COMPOUND RENEWAL PROCESSES

Compound renewal processes (CRPs) are among the most ubiquitous models used in applications of probability. At the same time, they are a natural generalization of random walks, the most well-studied classical objects in probability theory. This monograph, written for researchers and graduate students, presents the general asymptotic theory and generalizes many well-known results concerning random walks. The book contains the key limit theorems for CRPs, functional limit theorems, integro-local limit theorems, large and moderately large deviation principles for CRPs in the state space and in the space of trajectories, including large deviation principles in boundary crossing problems for CRPs, with an explicit form of the rate functionals, and an extension of the invariance principle for CRPs to the domain of moderately large and small deviations. Applications establish the key limit laws for Markov additive processes, including limit theorems in the domains of normal and large deviations.

Encyclopedia of Mathematics and Its Applications

This series is devoted to significant topics or themes that have wide application in mathematics or mathematical science and for which a detailed development of the abstract theory is less important than a thorough and concrete exploration of the implications and applications.

Books in the **Encyclopedia of Mathematics and Its Applications** cover their subjects comprehensively. Less important results may be summarized as exercises at the ends of chapters. For technicalities, readers can be referred to the bibliography, which is expected to be comprehensive. As a result, volumes are encyclopedic references or manageable guides to major subjects.

All the titles listed below can be obtained from good booksellers or from Cambridge University Press. For a complete series listing visit

www.cambridge.org/mathematics.

134 Y. Crama and P. L. Hammer (eds.) *Boolean Models and Methods in Mathematics, Computer Science, and Engineering*
135 V. Berthé and M. Rigo (eds.) *Combinatorics, Automata and Number Theory*
136 A. Kristály, V. D. Rădulescu and C. Varga *Variational Principles in Mathematical Physics, Geometry, and Economics*
137 J. Berstel and C. Reutenauer *Noncommutative Rational Series with Applications*
138 B. Courcelle and J. Engelfriet *Graph Structure and Monadic Second-Order Logic*
139 M. Fiedler *Matrices and Graphs in Geometry*
140 N. Vakil *Real Analysis through Modern Infinitesimals*
141 R. B. Paris *Hadamard Expansions and Hyperasymptotic Evaluation*
142 Y. Crama and P. L. Hammer *Boolean Functions*
143 A. Arapostathis, V. S. Borkar and M. K. Ghosh *Ergodic Control of Diffusion Processes*
144 N. Caspard, B. Leclerc and B. Monjardet *Finite Ordered Sets*
145 D. Z. Arov and H. Dym *Bitangential Direct and Inverse Problems for Systems of Integral and Differential Equations*
146 G. Dassios *Ellipsoidal Harmonics*
147 L. W. Beineke and R. J. Wilson (eds.) with O. R. Oellermann *Topics in Structural Graph Theory*
148 L. Berlyand, A. G. Kolpakov and A. Novikov *Introduction to the Network Approximation Method for Materials Modeling*
149 M. Baake and U. Grimm *Aperiodic Order I: A Mathematical Invitation*
150 J. Borwein et al. *Lattice Sums Then and Now*
151 R. Schneider *Convex Bodies: The Brunn-Minkowski Theory (Second Edition)*
152 G. Da Prato and J. Zabczyk *Stochastic Equations in Infinite Dimensions (Second Edition)*
153 D. Hofmann, G. J. Seal and W. Tholen (eds.) *Monoidal Topology*
154 M. Cabrera García and Á. Rodríguez Palacios *Non-Associative Normed Algebras I: The Vidav-Palmer and Gelfand-Naimark Theorems*
155 C. F. Dunkl and Y. Xu *Orthogonal Polynomials of Several Variables (Second Edition)*
156 L. W. Beineke and R. J. Wilson (eds.) with B. Toft *Topics in Chromatic Graph Theory*
157 T. Mora *Solving Polynomial Equation Systems III: Algebraic Solving*
158 T. Mora *Solving Polynomial Equation Systems IV: Buchberger Theory and Beyond*
159 V. Berthé and M. Rigo (eds.) *Combinatorics, Words and Symbolic Dynamics*
160 B. Rubin *Introduction to Radon Transforms: With Elements of Fractional Calculus and Harmonic Analysis*
161 M. Ghergu and S. D. Taliaferro *Isolated Singularities in Partial Differential Inequalities*
162 G. Molica Bisci, V. D. Radulescu and R. Servadei *Variational Methods for Nonlocal Fractional Problems*
163 S. Wagon *The Banach-Tarski Paradox (Second Edition)*
164 K. Broughan *Equivalents of the Riemann Hypothesis I: Arithmetic Equivalents*
165 K. Broughan *Equivalents of the Riemann Hypothesis II: Analytic Equivalents*
166 M. Baake and U. Grimm (eds.) *Aperiodic Order II: Crystallography and Almost Periodicity*
167 M. Cabrera García and Á. Rodríguez Palacios *Non-Associative Normed Algebras II: Representation Theory and the Zelmanov Approach*
168 A. Yu. Khrennikov, S. V. Kozyrev and W. A. Zúñiga-Galindo *Ultrametric Pseudodifferential Equations and Applications*
169 S. R. Finch *Mathematical Constants II*
170 J. Krajíček *Proof Complexity*
171 D. Bulacu, S. Caenepeel, F. Panaite and F. Van Oystaeyen *Quasi-Hopf Algebras*
172 P. McMullen *Geometric Regular Polytopes*
173 M. Aguiar and S. Mahajan *Bimonoids for Hyperplane Arrangements*
174 M. Barski and J. Zabczyk *Mathematics of the Bond Market: A Lvy Processes Approach*
175 T. R. Bielecki, J. Jakubowski and M. Niewęgłowski *Structured Dependence between Stochastic Processes*
176 A. A. Borovkov, V. V. Ulyanov and Mikhail Zhitlukhin *Asymptotic Analysis of Random Walks: Light-Tailed Distributions*
177 Y.-K. Chan *Foundations of Constructive Probability Theory*
178 L. W. Beineke, M. C. Golumbic and R. J. Wilson (eds.) *Topics in Algorithmic Graph Theory*
179 H.-L. Gau and P. Y. Wu *Numerical Ranges of Hilbert Space Operators*
180 P. A. Martin *Time-Domain Scattering*
181 M. D. de la Iglesia *Orthogonal Polynomials in the Spectral Analysis of Markov Processes*
182 A. E. Brouwer and H. Van Maldeghem *Strongly Regular Graphs*
183 D. Z. Arov and O. J. Staffans *Linear State/Signal Systems*

Encyclopedia of Mathematics and Its Applications

Compound Renewal Processes

A. A. BOROVKOV
Sobolev Institute of Mathematics, Russia

Translated by
ALEXEY ALIMOV
Steklov Institute of Mathematics, Moscow

CAMBRIDGE
UNIVERSITY PRESS

CAMBRIDGE
UNIVERSITY PRESS

University Printing House, Cambridge CB2 8BS, United Kingdom

One Liberty Plaza, 20th Floor, New York, NY 10006, USA

477 Williamstown Road, Port Melbourne, VIC 3207, Australia

314–321, 3rd Floor, Plot 3, Splendor Forum, Jasola District Centre,
New Delhi – 110025, India

103 Penang Road, #05–06/07, Visioncrest Commercial, Singapore 238467

Cambridge University Press is part of the University of Cambridge.

It furthers the University's mission by disseminating knowledge in the pursuit of education, learning, and research at the highest international levels of excellence.

www.cambridge.org
Information on this title: www.cambridge.org/9781009098441
DOI: 10.1017/9781009093965

First published 2022

A catalogue record for this publication is available from the British Library.

Library of Congress Cataloging-in-Publication Data
Names: Borovkov, A. A. (Aleksandr Alekseevich), 1931- author. |
Alimov, Alexey, translator.
Title: Compound renewal processes / A.A. Borovkov ; translated by Alexey Alimov.
Description: Cambridge ; New York : Cambridge University Press, 2022. |
Series: Encyclopedia of mathematics and its applications |
Includes bibliographical references and index.
Identifiers: LCCN 2021054195 (print) | LCCN 2021054196 (ebook) |
ISBN 9781009098441 (hardback) | ISBN 9781009093965 (epub)
Subjects: LCSH: Limit theorems (Probability theory) | Deviation
(Mathematics) | BISAC: MATHEMATICS / Probability & Statistics / General
Classification: LCC QA273.67 .B67 2022 (print) | LCC QA273.67 (ebook) |
DDC 519.2–dc23/eng20220215
LC record available at https://lccn.loc.gov/2021054195
LC ebook record available at https://lccn.loc.gov/2021054196

ISBN 978-1-009-09844-1 Hardback

Contents

Introduction

The Objects of Study

Compound renewal processes (CRPs) are among the most common mathematical models in many applications of probability theory, such as queuing theory, insurance theory, risk theory, and others. They are also used in theoretical research, for example, in the study of Markov additive processes (see [95, 96]). At the same time they are a natural generalization of random walks, the most well-studied classical objects in probability theory. The new general asymptotic theory of CRPs constructed in this monograph is therefore both of applied interest and generalizes many well-known results of probability theory related to random walks (see, e.g., A.A. Borovkov's monograph [23]). Some of the results obtained in this monograph for CRPs turned out to be new for the special case of random walks as well (see, e.g., Chapter 7).

Suppose we are given a random vector (τ_1, ζ_1) and a sequence of independent identically distributed random vectors (τ, ζ), (τ_2, ζ_2),... independent of (τ_1, ζ_1), where $\tau_1 \geq 0$, $\tau > 0$. We set

$$T_n := \sum_{j=1}^{n} \tau_j, \qquad Z_n := \sum_{j=1}^{n} \zeta_j \quad \text{for } n \geq 1, \qquad T_0 = Z_0 = 0. \tag{1}$$

For $t \geq 0$, we put

$$\eta(t) := \min\{k \geq 0 : T_k > t\}, \qquad \nu(t) := \eta(t) - 1. \tag{2}$$

Clearly,

$$\nu(t) = \max\{k \geq 0 : T_k \leq t\}$$

for all $t \geq 0$.

The random processes $\eta(t)$ and $\nu(t)$ are called *renewal processes* (or *simple renewal processes*).

The term "renewal process" first appeared in connection with "technical" applied problems in which failures and renewals of some devices such as electrical gadgets are present, the lifetimes of the devices usually being random (see, e.g., [53]). Let $\tau_1, \tau_2, \ldots$ be the trouble-free operation times of the devices. After time τ_j, a *failure* occurs and the faulty device is subject to *renewal* (or replacement). Assume that renewal (replacement) occurs instantly. Then $\nu(t)$ will be the *number of renewals* that occur before time t if we do not count the "renewal" at time $t = 0$. The number of renewals will be equal to $\eta(t)$ if we assume that a renewal occurs at $t = 0$.

In almost all applied problems, the properties of the distribution of τ_1 depend on when we start observing the system. If we know that a renewal occurs at time $t = 0$, then we can assume that $\tau_1, \tau_2, \ldots$ are identically distributed. But if we start observing at some time and we do not know when the last renewal occurred, then it is natural to assume that the time τ_1 until the first renewal after the beginning of observation has in general a distribution different from the distribution of the intervals $\tau_2, \tau_3, \ldots$ between subsequent renewals.

We will now introduce a wider class of processes.

Definition 1 The process

$$Z(t) := Z_{\nu(t)}, \qquad t \geq 0 \tag{3}$$

is called a *compound* (or *generalized*) *renewal process* (a *CRP*).

The sequence $\{(\tau_j, \zeta_j)\}$ will be called the *governing sequence of the CRP*.

As we have already mentioned, CRPs arise as a mathematical model in many applied problems, for example, in queuing theory and insurance theory. The standard generally accepted model of a CRP assumes that the time τ_1 of the first jump and the size ζ_1 of this jump have a joint distribution in general different from the joint distribution of (τ, ζ) (see, e.g., [53], [5]). This is the case, for example, for a class of CRPs important from an applications point of view, those with *stationary increments* (see §1.1.2). If $(\tau_1, \zeta_1) \underset{d}{=} (\tau, \zeta)$, then the process $Z(t)$ is called a *homogeneous CRP*; otherwise, it is called *inhomogeneous*.

The trajectories of $Z(t)$ on $[0, \infty)$ for $\tau_1 > 0$ have the following form:

$$Z(t) = \begin{cases} 0 & \text{if} \quad t \in [0, \tau_1), \\ \zeta_1 & \text{if} \quad t \in [\tau_1, T_2), \\ Z_2 & \text{if} \quad t \in [T_2, T_3) \quad \text{and so on;} \end{cases} \tag{4}$$

they are right-continuous. If $\tau_1 = 0$ (this is not excluded since $\tau_1 \geq 0$ by assumption), then the set $[0, \tau_1)$ is empty, $Z(t) = \zeta_1$ for $t \in [0, \tau_2)$, and so, the process $Z(t)$ on the event $\{\tau_1 = 0\}$ can be regarded as a homogeneous CRP with initial value $Z(0) = \zeta_1$.

Let us give two examples in which the process $Z(t)$ plays a key role in describing the work of the system under study. First, we consider the simplest problem in queuing theory. Customers with service times $\zeta_1, \zeta_2, \ldots$ arrive at the queueing system at times $T_1, T_2, \ldots$, respectively. For example, the customers could be airplanes arriving for landing at a busy airport: They arrive at times $T_1, T_2, \ldots$, and the time needed for the landing of the jth plane is equal to ζ_j. Or the context could be an information processing system, which receives information in packets at times $T_1, T_2, \ldots$, the jth packet requiring time ζ_j for its processing.

In such systems, $Z(t) = Z_{\nu(t)}$ is the time required for serving the customers received by time t. Suppose that the system under consideration is a queuing system (the customers that find the system busy join a queue). An important characteristic of such a system is the "virtual" waiting time $W(t)$ before the customer that arrives at time t starts to be serviced. It is not hard to see that $W(t)$ satisfies the equation

$$dW(t) = \begin{cases} dZ(t) - dt & \text{if} \quad W(t) > 0, \\ dZ(t) & \text{if} \quad W(t) = 0, \end{cases}$$

which has an explicit solution

$$W(t) = Z(t) - t - \inf_{0 \le u \le t} (0, Z(u) - u) \tag{5}$$

(the trajectory of $W(t)$ is obtained from the trajectory of $Z(t) - t$ by means of a "stopping" barrier at point 0). If, for example, τ_j and ζ_j are independent, then (5) implies that the limiting distribution of $W(t)$ as $t \to \infty$ coincides with the distribution of $\sup_{0 \le t < \infty} (Z^{(\mathrm{st})}(t) - t)$ (see, e.g., [17, §6]), where $Z^{(\mathrm{st})}(t)$ is a CRP with stationary increments, i.e., the process $Z(t)$ with a specially chosen distribution of τ_1 (see §1.1.2).

The second example is concerned with the operations of an insurance company. Let $T_1, T_2, \ldots$ be the times of significant claim payouts and let $\zeta_1, \zeta_2, \ldots$ be the amounts of these payments, respectively. Further, let r be the premium rate (the amount received by the company from insured customers per time unit). If x is the company's initial surplus, then its surplus at time t is equal to $x + rt - Z(t)$. This means that if $\inf_{u \le t} (x + ru - Z(u)) < 0$, then the company will go bankrupt by time t. In other words, the probability of ruin prior to time t is equal to $\mathbf{P}(\sup_{u \le t} (Z(u) - ru) > x)$. This is the classical ruin probability problem, which is the subject of many publications including monographs (see, e.g., [9], [4]–[6], [63]). It is considered in §6.7.

In both examples, the objects of study are the probabilities that the trajectory of the process $Z(u)$ crosses some boundary prior to time t. Problems of this kind are called boundary crossing problems for the CRP. They are considered in §1.6 and in Chapters 4 and 6.

CRPs also appear in theoretical research. For example, they emerge when studying the asymptotic laws for Markov additive processes (sums of random variables defined on the states of Markov chains). If the chain is Harris, then it has a positive atom, sometimes an "artificial" one. If we construct cycles (of respective lengths $\tau_1, \tau_2, \ldots$) generated by the returns of the chain to a positive atom and denote by $\zeta_1, \zeta_2, \ldots$ the increments of the sums on these cycles, then we obtain independent identically distributed vectors (τ_j, ζ_j), which define the corresponding CRP $Z(n)$ (time $t = n$ is discrete in this case). Using this process and the results presented in this monograph, one can obtain all the main limit laws for Markov additive processes (see §§1.8, 2.5, 5.7, and the references therein).

Alongside the CRP $Z(t)$, we will also consider stochastic processes

$$Y(t) := Z_{\eta(t)} = Z_{\nu(t)} + \zeta_{\eta(t)}, \qquad t \ge 0. \tag{6}$$

They will also be referred to as CRPs. The trajectories of $Y(t)$ on $[0, \infty)$ have the following form for $\tau_1 > 0$:

$$Y(t) = \begin{cases} \zeta_1 & \text{if} \quad t \in [0, \tau_1), \\ Z_2 & \text{if} \quad t \in [\tau_1, T_2) \quad \text{and so on.} \end{cases}$$

When $\tau_1 = 0$, there are changes similar to those mentioned after (4). We will show that the limit laws in the domain of normal deviations to be studied in Chapters 1 and 2, under appropriate conditions, are the same for the CRPs $Z(t)$ and $Y(t)$. In the domain of large deviations (see Chapters 3 and 6), this is not always the case.

Since $\eta(t)$ is a Markov time, the processes $Y(t) = Z_{\eta(t)}$ have a somewhat simpler structure, and, in a number of cases, it is more convenient to study these processes.

If $\tau_1 \equiv \tau \equiv 1$, then the CRP

$$Z(t) = Z([t]) = Z_{[t]}$$

becomes the *sequence of partial sums* $Z_{[t]}$ of the random variables ζ_j, i.e., a *random walk*. This object has been extensively studied.

If $\zeta_1 \equiv \zeta \equiv 1$, then $Z(t) = \nu(t) = \eta(t) - 1$, where $\eta(t)$ is a *simple renewal process*. For this process, $\mathbf{P}(\eta(t) > n) = \mathbf{P}(T_n \le t)$, and the problem of studying the distribution of $\eta(t)$ reduces to studying the distribution of sums of random variables – this time, of the sums T_n. Clearly, similar observations are also valid for the CRP $Y(t)$.

If $(\tau_1, \zeta_1) \underset{d}{=} (\tau, \zeta)$, while τ and ζ are independent,

$$\mathbf{P}(\tau > v) = e^{-\lambda v}, \qquad v \ge 0, \quad \lambda > 0, \tag{7}$$

then the process $Z(t)$ becomes a *compound Poisson process*, i.e., a process with independent increments.

If, instead of (7), we have

$$\mathbf{P}(\tau = k) = (1 - q)q^k, \quad q \in (0, 1), \quad k = 0, 1, \ldots,$$

or $\mathbf{P}(\tau = 1) = 1$, then, as in the case of (7), the sequence $Z(k)$ is a process with independent increments but in discrete time, i.e., a sequence of partial sums of independent identically distributed random variables (a random walk).

A Short History and the Contents of the Book

The study of CRPs is the topic of many publications. A number of general results are known such as the strong law of large numbers, the central limit theorem (see, e.g., the textbook by A.A. Borovkov [21, §10.6]), the law of the iterated logarithm, the invariance principle (see M. Csörgo, L. Hervatt, and J. Steinebach [55], [112]; A. Gut [72, Chapter 5]). The proofs in these publications rely upon a very complex technique and are simplified in Chapter 1. In the monograph [38, Chapter 16] by A.A. Borovkov and K.A.Borovkov, large deviation probabilities for CRPs and their trajectories are studied in the case when the jumps of the process have distributions regularly varying at infinity.

A significant portion of the published work related to CRPs concerns applications or rather special problem formulations. The body of work is fragmented in nature and does not deal with the main directions of interest. From 2008 to 2019, a major cycle of works by the author, K.A. Borovkov, A.A. Mogul'skii, and E.I. Prokopenko appeared (many of which are joint works) devoted to limit laws for CRPs. These works served as the basis of this book.

The goal of this monograph is a systematic exposition of the asymptotic theory of CRPs in its general form. It comprises analogs and generalizations of all the main limit laws established for random walks presented, for example, in [23]. This is the first presentation of the theory in monographic literature. It is of both theoretical and applied interest.

The theory contains:

- the basic limit theorems for CRPs in the domain of normal deviations (with the functional limit theorems), including the case of infinite variance of the jumps of the process; the law of the iterated logarithm and its analogs (Chapter 1);
- integro-local limit theorems for CRPs in the domains of normal, moderately large and large deviations (Chapters 2 and 5);
- large and moderately large deviation principles for CRPs in the state space and in the space of trajectories, including large deviation principles in boundary crossing problems for CRPs, with an explicit form of deviation function (rate function) (Chapters 3 and 4);
- limit theorems describing the sharp asymptotics in boundary crossing problems for CRPs (Chapter 6);
- extension of the invariance principle for CRPs to the domain of moderately large and small deviations (Chapter 7); the results of Chapter 7 turn out to be new for random walks as well.

We apply the theory to establish the main limit laws for Markov additive processes including functional limit theorems in the domains of normal and large deviations (§§1.8, 3.6, and 5.7).

The above suggests that a significant portion of the monograph (Chapters 3–7) is concerned with studying large deviation probabilities for CRPs. Mathematically, this is the most content-rich and difficult part of the theory we present. Note that an essential role here is played by the following circumstance. It turns out that there exists a function that encapsulates all information about the asymptotic behavior of the distribution of the CRP on increasing time intervals. We found and studied it, and call it the *fundamental* function (see §3.5). For a random walk (which is a special case of CRP), the fundamental function is equal to the logarithm of the Laplace transform of the jump distribution. In the general case, the fundamental function plays the same role as this transform, but instead of explicit equalities there will be analogous asymptotic relations as time grows.

A vast literature including several monographs is devoted to large deviations problems for random processes. Its main concern is the study of "rough" asymptotics (asymptotics of the logarithms) of the probabilities of the corresponding rare events for a wide class of processes. The main approaches and a survey of the results can be found, for example, in the monographs by J. Feng and T.G. Kurtz [64] and A. Dembo and O. Zeitouni [57] (see also the bibliography therein). In our case, in studying CRPs, we use more effective and more constructive direct approaches based on knowledge of the specific nature of CRPs, defined in (3), which makes it possible to find the fundamental function. They enable us to obtain in Chapters 3–7 much more advanced results including (a) the large deviation principle (LDP) in Chapters 3 and 4, with an explicit form of the deviation rate functional which is defined by the Legendre transform of the fundamental function; (b) in Chapters 5–7, solutions to more difficult problems on the sharp asymptotics of the large deviation probabilities.

The exposition in §§1.5–1.7 and Chapters 2, 6, and 7 is based on the author's papers [25], [30]–[36]. The exposition in Chapters 3–5 relies on the author's papers [26, 34] and on joint work with A.A. Mogul'skii [39], [44]–[48] (with E.I. Prokopenko as coauthor in [48] as well). The monograph also contains a number of previously unpublished results.

In the study of the large deviation probabilities of CRPs in Chapters 3–7, we assume that Cramér's moment condition is met (fast decay at infinity of the jump distributions). For the case where the jump distributions vary regularly at infinity (slow decay), in Chapter 8, for the sake of completeness, we present a number of results from [38, Chapter 16] without proofs.

The creation of a general asymptotic theory of CRPs under Cramér's condition became possible owing to the previous work in the following three areas:

- limit theorems on the asymptotics of the renewal measure in the domain of large deviations for multidimensional random walks (see A. A. Borovkov and A. A. Mogul'skii [39, 44]);
- large deviations of multidimensional random walks (see, e.g., A.A. Borovkov's monograph [23]); in applications to CRPs, approaches related to so-called *local* large deviation principles are very useful;
- Stone's integro-local theorems for random walks (see [114, 115]; they are particularly important as research tools).

The possession of the results and techniques in these three areas is necessary for a sufficiently complete asymptotic analysis of CRPs.

Acknowledgement The author is grateful to T.V. Belyaeva for her great help in preparing the manuscript.

1

Main Limit Laws in the Normal Deviation Zone

1.1 Preliminary Results

1.1.1 Convergence of Distributions and Moments of Some Functionals of CPRs

Compound renewal processes (CRPs) $Z(t)$, $Y(t)$ were defined in the introduction (see (1)–(6)). Their definition is based on the governing sequence of vectors $\{\tau_j, \zeta_j\}$, the sums

$$T_n = \sum_{j=1}^{n} \tau_j, \qquad Z_n = \sum_{j=1}^{n} \zeta_j,$$

and the renewal processes $\eta(t) = \min\{k : T_k > t\}$ and $\nu(t) = \max\{k : T_k \le t\} = \eta(t) - 1$. In this notation,

$$Z(t) = Z_{\nu(t)}, \qquad Y(t) = Z_{\eta(t)}.$$

To describe the properties of the CRPs $Z(t)$, $Y(t)$, we need some more notation. Let $\chi(t)$ be the first overshoot over level t by the random walk $\{T_k\}_{k=1}^{\infty}$,

$$\chi(t) := T_{\eta(t)} - t, \tag{1.1.1}$$

and let

$$\gamma(t) := t - T_{\nu(t)} \tag{1.1.2}$$

be the corresponding undershoot. We also set

$$\zeta(t) := \zeta_{\eta(t)}, \qquad \tau(t) := \tau_{\eta(t)},$$

so that

$$\gamma(t) + \chi(t) = \tau(t).$$

In what follows, it is always assumed that

$$\mathbf{E}\tau =: a_\tau \quad \text{and} \quad \mathbf{E}\zeta =: a_\zeta$$

both exist, and hence the "mean drift"

$$a := \frac{a_\zeta}{a_\tau}$$

of the CRP is defined. The name "mean drift" is reasonable, because we shall show that

$$\frac{\mathbf{E}Z(t)}{t} \to a, \qquad \frac{Z(t)}{t} \underset{\text{a.s.}}{\to} a \quad \text{as } t \to \infty.$$

The same relations also hold for $Y(t)$. *In what follows, the assumption that a_τ, a_ζ exist and are finite will generally not be repeated.*

The distribution of (τ_1, ζ_1) can be arbitrary; the conditions related to this vector will be specified if necessary.

Below, we will also use the *renewal function* corresponding to the sequence $\{T_k\}$; in the homogeneous case, we denote it by

$$H(t) = \sum_{k=0}^{\infty} \mathbf{P}(T_k \le t).$$

In the inhomogeneous case $\tau_1 \underset{d}{\neq} \tau$, it is denoted by $\widetilde{H}(t)$.

We have

$$\{\eta(t) > k\} = \{T_k \le t\},$$

and hence, in the homogeneous case,

$$\mathbf{E}\eta(t) = \sum_{k=1}^{\infty} \mathbf{P}(\eta(t) \ge k) = \sum_{k=0}^{\infty} \mathbf{P}(\eta(t) > k) = \sum_{k=0}^{\infty} \mathbf{P}(T_k \le t) = H(t), \tag{1.1.3}$$

$$T_{\eta(t)} = t + \chi(t)$$

and so, by the Wald identity,

$$\begin{aligned} \mathbf{E}T_{\eta(t)} &= a_\tau \mathbf{E}\eta(t) = t + \mathbf{E}\chi(t), \\ \mathbf{E}\eta(t) &= H(t) = \frac{t + \mathbf{E}\chi(t)}{a_\tau}. \end{aligned} \tag{1.1.4}$$

Lemma 1.1.1 *Let the distribution of τ be nonlattice. Then*

(i) *The proper limit distributions*

$$\lim_{t\to\infty} \mathbf{P}(\gamma(t) \ge u, \chi(t) \ge v, \zeta(t) \ge w) = \frac{1}{a_\tau} \int_u^\infty \mathbf{P}(\tau \ge y + v, \ \zeta \ge w)\, dy, \tag{1.1.5}$$

$$\lim_{t\to\infty} \mathbf{P}(\zeta(t) \ge w) = \frac{\mathbf{E}(\tau;\ \zeta \ge w)}{a_\tau} \tag{1.1.6}$$

always exist.

(ii) *In the homogeneous case, there exist constants $c_1 \in (0, \infty)$ and $c_2 \in (1/a_\tau, \infty)$ such that*

$$\sup_t \mathbf{P}(\gamma(t) \ge u, \chi(t) \ge v, \zeta(t) \ge w)$$
$$\le c_1 \mathbf{P}(\tau \ge u + v, \zeta \ge w) + c_2 \int_u^\infty \mathbf{P}(\tau > y + v, \zeta \ge w)\, dy. \qquad (1.1.7)$$

In the inhomogeneous case, the term $\mathbf{P}(\tau_1 \ge u + v, \zeta_1 \ge w)$ *is added to the right-hand side of* (1.1.7).

(iii) *If* $\mathbf{E}\tau_1^{k-1} < \infty$ *and* $\mathbf{E}\tau^k < \infty$, *then*

$$\mathbf{E}\gamma^k(t) = o(t), \qquad \mathbf{E}\chi^k(t) = o(t) \quad as \quad t \to \infty. \qquad (1.1.8)$$

If $\mathbf{E}\tau_1^k < \infty$ *and* $\mathbf{E}\tau^{k+1} < \infty$, *then for nonlattice* τ,

$$\mathbf{E}\gamma^k(t) \to \frac{\mathbf{E}\tau^{k+1}}{(k+1)a_\tau}, \ \mathbf{E}\chi^k(t) \to \frac{\mathbf{E}\tau^{k+1}}{(k+1)a_\tau} \quad as \quad t \to \infty. \qquad (1.1.9)$$

If $\mathbf{E}e^{\lambda\tau_1} < \infty$ *and* $\mathbf{E}e^{\lambda\tau} < \infty$ *for* $\lambda > 0$, *then*

$$\mathbf{E}e^{\lambda\gamma(t)} \to \frac{\mathbf{E}e^{\lambda\tau} - 1}{a_\tau \lambda} \quad as \quad t \to \infty. \qquad (1.1.10)$$

The same relation also holds for $\chi(t)$.

If f *is a measurable function and* $\mathbf{E}|f(\zeta_1)| < \infty$, $\mathbf{E}\tau|f(\zeta)| < \infty$, *then*

$$\mathbf{E}f(\zeta(t)) \to \frac{\mathbf{E}(\tau f(\zeta))}{a_\tau} \quad as \quad t \to \infty. \qquad (1.1.11)$$

The lemma shows that the limit values of the moments of the random variables $\gamma(t)$, $\chi(t)$, $\zeta(t)$ coincide with the moments of the variables γ_∞, χ_∞, ζ_∞, whose joint distribution is given by

$$\mathbf{P}(\gamma_\infty \ge u, \chi_\infty \ge v, \ \zeta_\infty \ge w) = \frac{1}{a_\tau} \int_u^\infty \mathbf{P}(\tau \ge y + v, \zeta \ge w)\, dy \qquad (1.1.12)$$

(see (1.1.5)).

If the distribution of τ is arithmetic, then the integrals in (1.1.12) are replaced by sums – this slightly changes the values of the right-hand sides in (1.1.9) and (1.1.10).

Proof of Lemma 1.1.1 (i) For homogeneous CRPs, in view of the main renewal theorem for nonlattice τ, we have

$$\mathbf{P}(\gamma(t) \ge u, \chi(t) \ge v, \zeta(t) \ge w) = \sum_{k=1}^\infty \int_0^{t-u} \mathbf{P}(T_k \in dy)\mathbf{P}(\tau \ge t - y + v, \zeta \ge w)$$
$$= \int_0^{t-u} dH(y)\mathbf{P}(\tau \ge t - y + v, \zeta \ge w) \to \frac{1}{\mathbf{E}\tau} \int_u^\infty \mathbf{P}(\tau \ge y + v, \zeta \ge w)\, dy \qquad (1.1.13)$$

as $t \to \infty$.

If the distribution of (τ_1, ζ_1) differs from the general distribution of (τ, ζ), then, for $t > u$, we have

$$\mathbf{P}(\gamma(t) \geq u, \chi(t) \geq v, \zeta(t) \geq w) = \mathbf{P}(\tau_1 \geq t + v, \zeta_1 \geq w) \\ + \int_0^{t-u} \mathbf{P}(\tau_1 \in ds)\mathbf{P}(\gamma(t-s) \geq u, \chi(t-s) \geq v, \zeta(t-s) \geq w), \quad (1.1.14)$$

where, for each fixed s, we have the convergence as $t \to \infty$ of the form (1.1.13) for the second factor under the integral sign in (1.1.14). This means that for the inhomogeneous CRPs the left-hand side of (1.1.14) also converges as $t \to \infty$ to the right-hand side of (1.1.13). This proves (1.1.5). Putting $u = v = 0$ in (1.1.13), (1.1.14), we get (1.1.6).

(ii) For the renewal function $H(t)$, there always exist constants $c_1 > 0$ and $c_2 > 1/a_\tau$ such that

$$H(t) \leq c_1 + c_2 t \quad \text{for all} \quad t \geq 0.$$

On the other hand, the integrand $\mathbf{P}(\tau \geq t - y + v, \zeta \geq w)$ in (1.1.13) is increasing with y. Hence the left-hand side in (1.1.13), which is 0 for $t \leq u$, is majorized by

$$c_1\mathbf{P}(\tau \geq t + v, \zeta \geq w) + c_2 \int_0^{t-u} \mathbf{P}(\tau \geq t - y + v, \zeta \geq w)\, dy \quad (1.1.15)$$

for $t \geq u$. This implies (1.1.7).

The additional term on the right of (1.1.7) in the inhomogeneous case is brought about by the appearance of the first term on the right of (1.1.14).

(iii) Assertion (1.1.8) follows from (1.1.15). If $\mathbf{E}\tau^{k+1} < \infty$, then by (1.1.12) the function

$$u^k\mathbf{P}(\gamma_\infty \geq u) = \frac{u^k}{a_\tau} \int_u^\infty \mathbf{P}(\tau \geq u)\, dy$$

is integrable. Hence by assertions (i), (ii) of the theorem, we can apply the dominated convergence theorem, which implies that

$$\mathbf{E}\gamma^k(t) \to \mathbf{E}\gamma_\infty^k = \frac{1}{a_\tau} \int_0^\infty y^k\mathbf{P}(\tau \geq y)\, dy = \frac{\mathbf{E}\tau^{k+1}}{(k+1)a_\tau}.$$

The proof of the remaining relations in (1.1.9) is similar. The last assertion follows from the equality

$$\mathbf{E}\zeta(t) = \mathbf{E}(\zeta(t);\ \zeta(t) \geq 0) + \mathbf{E}(\zeta(t);\ \zeta(t) < 0) \\ = \int_0^\infty \mathbf{P}(\zeta(t) \geq w) dw - \int_{-\infty}^0 \mathbf{P}(\zeta(t) \leq w) dw.$$

Lemma 1.1.1 is proved. □

In studying the limit distributions of the variables $\chi(t), \zeta(t)$ as $t \to \infty$, we can also consider the "triangular array scheme," when the distribution of the "inhomogeneous" vector (τ_1, ζ_1) depends on some parameter N. Such schemes appear, for example, when considering the processes

$$Z_N(t) = Z(N+t) - Z(N). \tag{1.1.16}$$

The role of the initial jumps for $Z_N(t)$ will be played by the variables $\chi(N)$ and $\zeta(N)$, whose distribution depends on N.

Corollary 1.1.2 *If in a partial triangular array scheme, the parameter* $N = N(t)$ *depends on* t *so that* $\tau_1 = o_p(t)$ *as* $t \to \infty$, *then assertions* (1.1.5) *and* (1.1.6) *of Lemma* 1.1.1 *remain valid. If, moreover,* $\mathbf{E}(\tau_1; \tau_1 \geq t) \to 0$ *and* $\mathbf{E}(\zeta_1; \tau_1 \geq t) \to 0$ *as* $t \to \infty$, *then assertions* (1.1.9) *and* (1.1.12) *for* $k = 1$, $f(z) = z$ *also remain valid.*

The first assertion of the corollary follows from (1.1.14). The second one follows from (1.1.9), (1.1.11) for $\mathbf{E}\chi(t)$ and $\mathbf{E}\zeta(t)$.

Corollary 1.1.3 *If* $\mathbf{E}\tau_1 < \infty$, *then as* $t \to \infty$

$$\widetilde{H}(t) = \frac{t + o(t)}{a_\tau}.$$

This result follows from (1.1.3), Lemma 1.1.1, and the relations

$$\begin{aligned}\widetilde{H}(t) = \mathbf{E}\eta(t) &= \mathbf{P}(\tau_1 > t) + \int_0^t \mathbf{P}(\tau_1 \in ds)\,[1 + \mathbf{E}\eta_0(t-s)] \\ &= \mathbf{P}(\tau_1 > t) + \int_0^t \mathbf{P}(\tau_1 \in ds)\Big(1 + \frac{t - s + \mathbf{E}\chi(t-s)}{a_\tau}\Big),\end{aligned}$$

where the first passage time $\eta_0(t)$ corresponds to the homogeneous sequence $\{T_k\}$.

Results similar to Lemma 1.1.1 can be found, for example, in [21, §§10.4, 10.6].

1.1.2 CRPs with Stationary Increments

Let us return to the special case (1.1.16). If the initial jumps $(\tau_{1,N}, \zeta_{1,N})$ of the process $Z_N(t)$ are labeled by the index N, then, as already pointed out,

$$(\tau_{1,N}, \zeta_{1,N}) \underset{d}{=} (\chi(N), \zeta(N)),$$

so that, by Lemma 1.1.1,

$$(\tau_{1,N}, \zeta_{1,N}) \Rightarrow (\chi_\infty, \zeta_\infty) \quad \text{as} \quad N \to \infty$$

(the sign "⇒" denotes weak convergence of distributions), where the distribution $(\chi_\infty, \zeta_\infty)$ is described in (1.1.12). Consider the CRP $Z^{(\mathrm{st})}(t)$ (the meaning of the index (st) will be explained in Definition 1.1.4) with initial jumps $(\tau^{(\mathrm{st})}, \zeta^{(\mathrm{st})})$, which have the same distribution as $(\chi_\infty, \zeta_\infty)$ (see (1.1.12)),

$$\mathbf{P}(\tau_1^{(\mathrm{st})} \geq v, \zeta_1^{(\mathrm{st})} \geq w) = \frac{1}{a_\tau} \int_v^\infty \mathbf{P}(\tau \geq y, \zeta \geq w)\, dy; \tag{1.1.17}$$

the successive jumps being (τ_2, ζ_2), (τ_3, ζ_3),... By the above interpretation, for such a process, the distribution of $(\chi^{(\mathrm{st})}(t), \zeta^{(\mathrm{st})}(t))$ (with obvious agreement in notation) will be the same for all $t \geq 0$ (the distribution of $(\chi^{(\mathrm{st})}(t'), \zeta^{(\mathrm{st})}(t'))$ can be looked upon as the limit distribution of the initial jump of the process $Z_{N'}(t) = Z_{N+t'}(t)$ with $N' = N + t' \to \infty$). This means that the distribution of the increments of $Z^{(\mathrm{st})}(u+t) - Z^{(\mathrm{st})}(u)$ will be the same for all $u \geq 0$.

Definition 1.1.4 The process $Z(t)$ with the initial jump distribution (1.1.17) is called a CRP *with stationary increments* and is denoted by $Z^{(\mathrm{st})}(t)$.

Let us find the form of the characteristic function $\varphi^{(\mathrm{st})}(\lambda, \mu)$ of the vector $(\tau_1^{(\mathrm{st})}, \zeta_1^{(\mathrm{st})})$. We set

$$\varphi(\lambda, \mu) = \mathbf{E} e^{i\lambda\tau + i\mu\zeta}.$$

Lemma 1.1.5 (i) *The following representation holds*:

$$\varphi^{(\mathrm{st})}(\lambda, \mu) = \frac{\varphi(\lambda, \mu) - \varphi(0, \mu)}{i\lambda a_\tau}. \tag{1.1.18}$$

(ii) *The process $Z(t) = Z^{(\mathrm{st})}(t)$ is a homogeneous CRP if and only if $Z(t)$ is a compound (generalized) Poisson process, that is, τ and ζ are independent, and* $\mathbf{P}(\tau > x) = e^{-x/a_\tau}$.

Proof of Lemma 1.1.5 (i) In view of (1.1.17) we have

$$\begin{aligned}
\varphi^{(\mathrm{st})}(\lambda, \mu) &= \int_0^\infty e^{i\lambda v} \int_{-\infty}^\infty e^{i\mu w} \mathbf{P}(\tau_1^{(\mathrm{st})} \in dv, \zeta_1^{(\mathrm{st})} \in dw) \\
&= \frac{1}{a_\tau} \int_0^\infty e^{i\lambda v} \Big[\int_{-\infty}^\infty e^{i\mu w} \mathbf{P}(\tau \geq v, \zeta \in dw) \Big] dv.
\end{aligned} \tag{1.1.19}$$

We denote by $U(v)$ the expression in square brackets on the right of (1.1.19) and let $V(v) = \frac{e^{i\lambda v}}{i\lambda}$. Hence, integrating by parts (1.1.19), we find that

$$\begin{aligned}
\varphi^{(\mathrm{st})}(\lambda, \mu) &= \frac{1}{a_\tau} U(v)V(v)\Big|_0^\infty + \frac{1}{i\lambda a_\tau} \int_0^\infty e^{i\lambda v + i\mu w} \mathbf{P}(\tau \in dv, \zeta \in dw) \\
&= \frac{1}{i\lambda a_\tau} [\varphi(\lambda, \mu) - \varphi^{(\zeta)}(\mu)],
\end{aligned}$$

where $\varphi^{(\zeta)}(\mu) = \mathbf{E} e^{i\mu\zeta}$.

(ii) The second assertion follows from the fact that by (1.1.18) the equality

$$\varphi^{(\mathrm{st})}(\lambda, \mu) = \varphi(\lambda, \mu)$$

is equivalent to the equality

$$\varphi(\lambda, \mu) = \frac{\varphi^{(\zeta)}(\mu)}{1 - i a_\tau \lambda}.$$

The lemma is proved. □

From Lemma 1.1.1 it follows that $\mathbf{E}\tau_1^{(\mathrm{st})} < \infty$ if $\mathbf{E}\tau^2 < \infty$ (see (1.1.9)), and $\mathbf{E}\zeta^{(\mathrm{st})} < \infty$ if $\mathbf{E}\tau^2 < \infty$, $\mathbf{E}\zeta^2 < \infty$ (see (1.1.11)). If τ and ζ are independent, then $\mathbf{E}\zeta^{(\mathrm{st})} < \infty$ if $\mathbf{E}\tau < \infty$, $\mathbf{E}|\zeta| < \infty$.

1.1.3 Strong Law of Large Numbers for a Simple Renewal Process $\eta(t)$

Lemma 1.1.6 *The almost sure convergence*

$$\frac{\eta(t)}{t} \underset{a.s.}{\longrightarrow} \frac{1}{a_\tau} \quad as \quad t \to \infty \tag{1.1.20}$$

always holds.

Proof of Lemma 1.1.6 Consider the function $T_t := T_{[t]}$ of the real variable t. For this function, by the strong law of large numbers

$$\frac{T_t}{t} \underset{\text{a.s.}}{\longrightarrow} a_\tau,$$

as $t \to \infty$. That is, for any $\varepsilon > 0$, there exists a (random) number $t_0 = t_0(\varepsilon) < \infty$ such that T_t lies between the rays $y = (a_\tau \pm \varepsilon)t$ for all $t > t_0$. This means that the function $\eta(y) = \min\{t : T_t > y\}$, which is the inverse of T_t, lies for all $y \geq t_0(a_\tau + \varepsilon)$ between the rays

$$t = \frac{y}{a_\tau \pm \varepsilon}.$$

But this means that (1.1.20) holds. Lemma 1.1.6 is proved. □

1.1.4 Almost Sure Convergence of Some Functionals of CRPs

Consider some measurable function $g(t, y) \geq 0$ and set, for brevity,

$$g = g(\tau, \zeta), \qquad g_n = g(\tau_n, \zeta_n), \qquad \overline{g}_n = \max_{k \leq n} g_k.$$

Let $V(x)$ be some nondecreasing regularly varying function on $[0, \infty)$, that is, a function with the representation

$$V(x) = x^\alpha l(x), \qquad x \geq 0, \quad \alpha > 1, \tag{1.1.21}$$

where $l(x)$ is a slowly varying function as $x \to \infty$. We let $V^{(-1)}(y)$ denote the inverse function of $V(x)$,

$$V^{(-1)}(y) := \inf\{x : V(x) \geq y\}.$$

The function $V^{(-1)}(y)$ is also a regularly varying function (with exponent $1/\alpha$); see, for example, Theorem 1.1.3 in [38].

Similarly, if

$$F(x) = x^{-\alpha} l(x) \tag{1.1.22}$$

is a regularly varying at infinity function with exponent $-\alpha$, then the function

$$\sigma(u) := F^{(-1)}(1/u) = \inf\{x : F(x) < 1/u\}$$

is also a regularly varying at infinity function (with exponent $1/\alpha$).

Lemma 1.1.7 (i) *If* (1.1.21) *holds and* $\mathbf{E}V(g) < \infty$, *then, as* $n \to \infty$, $t \to \infty$,

$$\frac{\overline{g}_n}{V^{(-1)}(n)} \underset{a.s.}{\to} 0, \qquad \frac{\overline{g}_{\eta(t)}}{V^{(-1)}(t)} \underset{a.s.}{\to} 0. \tag{1.1.23}$$

(ii) *If* $\mathbf{P}(g \geq x) \leq cF(x)$, $c = \text{const}$, $F(x) = x^{-\alpha}l(x)$, *then, for any* $\theta > 1/\alpha$, $n \to \infty$, $t \to \infty$,

$$\frac{\overline{g}_n}{\sigma(n)(\ln n)^\theta} \underset{a.s.}{\to} 0, \qquad \frac{\overline{g}_{\eta(t)}}{\sigma(t)(\ln t)^\theta} \underset{a.s.}{\to} 0. \tag{1.1.24}$$

(iii) *If* $\mathbf{P}(g \geq x) = o(F(x))$ *as* $x \to \infty$, *then*

$$\frac{\overline{g}_n}{\sigma(n)} \underset{p}{\to} 0, \qquad \frac{\overline{g}_{\eta(t)}}{\sigma(t)} \underset{p}{\to} 0 \tag{1.1.25}$$

as $n \to \infty$, $t \to \infty$, *respectively.*

Proof of Lemma 1.1.7 (i) We first consider the homogeneous case. Let us show that

$$\frac{g_n}{V^{(-1)}(n)} \underset{a.s.}{\to} 0 \quad as \quad n \to \infty. \tag{1.1.26}$$

For any $\varepsilon > 0$,

$$\sum_{n=1}^{\infty} \mathbf{P}(g_n \geq \varepsilon V^{(-1)}(n)) \leq \int_0^\infty \mathbf{P}(g \geq \varepsilon V^{(-1)}(t))\,dt = \int_0^\infty \mathbf{P}\left(V\left(\frac{g}{\varepsilon}\right) \geq t\right)dt = \mathbf{E}V\left(\frac{g}{\varepsilon}\right). \tag{1.1.27}$$

Since $V(g/\varepsilon) \sim \varepsilon^{-\alpha}V(g)$ as $g \to \infty$, we have

$$\mathbf{E}V\left(\frac{g}{\varepsilon}\right) \leq c + 2\varepsilon^{-\alpha}\mathbf{E}V(g) < \infty, \quad c = \text{const}. \tag{1.1.28}$$

Now (1.1.26) follows from (1.1.27), (1.1.28) and the Borel–Cantelli lemma, because only a finite number of events $\{g_n \geq \varepsilon V^{(-1)}(n)\}$ occur with probability 1.

Let us now prove the first relation in (1.1.23). From (1.1.26) it follows that there exists a random number $n_0 = n_0(\varepsilon)$ such that

$$g_n < \varepsilon V^{(-1)}(n) \quad \text{for all} \quad n \geq n_0. \tag{1.1.29}$$

Moreover, there always exists a number $m_0 \geq n_0$ such that $\overline{g}_{n_0} < \varepsilon V^{(-1)}(m_0)$, and hence, by (1.1.29),

$$\overline{g}_{n_0+1} < \varepsilon V^{(-1)}(m_0), \ldots, \overline{g}_{m_0} < \varepsilon V^{(-1)}(m_0).$$

Another appeal to (1.1.29) shows that $\overline{g}_n < \varepsilon V^{(-1)}(n)$ for all $n \geq m_0$. But this means that

$$\frac{\overline{g}_n}{V^{(-1)}(n)} \underset{\text{a.s.}}{\to} 0 \quad \text{as} \quad n \to \infty. \tag{1.1.30}$$

We have $\eta(t) \underset{\text{a.s.}}{\to} \infty$ as $t \to \infty$, and now from (1.1.30) we get

$$\frac{g_{\eta(t)}}{V^{(-1)}(\eta(t))} \underset{\text{a.s.}}{\to} 0 \quad \text{as} \quad t \to \infty.$$

Moreover, by Lemma 1.1.6,

$$\frac{\eta(t)}{t} \underset{\text{a.s.}}{\to} \frac{1}{a_\tau}, \qquad \frac{V^{(-1)}(\eta(t))}{V^{(-1)}(t)} \underset{\text{a.s.}}{\to} a_\tau^{-1/\alpha}.$$

This proves the second relation in (1.1.23).

(ii) For any $\delta > 0$ and sufficiently large $n = n_\delta$, using, for example, Theorem 1.1.3 in [38], we have, for all $n \ge n_\delta$,

$$l(\sigma(n)(\ln n)^\theta) \le (\ln n)^{\theta\delta} l(\sigma(n)).$$

Hence, for $n \ge n_\delta$,

$$\begin{aligned} F(\sigma(n)(\ln n)^\theta) &= \sigma(n)^{-\alpha}(\ln n)^{-\theta\alpha} l(\sigma(n)(\ln(n))^\theta) \\ &\le \sigma(n)^{-\alpha}(\ln n)^{-\theta(\alpha-\delta)} l(\sigma(n)) = (\ln n)^{-\theta(\alpha-\delta)} F(\sigma(n)) = \frac{(\ln n)^{-\theta(\alpha-\delta)}}{n}. \end{aligned}$$

A similar analysis shows that, for $\alpha\theta > 1$ and $\delta < \frac{\alpha\theta-1}{\theta}$, we have $\theta(\alpha-\delta) > 1$ and

$$\sum_{n=n_\delta}^{\infty} \mathbf{P}(g_n > \sigma(n)(\ln n)^\theta) \le c \sum_{n=n_\delta}^{\infty} F(\sigma(n)(\ln n)^\theta) \le c \sum_{n=n_\delta}^{\infty} \frac{(\ln n)^{-\theta(\alpha-\delta)}}{n} < \infty.$$

Therefore,

$$\frac{g_n}{\sigma(n)(\ln n)^\theta} \underset{\text{a.s.}}{\to} 0 \quad \text{for} \quad \theta > \frac{1}{\alpha}, \quad n \to \infty. \tag{1.1.31}$$

The remaining part of the proof, in which (1.1.31) is used to derive (1.1.24), is similar to the proof of assertion (i) of the lemma.

(iii) For any $\varepsilon > 0$,

$$\begin{aligned} \mathbf{P}(\overline{g}_n > \varepsilon\sigma(n)) \le n\mathbf{P}(g > \varepsilon\sigma(n)) &= o(nF(\varepsilon\sigma(n))) \\ &= o(nF(\sigma(n))) = o(1) \quad \text{as} \quad n \to \infty. \end{aligned}$$

This proves the first relation in (1.1.25). The second relation follows from the fact that

$$\frac{\overline{g}_{\eta(t)}}{\sigma(\eta(t))} \underset{p}{\to} 0 \quad \text{and} \quad \frac{\eta(t)}{t} \underset{\text{a.s.}}{\to} \frac{1}{a_\tau} \quad \text{as} \quad t \to \infty.$$

It is easily seen that the above arguments do not change if an arbitrary fixed random vector is considered in place of (τ_1, ζ_1). Lemma 1.1.7 is proved. □

From Lemma 1.1.7 we have the following result. We set

$$\overline{\chi}(t) = \max_{u \le t} \chi(u), \qquad \overline{\zeta}(t) = \max_{u \le t} \zeta(u).$$

Corollary 1.1.8 (i) $\dfrac{\overline{\chi}(t)}{t} \underset{a.s.}{\to} 0$, $\dfrac{\overline{\zeta}(t)}{t} \underset{a.s.}{\to} 0$ *as* $t \to \infty$.

(ii) *The following relations hold*:

If $\mathbf{E}\tau^2 < \infty$, *then* $\dfrac{\overline{\chi}(t)}{\sqrt{t}} \underset{a.s.}{\to} 0$.

If $\mathbf{E}\zeta^2 < \infty$, *then* $\dfrac{\overline{\zeta}(t)}{\sqrt{t}} \underset{a.s.}{\to} 0$.

Proof We have $\mathbf{E}\tau < \infty$, $\mathbf{E}|\zeta| < \infty$, and hence the first assertion follows from the first assertion of Lemma 1.1.7 with $g(t, y) = t$, $g(t, y) = |y|$, and $V(x) = x$. The second assertion follows similarly from Lemma 1.1.7 with $V(x) = x^2$. □

1.2 First Moments of the Processes $Z(t)$ and $Y(t)$. Strong Laws of Large Numbers

1.2.1 Asymptotics for First- and Second-Order Moments of $Z(t)$ and $Y(t)$

We set

$$\xi_i = \zeta_i - a\tau_i, \qquad S_n = \sum_{i=1}^{n} \xi_i = Z_n - aT_n, \tag{1.2.1}$$

so that ξ_i for $i \geq 2$ are independent copies of the random variable

$$\xi = \zeta - a\tau, \qquad \mathbf{E}\xi = 0. \tag{1.2.2}$$

Theorem 1.2.1 I. *Let* $Z(t)$, $Y(t)$ *be homogeneous CRPs.*

(i) *The following relations hold*:

$$\mathbf{E}Y(t) = a(t + \mathbf{E}\chi(t)) = at + r_Y(t), \quad r_Y(t) = o(t), \tag{1.2.3}$$

$$\mathbf{E}Z(t) = a(t + \mathbf{E}\chi(t)) - \mathbf{E}\zeta(t) = at + r_Z(t), \; r_Z(t) = o(t), \tag{1.2.4}$$

as $t \to \infty$.

(ii) *If* $\mathbf{E}\tau^2 < \infty$, *then in* (1.2.3) *the asymptotic expansion holds, in which, in the nonlattice case,*

$$r_Y(t) = \frac{a_\zeta \mathbf{E}\tau^2}{2a_\tau^2} + o(1). \tag{1.2.5}$$

If, in addition $\mathbf{E}|\tau\zeta| < \infty$, *then in* (1.2.4)

$$r_Z(t) = \frac{a_\zeta \mathbf{E}\tau^2}{2a_\tau^2} - \frac{\mathbf{E}\tau\zeta}{a_\tau} + o(1). \tag{1.2.6}$$

(iii) *If* $\mathbf{E}\tau^2 < \infty$, $\sigma_\xi^2 := \mathbf{D}\xi$, *then as* $t \to \infty$

$$DY(t) = \frac{\sigma_\xi^2 t}{a_\tau} + o(t), \tag{1.2.7}$$

$$DZ(t) = \frac{\sigma_\xi^2 t}{a_\tau} + o(t). \tag{1.2.8}$$

If, in addition $\mathbf{E}\tau^3 < \infty$, *then*

$$\mathbf{D}Y(t) = \frac{\sigma_\xi^2 t}{a_\tau} + O(\sqrt{t}). \tag{1.2.9}$$

The same representation also holds for $\mathbf{D}Z(t)$.

II. *In the inhomogeneous case,* (1.2.3) *and* (1.2.4) *remain valid if* $\mathbf{E}\tau_1 < \infty$ *and* $\mathbf{E}|\zeta_1| < \infty$. *Moreover,* (1.2.7) *and* (1.2.8) *remain valid if* $\mathbf{E}\tau_1^2 < \infty$ *and* $\mathbf{E}\zeta_1^2 < \infty$.

Proof of Theorem 1.2.1 I. We first consider the homogeneous case $(\tau_1, \zeta_1) \underset{d}{=} (\tau, \zeta)$.

(i), (ii) Using the Wald identity and (1.1.4), we find that

$$\mathbf{E}Y(t) = \mathbf{E}Z_{\eta(t)} = a_\zeta \mathbf{E}\eta(t) = a(t + \mathbf{E}\chi(t)).$$

This proves (1.2.3) in view of Lemma 1.1.1.

Similarly,

$$\mathbf{E}Z(t) = \mathbf{E}Y(t) - \mathbf{E}\zeta(t),$$

which verifies (1.2.4). In the cases $\mathbf{E}\tau^2 < \infty$, $\mathbf{E}|\tau\zeta| < \infty$, the values of $\mathbf{E}\chi(t)$, $\mathbf{E}\zeta(t)$ are given, respectively, by the formulas (1.1.9) and (1.1.11). This proves (1.2.5) and (1.2.6).

The values of $r_Y(t)$, $r_Z(t)$ in the arithmetic case can also be found (see Lemma 1.1.1).

(iii) We have

$$\begin{aligned}\mathbf{D}Y(t) = \mathbf{E}\big[Y(t) - a(t + \mathbf{E}\chi(t))\big]^2 &= \mathbf{E}\big[Z_{\eta(t)} - aT_{\eta(t)} + a(\chi(t) - \mathbf{E}\chi(t))\big]^2 \\ &= \mathbf{E}\big[S_{\eta(t)} + a(\chi(t) - \mathbf{E}\chi(t))\big]^2.\end{aligned}$$

By the Wald identity,

$$\mathbf{E}S_{\eta(t)}^2 = \sigma_\xi^2 \mathbf{E}\eta(t) = \frac{\sigma_\xi^2(t + \mathbf{E}\chi(t))}{a_\tau}$$

(see, e.g., Theorem 15.2.5 in [21]). Therefore,

$$\mathbf{D}Y(t) = \frac{\sigma_\xi^2(t + \mathbf{E}\chi(t))}{a_\tau} + a^2\mathbf{D}\chi(t) + 2a\mathbf{E}S_{\eta(t)}(\chi(t) - \mathbf{E}\chi(t)). \tag{1.2.10}$$

Since $\mathbf{E}\tau^2 < \infty$, we have $\mathbf{D}\chi(t) = \mathbf{E}\chi^2(t) - (\mathbf{E}\chi(t))^2 = o(t)$ as $t \to \infty$, and hence, using the Cauchy–Schwarz inequality for $|\mathbf{E}S_{\eta(t)}(\chi(t) - \mathbf{E}\chi(t))|$, we get

$$\mathbf{D}Y(t) = \frac{\sigma_\xi^2 t}{a_\tau} + o(t) \quad \text{as} \quad t \to \infty. \tag{1.2.11}$$

This proves (1.2.7). A similar analysis establishes (1.2.8). If $\mathbf{E}\tau^3 < \infty$, then the value of $\mathbf{D}\chi(t)$ is bounded uniformly in t, and so we get (1.2.9).

II. Let us now consider the inhomogeneous CRPs, which we denote by $\widetilde{Y}(t)$ and $\widetilde{Z}(t)$. We have

$$\begin{aligned}\mathbf{E}\widetilde{Y}(t) &= \mathbf{E}(\zeta_1;\ \tau_1 > t) + \int_0^t \mathbf{E}(\zeta_1 + Y(t-s);\ \tau_1 \in ds)\\ &= \mathbf{E}\zeta_1 + \int_0^t \mathbf{E}Y(t-s)\mathbf{P}(\tau_1 \in ds),\end{aligned}$$

where $Y(u)$ is a homogeneous CRP not depending on (τ_1, ζ_1). Hence by assertion I(i), we have, as $t \to \infty$,

$$\begin{aligned}\mathbf{E}\widetilde{Y}(t) &= \mathbf{E}\zeta_1 + \int_0^{t/2} \mathbf{E}Y(t-s)\mathbf{P}(\tau \in ds) + \mathbf{P}(\tau_1 > t/2)O(t)\\ &= \mathbf{E}\zeta_1 + at - a\mathbf{E}\tau_1 + o(t).\end{aligned}$$

A relation of the same form also holds for $\mathbf{E}\widetilde{Z}(t)$. This proves the first claim in assertion II. The second claim is proved similarly, but the argument is somewhat more tedious. Theorem 1.2.1 is proved. □

The parameter

$$\sigma^2 := \frac{\sigma_\xi^2}{a_\tau}$$

in (1.2.7), (1.2.8) can be interpreted as the "specific" (per time unit) asymptotic variance of the processes $Y(t)$, $Z(t)$.

Note that in the case $\mathbf{E}\tau^2 = \infty$ we have $\mathbf{D}\chi(t) = \infty$, and for $a \neq 0$, from (1.2.10) we get $\mathbf{D}Y(t) = \infty$. At the same time, we shall show in Theorem 1.3.3 that the distribution of $Y(t)$ for $\sigma_\xi < \infty$ is asymptotically normal (the relations $\mathbf{E}\tau^2 = \infty$ and $\mathbf{D}\xi < \infty$ are compatible if, for example, $a = 0$ or $\zeta = a\tau + \omega$, $\mathbf{E}\omega^2 < \infty$).

If $a = 0$, then for nonlattice τ, in the homogeneous case

$$\mathbf{E}Y(t) = 0, \qquad \mathbf{D}Y(t) = \sigma^2\left(t + \frac{\mathbf{E}\tau^2}{2a_\tau} + o(1)\right) \quad \text{as} \quad t \to \infty.$$

Remark 1.2.2 It is easily checked that (1.2.3), (1.2.4) and (1.2.7), (1.2.8) also hold in the cases when the distributions of τ_1, ζ_1 depend on t and, respectively, $\mathbf{E}\tau_1 = o(t)$, $\mathbf{E}|\zeta_1| = o(t)$ and $\mathbf{E}\tau_1^2 = o(t)$, $\mathbf{E}\zeta_1^2 = o(t)$ as $t \to \infty$. We will not enlarge on this, because it will not be required in what follows.

1.2.2 Strong Laws of Large Numbers

Theorem 1.2.3 *The following relations hold*:

$$\frac{Y(t)}{t} \underset{a.s.}{\to} a, \qquad \frac{Z(t)}{t} \underset{a.s.}{\to} a \quad as \quad t \to \infty. \tag{1.2.12}$$

Proof of Theorem 1.2.3 We have

$$\frac{Y(t)}{t} = \frac{Z_{\eta(t)}}{T_{\eta(t)} - \chi(t)} = \frac{Z_{\eta(t)}}{\eta(t)} \left(\frac{T_{\eta(t)}}{\eta(t)} - \frac{\chi(t)}{t} \frac{t}{\eta(t)} \right)^{-1}. \tag{1.2.13}$$

Here by the strong law of large numbers for sums of random variables and since $\eta(t) \underset{\text{a.s.}}{\to} \infty$ as $t \to \infty$, we have

$$\frac{Z_{\eta(t)}}{\eta(t)} \underset{\text{a.s.}}{\to} a_\zeta, \qquad \frac{T_{\eta(t)}}{\eta(t)} \underset{\text{a.s.}}{\to} a_\tau \quad \text{as} \quad t \to \infty.$$

Moreover, $\frac{t}{\eta(t)} \underset{\text{a.s.}}{\to} a_\tau$ by Lemma 1.1.6 and $\frac{\chi(t)}{t} \underset{\text{a.s.}}{\to} 0$ as $t \to \infty$ by Corollary 1.1.8. Now (1.2.12) follows from (1.2.13). For the process $Z(t)$ the proof is similar. Theorem 1.2.3 is proved. □

1.3 Central Limit Theorem and the Law of the Iterated Logarithm

1.3.1 Anscombe's Theorem

The proofs of the main results in this and some other sections will be based on Anscombe's theorem (see [2]) on the possibility of replacing in limit theorems a deterministic growing time parameter by a random growing parameter. Below we present slightly different statements and proofs of this theorem, which are (in our opinion) simpler than those in the original paper [2]. Let G be some distribution function; the sign "$\Rightarrow$" denotes the weak convergence.

Theorem 1.3.1 (Anscombe) *Let $s(n)$ be a sequence of random variables such that:*

(1)
$$\mathbf{P}(s(n) < v) \Rightarrow G(v) \quad as \quad n \to \infty, \tag{1.3.1}$$

(2)
$$\max_{|k|<\delta n} |s(n+k) - s(n)| \underset{P}{\to} 0 \quad as \quad n \to \infty, \quad \delta = \delta(n) \to 0. \tag{1.3.2}$$

Let, further, $\theta(t)$ be a sequence of integer-valued random variables defined on the same probability space as $s(n)$ and depending on the increasing parameter t such that

(3) *there exists a function $h(t) \to \infty$ as $t \to \infty$ such that*

$$\frac{\theta(t)}{h(t)} \underset{P}{\to} 1. \tag{1.3.3}$$

Under these conditions,

$$\mathbf{P}(s(\theta(t)) < v) \Rightarrow G(v) \quad as \quad t \to \infty.$$

Remark 1.3.2 Condition (1.3.2) of Theorem 1.3.1 is a slightly simpler version of the so-called "Anscombe's condition," which was used in the original paper [2] (see also §1.3 in [72]). Condition (1.3.2) is slightly weaker than the original Anscombe's condition.

Proof of Theorem 1.3.1 By (1.3.3), there exists a sequence ε_t that converges to zero sufficiently slowly as $t \to \infty$ and such that

$$\mathbf{P}(|\,\theta(t) - h(t)| \geq \varepsilon_t h(t)) \to 0.$$

Hence, assuming without loss of generality that the function $h(t)$ is integer-valued, we have, for any point v of continuity of the function G,

$$\begin{aligned}\mathbf{P}\big(s(\theta(t)) < v\big) &= o(1) + \mathbf{P}\big(s(\theta(t)) < v;\ |\theta(t) - h(t)| < \varepsilon_t h(t)\big)\\ &= o(1) + \mathbf{P}\big(s(h(t)) + \rho(t) < v;\ |\theta(t) - h(t)| < \varepsilon_t h(t)\big),\end{aligned} \tag{1.3.4}$$

where $\rho(t) = s(\theta(t)) - s(h(t))$. Moreover, on the set $|\theta(t) - h(t)| < \varepsilon_t h(t)$ in view of (1.3.2) we have

$$|\rho(t)| \leq \max_{|k|<\varepsilon_t h(t)} \big|s(h(t) + k) - s(h(t))\big| \underset{p}{\to} 0$$

as $t \to \infty$. Hence from (1.3.1), (1.3.4) it follows that

$$\mathbf{P}\big(s(\theta(t)) < v\big) = o(1) + \mathbf{P}\big(s(h(t)) + \rho(t) < v\big) \Rightarrow G(v)$$

as $t \to \infty$. Theorem 1.3.1 is proved. □

1.3.2 Central Limit Theorem

The central limit theorem for the GRPs $Y(t)$, $Z(t)$ can be obtained either directly (see, e.g., [21], Chapter 10) or using Anscombe's theorem – this approach is employed below under slightly more general (than usually) conditions.

Consider the sequence

$$y(t) := \frac{Y(t) - at}{\sigma\sqrt{t}}, \tag{1.3.5}$$

where, as before $a = a_\zeta / a_\tau$, $\sigma^2 = \sigma_\xi^2 a_\tau^{-1}$, $\xi = \zeta - a\tau$, and it is assumed that $\sigma_\xi^2 = \mathbf{E}\xi^2 < \infty$. Below, as in Corollary 1.1.2 and in Remark 1.2.2, we assume that for a partial triangular array scheme the distribution of (τ_1, ζ_1) may depend on t.

Theorem 1.3.3 *If $\sigma_\xi < \infty$, $\tau_1 = o_p(\sqrt{t})$, $\zeta_1 = o_p(\sqrt{t})$ as $t \to \infty$, then*

$$y(t) \Leftrightarrow \mathbf{\Phi}, \tag{1.3.6}$$

where $\mathbf{\Phi}$ is the standard normal distribution. The same result also holds for $z(t) = \frac{Z(t)-at}{\sigma\sqrt{t}}$.

The symbol "$\Leftrightarrow$" in (1.3.6) means weak convergence as $t \to \infty$ of the distributions of the sequence of random variables $y(t)$ to the distribution $\mathbf{\Phi}$.

Note that in the theorem we do not suppose that σ_τ is finite – it suffices to assume only that σ_ξ is finite; so if, for example, $a = a_\zeta = 0$, then $\xi = \zeta$, and for (1.3.6) it suffices that only $a_\tau = \mathbf{E}\tau$ and $\sigma_\zeta^2 = \mathbf{D}\zeta$ exist.

The next result is required in order to be able to employ Anscombe's theorem in the proof of Theorem 1.3.3.

Lemma 1.3.4 *The normalized sums*

$$s(n) = \frac{S_n - an}{\sigma_\xi \sqrt{n}} \tag{1.3.7}$$

satisfy condition 2) *of Theorem* 1.3.1 (*see* (1.3.2), *where* S_n *are defined in* (1.2.1)).

Proof We have for $m = n + k$,

$$s(m) - s(n) = \frac{S_m - am}{\sigma_\xi \sqrt{m}} - \frac{S_n - an}{\sigma_\xi \sqrt{n}} \pm \frac{S_n - an}{\sigma_\xi \sqrt{m}}, \tag{1.3.8}$$

where

$$\max_{|k|<\delta n} \left| \frac{S_n - an}{\sigma_\xi \sqrt{n}} - \frac{S_n - an}{\sigma_\xi \sqrt{m}} \right| \le \left| \frac{S_n - an}{\sigma_\xi \sqrt{n}} \right| \max \left| 1 - \frac{1}{\sqrt{1 \pm \delta}} \right| \underset{p}{\to} 0 \tag{1.3.9}$$

as $n \to \infty$, $\delta = \delta(n) \to 0$. Next, assume, for definiteness, that $m > n$ $(k > 0)$. Then

$$\max_{k<\delta n} \left| \frac{S_m - am}{\sigma_\xi \sqrt{m}} - \frac{S_n - an}{\sigma_\xi \sqrt{m}} \right| \underset{p}{\le} \sqrt{\delta} \max_{k<\delta n} \left| \frac{S_k - ak}{\sigma_\xi \sqrt{\delta n}} \right|, \tag{1.3.10}$$

where $\max_{k<\delta n} |\frac{S_k - ak}{\sigma_\xi \sqrt{\delta n}}|$ converges in distribution as $\delta n \to \infty$ to a proper random variable. Hence the right-hand side in (1.3.10) converges in probability to zero as $n \to \infty$, $\delta = \delta(n) \to 0$. The case $m < n$ is considered similarly. (Note that the convergence of the right-hand side in (1.3.10) in probability to zero as $\delta \to 0$ can be also derived from the Kolmogorov–Doob inequality for the distribution of the right-hand side in (1.3.10) (see, e.g., Theorem 15.3.3 in [21])).

Now (1.3.2) follows from (1.3.8) and (1.3.9). Lemma 1.3.4 is proved. □

Proof of Theorem 1.3.3 Using the notation of §1.1.1, in the homogeneous case we can write

$$y(t) = \frac{S_{\eta(t)}}{\sigma \sqrt{\eta(t)}} \cdot \sqrt{\frac{\eta(t)}{t}} + \frac{a\chi(t)}{\sigma \sqrt{t}} = s(\eta(t)) \sqrt{\frac{\eta(t) a_\tau}{t}} + \frac{a\chi(t)}{\sigma \sqrt{t}}, \tag{1.3.11}$$

where by Anscombe's theorem, $s(\eta(t)) \Leftrightarrow \mathbf{\Phi}$ as $t \to \infty$. By Lemmas 1.1.6 and 1.1.1,

$$\sqrt{\frac{\eta(t) a_\tau}{t}} \underset{\text{a.s.}}{\to} 1, \qquad \frac{\chi(t)}{\sqrt{t}} \underset{p}{\to} 0 \quad \text{as} \quad t \to \infty,$$

and hence from (1.3.11) we get (1.3.6), thereby proving Theorem 1.3.3 in the homogeneous case.

In the inhomogeneous case, for the CRP $\widetilde{Y}(t)$ we have

$$\widetilde{Y}(t) = \begin{cases} \zeta_1 & \text{for} \quad \tau_1 < t, \\ \zeta_1 + Y(t - \tau_1) & \text{for} \quad \tau_1 \geq t, \end{cases}$$

where $Y(t)$ is a homogeneous CRP not depending on (τ_1, ζ_1). If τ_1, ζ_1 are nonrandom, $\tau_1 = o(\sqrt{t})$, $\zeta_1 = o(\sqrt{t})$ as $t \to \infty$, then (1.3.6) clearly holds also for the process $\widetilde{Y}(t)$. It is easily seen that this result also holds in the case $\tau_1 = o_p(\sqrt{t})$, $\zeta_1 = o_p(\sqrt{t})$. This proves Theorem 1.3.3 for the process $Y(t)$. Note that we have proved this also for the process $Z(t)$, because

$$Z(t) = Y(t) - \zeta(t), \qquad \frac{\zeta(t)}{\sqrt{t}} \underset{\text{a.s.}}{\to} 0 \quad \text{as} \quad t \to \infty. \tag{1.3.12}$$

Theorem 1.3.3 is proved. □

The asymptotic normality of the CRP in the case $a = 0$ (that is, the normality of $\frac{S_{\eta(t)}}{\sigma\sqrt{t}}$) was proved in [102] (see also p. 261 in [50] and Chapter 1 in [72]) also with the help of Anscombe's theorem. The asymptotic normality of the CRP in the general case was established in § 10.6 of [21], as was already noted, via "direct" approaches.

1.3.3 Law of the Iterated Logarithm

Theorem 1.3.5 *Let $\sigma_\tau < \infty$, $\sigma < \infty$, let τ_1, ζ_1 be arbitrary fixed random variables, and let*

$$L(t) := \sigma\sqrt{2t \ln \ln t} \quad \textit{for} \quad t \geq 3. \tag{1.3.13}$$

Then, with probability 1,

$$\overline{y} := \overline{\lim_{t\to\infty}} \frac{Y(t) - at}{L(t)} = 1, \qquad \underline{\lim_{t\to\infty}} \frac{Y(t) - at}{L(t)} = -1. \tag{1.3.14}$$

If $a = 0$, then the condition $\sigma_\tau < \infty$ is superfluous.
The same result also holds for the process $Z(t)$.

The law of the iterated logarithm for CRPs with $a = 0$ (that is, for the process $S_{\eta(t)}$) was established in [117] (and follows from Chapters 1 and 5 in [72]).

Proof of Theorem 1.3.5 It suffices to prove (1.3.14). Since $t = T_{\eta(t)} - \chi(t)$, we have

$$\overline{y} = \overline{\lim_{t\to\infty}} \left(\frac{S_{\eta(t)}}{L(t)} + \frac{a\chi(t)}{L(t)} \right). \tag{1.3.15}$$

By Corollary 1.1.8 for $\sigma_\tau < \infty$ with probability 1,

$$\overline{\lim_{t\to\infty}} \frac{a\chi(t)}{L(t)} = 0.$$

If $a = 0$, then this relation always holds. Moreover, with probability 1,

$$\lim_{t\to\infty} \frac{a_\tau \eta(t)}{t} = 1, \qquad \lim_{t\to\infty} \frac{L(t)}{L(a_\tau \eta(t))} = 1,$$

and hence, with probability 1,

$$\overline{y} = \overline{\lim_{t\to\infty}} \frac{S_{\eta(t)}}{L(t)} = \overline{\lim_{t\to\infty}} \frac{S_{\eta(t)}}{L(a_\tau \eta(t))}.$$

With increasing t the function $\eta(t)$ goes through all values $1, 2, \ldots$. Hence, since $\sigma\sqrt{a_\tau} = \sigma_\xi$ and by the law of the iterated logarithm for $\{S_n\}$, we have

$$\overline{y} = \overline{\lim_{n\to\infty}} \frac{S_n}{L(a_\tau n)} = \overline{\lim_{n\to\infty}} \frac{S_n}{\sqrt{a_\tau} L(n)} = 1.$$

By (1.3.12) the same result also holds for the process $Z(t)$. Theorem 1.3.5 is proved. □

1.4 Convergence to a Stable Law. Analog of the Law of the Iterated Logarithm

1.4.1 Convergence to a Stable Law

In this section, we consider the case $\sigma_\xi^2 = \mathbf{D}\xi = \infty$ and assume that the condition of regular decay of the distribution of $\xi = \zeta - a\tau$ at infinity is met. We set

$$F_-(t) = \mathbf{P}(\xi \le -t), \quad F_+(t) = \mathbf{P}(\xi \ge t),$$
$$F(t) = F_-(t) + F_+(t) = \mathbf{P}(|\xi| \ge t).$$

We assume the following condition.

$[\mathbf{R}_{\alpha,\beta}]$ *The function $F(t)$ is regularly varying at infinity, that is, $F(t)$ has the form*

$$F(t) = t^{-\alpha} l(t), \qquad \alpha \in (1, 2),$$

where $l(t)$ is a slowly varying at infinity function. Moreover, the limit

$$\lim_{t\to\infty} \frac{F_+(t)}{F(t)} =: \beta_+ \in [0, 1]$$

exists, and we put $\beta := 2\beta_+ - 1$.

Next, let $F^{(-1)}(u)$ be the generalized inverse function for $F(t)$,

$$F^{(-1)}(u) = \inf\{t : F(t) < u\}$$

and

$$\sigma_\xi(n) := F^{(-1)}(1/n). \tag{1.4.1}$$

The function $\sigma_\xi(n)$ reads as $n^{1/\alpha} l_{\sigma_\xi}(n)$, where $l_{\sigma_\xi}(n)$ is a slowly varying sequence (see, e.g., §8.8 in [38]).

Under condition $[\mathbf{R}_{\alpha,\beta}]$, the normalized sums $s(n) := S_n/\sigma_\xi(n)$ (see (1.2.1)) weakly converge in distribution to the stable law $\mathbf{\Phi}_{\alpha,\beta}$ with parameters (α, β). We set

$$\sigma(t) = \sigma_\xi(t) a_\tau^{-1/\alpha}.$$

Theorem 1.4.1 *If $\xi = \zeta - a\tau$ satisfies condition $[\mathbf{R}_{\alpha,\beta}]$, and if $\tau_1 = o_p(\sigma(t))$, $\zeta_1 = o_p(\sigma(t))$, then, as $t \to \infty$,*

$$y(t) := \frac{Y(t) - at}{\sigma(t)} \Leftrightarrow \Phi_{\alpha,\beta}. \tag{1.4.2}$$

The same result also holds for the process $z(t) := \frac{Z(t)-at}{\sigma(t)}$.

(Recall that the notation $y(t) \Leftrightarrow \Phi_{\alpha,\beta}$ means the weak convergence of the distributions of $y(t)$ to the distribution $\mathbf{\Phi}_{\alpha,\beta}$ with the distribution function $\Phi_{\alpha,\beta}$.)

Assertion (1.4.2) in the case $a = 0$ (that is, for the stopped sums $S_{\eta(t)}/\sigma(t)$) was announced (without proof) in [72] (Theorem 1.3.2).

The scheme of the proof of Theorem 1.4.1 is the same as in the central limit theorem (see §1.3.2). We first show that the normalized sums

$$s(n) = \frac{S_n - an}{\sigma_\xi(n)}$$

satisfy condition 2) of Theorem 1.3.1, that is, we establish, in the homogeneous case, the following analog of Lemma 1.3.4.

Lemma 1.4.2 *Let a random variable $\xi = \zeta - a\tau$ satisfy condition $[\mathbf{R}_{\alpha,\beta}]$, $\alpha \in (1, 2)$. Then:*

$$\sup_{|k|<\delta n} |s(n+k) - s(n)| \xrightarrow[p]{} 0 \tag{1.4.3}$$

as $n \to \infty$, $\delta = \delta(n) \to 0$.

Proof of Lemma 1.4.2 As in (1.3.8), (1.3.9), we have, for $m = n + k$,

$$\begin{aligned}\sup_{|k|<\delta n} |s(n+k) - s(n)| &= \sup_{|k|<\delta n} \left| \frac{S_m - am}{\sigma_\xi(m)} - \frac{S_n - an}{\sigma_\xi(n)} \pm \frac{S_n - an}{\sigma_\xi(m)} \right| \\ &\le \sup_{|k|<\delta n} \frac{|S_k - ak|}{\sigma_\xi(m)} + |s(n)| \sup_{|k|<\delta n} \left| \frac{\sigma_\xi(n)}{\sigma_\xi(n+k)} - 1 \right|. \end{aligned} \tag{1.4.4}$$

Next, $\frac{\sigma_\xi(n)}{\sigma_\xi(n+k)} \to 1$ as $n \to \infty$, $k = o(n)$, and hence the second term on the right of (1.4.4) is, clearly, $o_p(1)$. For the first term on the right of (1.4.4), for $k > 0$ we have

$$\max_{k<\delta n} \left| \frac{S_k - ak}{\sigma_\xi(m)} \right| = \frac{\sigma_\xi(\delta n)}{\sigma_\xi(m)} \max_{k<\delta n} \left| \frac{S_k - ak}{\sigma_\xi(\delta n)} \right|. \tag{1.4.5}$$

From Theorem 3.1.1 and Corollary 3.1.2 in [38], we have, as $n \to \infty$ and for large v,

$$\mathbf{P}\left(\max_{k<\delta n} |S_k - ak| > v\sigma_\xi(\delta n)\right) \le 2\delta n F(v\sigma_\xi(\delta n)). \tag{1.4.6}$$

Taking into account that $\frac{\sigma(\delta n)}{\sigma(n)} \sim \delta^{1/\alpha}$ as $n \to \infty$, we set $v = \varepsilon\delta^{-1/\alpha}$ for any fixed $\varepsilon > 0$. Then the right-hand side in (1.4.6) is majorized by

$$3\delta n\varepsilon^{-\alpha}\delta F(\sigma(\delta n)) \sim 3\varepsilon^{-\alpha}\delta \to 0$$

as $\delta \to 0$. This means that the left-hand side in (1.4.5) converges in probability to 0 as $n \to \infty$, $\delta = \delta(n) \to 0$. As a result, we get (1.3.2). The case $k < 0$ is dealt with similarly. Lemma 1.4.2 is proved. □

Proof of Theorem 1.4.1 Consider the homogeneous case. The sequence $s(n)$ satisfies condition 1) of Theorem 1.3.1 with $G = \Phi_{\alpha,\beta}$. By Lemma 1.4.2, it also satisfies condition 2) of this theorem. Finally, as we have already seen, the sequence $\eta(t)$ satisfies condition 3) of Theorem 1.3.1. Hence by Theorem 1.3.1,

$$s(\eta(t)) \Leftrightarrow \Phi_{\alpha,\beta} \quad \text{as} \quad t \to \infty.$$

We have $\frac{a_\tau^{1/\alpha}\sigma_\xi(\eta(t))}{\sigma_\xi(t)} \underset{p}{\to} 1$, and hence $\frac{\sigma_\xi(\eta(t))}{\sigma(t)} \underset{p}{\to} 1$ as $t \to \infty$. Moreover, by Lemma 1.1.1, $\frac{\chi(t)}{\sigma(t)} \underset{p}{\to} 0$. Therefore,

$$y(t) = \frac{Y(t) - at}{\sigma(t)} = \frac{S_{\eta(t)}}{\sigma_\xi(\eta(t))}\,\frac{\sigma_\xi(\eta(t))}{\sigma(t)} + \frac{a\chi(t)}{\sigma(t)} = s(\eta(t)) + o_p(1) \Leftrightarrow \Phi_{\alpha,\beta}.$$

The extension of this result to the process $z(t) = \frac{Z(t)-at}{\sigma(t)}$ and the inhomogeneous case proceeds as in Theorem 1.3.3. Theorem 1.4.1 is proved. □

1.4.2 Analog of the Law of the Iterated Logarithm

For $\sigma = \infty$, an analog of the law of the iterated logarithm for the process $Y(t) - at$ is as follows.

Theorem 1.4.3 *Let ξ satisfy condition $[\mathbf{R}_{\alpha,\beta}]$ with $\alpha \in (1, 2)$, $\beta > -1$, let τ_1, ζ_1 be arbitrary fixed random variables, and let at least one of the following two conditions hold*:

$$a = 0 \ \text{or} \ F_\tau(t) := \mathbf{P}(\tau \geq t) \leq cF(t), \quad c = \text{const}, \quad F(t) = \mathbf{P}(|\xi| \geq t). \qquad (1.4.7)$$

Then:

(i) *For any $\theta > 1/\alpha$,*

$$\overline{y}_\theta := \varlimsup_{t\to\infty} \frac{Y(t) - at}{\sigma(t)(\ln t)^\theta} = 0, \qquad (1.4.8)$$

and, for any $\theta < 1/\alpha$,

$$\overline{y}_\theta = \infty \qquad (1.4.9)$$

with probability 1.

(ii) *Assertion* (i) *remains valid for the process* $Z(t)$ *if the second condition in* (1.4.7) *is augmented with the condition*

$$F_\zeta(t) := \mathbf{P}(\zeta \geq t) \leq cF(t), \quad c = \text{const}, \quad t > 0. \tag{1.4.10}$$

Theorem 1.4.3 can be considered as the "law of the single logarithm". This result can also be phrased in terms of the iterated logarithm.

Corollary 1.4.4 *Under conditions of assertion* (i) *of Theorem* 1.4.3,

$$\overline{\lim_{t\to\infty}} \frac{\ln(Y(t) - at)^+ - \ln \sigma(t)}{\ln \ln t} = \frac{1}{\alpha}, \tag{1.4.11}$$

with probability 1, *where* $(x)^+ = \max(0, x)$.

Proof of Corollary 1.4.4 We set

$$y_{(\theta)}(t) := \frac{Y(t) - at}{\sigma(t)(\ln t)^\theta}.$$

Hence by (1.4.8) with $\theta = \frac{1}{\alpha} + \varepsilon$, $\varepsilon > 0$, for all sufficiently large t, with probability 1

$$\begin{gathered} y_{(\theta)}(t) < 1, \quad \ln \left(y_{(\theta)}(t)\right)^+ < 0, \\ \frac{\ln(Y(t) - at)^+ - \ln \sigma(t)}{\ln \ln t} < \theta = \frac{1}{\alpha} + \varepsilon. \end{gathered} \tag{1.4.12}$$

We now get an estimate from below. From (1.4.9) it follows that, for almost all elementary events from the probability space on which the process $Y(t)$, $t \geq 0$ (or the sequence $\{\tau_j, \zeta_j\}_{j=1}^\infty$) is defined, for any $\theta = \frac{1}{\alpha} - \varepsilon$, $\varepsilon > 0$, there exists a sequence $\{t_k\}$, $t_k \to \infty$ as $k \to \infty$ such that $y_\theta(t_k) \to \infty$ as $k \to \infty$. Hence, arguing as above, we get

$$\begin{gathered} y_{(\theta)}(t_k) > 1, \quad \ln y_{(\theta)}(t_k) > 0, \\ \frac{\ln(Y(t_k) - at_k) - \ln \sigma(t_k)}{\ln \ln t_k} > \theta = \frac{1}{\alpha} - \varepsilon \quad \text{as} \quad k \to \infty. \end{gathered}$$

Since $\varepsilon > 0$ is arbitrary, this together with (1.4.12) yields (1.4.11), which proves Corollary 1.4.4. □

From the proof of Corollary 1.4.4 it follows that there exists a sequence $\varepsilon_t \to 0$ as $t \to \infty$ such that

$$\mathbf{P}(Y(t) - at < \sigma(t)(\ln t)^{1/\alpha + \varepsilon_t} \text{ for all } t \geq T) \to 1 \quad \text{as} \quad T \to \infty.$$

Results similar to Theorem 1.4.3 and Corollary 1.4.4 also hold for the process $Z(t)$ and negative deviations of $Y(t) - at$ and $Z(t) - at$.

Proof of Theorem 1.4.3 The scheme of the proof of the first assertion is similar to that used in the proof of Theorem 1.3.5. The following modifications are required. Instead of (1.3.15), we will employ the equality

$$\overline{y}_\theta = \overline{\lim_{t\to\infty}} \left(\frac{S_{\eta(t)}}{\sigma(t)(\ln t)^\theta} + \frac{a\chi(t)}{\sigma(t)(\ln t)^\theta} \right), \tag{1.4.13}$$

in which by condition (1.4.7) and the second assertion of Lemma 1.1.7,

$$\lim_{t\to\infty} \frac{a\chi(t)}{\sigma(t)(\ln t)^\theta} = 0 \quad \text{a.s.} \tag{1.4.14}$$

To estimate the principal term on the right of (1.4.13), we need the following result (an analog of the "law of the single logarithm") for sums of random variables; see Theorem 3.9.1 in [38]):

Under the conditions of Theorem 1.4.3,

$$\overline{s}_\theta := \overline{\lim_{n\to\infty}} \frac{S_n}{\sigma(n)(\ln)^\theta} = 0 \quad \textit{a.s.} \tag{1.4.15}$$

for any $\theta > 1/\alpha$, *and* $\overline{s}_\theta = \infty$ *a.s. for any* $\theta < 1/\alpha$.

Since

$$\frac{\eta(t)}{t} \underset{\text{a.s.}}{\rightarrow} a_\tau^{-1} \quad \text{as} \quad t \to \infty,$$

we have

$$\overline{\lim_{t\to\infty}} \frac{S_{\eta(t)}}{\sigma(t)(\ln t)^\theta} = a_\tau^{-1} \overline{\lim_{t\to\infty}} \frac{S_{\eta(t)}}{\sigma(\eta(t))(\ln \eta(t))^\theta}.$$

In view of (1.4.15) this implies the first assertion of the theorem.

(ii) The second assertion can be proved in the same way, but since $Z(t) = Y(t) - \zeta(t)$, the condition

$$\lim_{t\to\infty} \frac{|\zeta(t)|}{\sigma(t)(\ln t)^\theta} = 0 \quad \text{a.s.} \tag{1.4.16}$$

should in addition be satisfied. This condition follows from condition (1.4.10) and Lemma 1.4.2. If $a = 0$, then $F_\zeta(t) = F(t)$ and condition (1.4.10) is superfluous, as well as condition (1.4.7). Theorem 1.4.3 is proved. □

1.5 Invariance Principle

1.5.1 Introduction

In this section, the behavior of CRP *trajectories* will be studied under the assumption that

$$\mathbf{E}\tau^2 < \infty, \qquad \mathbf{E}\zeta^2 < \infty.$$

It will be convenient to consider the processes $Y(t)$, $Z(t)$ on an increasing time interval $[0, u_0T]$ as $T \to \infty$ and with fixed $u_0 > 0$. Compressing the time by the coefficient T, we consider the processes

$$y_T(u) = \frac{Y(uT) - auT}{\sigma\sqrt{T}} \quad \text{and} \quad z_T(u) = \frac{Z(uT) - auT}{\sigma\sqrt{T}} \tag{1.5.1}$$

defined on the interval $[0, u_0]$.

The invariance principle for the process $y_T(u)$ was established in [55, 112] with the help of quite complicated constructions on the common probability space and as a corollary of strong approximation of the processes y_T by a Wiener process w; see also [66]. For the convergence rate in the invariance principle, see, for example, [37], [56] and the references given therein.

From the proof of Theorem 1.5.3 on the invariance principle for $y_T(u)$ it will follow that the principal contribution to the limit distribution of y_T comes from the distribution of the stopped normalized sums $S_{\nu(uT)}$. Functional limit theorems for the stopped sums $\frac{S_{\nu(uT)}}{\sigma_\xi \sqrt{T}}$ (that is, theorems for the processes $z_T(u)$ with $a = 0$) are consequences of the results of [72, Chapter 5].

1.5.2 Analog of Anscombe's Theorem in the Case of Convergence to a Continuous Process

Let $s_T(u)$ be a sequence of processes on the interval $u \in [0, u_0]$ defined in a measurable space $(\mathbb{D}(0, u_0), \mathfrak{B}_\mathbb{D})$, where $\mathbb{D}(0, u_0)$ is the space of functions on $[0, u_0]$ without discontinuities of the second kind and which are right-continuous at each point, and let $\mathfrak{B}_\mathbb{D}$ be the σ-algebra of sets from $\mathbb{D}$ generated by cylindrical sets. Next, let $\mathbb{C}(0, u_0)$ be the space of continuous functions on $[0, u_0]$, let $\mathfrak{B}_\mathbb{C}$ be the σ-algebra of sets from $\mathbb{C}$ generated by cylindrical sets, and let $\rho_{\mathbb{C},u_0}$ be the uniform metric: for $g_i \in \mathbb{D}(0, u_0)$, $i = 1, 2$,

$$\rho_{\mathbb{C},u_0}(g_1, g_2) = \sup_{u \le u_0} |g_1(u) - g_2(u)|.$$

We say that a sequence of processes $s_T(u)$, defined on the space $(\mathbb{D}(0, u_0), \mathfrak{B}_\mathbb{D})$, $\mathbb{C}$-*converges* (in distribution) as $T \to \infty$ to a process $s(u)$, $u \in [0, u_0]$, defined on the space $(\mathbb{C}(0, u_0), \mathfrak{B}_\mathbb{C})$ if

$$f(s_T) \Rightarrow f(s) \quad \text{as} \quad T \to \infty, \tag{1.5.2}$$

for any measurable in $(\mathbb{D}(0, u_0), \mathfrak{B}_\mathbb{D})$ functional f continuous in the uniform metric at "points" of the space $\mathbb{C}(0, u_0)$. Here the symbol "$\Rightarrow$" means the weak convergence of distributions. The $\mathbb{C}$-convergence will be denoted by

$$s_T \underset{\mathbb{C}}{\Rightarrow} s \quad \text{as} \quad T \to \infty.$$

The following result is an analog of Anscombe's theorem for random processes.

Theorem 1.5.1 *Let $s_T(v)$ be random process in $(\mathbb{D}(0, v_0), \mathfrak{B}_\mathbb{D})$ that $\mathbb{C}$-converge on $[0, v_0]$ to a process $s(v)$ continuous with probability* 1.

Next, let $\theta_T(u)$ be nondecreasing random processes in $(\mathbb{D}(0, v_0/b), \mathfrak{B}_\mathbb{D})$ defined on the same probability space as s_T and such that

$$\max_{u \le v_0/b} |\theta_T(u) - bu| \underset{p}{\to} 0 \quad \textit{as} \quad T \to \infty \tag{1.5.3}$$

for a given $b > 0$.

Then, for any $u_0 < v_0/b$,

$$\mathbf{P}\{\theta_T(u_0) < v_0\} \to 1 \quad as \quad T \to \infty, \tag{1.5.4}$$

and the processes $s_T(u)$ *and* $s(u)$ *can be defined on the same probability space so that*

$$\rho_{\mathbb{C},u_0}(s_T(\theta_T(u)), s(bu)) \underset{p}{\to} 0 \quad as \quad T \to \infty. \tag{1.5.5}$$

Moreover, the processes $s_T(\theta_T(u))$ $\mathbb{C}$*-converge on* $[0, u_0]$ *to the process* $s(bu)$.

Note that in place of the function bu in (1.5.3) one may consider any continuous strictly increasing function $b(u)$, $b(0) = 0$ and choose $u_0 < b^{(-1)}(v_0)$. In this case, the role of the limiting process is played by the process $s(b(u))$.

The case when $\theta_T(u) = \frac{N(Tu)}{T}$, where $\frac{N(t)}{t} \underset{a.s.}{\to} b$ as $t \to \infty$ in the context of stopped random walks was considered in [72].

In Theorem 1.5.1, the assumption that the limiting process $s(v)$ is continuous (which implies that the processes $s_T(v)$ are also in a sense continuous) is an analog of the continuity condition 2) in Anscombe's theorem.

In what follows, in parallel with the metric $\rho_{\mathbb{C},u_0}$, we will also consider some other metrics on the space $\mathbb{D}$. The following simple result will be useful.

Lemma 1.5.2 *Let* ρ *be a metric on the space* $\mathbb{D}(0, u_0)$ *and let* f *be a measurable functional in* $(\mathbb{D}(0, u_0), \mathfrak{B}_{\mathbb{D}})$ *continuous with respect to the metric* ρ. *Next, let* s_T *and* s *be random processes in* $(\mathbb{D}(0, u_0), \mathfrak{B}_{\mathbb{D}})$ *defined on the same probability space and such that* $\rho(s_T, s) \underset{p}{\to} 0$ *as* $T \to \infty$. *Then*

$$f(s_T) \Rightarrow f(s) \quad as \quad T \to \infty.$$

Proof of Lemma 1.5.2 Setting $B_\delta = \{\rho(s_T, s) < \delta\}$, we have $\mathbf{P}(\overline{B}_\delta) \to 0$ as $T \to \infty$ and any $\delta > 0$. This means that there exists a sequence $\delta = \delta_T$ converging sufficiently slowly to 0 as $T \to \infty$ and such that the convergence $\mathbf{P}(\overline{B}_\delta) \to 0$ still takes place. Next, for $\delta = \delta_T$ we have

$$\mathbf{P}(f(s_T) < v) = \mathbf{P}(f(s_T) < v, B_\delta) + \mathbf{P}(f(s_T) < v, \overline{B}_\delta),$$

where the second term on the right is $o(1)$ as $T \to \infty$. Moreover, since f is continuous, we have, for $\delta = \delta_T$,

$$\begin{aligned}\mathbf{P}(f(s_T) < v, B_\delta) &= \mathbf{P}(f(s) + o_p(1) < v, B_\delta)\\ &= \mathbf{P}(f(s) < v + o_p(1)) - \mathbf{P}(f(s) < v + o_p(1), \overline{B}_\delta)\\ &= \mathbf{P}(f(s) < v + o_p(1)) + o(1).\end{aligned}$$

It follows that

$$\mathbf{P}(f(s_T) < v) \to \mathbf{P}(f(s) < v) \quad \text{as} \quad T \to \infty$$

at any point v of continuity of the distribution function of the random variable $f(s)$. The lemma is proved. □

Proof of Theorem 1.5.1 It is known that the processes $s_T(v)$ and $s(v)$ can be defined on the same probability space so as to have

$$\rho_{\mathbb{C},v_0}(s_T(v), s(v)) \underset{p}{\to} 0 \quad \text{as} \quad T \to \infty. \tag{1.5.6}$$

(For a general theorem here, see, e.g., [10], Theorem 1.6.7.) Hence by Lemma 1.5.2 to prove the theorem it suffices to verify (1.5.4), (1.5.5). We set

$$A_{T,\varepsilon} = \Big\{ \max_{u \le u_0} |\theta_T(u) - bu| \le \varepsilon \Big\}.$$

Now condition (1.5.3) implies that there exists a sequence $\varepsilon = \varepsilon_T$ converging sufficiently slowly to 0 as $T \to \infty$ and such that

$$\mathbf{P}(\overline{A}_{T,\varepsilon}) \to 0 \quad \text{as} \quad T \to 1. \tag{1.5.7}$$

It is clear that

$$A_{T,\varepsilon} \subset \{\theta_T(u_0) \le v_0\} \quad \text{for any} \quad u_0 < v_0/b,$$

which verifies (1.5.4).

Let us now check (1.5.5). For any $h > 0$, we have

$$\begin{aligned}
&\mathbf{P}\Big(\rho_{\mathbb{C},u_0}(s_T(\theta_T(u)), s(bu) > 2h\Big) \\
&\qquad \le \mathbf{P}(\overline{A}_{T,\varepsilon}) + \mathbf{P}\Big(\rho_{\mathbb{C},u_0}(s_T(\theta_T(u)), s(\theta_T(u))) > h; A_{T,\varepsilon}\Big) \\
&\qquad\quad + \mathbf{P}\Big(\rho_{\mathbb{C},u_0}(s(\theta_T(u)), s(bu)\Big) > h; A_{T,\varepsilon}\Big).
\end{aligned} \tag{1.5.8}$$

On the set $A_{T,\varepsilon}$, $\rho_{\mathbb{C},u_0}(s_T(\theta_T(u)), s(\theta_T(u)))$ on the right of (1.5.8) is not greater than

$$\rho_{\mathbb{C},v_0}(s_T(v), s(v))$$

and hence, by (1.5.6), the second term on the right of (1.5.8) converges to 0 as $T \to \infty$. The convergence to 0 of the third term on the right of (1.5.8) follows from the definition of the set $A_{T,\varepsilon}$ and the continuity of the function $s(v)$ with probability 1. Now (1.5.5) follows, since $\mathbf{P}(\overline{A}_{T,\varepsilon}) \to 0$ as $T \to \infty$. Theorem 1.5.1 is proved. □

1.5.3 Invariance Principle for Compound Renewal Processes

Theorem 1.5.3 *If* $\mathbf{E}\tau^2 < \infty$, $\mathbf{E}\zeta^2 < \infty$, $\tau_1 = o_p(\sqrt{T})$, $\zeta_1 = o_p(\sqrt{T})$, *then the sequences of processes* $y_T(u)$, $z_T(u)$, *defined in* (1.5.1), $\mathbb{C}$*-converge for any* $u_0 > 0$ *as* $T \to \infty$ *to a standard Wiener process* $w(u)$, $u \in [0, u_0]$ *(that is,* (1.5.2) *holds with* s_T *replaced by* y_T *and* z_T *and if* w *is substituted for* s*).*

Note that for $u_0 = 1$ by Corollary 1.1.8 we have

$$\frac{\overline{\zeta}(T)}{\sqrt{T}} \underset{\text{a.s.}}{\to} 0 \quad \text{as} \quad T \to \infty. \tag{1.5.9}$$

Next,

$$\rho_{\mathbb{C},1}(y_T, z_T) \le \frac{1}{\sigma\sqrt{T}}\,\overline{\zeta}(T),$$

and hence

$$\rho_{\mathbb{C},1}(y_T, z_T) \underset{\text{a.s.}}{\to} 0 \quad \text{as} \quad T \to \infty,$$

which shows that the processes y_T and z_T are asymptotically equivalent. It is clear that the same results also hold for any $u_0 > 0$. Hence in the proof of Theorem 1.5.3 we can consider only the processes y_T.

Proof of Theorem 1.5.3 Consider the processes

$$s_T(v) = \frac{S_{vT}}{\sigma_\xi\sqrt{T}}, \quad \text{where} \quad S_{vT} := S_{[vT]}, \quad S_k = \sum_{j=1}^{k} \xi_j. \tag{1.5.10}$$

It is known that there exists a standard Wiener process $w(v)$ defined on the same probability space as $s_T(v)$ such that (see (1.5.6))

$$\rho_{\mathbb{C},v_0}(s_T(v), w(v)) \underset{p}{\to} 0 \quad \text{as} \quad T \to \infty. \tag{1.5.11}$$

The probability space on which the processes $s_T(v)$ and $w(v)$ are defined can always be extended so as to accommodate random variables τ_i and ζ_i so that ξ_i will be the difference of the variables ζ_i and $a\tau_i$ for all i. As a result, the processes $Z(t)$ and $Y(t)$ are defined on the entire half-line, and moreover, the random variables $\eta(t)$ are such that $\frac{\eta(t)}{t} \underset{\text{a.s.}}{\to} \frac{1}{a_\tau}$ as $t \to \infty$. Therefore,

$$\theta_T(u) := \frac{\eta(uT)}{T} \underset{\text{a.s.}}{\to} \frac{u}{a_\tau} \quad \text{as} \quad uT \to \infty.$$

It follows that, for any u_0,

$$\sup_{u \le u_0} \left| \theta_T(u) - \frac{u}{a_\tau} \right| = \rho_{\mathbb{C},u_0}(\theta_T(u), ub) \underset{\text{a.s.}}{\to} 0$$

as $T \to \infty$, and so in our case condition (1.5.3) is met and relation (1.5.4) of Theorem 1.5.1 holds with $b = 1/a_\tau$.

From Theorem 1.5.1 it follows that (see (1.5.5))

$$\rho_{\mathbb{C},u_0}(s_T(\theta_T(u)), w(bu)) \underset{p}{\to} 0 \text{ for any } u_0 < \frac{v_0}{b} \text{ and } T \to \infty. \tag{1.5.12}$$

Now we consider the processes

$$y_T(u) = \frac{Y(uT) - auT}{\sigma\sqrt{T}} = \frac{S_{\eta(uT)} + a\chi(uT)}{\sigma\sqrt{T}}$$

(see (1.5.1)) and put

$$y_T^0(u) := \frac{S_{\eta(uT)}}{\sigma\sqrt{T}} = \sqrt{a_\tau}\, s_T\!\left(\frac{\eta(uT)}{T}\right) = \sqrt{a_\tau}\, s_T(\theta_T(u)). \tag{1.5.13}$$

Consider the standard Wiener process

$$w^*(u) = \frac{w(bu)}{\sqrt{b}}.$$

By Lemma 1.5.2, to prove the theorem it suffices to verify that

$$\rho_{\mathbb{C},v_0}(y_T(u), w^*(u)) \underset{P}{\to} 0 \quad \text{as} \quad T \to \infty.$$

We have

$$\rho_{\mathbb{C},u_0}(y_T, w^*) \le \rho_{\mathbb{C},u_0}(y_T, y_T^0) + \rho_{\mathbb{C},u_0}(y_T^0, w^*),$$

where

$$\rho_{\mathbb{C},u_0}(y_T, y_T^0) = \frac{|a|}{\sigma} \sup_{u \le u_0} \frac{\chi(uT)}{\sqrt{T}}, \qquad \sup_{u \le u_0} \frac{\chi(uT)}{\sqrt{T}} \underset{P}{\to} 0 \text{ as } T \to \infty \tag{1.5.14}$$

(see Corollary 1.1.8), and hence, $\rho_{\mathbb{C},u_0}(y_T, y_T^0) \underset{P}{\to} 0$ as $T \to \infty$. So it suffices to verify that, for $b = 1/a_T$,

$$\begin{aligned} \rho_{\mathbb{C},u_0}(y_T(u), w^*(u)) &= \sqrt{a_T}\rho_{\mathbb{C},u_0}\Big(s_T(\theta_T(u)), w^*(u)/\sqrt{a_T}\Big) \\ &= \sqrt{a_T}\rho_{\mathbb{C},u_0}\Big(s_T(\theta_T(u)), w(ub)\Big) \underset{P}{\to} 0 \quad \text{as} \quad T \to \infty. \end{aligned} \tag{1.5.15}$$

But this was already established in (1.5.12). This proves Theorem 1.5.3 in the homogeneous case.

In the inhomogeneous case,

$$\widetilde{Y}(t) = \begin{cases} \zeta_1 & \text{for} \quad t < \tau_1, \\ \zeta_1 + Y(t - \tau_1) & \text{for} \quad t \ge \tau_1, \end{cases}$$

where $Y(t)$ is a homogeneous CRP not depending on τ_1 and ζ_1. For nonrandom $\tau_1 = o(\sqrt{T})$, $\zeta_1 = o(\sqrt{T})$, the conclusion of the theorem clearly holds, because $\zeta_1/\sqrt{T} \underset{P}{\to} 0$, and the replacement of the time interval $[0, u_0T]$ of length u_0T by an interval of length $u_0T - o(\sqrt{T})$ does not introduce any changes in the above arguments. The change to random $\tau_1 = o_p(\sqrt{T})$, $\zeta_1 = o_p(\sqrt{T})$ is also clear. Theorem 1.5.3 is proved. □

Remark 1.5.4 Theorem 1.5.3 can also be proved via the "classical" approach, according to which

A sufficient condition for the $\mathbb{C}$*-convergence* $y_T \underset{\mathbb{C}}{\Rightarrow} w$ *is that*

(1) *the finite-dimensional distributions of* y_T *should weakly converge to the corresponding distributions of* w*;*

(2) *for any* $\varepsilon > 0$,

$$\lim_{\Delta \to 0} \overline{\lim_{T \to \infty}} \mathbf{P}(\omega(y_T, \Delta) > \varepsilon) = 0, \tag{1.5.16}$$

where, for $g \in \mathbb{D}(0,1)$, the functional

$$\omega(g, \Delta) = \sup_{(t,u)\in B_\Delta} |g(u+v) - g(u)|$$

is the modulus of continuity of the function g,

$$B_\Delta := \{(u, v) : v \in (0, \Delta), u \in [0, u_0], u + v \le u_0\}$$

(*see, e.g., Chapter* 2 *in* [10] *or Theorem* 1.2.1 *in* [18]).

However, in the classical approach one would have to repeat in the more complicated setting the proof of (1.5.16), which is known for processes $s_T(u)$ generated by random walks, and then verify the convergence of finite-dimensional distributions.

In our opinion, the approach applied in the proof of Theorem 1.5.3, which is based on the already verified invariance principle for random walks and the analog of Anscombe's theorem for random processes, is the most simple and transparent.

Remark 1.5.5 It can easily be checked that the proof of Theorem 1.5.3 will not change if in the definition of the $\mathbb{C}$-convergence (1.5.3) one assumes that the functional f is continuous, not at all points of the space $\mathbb{C}$, but only at points of the set $\mathbb{C}_f \subset \mathbb{C}$ such that $\mathbf{P}(s \in \mathbb{C}_f) = 1$ (or $\mathbf{P}(w \in \mathbb{C}_f) = 1$ in the invariance principle). Such a functional is known as a functional continuous with probability 1. For example, the functional $f(s) = I\{s(1/2) > 1\}$ is continuous with w-probability 1, inasmuch as $\mathbf{P}(w(1/2) = 1) = 0$. For this f, the subspace $\mathbb{C}_f$ is $\mathbb{C}_f = \{s \in \mathbb{C} : s(1/2) \neq 1\}$. For such generalization, one can equivalently formulate the invariance principle in terms of the convergence of probabilities

$$\mathbf{P}(y_T \in B) \to \mathbf{P}(w \in B)$$

for the so-called sets B with w-zero boundary

$$\mathbf{P}(w \in \partial B) = 0 \tag{1.5.17}$$

(the boundary of measure 0 with respect to the distribution of w).

Indeed, under (1.5.17), consider the functional

$$f_B(s) = \begin{cases} 1 & \text{for } s \in (B), \\ 1/2 & \text{for } s \in \partial B, \\ 0 & \text{for } s \notin [B]. \end{cases}$$

This functional is continuous with probability 1 with respect to the distribution of w, and hence,

$$\mathbf{P}(y_T \in (B)) = \mathbf{E}f(y_T) \to \mathbf{E}f(w) = \mathbf{P}(w \in (B)) = \mathbf{P}(w \in [B]).$$

Conversely, if f is a functional continuous with w-probability 1, then the set $B_t = \{s : f(s) < t\}$, where t is any continuity point of the distribution of $f(w)$, is a set with w-zero boundary. Hence

$$\mathbf{P}(f(y_T) < f) = \mathbf{P}(y_T \in B_t) \to \mathbf{P}(w \in B_t) = \mathbf{P}(f(w) < t),$$

which is equivalent to the $\mathbb{C}$-convergence.

In Chapter 7 we will study the possibility of extending the invariance principle to the zones of large and small deviations in terms of the asymptotic equivalence of the small probabilities $\mathbf{P}(y_T \in B)$ and $\mathbf{P}(w \in B)$ for a broad class of sets B depending on T with "w-thin" boundary.

1.6 Convergence of Normalized Compound Renewal Processes to Stable Processes in the Case when ξ Has Infinite Variance

In this section, we consider the case when $\sigma_\xi^2 = \mathbf{D}\xi = \infty$ and if conditions $[\mathbf{R}_{\alpha,\beta}]$ of regular decay of the distribution of $\xi = \zeta - a\tau$ at infinity (see § 1.5) are met.

1.6.1 S-Convergence to Stable Processes

Assume that processes $s_T(u)$ and $s(u)$ are defined on the space $(\mathbb{D}(0,u_0), \mathfrak{B}_\mathbb{D})$ equipped with the Skorokhod metric

$$\rho_{S,u_0}(g_1, g_2) := \inf_h \Big[\sup_{u\in[0,u_0]} |g_1(u) - g_2(h(u))| + \sup_{u\in[0,u_0]} |h(u) - u| \Big], \qquad (1.6.1)$$

where the infimum is taken over all continuous increasing functions $h(u)$ such that $h(0) = 0$, $h(u_0) = u_0$.

We say that a sequence of processes $s_T(u)$ *S-converges* (in distribution) *as $T \to \infty$ to a process $s(u)$ on the interval* $[0, u_0]$ (written $s_T \underset{S}{\Rightarrow} s$) if

$$f(s_T) \Rightarrow f(s) \quad \text{as} \quad T \to \infty \qquad (1.6.2)$$

for any $(\mathbb{D}(0,u_0), \mathfrak{B}_\mathbb{D})$-measurable functional f continuous in the Skorokhod metric.

For conditions of S-convergence of processes, see, for example, § 6.5 in [68], [10], [18], and [10]. With the help of these results it was shown, in particular, that under conditions $[\mathbf{R}_{\alpha,\beta}]$ the normalized process

$$s_T(u) = \frac{S_{uT}}{\sigma_\xi(T)}, \qquad u \le u_0 \qquad (1.6.3)$$

S-converges as $T \to \infty$ to a stable process $w_{\alpha,\beta}(u)$ with parameters α, β: $s_T \underset{S}{\Rightarrow} w_{\alpha,\beta}$ ($\sigma_\xi(T)$ is defined in (1.4.1)).

We set, as before,

$$\sigma(t) = \sigma_\xi(t) a_\tau^{-1/\alpha}$$

and consider the sequence of processes

$$y_T(u) = \frac{Y(u) - auT}{\sigma(T)} = \frac{Z_{\eta(uT)} - auT}{\sigma(T)} = \frac{S_{\eta(uT)}}{\sigma(T)} + \frac{a\chi(uT)}{\sigma(T)}, \qquad u \in [0, u_0] \qquad (1.6.4)$$

(here we used the equality $t = T_{\eta(t)} - \chi(t)$).

The question is whether under conditions $[\mathbf{R}_{\alpha,\beta}]$ the distributions of the processes y_T converge to the distribution of the stable process $w_{\alpha,\beta}(u)$, $u \in [0, u_0]$. It turns out that this convergence does not always hold.

The proof of the S-convergence of a CRP to the process $w_{\alpha,\beta}$ faces fundamental difficulties due to the fact that the trajectory of the process $y_T(u)$ has the form (1.6.4) and contains the "irregular" sawtooth term $\frac{a\chi(uT)}{\sigma(T)}$, which in some cases can become prominent. Hence the convergence $y_T \underset{S}{\Rightarrow} w_{\alpha,\beta}$ will take place only under additional conditions.

Theorem 1.6.1 *Let $\xi = \zeta - a\tau$ satisfy conditions $[\mathbf{R}_{\alpha,\beta}]$, and let, in addition, at least one of the conditions*

$$a = 0 \quad \text{or} \quad F^{(\tau)}(t) := \mathbf{P}(\tau \ge t) = o\big(F(t)\big) \quad \text{as} \quad t \to \infty \tag{1.6.5}$$

holds. Then, for any $u_0 > 0$,

$$y_T(u) \underset{S}{\Rightarrow} w_{\alpha,\beta}(u) \quad \text{on} \quad [0, u_0]. \tag{1.6.6}$$

The result of Theorem 1.6.1 for $a = 0$ was established in Chapter 5 of [72]. It is easily seen, for independent τ and ζ, that the second condition in (1.6.5) is met if

$$F^{(\tau)}(t) = o\big(F^{(|\zeta|)}(t)\big) \quad \text{as} \quad t \to \infty;$$

for example, this condition is always satisfied for the compound Poisson processes.

If condition (1.6.5) is not met, then the S-convergence (1.6.6) is in general absent, even though the finite-dimensional distributions converge. For more details, see §1.6.2 below.

Proof of Theorem 1.6.1 We set, as before, $\theta_T(u) = \frac{\eta(uT)}{T}$. Hence by (1.6.3)

$$y_T(u) = a_\tau^{1/\alpha} s_T(\theta_T(u)) + \frac{a\chi(uT)}{\sigma(T)}. \tag{1.6.7}$$

We claim that under condition (1.6.5) the second term on the right of (1.6.7) is negligibly small. This claim is obvious for $a = 0$. If $a \neq 0$, then we shall show that

$$\rho_{\mathbb{C},u_0}(y_T(u), a_\tau^{1/\alpha} s_T(\theta_T(u))) \underset{p}{\to} 0 \quad \text{as} \quad T \to \infty \tag{1.6.8}$$

or, equivalently,

$$\sup_{u \le u_0} \frac{\chi(uT)}{\sigma(T)} \underset{p}{\to} 0 \quad \text{as} \quad T \to \infty. \tag{1.6.9}$$

Indeed, on the set $A_T = \{\theta_T(u_0) < v_0\}$, $u_0 < bv_0$, for $b = 1/a_\tau$ we have $P(A_T) \to 1$,

$$\max_{u \le u_0} \chi(uT) \le \max_{i \le v_0 T} \tau_i =: \overline{\tau}_T,$$

where, for any $\varepsilon > 0$,

$$\begin{aligned}\mathbf{P}(\overline{\tau}_T > \varepsilon\sigma(T)) &= o(1) + \mathbf{P}(\overline{\tau}_T > \varepsilon\sigma(T), A_T) \le o(1) + v_0 T F^{(\tau)}(\varepsilon\sigma(T)) \\ &= o(1) + o(TF(\sigma_\xi(T))) = o(1)\end{aligned}$$

as $T \to \infty$. This proves (1.6.9) and (1.6.8).

Thus, under the above conditions, the processes $y_T(u)$ and $a_T^{\frac{1}{\alpha}} s_T(\theta_T(u))$ are asymptotically equivalent.

The convergence of the processes $a^{1/\alpha} s_T(\theta_T(u))$ to $w_{\alpha,\beta}(u)$, as we already noted, was established in [72] (Theorem 5.2.2). At this point, we could have completed the proof of Theorem 1.6.1. However, the proof of the required convergence in [72] contains many steps and relies on other, sometimes more involved, results. At the same time, there is a direct proof of the required convergence, which we now present.

Without loss of generality, we set $a_\tau = 1$, $u_0 = 1$, $\rho_{S,u_0} = \rho_S$. In view of Theorem 1.6.7 in [10], we can define the processes $s_T(u)$ and $w_{\alpha,\beta}(u)$ on the same probability space so as to have

$$\rho_S(s_T(u), w_{\alpha,\beta}(u)) \underset{p}{\to} 0 \quad \text{as} \quad T \to \infty. \tag{1.6.10}$$

As in the proof of Theorem 1.5.3, we enlarge the probability space underlying the processes $s_T(u)$ and $w_{\alpha,\beta}(u)$ so that the entire sequence (τ_i, ζ_i) would be defined on it, and therefore the processes $Z(t)$ and $Y(t)$. Hence the random variables $\eta(t)$, $\theta_T(u) = \frac{\eta(uT)}{T}$ are also defined on this space. Next,

$$\rho_S\big(s_T(\theta_T(\cdot)), w_{\alpha,\beta}(\cdot)\big) \le \rho_S\big(s_T(\theta_T(\cdot)), s_T(\cdot)\big) + \rho_S(s_T(\cdot), w_{\alpha,\beta}(\cdot)), \tag{1.6.11}$$

where the second term on the right of (1.6.11) converges in probability to 0 as $T \to \infty$. Therefore, it suffices to establish the convergence to 0 of the first term, that is, the asymptotic equivalence of $s_T(\theta_T(u))$ and $s_T(u)$. We have

$$\rho_S\big(s_T(\theta_T(\cdot)), s_T(\cdot)\big) \le \sup_{u \le 1} \big|s_T(\theta_T(u)) - s_T(h(u))\big| + \sup_{u \le 1} |h(u) - u| \tag{1.6.12}$$

for any continuous increasing function $h(u)$, $h(0) = 0$, $h(1) = 1$, and hence it suffices to choose a function $h(u) = h_T(u)$ for which both terms on the right of (1.6.12) converge in probability to 0 as $T \to \infty$.

Given a small $\varepsilon > 0$, consider the set

$$A_{T,\varepsilon} = \big\{|\eta(T) - T| < \varepsilon T + 1\big\}.$$

It is clear that

$$P(A_{T,\varepsilon}) \to 1 \quad \text{for any} \quad \varepsilon > 0 \quad \text{as} \quad T \to \infty. \tag{1.6.13}$$

Hence there exists a sequence $\varepsilon = \varepsilon_T$ converging to 0 sufficiently slowly as $T \to \infty$ and such that (1.6.13) still holds.

We choose as h the smoothed version

$$h(u) = \widetilde{\theta}_T(u)$$

of the function $\theta_T(u)$, which is defined on the set $A_{T,\varepsilon}$ as the continuous polygon passing through the nodes

$$\left(t_k = \frac{T_k}{T}, u_k = \frac{k}{T}\right) \quad \text{for} \quad k = 0, 1, \ldots, k_T := [T(1 - \varepsilon)],$$

and through the point (1, 1), so that

$$\theta_T(t_k) = \widetilde{\theta}_T(t_k) \tag{1.6.14}$$

for $k \le k_T$.

The function $s_T(u)$ is piecewise continuous (it takes the same value on (u_{k-1}, u_k), $k \ge 1$), and hence

$$s_T(\theta_T(u)) = s_T(\widetilde{\theta}_T(u)) \quad \text{for} \quad u \le t_{k_T}. \tag{1.6.15}$$

For the first term on the right of (1.6.12), on the set $A_{T,\varepsilon}$ we have $u_{k_T} < 1$, and

$$\sup_{u\le 1} |s_T(\theta_T(u)) - s_T(\widetilde{\theta}_T(u))| \le d_1 + d_2,$$

where by (1.6.15), (1.6.14)

$$\begin{aligned} d_1 &= \sup_{u\in[u_{k_T},1]} \big|s_T(\theta_T(u)) - s_T(\theta_T(u_{k_T}))\big| \\ &\underset{p}{=} \frac{1}{\sigma(T)} \max_{k\le \eta(T)-T(1-\varepsilon)} |S_k^*| \le \frac{1}{\sigma(T)} \max_{k\le 2\varepsilon T} |S_k^*|, \\ d_2 &= \sup_{u\in[u_{k_T},1]} \big|s_T(\widetilde{\theta}_T(u)) - s_T(\widetilde{\theta}_T(u_{k_T}))\big|. \end{aligned}$$

Here the sequence $\{S_k^*\}$ has the same distribution as $\{S_k\}$, and the variable

$$\frac{1}{\sigma(2\varepsilon T)} \max_{k\le 2\varepsilon T} |S_k^*|$$

has proper limit distribution when $\varepsilon = \varepsilon_T$, $\varepsilon T \to \infty$; moreover, $\frac{\sigma(2\varepsilon T)}{\sigma(T)} \to 0$. Hence, for any $\delta > 0$,

$$\mathbf{P}(d_1 > \delta, A_{T,\varepsilon}) \to 0 \quad \text{as} \quad T \to \infty.$$

The variable d_2 is estimated in the same way.

Next, for the second term on the right of (1.6.12), we have, on the set $A_{T,\varepsilon}$,

$$d_3 := \sup_{u\le 1} |h(u) - u| = \frac{1}{T} \sup_{k\le k_T} |T_k - k|,$$

$$\mathbf{P}(d_3 > \delta, A_{T,\varepsilon}) \to 0 \quad \text{as} \quad T \to \infty.$$

Therefore, by (1.6.12),

$$\begin{aligned} &\mathbf{P}\Big(\rho_S(s_T(\theta_T(\cdot)), s_T(\cdot)) > \delta\Big) \\ &\quad = o(1) + \mathbf{P}\Big(\rho_S(s_T(\theta_T(\cdot)), s_T(\cdot)) > \delta, A_{T,\varepsilon}\Big) = o(1) \quad \text{as} \quad T \to \infty. \end{aligned}$$

Theorem 1.6.1 is proved. □

1.6.2 Absence of S-Convergence without Condition (1.6.5)

Let

$$a = 1, \qquad F^{(\tau)}(t) \gg F(t) \quad \text{as} \quad t \to \infty, \tag{1.6.16}$$

so that condition (1.6.5) is not met. This case is realized, for example, if $\zeta = \tau + \omega$, where $\mathbf{E}\omega = 0$, the variable $\omega = \xi$ satisfies condition $[\mathbf{R}_{\alpha,\beta}]$ and does not depend on τ, τ satisfies condition $[\mathbf{R}_{\alpha',1}]$ with $1 < \alpha' < \alpha$, so that the possible jumps of τ are "greater" than the jumps of $\xi = \omega$. Let us show that in this case the convergence $y_T \underset{S}{\Rightarrow} w_{\alpha,\beta}$ is impossible. We first note that the variable

$$h_{T,\tau} := \frac{\sup \chi(uT)}{\sigma_\tau(T)},$$

where $\sigma_\tau(T) = (F^{(\tau)})^{(-1)}(1/T)$, has proper limit distribution as $T \to \infty$. Indeed, it can be easily shown (again with the help of the sets $A_{T,\varepsilon}$) that $h_{T,\tau}$ has the same limit distribution as $\frac{\overline{\tau}_T}{\sigma_\tau(T)}$, where $\overline{\tau}_T = \max_{k \le T} \tau_k$. But $\frac{\overline{\tau}_T}{\sigma_\tau(T)}$ has proper nondegenerate limit distribution as $T \to \infty$ because (for simplicity, we assume that T is integer)

$$\mathbf{P}(\overline{\tau}_T < v\sigma_\tau(T)) = \left(1 - F^{(\tau)}(v\sigma_\tau(T))\right)^T, \tag{1.6.17}$$

where

$$F^{(\tau)}(v\sigma_\tau(T)) \sim \frac{v^{-\alpha'}}{T} \quad \text{as} \quad T \to \infty.$$

Hence the left-hand side in (1.6.17) reads as

$$\left(1 - \frac{v^{-\alpha'}(1 + o(1))}{T}\right)^T \to \exp\{-v^{-\alpha'}\} \quad \text{as} \quad T \to \infty$$

(see also [111, 70]). We have $\sigma(T) \ll \sigma_\tau(T)$ as $T \to \infty$, and hence by the above

$$h_T := \frac{\sup \chi(uT)}{\sigma(T)} \to \infty \quad \text{as} \quad T \to \infty. \tag{1.6.18}$$

At the same time, $s_T(\theta_T(\cdot)) \underset{S}{\Rightarrow} w_{\alpha,\beta}(\cdot)$, and hence,

$$\inf_{u \le 1} s_T(\theta_T(u)) \Rightarrow \inf_{u \le 1} w_{\alpha,\beta}(u).$$

Since the right-hand side in this relation is finite with probability 1, it follows by (1.6.7), (1.6.18) that

$$\sup_{u \le 1} y_T(u) \underset{p}{\to} \infty \quad \text{as} \quad T \to \infty. \tag{1.6.19}$$

But the functionals $f(s) = \inf_{u \le 1} s(u)$ and $f(s) = \sup_{u \le 1} s(u)$ are ρ_S-continuous. Hence (1.6.19) means the absence of the S-convergence $y_T \underset{S}{\Rightarrow} w_{\alpha,\beta}$.

We also note that, in addition, for any $u \le 1$, we have

$$\frac{\chi(uT)}{\sigma(T)} \underset{p}{\to} 0 \quad \text{as} \quad T \to \infty$$

and the finite-dimensional distributions $\frac{\chi(uT)}{\sigma(T)}$ correspond to the zero limiting process.

If τ and ζ are independent, $a \neq 0$, then the distribution F is formed as the convolution of the distributions of τ and ζ, and instead of (1.6.16) only the inequality

$$F^{(\tau)}(t) \geq cF(t) \quad \text{as} \quad t \to \infty, \quad c = \text{const}$$

is possible. In this case, in analogy with the above, the limiting process for y_T in (1.6.7) is formed as the sum of the process $w_{\alpha,\beta}$ and the *sawtooth* term $\frac{a\chi(uT)}{\sigma(T)}$, whose jumps for $a \neq 0$ are comparable with 1, so that in this case, in addition to the vertical jumps of the process $s_T(\theta_T(u))$ in (1.6.7), the "*slanted* jumps" of the process $-\frac{a\chi(uT)}{\sigma(T)}$ will be added, and the S-convergence $y_T \underset{S}{\Rightarrow} w_{\alpha,\beta}$ will also be absent. This is because the metric ρ_S is incapable of describing the proximity of processes of which one has vertical jumps and the other has jumps of the same size, but slanted.

To avoid this nuisance, in the next section we consider the metric $\rho_{\mathbb{D}}$, which is free from this defect – this metric is close to ρ_S and is slightly weaker than ρ_S. But even for this metric, we will need additional conditions under which the processes y_T, z_T weakly converge to stable processes.

Remark 1.6.2 In the proof of Theorem 1.6.1 one can skip the proof of the intermediate assertion on the convergence to zero of $\rho_S(s_T(\theta_T(\cdot)), w_{\alpha,\beta})$ (see (1.6.11)) and instead invoke Theorem 5.2.2 in [72] on the S-convergence of $s_T(\theta_T(\cdot))$ to $w_{\alpha,\beta}(\cdot)$. However, in our opinion, the above proof is simpler, more transparent, and has no references to a number of other works.

1.6.3 $\mathbb{D}$-Convergence to Stable Processes

In order to widen the conditions under which the distributions of CRPs y_T, z_T weakly converge in $(\mathbb{D}(0, u_0), \mathfrak{B}_{\mathbb{D}})$ to stable processes, we introduce one more metric $\rho_{\mathbb{D},u_0}$, which is close to ρ_{S,u_0} but weaker.

For simplicity, assume that $u_0 = 1$, $\rho_{\mathbb{D},u_0} = \rho_{\mathbb{D}}$, $\mathbb{D} = \mathbb{D}(0, 1)$.

The metric $\rho_{\mathbb{D}}$ is defined as follows. For any function $g \in \mathbb{D}$, we define the *graph* of this function as the simply connected set Γg in $[0, 1] \times \mathbb{R}$ coinciding with the curve $(t, g(t))$ everywhere except at the discontinuity points of the function g, at which the points $(t, g(t-0))$, $(t, g(t+0)) \in \mathbb{R}^2$ are joined by the straight line segment. The distance $\rho_{\mathbb{D}}(g_1, g_2)$ is defined as the Hausdorff distance between the sets Γg_1 and Γg_2 in $\mathbb{R}^2$, that is, we write $\rho_{\mathbb{D}}(g_1, g_2) < \varepsilon$ if simultaneously $\Gamma g_1 \in (\Gamma g_2)_\varepsilon$ and $\Gamma g_2 \in (\Gamma g)_\varepsilon$, where the ε-neighborhoods in $\mathbb{R}^2$ are defined with respect to the Euclidean metric. In other words,

$$\begin{aligned}\rho_{\mathbb{D}}(g_1, g_2) &:= \max\{r(g_1, g_2), r(g_2, g_1)\}, \\ r(g_1, g_2) &:= \max_{\boldsymbol{v}\in\Gamma g_1} \min_{\boldsymbol{u}\in\Gamma g_2} |\boldsymbol{u} - \boldsymbol{v}|,\end{aligned} \tag{1.6.20}$$

where $\boldsymbol{u} = (u_1, u_2)$, $\boldsymbol{v} = (v_1, v_2)$, $u_1, v_1 \in [0, 1]$, $u_2, v_2 \in \mathbb{R}$, and $|\cdot|$ means the Euclidean norm in $\mathbb{R}^2$. It is clear that

$$\rho_{\mathbb{D}}(g_1, g_2) \le \rho_{\mathbb{C}}(g_1, g_2) := \sup_{0 \le t \le 1} |g_1(t) - g_2(t)|. \tag{1.6.21}$$

Let us compare the distances ρ_S and $\rho_{\mathbb{D}}$. Given functions g_1, g_2 from $\mathbb{D}$, assume, for simplicity, that there is a function $h_0(t)$ on which the infimum $\inf_h$ in (1.6.1) is attained,

$$\rho_S(g_1, g_2) = \rho_1 + \rho_2, \text{ where } \rho_1 = \sup_u |g_1(u) - g_2(h_0(u))|, \ \rho_2 = \sup |h_0(u) - u|.$$

Then, for each point u, the distance in $\mathbb{R}^2$ between the points $(u, g_1(u))$ and $(h_0(u), g_2(h_0(u)))$ is at most

$$\sqrt{\rho_1^2 + \rho_2^2} \le \rho_1 + \rho_2 = \rho_S(g_1, g_2).$$

The same inequality also holds for the pair of points

$$(u, g_2(u)) \quad \text{and} \quad (h_1(u), g_1(h_1(u)))$$

with a suitable function $h_1(u)$. It follows that

$$\rho_{\mathbb{D}}(g_1, g_2) \le \rho_S(g_1, g_2).$$

Moreover, for the functions

$$g(u) = \begin{cases} 0, & u \in [0, p), \ p \in (0, 1), \\ 1, & u \in [p, 1]; \end{cases}$$

$$g_n(u) = \begin{cases} 0, & u \in [0, p), \\ n(u - p), & u \in [p, p + 1/n), \\ 1, & u \in [p + 1/n, 1], \end{cases}$$

we have

$$\rho_{\mathbb{D}}(g, g_n) \le \frac{1}{n} \to 0 \quad \text{as} \quad n \to \infty,$$
$$\rho_S(g, g_n) = 1 \quad \text{for all } n.$$

This means that the *metric* $\rho_{\mathbb{D}}$ *is weaker than the metric* ρ_S, and so the class of functionals f continuous with respect to $\rho_{\mathbb{D}}$ is slightly narrower than the class of functionals continuous with respect to ρ_S. Therefore, the convergence $s_T \underset{S}{\Rightarrow} s$ implies the convergence $s_T \underset{\mathbb{D}}{\Rightarrow} s$ (see below), but not conversely. The difference between $\rho_{\mathbb{D}}$ and ρ_S is small, but we are unaware of examples of functionals with interesting applications which are continuous with respect to ρ_S, but discontinuous with respect to $\rho_{\mathbb{D}}$. The class of $\rho_{\mathbb{D}}$-continuous functionals includes the widely applicable functionals like $\max_{u \in [0,1]} (g(u) - g_0(u))$, where $g_0(u)$ is continuous, as well as integral functionals, and so on. The metric $\rho_{\mathbb{D}}$ was used with success in [43], [23] to establish the extended large deviation principle for random walks.

The metric $\rho_{\mathbb{D}}$ was introduced and studied in [16] for the space $\mathbb{F}$, which is broader than $\mathbb{D}$ (see [16]). The topology induced by this metric coincides with the Skorokhod topology M_2 described in [110]. The metric $\rho_{\mathbb{D}}$ can be also looked upon as an extension of the Lévy metric in the space of nondecreasing functions (see, e.g., [69]) to the wider space $\mathbb{D}$.

Let us return to CRPs on $[0, u_0]$ with an arbitrary $u_0 > 0$. Let the distance $\rho_{\mathbb{D},u_0}$ be defined by (1.6.20), where g_1, g_2 are functions on $[0, u_0]$.

In analogy with the above, we write

$$s_T \underset{\mathbb{D}}{\Rightarrow} s \quad \text{as} \quad T \to \infty$$

if (1.6.2) holds for any functional f in $(\mathbb{D}(0, u_0), \mathfrak{B}_{\mathbb{D}})$ continuous with respect to $\rho_{\mathbb{D},u_0}$.

The main result of this section is as follows.

Theorem 1.6.3 *Let $\xi = \zeta - a\tau$ satisfy conditions $[\mathbf{R}_{\alpha,\beta}]$. If at least one of the conditions*

$$a = 0 \quad or \quad \mathbf{P}(\tau > t, a\zeta > t) = o(F(t)) \quad as \quad t \to \infty \tag{1.6.22}$$

is satisfied, then, for any $u_0 > 0$,

$$y_T(u) \underset{\mathbb{D}}{\Rightarrow} w_{\alpha,\beta}(u) \quad on \quad [0, u_0]. \tag{1.6.23}$$

If τ and ζ are independent, then the convergence in (1.6.23) *always holds.*

The same assertions also hold for z_T.

Proof of Theorem 1.6.3 If $a = 0$, then (1.6.23) follows from Theorem 1.6.1.

Suppose now that $a \neq 0$, and for simplicity (and without loss of generality) we put

$$u_0 = 1, \quad a_\tau = 1.$$

For brevity, we set $\rho_{\mathbb{D},u_0} = \rho_{\mathbb{D}}$, $\rho_{S,u_0} = \rho_S$. In the proof of Theorem 1.6.1, we have shown (see (1.6.11)) that the processes $s_T(u)$, $\theta_T(u) = \frac{\eta(uT)}{T}$, $y_T(u)$, and $w_{\alpha,\beta}(u)$ can be defined on the same probability space so that

$$\rho_S\big(s_T(\theta_T(\cdot)), w_{\alpha,\beta}(\cdot)\big) \underset{p}{\to} 0 \quad \text{as} \quad T \to \infty. \tag{1.6.24}$$

We have

$$\rho_{\mathbb{D}}(y_T, w_{\alpha,\beta}) \le \rho_{\mathbb{D}}\big(y_T(\cdot), s_T(\theta_T(\cdot))\big) + \rho_{\mathbb{D}}\big(s_T(\theta_T(\cdot)), w_{\alpha,\beta}(\cdot)\big). \tag{1.6.25}$$

Next, $\rho_{\mathbb{D}} \le \rho_S$, and so by (1.6.24) the second term on the right of (1.6.25) converges in probability to 0 as $T \to \infty$. Hence by Lemma 1.5.2 it suffices to verify that

$$\rho_{\mathbb{D}}\big(y_T(\cdot), s_T(\theta(\cdot))\big) \underset{p}{\to} 0 \quad \text{as} \quad T \to \infty. \tag{1.6.26}$$

We split the half-open interval $[0, 1 + \frac{\chi(T)}{T})$ into half-open intervals $[t_{i-1}, t_i) = [T_{i-1}/T, T_i/T)$, $i = 1, 2, \ldots, \eta(T)$, and consider the projections of the trajectories $y_T(u)$ and $s_T(\theta_T(u))$ to these half-open intervals.

The trajectory of $y_T(u)$ on $[t_{i-1}, t_i)$ consists of the vertical jump $\frac{\zeta_i}{\sigma(T)}$ at the point t_{i-1} and the slanted line $y_T(t_{i-1}+0) - \frac{auT}{\sigma(T)}$, $u < \tau_i$, on the interval (t_{i-1}, t_i) so that

$$y_T(t_i - 0) = y_T(t_{i-1} - 0) + \frac{\zeta_i - a\tau_i}{\sigma(T)}, \qquad \zeta_i - a\tau_i = \xi_i.$$

The trajectory of $s_T(\theta(u))$ on $[t_{i-1}, t_i)$ consists of the vertical jump ξ_i at the point t_{i-1} and the horizontal part on (t_{i-1}, t_i), so that at the points t_i the values of the processes $y_T(\cdot)$ and $s_T(\theta_T(\cdot))$ coincide.

The distances $\rho_{\mathbb{D}}^{(i)}$ between these two *projections* of the processes $y_T(\cdot)$ and $s_T(\theta(\cdot))$ on $[t_{i-1}, t_i)$ in the metric $\rho_{\mathbb{D}}$ is clearly at most $\frac{\tau_i}{T}$ if ζ_i and ξ_i have the same sign; these distances are majorized by $\max\left(\frac{\tau_i}{T}, \frac{|\zeta_i|}{\sigma(T)}\right)$ if ζ_i and ξ_i have different signs.

More precisely, let $a > 0$. Then

$$\rho_{\mathbb{D}}^{(i)} = \frac{\tau_i}{T} I(\{\zeta_i \geq 0,\ \xi_i \geq 0\} \bigcup \{\zeta_i \leq 0, \xi_i \leq 0\})$$
$$+ \max\left(\frac{\tau_i}{T}, \frac{\zeta_i}{\sigma(T)}\right) I(\zeta_i > 0, \xi_i < 0)$$

(for $a > 0$ the event $\{\zeta_i < 0, \xi_i > 0\}$ is impossible because $\xi_i < \zeta_i$). Therefore,

$$\rho_{\mathbb{D}}^{(i)} \leq \max\left(\frac{\tau_i}{T}, \frac{X_i^+}{\sigma(T)}\right),$$

where

$$X_i^+ := \zeta_i I(\zeta_i > 0, \xi_i < 0),$$

and for $v > 0$,

$$\begin{aligned}\mathbf{P}(X_i^+ > v) &= \mathbf{P}(\zeta_i > v,\ \zeta_i - a\tau_i < 0)\\ &= \mathbf{P}\left(\tau > \frac{\zeta_i}{a}, \zeta > v\right) \leq \mathbf{P}\left(\tau > \frac{v}{a}, \zeta > v\right).\end{aligned} \tag{1.6.27}$$

If $a < 0$, then a similar analysis shows that

$$\rho_{\mathbb{D}}^{(i)} \leq \max\left(\frac{\tau_i}{T}, -\frac{X_i^-}{\sigma(T)}\right),$$

where

$$X_i^- := \zeta_i I(\zeta_i < 0, \xi_i > 0),$$

so that, for $v > 0$,

$$\begin{aligned}\mathbf{P}(X_i^- < -v) &= \mathbf{P}(\zeta_i < -v,\ \zeta_i - a\tau_i > 0)\\ &= \mathbf{P}\left(\tau > \frac{\zeta_i}{a}, \zeta < -v\right) \leq \mathbf{P}\left(\tau > \frac{v}{|a|}, \mathrm{e}_a\zeta > v\right), \quad \mathrm{e}_a = \frac{a}{|a|}\end{aligned} \tag{1.6.28}$$

and the right-hand side in the case $a > 0$ coincides with the right-hand side of (1.6.27). As a result, we get

$$\mathbf{P}(\rho_{\mathbb{D}}^{(i)} > \delta) \leq \mathbf{P}(\tau > \delta T) + \mathbf{P}\left(\tau > \frac{\delta\sigma(T)}{|a|}, \mathrm{e}_a\zeta > \delta\sigma(T)\right) \tag{1.6.29}$$

for any $\delta > 0$,

$$\rho_{\mathbb{D}}\Big(y_T(\cdot), s_T(\theta(\cdot))\Big) \leq \max_{i \leq \eta(T)} \rho_{\mathbb{D}}^{(i)}.$$

Consider the set

$$A_{T,\varepsilon} = \{\theta_T(1) < 1 + \varepsilon\} = \{\eta(T) < T(1 + \varepsilon)\}.$$

It is clear that $\mathbf{P}(A_{T,\varepsilon}) \to 1$ as $T \to \infty$ and any $\varepsilon > 0$. On this set

$$\rho_{\mathbb{D}}\Big(y_T(\cdot), s_T(\theta(\cdot))\Big) \leq \max_{i \leq T(1+\varepsilon)} \rho_{\mathbb{D}}^{(i)}$$

and

$$\begin{aligned}
&\mathbf{P}\Big(\rho_{\mathbb{D}}(y_T(\cdot), s_T(\theta(\cdot))) > \delta\Big) \\
&\qquad \leq \mathbf{P}(\overline{A}_{T,\varepsilon}) + \mathbf{P}\Big(\rho_{\mathbb{D}}(y_T(\cdot), s_T(\theta(\cdot)))\Big) > \delta, A_{T,\varepsilon}\Big) \\
&\qquad \leq \mathbf{P}(\overline{A}_{T,\varepsilon}) + T(1+\varepsilon)\mathbf{P}(\rho_{\mathbb{D}}^{(i)} > \delta).
\end{aligned} \tag{1.6.30}$$

We have $\mathbf{P}(\overline{A}_{T,\varepsilon}) \to 0$ as $T \to \infty$, and hence for the proof of (1.6.26) it suffices in view of (1.6.29) to verify that

$$T\mathbf{P}(\tau > \delta T) \to 0, \qquad T\mathbf{P}\Big(\tau > \frac{\delta\sigma(T)}{|a|}, \mathrm{e}_a\zeta > \delta\sigma(T)\Big) \to 0 \tag{1.6.31}$$

as $T \to \infty$. The first relation is clear, because $\mathbf{E}\tau$ exists. To verify the second relation, we employ condition (1.6.22), which implies that

$$\begin{aligned}
\mathbf{P}\Big(\tau > \frac{\delta\sigma(T)}{|a|}, \mathrm{e}_a\zeta > \delta\sigma(T)\Big) &= o\Big(F\Big(\min\Big(|a|, \frac{1}{|a|}\Big)\Big)\delta\sigma(T)\Big) \\
&= o\Big(F(\sigma(T))\Big) = o\Big(\frac{1}{T}\Big)
\end{aligned}$$

as $T \to \infty$. Hence the second estimated value in (1.6.31) is $o(1)$ as $T \to \infty$. This proves (1.6.26), and therefore, (1.6.23).

If τ and ζ are independent, then

$$\mathbf{P}(\tau > t, a\zeta > t) = \mathbf{P}(\tau > t)\mathbf{P}(a\zeta > t) = o(t^{-2}) \quad \text{as} \quad t \to \infty$$

and condition (1.6.22) obviously holds, which implies (1.6.23).

The argument for the processes z_T is similar. Theorem 1.6.3 is proved. □

Remark 1.6.4 The assumptions in Theorem 1.6.3 are close to the minimal ones. If condition (1.6.22) (which is broader than (1.6.5)) is violated, then the S- and $\mathbb{D}$-convergences of the distributions of the CRPs y_T, z_T to the process $w_{\alpha,\beta}$ are generally speaking absent. This is confirmed by the example given in §1.6.2, when $\zeta = \tau + \omega$, $\mathbf{E}\omega = 0$, τ, and ω are independent, $a = 1$. In this example, not only condition (1.6.5) is violated, but also the broader condition (1.6.22) is not met. Indeed, in this case $p := \mathbf{P}(\omega > 0) > 0$, and so

$$\begin{aligned}\mathbf{P}(\tau > t, \zeta > t) &\geq \mathbf{P}(\tau > t,\ \tau + \omega > t, \omega > 0)\\ &= \mathbf{P}(\tau > t, \omega > 0) = p\mathbf{P}(\tau > t) \gg F(t) \quad \text{as} \quad t \to \infty.\end{aligned}$$

In this example, we have $\sup_{u\leq 1} y_T(u) \underset{p}{\to} \infty$ as $T \to \infty$ (see §1.6.2), and hence the $\mathbb{D}$-convergence $y_T \underset{D}{\Rightarrow} w_{\alpha,\beta}$ is absent for the same reason as for the S-convergence fails to hold. (The functionals $f(s) = \inf_{u\leq 1} s(u)$ and $f(s) = \sup_{u\leq 1} s(u)$ are $\rho_{\mathbb{D}}$-continuous.) Note that the convergence is also absent in the metric space $\mathbb{D}$ equipped with any other metric ρ with respect to which the functional $f(s)= \sup_{u\leq 1} s(u)$ is continuous.

1.7 Limit Theorems for the First Passage Time of an Arbitrary Boundary by a Compound Renewal Process

1.7.1 Introduction

Let $g(t) = g_x(t)$, $t > 0$, be an arbitrary function (boundary) depending on the parameter $x \to \infty$. The main goal of the present section is to find the limit laws for the distribution of the first passage time η_g of the boundary $g_x(t)$ by a CRP $Y(t)$, $(Z(t))$,

$$\eta_g := \inf\{t > 0 : Y(t) > g_x(t)\}. \tag{1.7.1}$$

The boundaries $g_x(t)$ are assumed to be "remote," that is, it is assumed that the value

$$t_g := \inf\{t > 0 : at > g_x(t)\}$$

is defined and

$$t_g \to \infty \qquad \text{as} \quad x \to \infty. \tag{1.7.2}$$

We shall consider two classes of problems. In the first class it will be assumed that the random variable $\xi = \zeta - a\tau$ has finite second moment, that is,

$$\sigma_\xi^2 := \mathbf{E}(\zeta - a\tau)^2 = \mathbf{D}\xi < \infty. \tag{1.7.3}$$

In the second class, we will assume that $\sigma_\xi^2 = \infty$ and the random variable ξ satisfies the attraction conditions to a stable law (for more details, see §§1.4, 1.6).

The main limit theorems on the distribution of η_g that follow were obtained in [25]. Similar results for much simpler processes (random walks) can be found in [72, §5.6], [27].

1.7.2 Case of Finite Variance

Let us study the distribution of the first passage time η_g (see (1.7.1)) of the boundary $g_x(t)$ by the trajectory of the CRP $Y(t) = Z_{\eta(t)}$. In Theorem 1.7.3, the results obtained will be carried over to the process $Z(t) = Z_{\nu(t)}$ and the processes with linear drift.

We shall assume that (1.7.2) is met and that the boundary $g_x(t)$ satisfies the following two conditions.

$[\mathcal{A}]_1$ *For some* $\varepsilon > 0$, *there exists a sequence* $b_x \to b < a$ *as* $x \to \infty$ *such that*

$$g_x(t) = at_g + (t - t_g)b_x + o(\sqrt{t_g}) \quad as \quad x \to \infty \tag{1.7.4}$$

for

$$|t - t_g| < 2\frac{(1+\varepsilon)L(t_g)}{a-b}, \qquad L(t) := \sigma\sqrt{2t \ln\ln t}. \tag{1.7.5}$$

Consider the point $t_g^- < t_g$ of intersection of the functions $g_x(t)$ and $at+(1+\varepsilon)L(t)$ which is closest to t_g,

$$t_g^- := \sup\{t < t_g : g_x(t) > at + (1+\varepsilon)L(t)\}. \tag{1.7.6}$$

The second condition on $g_x(t)$ is as follows.

$[\mathcal{A}]_2$ $g_x(t) \to \infty$ *for any fixed t and* $x \to \infty$,

$$g_x(t) \geq at + (1+\varepsilon)L(t) \quad for \quad t \leq t_g^-. \tag{1.7.7}$$

It is clear that

$$t_g^- \geq t_g - \frac{2L(t_g)}{a-b} \tag{1.7.8}$$

for sufficiently large x.

The meaning of conditions $[\mathcal{A}]_1$, $[\mathcal{A}]_2$ is clear: the boundary $g_x(t)$ should be separated from the ray at (and its neighborhood) up to time t_g. Near the point t_g of the intersection of the ray at and the boundary $g_x(t)$, this boundary should be in a sense close to the linear boundary with angular coefficient $b < a$.

We now give the following two examples, in which conditions $[\mathcal{A}]_1$, $[\mathcal{A}]_2$ are met.

(1) Let $f(t)$ be a fixed (not depending on x) function and let $g_x(t)$ be obtained from $f(t)$ by stretching by the coefficient x in two coordinate axes,

$$g_x(t) = xf\left(\frac{t}{x}\right). \tag{1.7.9}$$

Assume that

$$u_f := \inf\{u : au \geq f(u)\}.$$

Then, clearly, $t_g = xu_f$. In order that conditions $[\mathcal{A}]_1$ and $[\mathcal{A}]_2$ be met, it suffices in this setting that the following conditions be satisfied:

1. $f(u)/u \to \infty$ as $u \downarrow 0$,
2. $f(u) > au + c\delta$ for $u \in [\delta, u_f - \delta]$, a sufficiently small δ, and some $c > 0$.
3. $f(u_f + u) = au_f + bu + O(|u|^{1+h})$ for $b = f'(u_f) < a$, $h > 0$, $|u| < \delta \to 0$.

The family (1.7.9) also includes the function

$$g_x(t) = xt^\alpha, \qquad \alpha \in [0, 1). \tag{1.7.10}$$

Indeed, if we put $f(t) = t^\alpha$, then in (1.7.9) we get $g_x(t) = x^{1-\alpha}t^\alpha$, and as a normalizing factor one should take $x^{1-\alpha}$, rather than x. For the function $g_x(t)$ from (1.7.10),

$$t_g = \left(\frac{x}{a}\right)^{1/(1-\alpha)}, \qquad b = \alpha a < a.$$

(2) In the second example, the function f is again fixed and

$$g_x(t) = x + f(t).$$

Here, the value t_g is defined if

$$\overline{\lim_{t\to\infty}} \,(at - f(t)) = \infty.$$

If, for example, $f(t) = bt + o(\sqrt{t})$ as $t \to \infty$, $b < a$, then $t_g = \frac{x+o(\sqrt{x})}{a-b}$ and conditions $[\mathcal{A}]_1$ and $[\mathcal{A}]_2$ are met. These conditions also hold in the case $f(t) = ct^\alpha$, $\alpha \in (0, 1)$.

Moreover, if in the above examples, in parallel with the functions $g_x(t)$, one considers the functions

$$g_x(t) + \sqrt{t}\,\delta_x(t), \tag{1.7.11}$$

where $\sup_t \delta_x(t) \to 0$ as $x \to \infty$, then these functions will also satisfy conditions $[\mathcal{A}]_1$ and $[\mathcal{A}]_2$, even though, as (1.7.11) shows, they can have arbitrary oscillations and discontinuities of order $o(\sqrt{t_g})$ near the point t_g.

Theorem 1.7.1 *Let $\sigma_\tau < \infty$, $\sigma < \infty$, let functions $g_x(t)$ be such that $t_g \to \infty$ as $x \to \infty$* (see (1.7.2)), *let $\tau_1 = o_p(\sqrt{t_g})$, $\zeta_1 = o_p(\sqrt{t_g})$, and let conditions $[\mathcal{A}]_1$, $[\mathcal{A}]_2$ be met. Then, as $x \to \infty$,*

$$\mathbf{P}\left(\frac{\eta_g - t_g}{\sigma\sqrt{t_g}} < v\right) \to \Phi((a-b)v),$$

where Φ is the distribution function of the normal law $\boldsymbol{\Phi}$.

If $a = 0$, then the condition $\sigma_\tau < \infty$ is superfluous.

Proof We will use Theorem 1.3.1 (Anscombe's theorem) and check its conditions for the process $y(t) = \frac{Y(t)-at}{\sigma\sqrt{t}}$ and the random time η_g.

The fulfillment of condition 1) of Theorem 1.3.1 for the process $y(t)$ follows from Theorem 1.3.3. Condition 2) is now a consequence of the following result.

Lemma 1.7.2 *If $\sigma_\tau < \infty$, $\sigma_\zeta < \infty$, $\tau_1 = o_p(\sqrt{t})$, $\zeta_1 = o_p(\sqrt{t})$, then the process $y(t)$ satisfies condition* 2) *of Theorem* 1.3.1, *that is,*

$$\sup_{|u|<\delta t} |y(t+u) - y(t)| \underset{P}{\to} 0 \tag{1.7.12}$$

as $t \to \infty$, $\delta = \delta(t) \to 0$.

If $a = 0$, then the condition $\sigma_\tau < \infty$ is superfluous.

Proof of Lemma 1.7.2 We first consider the homogeneous case. We have

$$y(t+u) - y(t) = \frac{Y(t+u) - Y(t) - au}{\sigma\sqrt{t+u}} - (Y(t) - at)\left(\frac{1}{\sigma\sqrt{t}} - \frac{1}{\sigma\sqrt{t+u}}\right),$$

$$\begin{aligned}\sup_{|u|<\delta t} |y(t+u) - y(t)| \le{}& \sup_{|u|<\delta t}\left|\frac{Y(t+u) - Y(t) - au}{\sigma\sqrt{t+u}}\right| \\ &+ \left|\frac{Y(t) - at}{\sigma\sqrt{t}}\right| \sup_{|u|<\delta t}\left|\sqrt{\frac{t}{t+u}} - 1\right|,\end{aligned} \tag{1.7.13}$$

where the second term on the right of (1.7.13), which is equal to $|y(t)|O(\delta)$, is clearly $o_p(1)$ as $\delta \to 0$. Since $Y(t) - at = S_{\eta(t)} + a\chi(t)$, the first term on the right of (1.7.13) can be written as

$$\begin{aligned}\sup_{|u|<\delta t}\left|\frac{S_{\eta(t+u)} - S_{\eta(t)} + a\chi(t+u) - a\chi(t)}{\sigma\sqrt{t+u}}\right| \le{}& \sup_{|u|<\delta t}\left|\frac{S_{\eta(t+u)} - S_{\eta(t)}}{\sigma\sqrt{t+u}}\right| \\ &+ \frac{|a|\chi(t)}{\sigma\sqrt{t(1-\delta)}} + \frac{|a|}{\sigma}\sup_{|u|<\delta t}\frac{\chi(t+u)}{\sqrt{t+u}}.\end{aligned} \tag{1.7.14}$$

The last two terms with $a = 0$ disappear, and for $a \neq 0$, $\sigma_\tau < \infty$, in view of Corollary 1.1.8, they converge to 0 as $t \to \infty$ with probability 1. It remains to consider the first term on the right of (1.7.14). By Lemma 1.1.6, for the trajectory $\eta(t)$ the strong law of large numbers holds, so that, for all sufficiently large t, the values of $\eta(t)$ for any $\delta > 0$ will lie between $a_\tau^{-1}t(1 \mp \delta)$. This means that, for $|u| < \delta t$ and for sufficiently small δ, all values of $\eta(t+u)$ will lie between $a_\tau^{-1}t(1 \mp 5\delta)$. Hence, for

$$n = [ta_\tau^{-1}] \tag{1.7.15}$$

sufficiently large t, the first term on the right of (1.7.14) is majorized by

$$\frac{2}{\sigma\sqrt{t(1-\delta)}} \max_{|k|<5\delta t} |S_{n+k} - S_n|. \tag{1.7.16}$$

Here we used the fact that

$$|S_{\eta(t+u)} - S_{\eta(t)}| \le |S_{\eta(t+u)} - S_n| + |S_{\eta(t)} - S_n|. \tag{1.7.17}$$

In other words, as $t \to \infty$, $\delta \to 0$, for the first term on the right-hand side of (1.7.14), which we denote by M_t, we have

$$\mathbf{P}(M_t > \varepsilon) \leq o(1) + \mathbf{P}\Big(\max_{|k| \leq 5\delta t} |S_{n+k} - S_n| > \frac{\varepsilon\sigma\sqrt{t}}{3}\Big). \tag{1.7.18}$$

If, under the probability sign on the right of (1.7.18), t is na_τ (see (1.7.15)), then we get the probability of the same form as was estimated in Lemma 1.3.4. In accordance with this lemma, the probabilities in (1.7.18) converge to 0 as $t \to \infty$, $\delta \to 0$. This together with (1.7.14), (1.7.13) completes the proof of (1.7.12).

In the inhomogeneous case, we may use the representation

$$\widetilde{Y}(t) = \zeta_1 + Y(t - \tau_1), \tag{1.7.19}$$

where $Y(t)$ is a homogeneous CRP, employ conditions $\tau_1 = o_p(\sqrt{t})$, $\zeta_1 = o_p(\sqrt{t})$, and repeat the arguments from the proof of Theorem 1.3.3 in the inhomogeneous case. Lemma 1.7.2 is proved. □

Let us verify condition 3) of Theorem 1.3.1. In parallel with t_g^- (see (1.7.6)), we consider the point $t_g^+ > t_g$ of intersection of the functions $g_x(t)$ and $at - (1+\varepsilon)L(t)$,

$$t_g^+ = \min\{t > t_g : at - (1+\varepsilon)L(t) \geq g_x(t)\},$$

where $L(t)$ is defined in (1.7.5).

By condition $[\mathcal{A}]_2$ and using the law of the iterated logarithm for the process $Y(t)$, we have

$$\mathbf{P}(\eta_g \in (t_g^-, t_g^+)) \to 1 \quad \text{as} \quad x \to \infty.$$

By condition (1.7.8), for sufficiently large x we have

$$t_g^+ - t_g^- \leq \frac{4L(t_g)}{a-b},$$

so that

$$\mathbf{P}\Big(|\eta_g - t_g| < \frac{4L(t_g)}{a-b}\Big) \to 1.$$

This means that

$$\frac{\eta_g}{t_g} \xrightarrow[p]{} 1 \quad \text{as} \quad x \to \infty \tag{1.7.20}$$

and that condition 3) of Theorem 1.3.1 for $\theta(x) = \eta_g$ and $h(x) = t_g$ is met. Hence from Theorem 1.3.1 we get

$$\mathbf{P}(y(\eta_g) < v) \Rightarrow \Phi(v) \quad \text{as} \quad x \to \infty, \tag{1.7.21}$$

where

$$y(\eta_g) = \frac{Y(\eta_g) - a\eta_g}{\sigma\sqrt{\eta_g}}, \qquad \frac{\sqrt{\eta_g}}{\sqrt{t_g}} \xrightarrow[p]{} 1 \quad \text{as} \quad x \to \infty. \tag{1.7.22}$$

Next,

$$Y(\eta_g) = g_x(\eta_g) + r_g, \tag{1.7.23}$$

where $|r_g| \le |\zeta(\eta_g)|$. We set

$$n_x^\pm := \frac{t_g^\pm}{a_\tau}(1 \pm \varepsilon),$$

for $\varepsilon > 0$. Consider the events

$$\begin{aligned} B_1 &:= \{t_g^- < \eta_g < t_g^+\}, \\ B_2 &:= \{\eta(t_g^-) > n_x^-\}, \\ B_3 &:= \{\eta(t_g^+) < n_x^+\}, \qquad B = B_1B_2B_3. \end{aligned} \tag{1.7.24}$$

It is clear that $\mathbf{P}(B_1) \to 1$ as $x \to \infty$ and by the law of large numbers, for $\eta(t)$ we have $\mathbf{P}(B_2B_3) \to 1$, so that $\mathbf{P}(B) \to 1$ as $x \to \infty$. Moreover, on the set B

$$\begin{aligned} n_x^- < \eta(t_g^-) < \eta(\eta_g) < \eta(t_g^+) < n_x^+, \\ |r_g| \le \max\left\{|\zeta(t)| : t_g^- < t < t_g^+\right\} \le \max\left\{|\zeta_k| : n_x^- < k < n_x^+\right\}. \end{aligned} \tag{1.7.25}$$

Hence, for any $\delta > 0$,

$$\mathbf{P}(|r_g| > \delta\sqrt{t_g};\, B) \le (n_x^+ - n_x^-)\mathbf{P}(|\zeta| > \delta\sqrt{t_g}).$$

Next, for sufficiently large x and small ε,

$$n_x^+ - n_x^- < \frac{2(t_g^+ - t_g^-)}{a_\tau} < \frac{8L(t_g)}{a_\tau(a-b)},$$

so that

$$\mathbf{P}(r_g > \delta\sqrt{t_g}) \le o(1) + \frac{8L(t_g)}{a_\tau(a-b)}\frac{\mathbf{E}\zeta^2}{\delta^2 t_g} = o(1) \quad \text{as } x \to \infty,$$

$r_g = o_p(\sqrt{t_g})$. Hence from (1.7.23) and condition $[\mathcal{A}]_1$, we get

$$\begin{aligned} Y(\eta_g) - a\eta_g &= g_x(\eta_g) - a\eta_g + o_p(\sqrt{t_g}) = (a - b_x)(t_g - \eta_g) + o_p(\sqrt{t_g}), \\ y(\eta_g) &= -\frac{(a-b_x)(\eta_g - t_g)}{\sigma\sqrt{t_g}} + o_p(1) \end{aligned} \tag{1.7.26}$$

and by (1.7.21),

$$\mathbf{P}\Big(-\frac{(a-b_x)(\eta_g - t_g)}{\sigma\sqrt{t_g}} < v\Big) = \mathbf{P}\Big(\frac{\eta_g - t_g}{\sigma\sqrt{t_g}} > -\frac{v}{a - b_x}\Big) \to \Phi(v) \tag{1.7.27}$$

as $x \to \infty$, or, equivalently,

$$\mathbf{P}\Big(\frac{\eta_g - t_g}{\sigma\sqrt{t_g}} < u\Big) \to \Phi((a-b)u).$$

Theorem 1.7.1 is proved. □

Let us now show that the conclusion of Theorem 1.7.1 remains valid for the CRP $Z(t) = Z_{\nu(t)}$ and the processes

$$Y^{(q)}(t) = Y(t) + qt, \qquad Z^{(q)}(t) = Z(t) + qt \tag{1.7.28}$$

with linear drift qt.

Theorem 1.7.3 *The conclusion of Theorem* 1.7.1 *remains valid if* $Y(t)$ *is replaced by* $Z(t)$*. This result also holds if* $Y(t)$ *is replaced by* $Y^{(q)}(t)$ *or by* $Z^{(q)}(t)$ *with* b *replaced by* $b^{(q)} := b - q$.

Proof We have

$$Y(t) - Z(t) = \zeta(t) \tag{1.7.29}$$

and besides, by Corollary 1.1.8,

$$\frac{|\zeta(t)|}{\sqrt{t}} \xrightarrow[\text{a.s.}]{} 0 \tag{1.7.30}$$

as $t \to \infty$; therefore,

$$\frac{|Z(t) - Y(t)|}{\sqrt{t}} \xrightarrow[\text{a.s.}]{} 0. \tag{1.7.31}$$

This implies the conclusion of Theorem 1.7.1 with the CRP $Z(t)$ (the boundary $g_x(t)$ is defined up to $o(\sqrt{t_g})$; see (1.7.4)).

The second assertion of the theorem is clear, because the problem under consideration can be reduced to the problem for the first passage time of the boundary $g(t)-qt$ by the processes $Y(t)$, $Z(t)$. Theorem 1.7.3 is proved. □

1.7.3 Case of Infinite Variance

If $\sigma = \sigma_\xi = \infty$, then, as in § 1.5, we shall assume that the condition of regular decay of the distribution $\xi = \zeta - a\tau$ at infinity is met. If, as before, we set

$$F_-(t) = \mathbf{P}(\xi \le -t), \quad F_+(t) = \mathbf{P}(\xi \ge t),$$
$$F(t) = F_-(t) + F_+(t) = \mathbf{P}(|\xi| \ge t),$$

then we assume that the following condition is satisfied.

[$\mathbf{R}_{\alpha,\beta}$] *The function* $F(t)$ *is regularly varying at infinity, that is,* $F(t)$ *can be written as*

$$F(t) = t^{-\alpha} l(t), \qquad \alpha \in (1, 2),$$

where $l(t)$ *is a slowly varying function at infinity. Moreover, the limit*

$$\lim_{t\to\infty} \frac{F_+(t)}{F(t)} =: \beta_+ \in [0, 1]$$

exists, and we put $\beta := 2\beta_+ - 1$.

Next, let $F^{(-1)}(u)$ be the generalized inverse function of $F(t)$,

$$F^{(-1)}(u) = \inf\{t : F(t) < u\}$$

and let

$$\sigma_\xi(n) := F^{(-1)}(1/n). \tag{1.7.32}$$

The function $\sigma_\xi(n)$ has the form $n^{1/\alpha} l_\sigma(n)$, where $l_\sigma(n)$ is a slowly varying sequence (see, e.g., [21, §8.8]).

Under condition $[\mathbf{R}_{\alpha,\beta}]$, the normalized sums

$$s(n) := \frac{S_n}{\sigma_\xi(n)}$$

weakly converge in distribution to the stable law $\mathbf{\Phi}_{\alpha,\beta}$ with the parameters (α, β).

Let, as before,

$$\sigma(t) = \sigma_\xi(t) a_\tau^{-1/\alpha}.$$

As in §1.7.2, we will assume here that the time t_g of the first intersection of the curves $g_x(t)$ and at is defined and that the boundary $g_x(t)$ satisfies conditions $[\mathcal{A}]_1$, $[\mathcal{A}]_2$, in which $o(\sqrt{t_g})$ in (1.7.4) should be replaced by $o(\sigma(t_g))$ and write $L(t)$ in place of $(1+\varepsilon)L(t)$ (with $L(t)$ defined in (1.7.5)), where now

$$L(t) = \sigma(t)(\ln t)^\theta \tag{1.7.33}$$

with any fixed $\theta > \frac{1}{\alpha}$. The remainder term $o(\sqrt{t_g})$ in (1.7.4) should be replaced by $o(\sigma(t_g))$. In this setting, all the comments to conditions $[\mathcal{A}]_1$, $[\mathcal{A}]_2$, from §1.7.2 remain valid.

Let, as before, η_g be defined in (1.7.1).

Theorem 1.7.4 *Let ξ satisfy condition $[\mathbf{R}_{\alpha,\beta}]$, $\alpha \in (1, 2)$, let functions $g_x(t)$ be such that $t_g \to \infty$ as $x \to \infty$, let $\tau_1 = o_p(\sigma(t_g))$, $\zeta_1 = o_p(\sigma(t_g))$, and let conditions $[\mathcal{A}]_1$, $[\mathcal{A}]_2$ hold with the changes given in (1.7.33). Let, in addition,*

$$\mathbf{P}(\tau \geq t) < cF(t), \qquad \mathbf{P}(|\zeta| \geq t) < cF(t) \tag{1.7.34}$$

with some $c < \infty$. Then, as $x \to \infty$,

$$\mathbf{P}\left(\frac{\eta_g - t_g}{\sigma(t_g)} < v\right) \to \Phi_{\alpha,\beta}((a-b)v), \tag{1.7.35}$$

where $\Phi_{\alpha,\beta}$ is the distribution function of a stable law with parameters α, β.

If $a = 0$, then the first of conditions (1.7.34) is superfluous.

The conclusion of the theorem holds verbatim for the processes Z, $Y^{(q)}$, and $Z^{(q)}$.

Proof of Theorem 1.7.4 The proof is quite similar to that of Theorem 1.7.1. Using Theorem 1.3.1, we will verify conditions 1)–3) for the sequence

$$y(t) = \frac{Y(t) - at}{\sigma(t)}$$

and the random time η_g. That condition 1) holds follows from Theorem 1.4.1.

Let us now show that $y(t)$ satisfies condition 2) of Theorem 1.3.1, that is, we establish the following analog of Lemma 1.7.2.

Lemma 1.7.5 *Let a random variable $\xi = \zeta - a\tau$ satisfy condition $[\mathbf{R}_{\alpha,\beta}]$, $\alpha \in (1, 2)$,*

$$\mathbf{P}(\tau \geq t) \leq cF(t), \qquad c = \text{const}, \tag{1.7.36}$$

$\tau_1 = o_p(\sigma(t))$, $\zeta_1 = o_p(\sigma(t))$. Then

$$\sup_{|u|<\delta t} |y(t+u) - y(t)| \underset{p}{\to} 0 \tag{1.7.37}$$

as $t \to \infty$, $\delta = \delta(t) \to 0$.

If $a = 0$, then condition (1.7.36) is superfluous.

Proof of Lemma 1.7.5 We first consider the homogeneous case. By analogy to (1.7.13), we have

$$\sup_{|u|<\delta t} |y(t+u) - y(t)| \leq \sup_{|u|<\delta t} \frac{|Y(t+u) - Y(t) - au|}{\sigma(t+u)} + |y(t)| \sup_{u<\delta t} \left| \frac{\sigma(t)}{\sigma(t+u)} - 1 \right|,$$

where the second term on the right is clearly $o_p(1)$. The second term is majorized by (cf. (1.7.14))

$$\sup_{|u|<\delta t} \left| \frac{S_{\eta(t+u)} - S_{\eta(t)}}{\sigma(t+u)} \right| + \frac{|a|\chi(t)}{\sigma(t-\delta t)} + \frac{|a|}{\sigma(t-\delta t)} \sup_{|u|<\delta t} \chi(t+u). \tag{1.7.38}$$

Here $\sigma(t - \delta t) \sim \sigma(t) \to \infty$ as $t \to \infty$,

$$\frac{\chi(t)}{\sigma(t-\delta t)} \underset{p}{\to} 0, \tag{1.7.39}$$

since the distribution of $\chi(t)$ converges as $t \to \infty$ to a proper distribution (see Lemma 1.1.1). Next,

$$\sup_{|u|<\delta t} \chi(t+u) \leq \chi(t-\delta t) + \mu_t, \tag{1.7.40}$$

where $\mu_t \underset{p}{\leq} \overline{\tau}(2\delta t) = \sup_{u \leq 2\delta t} \tau(u)$, $\tau(u) = \tau_{\eta(u)}$, and for $\varepsilon > 0$, $m = [\frac{3\delta t}{a_\tau}] \to \infty$ we have

$$\mathbf{P}(\overline{\tau}(2\delta t) > \varepsilon\sigma(t)) \leq \mathbf{P}(\eta(2\delta t) > m) + \mathbf{P}(\overline{\tau}_m > \varepsilon\sigma(t)) = o(1) + m\mathbf{P}(\tau > \varepsilon\sigma(t)).$$

However, by condition (1.7.36),

$$m\mathbf{P}(\tau > \varepsilon\sigma(t)) \leq mcF(\varepsilon a_\tau^{-1/\alpha}\sigma_\xi(t)) \sim mc\,\varepsilon^{-\alpha} a_\tau F(\sigma_\xi(t)) \sim 3\delta c\,\varepsilon^{-\alpha} = o(1).$$

Hence $\frac{\mu_t}{\sigma(t)} = o_p(1)$ and so $\frac{\chi(t-\delta t)}{\sigma(t-\delta t)} = o_p(1)$ by (1.7.39). This means that the last two terms in (1.7.38) are $o_p(1)$.

If $a = 0$, then the last two terms in (1.7.38) disappear, and in this case, the above estimates are not necessary.

It remains to estimate the first term in (1.7.38). Here we use the same argument as for the analogous term in (1.7.14) in the proof of Lemma 1.7.2. Hence the first term in (1.7.38) with

$$n = [ta_\tau^{-1}] \tag{1.7.41}$$

and for sufficiently large t will not exceed

$$\frac{2}{\sigma(t-\delta t)} \max_{|k|<5\delta t} |S_{n+k} - S_n|$$

on a set of high probability (cf. (1.7.16)). In other words, as $t \to \infty$, $\delta \to 0$, for the first term in (1.7.38), which we denote by M_t, for any $\varepsilon > 0$ we have (cf. (1.7.18))

$$\mathbf{P}(M_t > \varepsilon) \le o(1) + \mathbf{P}\Big(\max_{|k|<5\delta t} |S_{n+k} - S_n| > \frac{\varepsilon\sigma(t)}{3}\Big). \tag{1.7.42}$$

If, under the probability sign on the right of (1.7.42), we replace t by na_τ (see (1.7.41)) and put $N := 5\delta n a_\tau \to \infty$ as $n \to \infty$, then we get probability estimates of the form

$$\mathbf{P}(\overline{S}_N > \varepsilon c\sigma(n)),$$

where $c = \text{const}$, $\overline{S}_N = \max_{k\le N} S_k$. Since $\frac{\sigma(N)}{\sigma(n)} \to 0$ as $n \to \infty$, these probabilities converge to 0 as $n \to \infty$ in view of the results of §3.1 in [38]. (One can also employ here the convergence of $\frac{S_{tN}}{\sigma(N)}$ in distribution to a stable process, which implies that $\frac{\overline{S}_N}{\sigma(N)}$ has a proper limit distribution as $N \to \infty$, $\frac{\overline{S}_N}{\sigma(n)} \xrightarrow[p]{} 0$.) This proves the convergence in (1.7.37) in the homogeneous case. In the inhomogeneous case, one should repeat (with clear modifications) the arguments in the proof of Lemma 1.7.2. Lemma 1.7.5 is proved. □

Let us now verify condition 3) of Theorem 1.3.1. In parallel with the value t_g^- (see (1.7.6) with $(1+\varepsilon)L(t)$ replaced by the function $L(t)$ defined in (1.7.33)), we consider the point t_g^+ of intersection of the lines $g_x(t)$ and $at - L(t)$ with $L(t)$ defined in (1.7.33),

$$t_g^+ := \inf\{t > t_g : at - L(t) \ge g_x(t)\}.$$

By condition $[\mathcal{A}]_2$ and using the analog of the law of the iterated logarithm (Theorem 1.4.3), we have

$$\mathbf{P}(\eta_g \in (t_g^-, t_g^+)) \to 1 \quad \text{as} \quad x \to \infty.$$

From relations of the form (1.7.8) for sufficiently large x we get

$$t_g^+ - t_g^- \le \frac{4L(t_g)}{a-b},$$

so that

$$\mathbf{P}\Big(|\eta_g - t_g| < \frac{4L(t_g)}{a-b}\Big) \to 1.$$

This means that

$$\frac{\eta_g}{t_g} \xrightarrow[p]{} 1 \quad \text{as} \quad x \to \infty,$$

which verifies condition 3) of Theorem 1.3.1 with $\theta(x) = \eta_g$, $h(x) = t_g$.

We can now invoke Theorem 1.3.1, according to which

$$y(\eta_g) \Leftrightarrow \mathbf{\Phi}_{\alpha,\beta} \quad \text{as} \quad x \to \infty,$$

where

$$y(\eta_g) = \frac{Y(\eta_g) - a\eta_g}{\sigma(\eta_g)}, \qquad \frac{\eta_g}{t_g} \xrightarrow[p]{} 1.$$

We set

$$n_x^{\pm} := (1 \pm \varepsilon) a_\tau^{-1} t_g^{\pm}, \qquad \varepsilon > 0,$$

and use representation (1.7.23). As in the proof of Theorem 1.7.1, consider the events B_1, B_2, B_3, B (which are defined in (1.7.24)), where, as before, $\mathbf{P}(B_1) \to 1$ as $x \to \infty$. Next, assuming for simplicity that $n_x^{\pm}$ are integers, we have

$$\mathbf{P}(\eta(t_g^-) < n_x^-) = \mathbf{P}(T_{n_x^-} \geq t_g^-) = \mathbf{P}(T_{n_x^-} - a_\tau n_x^- \geq \varepsilon t_g^-). \tag{1.7.43}$$

By the law of large numbers, the probability in (1.7.43) converges to 0 as $x \to \infty$, so that $\mathbf{P}(B_2) \to 1$. A similar analysis shows that $\mathbf{P}(B_3) \to 1$, $\mathbf{P}(B) \to 1$ as $x \to \infty$.

Next, in view of (1.7.23), (1.7.25), on the set B we have

$$\eta(\eta_g) \in (n_x^-, n_x^+), \qquad |r_g| \leq \max\{|\zeta_k| : n_x^- < k < n_x^+\}. \tag{1.7.44}$$

Let us estimate $|r_g|$ in (1.7.23). For any fixed $\delta > 0$, by (1.7.44) we have

$$\mathbf{P}(|r_g| > \delta\sigma_\xi(t_g);\ B) \leq (n_x^+ - n_x^-)\mathbf{P}(|\zeta| > \delta\sigma_\xi(t_g)).$$

Next, for sufficiently large x and small ε, we get

$$n_x^+ - n_x^- < 2a_\tau^{-1}(t_g^+ - t_g^-) < \frac{8L(t_g)}{a_\tau(a-b)},$$

and hence

$$\mathbf{P}(|r_g| > \delta\sigma_\xi(t_g)) \leq o(1) + \frac{8L(t_g)}{a_\tau(a-b)}\mathbf{P}(|\zeta| > \delta\sigma_\xi(t_g)). \tag{1.7.45}$$

However, by (1.7.34), as $x \to \infty$,

$$\mathbf{P}(|\zeta| > \delta\sigma_\xi(t_g)) \leq c\delta^{-\alpha} F(\sigma_\xi(t_g))(1 + o(1)) \sim \frac{c\delta^{-\alpha}}{t_g}.$$

We have $L(t) = o(t)$, and hence in view of (1.7.45) we get that $r_g = o_p(\sigma(t_g))$. Returning to (1.7.23), we have by condition $[\mathcal{A}]_1$,

$$Y(\eta_g) - a\eta_g = g_x(\eta_g) - a\eta_g + o_p(\sigma(t_g)) = (a-b)(t_g - \eta_g) + o_p(\sigma(t_g)). \tag{1.7.46}$$

The further arguments, as well as (1.7.46), repeat those given in the derivation of (1.7.26) and (1.7.27) in the proof of Theorem 1.7.1 with $\sigma\sqrt{t_g}$ replaced by $\sigma(t_g)$ and $\Phi_{\alpha,\beta}$ substituted for Φ. Theorem 1.7.4 is proved. □

The proof pertaining to the processes $Y^{(q)}$, $Z^{(q)}$ repeats the arguments given in Theorem 1.7.3.

1.8 Main Limit Laws for Markov Additive Processes (for Sums of Random Variables Defined on States of a Markov Chain)

1.8.1 Ergodic Theorems for Harris Markov Chains

Let $\{\mathrm{x}_k\}_{k=0}^{\infty}$ be a Markov chain with states $\mathrm{e} \in E$ in an arbitrary measurable space $(E, \mathcal{E})$ with σ-algebra $\mathcal{E}$ and the transition probability $P_{\mathrm{e}}(A)$, $A \in \mathcal{E}$. We shall consider Harris Markov chains, which are defined as follows. Let $\mathrm{x}_n(\mathrm{e})$ be the state of the chain at time n with the initial state e and let $P_{\mathrm{e},n}(A) := \mathbf{P}(\mathrm{x}_n(\mathrm{e}) \in A) = \mathbf{P}(\mathrm{x}_n \in A \,|\, \mathrm{x}_0 = \mathrm{e})$ be the n-step transient probability. For some fixed set $V \in \mathcal{E}$, we define the random variable

$$\tau_V(\mathrm{e}) = \min\{k \geq 1, \mathrm{x}_k(\mathrm{e}) \in V\}, \qquad \mathrm{e} \in E,$$

which is the first entrance time of the chain from the state $\mathrm{e} \in E$ into the set V ($\tau_V(\mathrm{e}) = \infty$ if $\mathrm{x}_k(\mathrm{e}) \notin V$).

A Markov chain $\{\mathrm{x}_n\}$ in $(E, \mathcal{E})$ is called a *Harris chain* (or *Harris irreducible chain*, $\boldsymbol{\varphi}$-irreducible chain) if there exist a set $V \in \mathcal{E}$, a probability measure $\boldsymbol{\varphi}$ on $(E, \mathcal{E})$, and numbers $n_0 \geq 1$, $p \in (0, 1)$ such that

$$(\mathrm{I}_0) \qquad \mathbf{P}(\tau_V(\mathrm{e}) < \infty) = 1 \quad \text{for all } \mathrm{e} \in E,$$

$$(\mathrm{II}) \qquad P_{\mathrm{e},n_0}(A) \geq p\boldsymbol{\varphi}(A) \quad \text{for all } \mathrm{e} \in V, \quad A \in \mathcal{E}.$$

Condition (I_0) can be looked upon as the condition for irreducibility of the chain: from any point $\mathrm{e} \in E$ the trajectory x_n will sooner or later enter the set V. Condition (II) guarantees that, after n_0 steps after hitting the set V, the "walking particle" will have the distribution which is "minorized" by the same "distribution" $p\boldsymbol{\varphi}(\cdot)$. This condition is sometimes called the "mixing condition" – this condition provides a "partial memory loss" about the past of the trajectory.

If the chain has a recurrent atom e_0, then conditions (I_0), (II) are always satisfied with $V = \{\mathrm{e}_0\}$, $n_0 = 1$, $p = 1$, $\boldsymbol{\varphi}(A) = P_{\mathrm{e}_0}(A)$, so that the chain is a Harris chain.

As a set V one usually considers "compact" sets (if $E = \mathbb{R}^k$, then these sets are bounded), otherwise, as a rule, it is not possible to ensure the fulfillment of inequality (II). If the space E itself is "compact" (a finite set or closed bounded set in $\mathbb{R}^k$), then condition (II) can be satisfied also for $V = E$ (in this case, condition (I_0) is always met). For example, if the chain $\{\mathrm{x}_n\}$ is finite, irreducible, and aperiodic, then there exists an n_0 such that $P_{\mathrm{e},n_0}(\mathrm{e}_1) \geq p > 0$ for all e, e_1 from E. Hence condition (II) with $V = E$ is met if one takes as $\boldsymbol{\varphi}$ the uniform distribution on E.

Condition (II) can be interpreted to say that all the distributions $P_{\mathrm{e},n_0}(\cdot)$ for $\mathrm{e} \in V$ have an absolutely continuous component with respect to the measure $\boldsymbol{\varphi}$

$$\inf_{\mathrm{e} \in V} \frac{P_{\mathrm{e},n_0}(dy)}{\boldsymbol{\varphi}(dy)} \geq p > 0.$$

For conditions ensuring conditions (I_0), (II), see, for example, §13.7 in [21].

We shall also require the condition of positive recurrency of the set V (or of positive recurrency of the chain)

(I) $$\sup_{e \in V} \mathbf{E}\tau_V(e) < \infty$$

and the aperiodicity condition, which we write as follows. Let $x_k(\varphi)$ be a Markov chain with initial distribution φ, where φ is from condition (II). We set

$$\tau_V(\varphi) = \min\{k \geq 0 : x_k(\varphi) \in V\}.$$

In view of condition (I_0) it is clear that $\tau_V(\varphi)$ is a proper random variable.

The chain aperiodicity condition is as follows:[1]

(III) $$\gcd\left\{k \geq 1 : \mathbf{P}(\tau_V(\varphi) = k) > 0\right\} = 1.$$

Condition (III) is always met if condition (II) is satisfied with $n_0 = 1$, $\varphi(V) > 0$ (in this case, $\mathbf{P}(\tau_V(\varphi) = k) > 0$ for all k).

It is clear that a finite irreducible aperiodic Markov chain always satisfies conditions (I_0), (I)–(III). Athreya, Ney, and Nummelin (see, e.g., [97], [83], [7], [19, Chapter 1], [21, §13.7]) proved a remarkable fact that any Harris chain $\{x_n\}$ can be "extended" to the chain $x_n^* = (x_n, y_n)$, where y_n are independent,

$$y_n = \begin{cases} 1 & \text{with probability } p, \\ 0 & \text{with probability } 1 - p, \end{cases}$$

so that the chain $\{x_n^*\}$ will have an "artificial" atom $V^* = (V, 1)$ with the transition probability

$$\mathbf{P}(x_{n+1}^* \in (B, 0 \cup 1) \mid x_n^* = V^*) = \varphi(B)$$

(for more details, see, e.g., [21, Chapter 13], [19, Chapter 1]).

Since the state space $(E, \mathcal{E})$ of the Harris chain is arbitrary, we can, without loss of generality, assume from the very beginning that the chain $\{x_n\}$ has a single-state atom $V = \{e_0\}$.

Hence the trajectory of the process $\{x_n\}$ after the first entrance into e_0 can be split into independent identically distributed blocks (cycles) generated by chain returns to the state e_0. This naturally leads to the renewal process with time intervals of length $\tau_1 = \min\{k \geq 1 : x_k = e_0\}$, $\tau_2, \tau_3, \ldots$ being independent copies of $\tau = \min\{k \geq 1 : x_k(e_0) = e_0\}$. The renewal process will be homogeneous if $x_0 = e_0$. The probability

$$Q(k, A) = \mathbf{P}(x_k \in A \mid x_0 = e_0, x_1 \neq e_0, \ldots, x_{k-1} \neq e_0)$$

is called the taboo probability of entering A from e_0 in k steps (avoiding a return to e_0).

We have the following ergodic theorem, which in essence is a simple corollary of the fact that in the case $a_\tau = \mathbf{E}\tau < \infty$ the sequence $\gamma(n) = n - T_{\nu(n)}$, $T_k = \sum_{j=1}^{k} \tau_j$ has a proper limit distribution (see Lemma 1.1.1) and

$$\lim_{n\to\infty} \mathbf{P}(\gamma(n) = k) = \frac{\mathbf{P}(\tau > k)}{a_\tau}, \qquad k = 0, 1, \ldots$$

[1] Here "gcd" stands for "the greatest common divisor."

Theorem 1.8.1 *If conditions* (I_0), (I)–(III) *are met, that is, if* $\{\mathrm{x}_n\}$ *is an aperiodic Harris Markov chain, then, for any* $\mathrm{e} \in E$, $A \in \mathcal{E}$,

$$P_{\mathrm{e},n}(A) \to \pi(A) := \frac{1}{a_\tau} \sum_{k=0}^{\infty} \mathbf{P}(\tau > k)\, Q(k, A) \quad \text{as } n \to \infty \tag{1.8.1}$$

and $\pi(\cdot)$ *is an invariant measure.*

Indeed, let e_0 be a positive atom of the chain. By Lemma 1.1.1,

$$P_{\mathrm{e},n}(A) = \sum_{l=0}^{n} \mathbf{P}(\gamma(n) = l) Q(l, A) \to \pi(A) \tag{1.8.2}$$

as $n \to \infty$. It is easily checked that π is an invariant measure and that (1.8.1) is preserved for any distribution of the initial value x_0.

There is a stronger result than (1.8.1) on the convergence of $P_{\mathrm{e},n}(A)$ to $\pi(A)$ in variation.

For a complete proof of Theorem 1.8.1 (which consists of the above arguments with the reverse transition from the chain $\{\mathrm{x}_n^*\}$ to $\{\mathrm{x}_n\}$), see, for example, Theorem 13.7.1 in [21]).

1.8.2 Markov Additive Process

A Markov additive process (MAP) is defined as follows (see, e.g., [95, 96]). Assume that in addition to a chain $\{\mathrm{x}_n\}_{n=0}^{\infty}$ we are given a sequence of random variables $\{\xi_n\}_{n=0}^{\infty}$ such that $\{(\mathrm{x}_n, \xi_n); n = 1, 2, \dots\}$ is a Markov chain on $(E \times \mathbb{R}, \mathcal{E} \times \mathfrak{B})$, where $\mathfrak{B}$ is the Borel σ-algebra on $\mathbb{R}$, and its transient probability has the property

$$\mathbf{P}\big((\mathrm{x}_{n+1}, \xi_{n+1}) \in A \times B \,|\, \mathcal{F}_n\big) = \mathbf{P}\big((\mathrm{x}_{n+1}, \xi_{n+1}) \in A \times B \,|\, \mathrm{x}_n\big) =: P_{\mathrm{x}_n}(A, B), \tag{1.8.3}$$

where $A \in \mathcal{E}$, $B \in \mathfrak{B}$, $\mathcal{F}_n$ is the σ-algebra generated by $(\mathrm{x}_0, \dots, \mathrm{x}_n; \xi_1, \dots, \xi_n)$. Next, let

$$\mathrm{X}_n = \mathrm{X}_0 + \sum_{k=1}^{n} \xi_k. \tag{1.8.4}$$

A pair $\{(\mathrm{x}_n, \mathrm{X}_n), n = 0, 1, \dots\}$ is called a *Markov additive process* (MAP) with the transition function $P_{\mathrm{e}}(A, B)$. For the main properties of MAPs, see, for example, [51, 52].

For simplicity, we assume that E is discrete (countable). Then, by definition of a MAP,

$$\mathbf{P}(\xi_{n+1} \in B \,|\, \mathrm{x}_n = \mathrm{e},\ \mathrm{x}_{n+1} = \mathrm{e}_1) = \frac{P_{\mathrm{e}}(\mathrm{e}_1, B)}{P_{\mathrm{e}}(\mathrm{e}_1, \mathbb{R})}, \tag{1.8.5}$$

so that the distribution of ξ_{n+1} depends in general on the *pair* of states $(\mathrm{x}_n, \mathrm{x}_{n+1})$ (that is, transitions of the chain).

We will use a more transparent and constructive definition of a MAP, which corresponds more closely to the title of §1.8. In this definition, the probability

$$\mathbf{P}(\xi_{n+1} \in B \,|\, \mathcal{F}_n, \mathrm{x}_{n+1}) = \mathbf{P}(\xi_{n+1} \in B \,|\, \mathrm{x}_{n+1}) \tag{1.8.6}$$

depends only on x_{n+1}. This construction does not impair the generality of (1.8.3), (1.8.4), since the process (1.8.3), (1.8.4) can be constructed also on the chain $\{\overset{\neg}{x}_n\}$ composed of the "transitions" $\overset{\neg}{x}_n = (x_{n-1}, x_n)$. For the chain $\{\overset{\neg}{x}_n, \xi_n\}$, (1.8.6) holds. It is clear that the limit laws for the MAPs (x_n, X_n) and $(\overset{\neg}{x}_n, X_n)$ will be essentially the same.

This construction assumes that we are given a family of distributions $\{\mathbf{F}_e\}$ depending on the parameter $e \in E$. If $F_e^{(-1)}(t)$ is the quantile transformation of the distribution $\{\mathbf{F}_e\}$ and ω is a random variable uniformly distributed on [0, 1], then

$$\xi(e) := F_e^{(-1)}(\omega)$$

has the distribution $\mathbf{F}_e$ (see, e.g., [21, §3.3]).

It is assumed that the mapping $\mathbf{F}_e$ of the space E into the set of distributions is such that the function $F_e^{(-1)}(t)$ on $E \times \mathbb{R}$ is measurable with respect to $\mathfrak{E} \times \mathfrak{B}$, and moreover, as a function of e, features, if necessary, some properties of a regular variation so that the integrals in the transition function $P_{e_1}(A, B) = \int_A P_{e_1}(de)\mathbf{P}(\xi(e) \in B)$ are defined. These conditions are not required if E is discrete (in this case, the integrals become series).

If ω_n are uniformly distributed on [0, 1] and independent, then the sequence

$$\xi(x_n) := F_{x_n}^{(-1)}(\omega_n), \qquad n = 0, 1, \ldots$$

is called a *sequence of random variables defined on the Markov chain* $\{x_n\}$.

In this section, we shall be mostly concerned with the asymptotic properties of the distribution of the sums

$$X_n := \sum_{k=0}^{n} \xi(x_k). \tag{1.8.7}$$

The two-dimensional process

$$\{x_n, X_n\}, \qquad n = 0, 1, \ldots$$

will be called a *Markov additive process*. As was already pointed out, this construction is equivalent in a sense to (1.8.3), (1.8.4). The asymptotic laws of MAPs have been extensively studied (see, e.g., [119], [120], [98], [1], [59], [71], [75], [84]–[96]). Apparently, the first limit theorems for MAPs in the normal deviation zone were given in [75], and in the large deviation zone, in [119, 120, 84]. For general approaches to the study of additive functionals of Markov processes, see [59]. The large deviation principle for MAPs defined on Harris Markov chains (that is, in essence, on chains with a positive atom) was established in [95, 96] via the construction of cycles generated by returns to a positive atom, which induces the corresponding CRP. Indeed, if $\zeta_1, \zeta_2, \ldots$ denotes the increments of the sums X_n on cycles of length $\tau_1, \tau_2, \ldots$, then the vectors (τ_j, ζ_j) will be independent and identically distributed for $j \geq 2$. These vectors define the CRP $Z(n)$. In this case, the time parameter is integer (τ_j is integer) and we denote it by n. Here the components τ_j and ζ_j are dependent except for the case when $\xi(e) = 0$ for $e \neq e_0$.

Theorem 13.8.5 in [21] describes the relationship between the moment characteristics a and σ for CRPs and the stationary distribution of the chain $\{\mathrm{x}_k\}$ with the moments of $\xi(\mathrm{e})$. For the parameter a, this relationship follows from the relation

$$\mathbf{E}\mathrm{X}_n \sim n \int \pi(d\mathrm{e})\mathbf{E}\xi(\mathrm{e}) \sim an,$$

so that

$$a = \frac{a_\zeta}{a_\tau} = \int \pi(d\mathrm{e})\mathbf{E}\xi(\mathrm{e}), \qquad a_\tau = \frac{1}{\pi(\mathrm{e}_0)}, \qquad a_\zeta = \frac{a}{\pi(\mathrm{e}_0)},$$

where π is the stationary distribution of the chain.

1.8.3 Main Limit Laws for Markov Additive Processes

From the results of the present chapter we have as corollaries all the main limit laws for $\{\mathrm{X}_n\}$. In what follows in this chapter, it is assumed that $\{\mathrm{x}_n\}$ is a Harris Markov chain.

Theorem 1.8.2 *If* $\mathbf{E}\tau < \infty$ *and* $\mathbf{E}\zeta < \infty$, *then the law of large numbers for* X_n *holds*

$$\frac{\mathrm{X}_n}{n} \underset{p}{\to} a \quad as \quad n \to \infty. \tag{1.8.8}$$

Indeed, for each fixed $\gamma(n) = l$, the variable

$$\mathrm{X}_n = Z(n) + \rho_n, \quad \text{where} \quad \rho_n = \sum_{k=n-\gamma(n)+1}^{n} \xi(\mathrm{x}_k) \tag{1.8.9}$$

is conditionally independent of $Z(n)$. As was already pointed out, if $\mathbf{E}\tau < \infty$, then there exists the proper limit distribution of $\gamma(n)$ as $n \to \infty$, and hence, the proper limit distribution of ρ_n as $n \to \infty$. Now (1.8.8) follows from the law of large numbers for $Z(n)$ (Theorem 1.2.3).

Theorem 1.8.3 *If, in addition to the conditions of Theorem* 1.8.2,

$$\mathbf{E}\widehat{\zeta} < \infty,$$

where

$$\widehat{\zeta} := \max_{1 \le k \le \tau} \mathrm{X}_k - \min_{1 \le k \le \tau} \mathrm{X}_k \tag{1.8.10}$$

(*for* $\mathrm{x}_0 = \mathrm{e}_0$), *then the strong law of large numbers holds*

$$\frac{\mathrm{X}_n}{n} \underset{a.s.}{\to} a \quad as \quad n \to \infty. \tag{1.8.11}$$

Indeed, if $\mathbf{E}\widehat{\zeta} < \infty$ and $\{\widehat{\zeta}_n\}$ is a sequence of independent copies of $\widehat{\zeta}$, where $\widehat{\zeta}_n$ corresponds to the nth cycle, then, for any $\varepsilon > 0$,

$$\varepsilon \sum_{n=1}^{\infty} \mathbf{P}(\widehat{\zeta}_n > \varepsilon n) \le \int_0^\infty \mathbf{P}(\widehat{\zeta} > t)dt = \mathbf{E}\widehat{\zeta} < \infty.$$

Hence from the Borel–Cantelli lemma only finitely many events $A_n = \{\widehat{\zeta}_n/n > \varepsilon\}$ occur with probability 1, because $\sum_{n=1}^{\infty} P(A_n) = \sum_{n=1}^{\infty} \mathbf{P}(\widehat{\zeta}_n > \varepsilon n) < \infty$. This means that

$$\frac{\widehat{\zeta}_n}{n} \underset{\text{a.s.}}{\longrightarrow} 0 \quad \text{as} \quad n \to \infty. \tag{1.8.12}$$

Next, on the half-open interval $(T_{n-1}, T_n]$ the trajectory $\{\mathrm{X}_n\}$ differs from $\{Z(n)\}$ by at most $\widehat{\zeta}_n$. Hence (1.8.11) follows from the strong law of large numbers for $Z(n)$ in view of (1.8.12).

Theorem 1.8.4 *If* $\mathbf{E}\tau^2 < \infty$, $\mathbf{E}\zeta^2 < \infty$, *then the central limit theorem for* X_n *holds*:

$$\frac{\mathrm{X}_n - an}{\sigma\sqrt{n}} \Longrightarrow \mathbf{\Phi} \quad \textit{as} \quad n \to \infty.$$

This result follows from Theorem 1.3.3, because $\frac{\rho_n}{\sqrt{n}} \underset{p}{\to} 0$ as $n \to \infty$ in (1.8.9).

Theorem 1.8.5 *If, in addition to the conditions of Theorem* 1.8.4, $\mathbf{E}\widehat{\zeta}^2 < \infty$, *then for the sequence*

$$\frac{\mathrm{X}_{nt} - ant}{\sigma\sqrt{n}}, \qquad t \in [0, 1], \tag{1.8.13}$$

the invariance principle holds.

Indeed, if $\mathbf{E}\widehat{\zeta}^2 < \infty$, then, for any $b > \frac{1}{a_\tau}$ and $\varepsilon > 0$,

$$\mathbf{P}(\eta(n) > bn) \to 0,$$

$$\mathbf{P}(\max_{k \le bn} \widehat{\zeta}_k > \varepsilon\sqrt{n}) \le bno\left(\frac{1}{\varepsilon^2 n}\right) = o(1) \text{ as } n \to \infty.$$

Hence

$$\frac{1}{\sqrt{n}} \rho_{\mathbb{C}}(\mathrm{X}_{[nt]}, Z(nt)) \underset{p}{\to} 0 \quad \text{as} \quad n \to \infty. \tag{1.8.14}$$

Now the invariance principle for processes (1.8.13) follows from Theorem 1.5.3.

Theorem 1.8.6 *Under the conditions of Theorem* 1.8.5, *the law of the iterated logarithm for the sequence* $\{\mathrm{X}_n\}$ *holds: the relations*

$$\overline{\lim_{n\to\infty}} \frac{\mathrm{X}_n - an}{L(n)} \to 1,$$

$$\underline{\lim_{n\to\infty}} \frac{\mathrm{X}_n - an}{L(n)} = -1$$

hold with probability 1, *where* $L(n) = \sigma\sqrt{2n \ln\ln n}$ *for* $n \ge 3$.

This fact follows from Theorem 1.3.5 and since

$$\frac{\widehat{\zeta}_n}{\sqrt{n}} \underset{\text{a.s.}}{\longrightarrow} 0 \quad \text{as} \quad n \to \infty,$$

which follows from Lemma 1.1.7.

For MAPs, an analog of Theorem 1.7.1 on the distribution of the first passage time by the $\{X_n\}$ of an arbitrary remote boundary $g_x(n)$ also holds. We shall consider boundaries $g_x(t) = g_x(n)$ (the time t for MAPs is discrete) of the same form as in §1.7.2 (see (1.7.5)–(1.7.7)). We study the asymptotic behavior of the first passage time

$$\eta_g := \min\{n : X_n \geq g_x(n)\}$$

(cf. (1.7.1)). Let

$$n_g := \min\{n \geq 1 : an \geq g_x(n)\}.$$

Theorem 1.8.7 *Let $\sigma_\tau < \infty$, $\sigma < \infty$, $\mathbf{E}\widehat{\zeta}^2 < \infty$, let a function $g_x(n)$ be such that $n_g \to \infty$ as $x \to \infty$, and let conditions $[\mathcal{A}]_1$, $[\mathcal{A}]_2$ from §1.7 be met. Then, as $x \to \infty$,*

$$\mathbf{P}\left(\frac{\eta_g - n_g}{\sigma\sqrt{t_g}} < v\right) \to \Phi((a-b)v). \tag{1.8.15}$$

Indeed, let η_g^Z be the first passage time of the boundary $g_x(n)$ by the CRP $Z(n)$. Then, clearly, $\eta_g \leq \eta_g^Z$. Next, for $n = 2t_g$ we have $\mathbf{P}(\eta_g > n) \to 0$ as $x \to \infty$, and further using (1.8.14), we get

$$\rho_{\mathbb{C}}(X_{[nt]}, Z([nt])) = o_p(\sqrt{t_g}) \quad \text{for} \quad t \in [0,1], \quad n \to \infty. \tag{1.8.16}$$

The boundary $g_x(n)$ is defined by conditions $[\mathcal{A}]_1$, $[\mathcal{A}]_2$ up to $o(\sqrt{t_g})$ (see (1.7.4)), and hence under (1.8.16) the problems of the first passage time of these boundaries for the processes $X_{[nt]}$ and $Z([nt])$ are indistinguishable, and hence, Theorem 1.8.7 follows from Theorem 1.7.1.

All these results can also be extended to the "inhomogeneous processes" for each fixed $x_0 \neq e_0$.

Remark 1.8.8 Theorems 1.8.3, 1.8.5, and 1.8.6 involve conditions on the moments of the random variable $\widehat{\zeta}$ defined in (1.8.10). Note that (a) the fulfillment of these conditions does not follow from analogous conditions on ζ, and (b) these conditions are essential, that is, if they are violated, then the corresponding results in general fail.

We illustrate this with the following example.

We first consider a finite ergodic Markov chain $\{x_n\}$ with states $0, 1, \ldots, N$ such that $x_0 = e_0 = 0$, $P_N(0) = 1$, and define the random variables $\xi(0) = 0$, $|\xi(j)| < c$ for $j \leq N$ so that the vector (τ, ζ) satisfies the Cramér condition. Next, we add the states $N+1, N+2, \ldots$ and from the chain $\{x_n\}$ we construct the new chain $\{x_n^*\}$ by replacing the probability $P_N(0) = 1$ by $P_N(0) = 0$, and setting for $k = 1, 2, \ldots$

$$P_N(N+2k-1) = \frac{c}{k^2}, \quad P_{N+2k-1}(N+2k) = 1, \quad P_{N+2k}(0) = 1.$$

Next, from x_n^* we construct the chain $\{x_n^*, \xi_k^*\}$ by putting

$$\xi^*(N + 2k - 1) = k^s, \quad \xi^*(N + 2k) = -k^s \text{ for } k > 0,$$
$$\xi^*(k) = \xi(k) \text{ for } k \le N.$$

Then, with obvious notational conventions (all the notation corresponding to the new chain $\{x_n^*, \xi_n^*\}$ will be equipped with the upper index "$*$"), we have

(a) $\tau^* = \tau + 2$, $\zeta^* = \zeta$, $X^*_{\tau^*-2} = \zeta + \zeta^{**}$, where ζ^{**} does not depend on ζ, $\mathbf{P}(\zeta^{**} = k^s) = ck^{-2}$, $\mathbf{E}|\zeta^{**}| = \infty$ for $s \ge 1$, $\mathbf{E}\widehat{\zeta}^* = \infty$ for $\mathbf{E}|\zeta^*| = \mathbf{E}|\zeta| < \infty$,

(b) it is clear that this makes the fulfillment of the above assertions for the process $\{X_n^*\}$ impossible, because infinitely many events $\{\widehat{\zeta}^* > n\}$ occur with probability 1. The sequence $\{X_n^*\}$ also fails to satisfy the law of the iterated logarithm, the invariance principle, and relation (1.8.15), even though the vector (τ^*, ζ^*) satisfies condition [**C**]. This follows from the following arguments. In each cycle, the probability that the maximal value x_j^* will exceed $n^{1/3}$ is comparable with $n^{-2/3}$. Hence

$$\mathbf{P}\big(\max_{k\le n} x_k^* > n^{1/3}\big) \to 1 \quad \text{as} \quad n \to \infty.$$

This means that for $s = 4$ at least one jump of the sums X_k^*, $k \le n$, with probability close to 1 will exceed $n^{4/3}$, and, for an appropriate $q > 0$,

$$\mathbf{P}\big(\max_{k\le n}(X_k^* - a^*k) > qn^{4/3}\big) \to 1 \quad \text{as} \quad n \to \infty.$$

It is clear that this makes it impossible that any of the above assertions for the process $\{X_n^*\}$ be satisfied.

2

Integro-Local Limit Theorems in the Normal Deviation Zone

In this chapter, assuming that $\mathbf{E}|\xi|^2 < \infty$ and $\xi = (\tau, \zeta)$, we will prove several results on the asymptotics of the distributions of nonnormalized CRPs $Z(t)$ and $Y(t)$. These results, known as integro-local theorems, are sharper than the central limit theorem and are concerned with the probabilities of $Z(t)$ and $Y(t)$ of hitting intervals of small length in the normal deviation zone. In Chapter 5 we shall establish integro-local theorems for a broader range, including the large deviation zone. In Chapter 6, these results will be applied to the study of the exact asymptotics of the large-deviation probabilities in boundary crossing problems for CRPs.

In §2.1, we obtain integro-local theorems in the homogeneous case for independent and linearly dependent τ and ζ under minimal finiteness conditions on the second moments. In §2.2, we present some refinements of Stone's integro-local theorems for multidimensional random walks. In §2.3, these results are used in deriving integro-local theorems for CRPs in the general case but under additional moment conditions. In §2.4, these results are extended to the inhomogeneous case. Applications to additive Markov processes are given in §2.5.

For arithmetic ξ, local theorems for CRPs can be derived in a similar way, but in this case the statements and proofs become substantially simpler.

The simplified form of the results of §2.1 for independent τ and ζ can be found in [21, Theorem 10.6.3]. The case of linearly dependent τ and ζ and the general case were studied in [30].

2.1 Integro-Local Limit Theorems in the Case of Independent or Linearly Dependent τ and ζ

The linear dependence of τ and ζ means that τ and ζ are related by

$$\zeta = h\tau + \omega, \qquad h = \text{const},$$

where τ and ω are independent (τ and ζ are independent if $h = 0$). Below, it will be shown that integro-local limit theorems for homogeneous CRPs $Z(t)$ and $Y(t)$ in the case of independent or linearly dependent τ and ζ can be derived from the integro-local theorem for homogeneous *random walks*. Since in what follows (see, e.g., §2.2) we shall need analogous theorems for random walks in the two-dimensional case, we first give here Stone's integro-local theorem for random walks in the general multidimensional case.

2.1.1 Integro-Local Theorem for Random Walks

Let $\boldsymbol{\xi}, \boldsymbol{\xi}_1, \boldsymbol{\xi}_2, \ldots$ be independent and identically distributed vectors in $\mathbb{R}^d$,

$$\boldsymbol{a} := \mathbf{E}\boldsymbol{\xi}, \qquad \sigma^2 := \mathbf{E}\boldsymbol{\xi}^*\boldsymbol{\xi}$$

be the matrix of second moments (the superscript "$*$" stands for transposition, so $\boldsymbol{\xi}^*$ is a column vector),

$$\mathbf{S}_n = \sum_{k=1}^{n} \boldsymbol{\xi}_k, \qquad n \geq 1$$

(as a rule, vectors will be denoted by boldfaced letters).

In the context of our CRP problems, it will frequently happen that $\boldsymbol{\xi} = (\tau, \zeta)$, so that in this case $\boldsymbol{a} = (a_\tau, a_\xi)$, $\mathbf{S}_n = (T_n, Z_n)$. We shall also use one-dimensional versions of the integro-local theorem.

It is assumed that the distribution of $\boldsymbol{\xi}$ is not degenerate, that is, the matrix σ^2 is positive definite and that by a suitable linear transformation (including a renumbering of coordinates and a shift of the origin) this distribution can be reduced to the following form: the first of $l \leq d$ coordinates $\xi_{(1)}, \ldots, \xi_{(l)}$ of the vector $\boldsymbol{\xi} = (\xi_{(1)}, \ldots, \xi_{(d)})$ have nonlattice distribution, that is, in the case $l \geq 1$ the characteristic function

$$\varphi(\boldsymbol{\lambda}) := \mathbf{E}e^{i\langle \boldsymbol{\lambda}, \boldsymbol{\xi} \rangle}, \qquad \boldsymbol{\lambda} = (\lambda_1, \ldots, \lambda_d)$$

($\langle \boldsymbol{\lambda}, \boldsymbol{\xi} \rangle$ is the inner product of $\boldsymbol{\lambda}$ and $\boldsymbol{\xi}$) has the property

$$|\varphi(\boldsymbol{\lambda})| < 1 \quad \text{for} \quad (\lambda_1, \ldots, \lambda_l) \neq 0.$$

The remaining $d-l$ coordinates $\xi_{(l+1)}, \ldots, \xi_{(d)}$ are arithmetic; that is, each coordinate is integer, and the greatest common divisor of the differences between its possible values equals 1.

Next, let $\boldsymbol{x} = (x_1, \ldots, x_d) \in \mathbb{R}^d$, and for $\Delta > 0$, consider the direct product of half-open intervals in $\mathbb{R}^d$,

$$\boldsymbol{\Delta}[\boldsymbol{x}) := \prod_{j=1}^{l} [x_j, x_j + \Delta) \prod_{j=l+1}^{d} [x_j, x_j + 1).$$

The following result holds.

Theorem 2.1.1 (Stone [114], [115]) *Suppose that the matrix σ^2 is positive definite. Then, as $n \to \infty$,*

$$\frac{\mathbf{P}(\mathbf{S}_n \in \boldsymbol{\Delta}[\boldsymbol{x}))}{\Delta^l} = n^{-d/2}\phi_\sigma\left(\frac{\boldsymbol{x} - \boldsymbol{a}n}{\sqrt{n}}\right) + o(n^{-d/2}), \tag{2.1.1}$$

where ϕ_σ is the normal density with parameters $(\mathbf{0}, \sigma^2)$,

$$\phi_\sigma(v) = \frac{1}{|\sigma|(2\pi)^{d/2}} \exp\left\{-\frac{v\sigma^{-2}v^*}{2}\right\}, \qquad v \in \mathbb{R}^d,$$

σ^{-2} is the inverse matrix of σ^2, $|\sigma|^2$ is the determinant of the matrix σ^2, and the remainder term $o(n^{-d/2})$ is uniform in all $\boldsymbol{x}$ and in Δ from every compact set disjoint from zero.

The proof of Theorem 2.1.1 in the one-dimensional case $d = 1$ can be found in textbooks (see [21, §8.7]).

Theorem 2.1.1 can be naturally looked upon as an *integro-local* theorem, since it contains some elements of both *integral* theorems (dealing with the probabilities that $\mathbf{S}_n$ enter a *set*) and *local* theorems (in which these sets are *small*). This name makes it possible to distinguish integro-local theorems of the form (2.1.1) from the classical local theorems for the distribution of $\mathbf{S}_n$, in which the densities are assumed to exist and which are concerned with the asymptotical behavior of the density of the sum $\mathbf{S}_n$, and from the classical integral theorems, which describe the asymptotics of the probability that $\mathbf{S}_n$ enters "large" sets and which are much rougher than Theorem 2.1.1. Integral theorems can be deduced from integro-local theorems but not vice versa.

If $j > l$, then the coordinate x_j is integer. If $l = 0$ (all components $\xi_{(j)}$ are arithmetic), then, without loss of generality, we may assume that $\boldsymbol{x}$ is a vector with integral coordinates and replace the left-hand side in (2.1.1) by $\mathbf{P}(\mathbf{S}_n = \boldsymbol{x})$. In this case, the integro-local theorem becomes a local theorem.

Stone's integro-local theorem is the most perfect and accurate version of the classical central limit theorem, which does not assume any additional conditions other than the existence of variances. It only distinguishes between the lattice and nonlattice cases. As was already pointed out, in the lattice (arithmetic) case this theorem becomes a local theorem. In the nonlattice case, Stone's integro-local theorem provides essentially the same accuracy as local limit theorems when there exists the density of $\boldsymbol{\xi}$, but it assumes nothing about the existence of densities.

Since in the nonlattice case Δ may take arbitrarily small values, the left-hand side in (2.1.1) can be called the "predensity" of the distribution of $\mathbf{S}_n$. The right-hand side of (2.1.1) is (up to $o(n^{-d/2})$) the normal density with parameters $(0, n\sigma^2)$.

Integro-local theorems are very effective and are often the most appropriate technical tool in the study of many problems of probability theory. They serve as a key element, for example, in constructing the most adequate theory of large deviations for sums of random vectors (see, e.g., Chapter 9 in [21]); in establishing integro-local theorems for compound renewal processes (see Chapter 5 below and Chapter 9 in [21]); in proving results about the first passage time of a remote curvilinear boundary by a random walk (see Chapter 6); and in many other problems. Therefore, the refinement and generalization of integro-local theorems are, in our opinion, of undoubted interest. The refinement of Theorem 2.1.1, which is obtained in §2.2, will allow us to prove an integro-local theorem for CRPs in the general case when τ and ζ are arbitrarily dependent.

In the one-dimensional *arithmetic* case, (2.1.1) was obtained by Gnedenko [69], and, in the multidimensional case, by Rvacheva [107]. In the arithmetic case, the local theorems for *nonidentically* distributed random variables and vectors were considered in [67, 85, 99, 101, 104].

Under the Cramér conditions (the moment condition and the condition on the characteristic function), asymptotic expansions in the powers of $1/\sqrt{n}$ in integro-local theorems and an unboundedly decreasing lower bound for the possible values of Δ were found in [40]. Integro-local theorems for nonidentically distributed random variables in a triangular array scheme, including the large deviation zone, were obtained in [22].

2.1.2 Integro-Local Theorems for Homogeneous CRPs in the Case of Independent or Linearly Dependent τ and ζ

In this section, we will prove the integro-local theorem in the case when ζ and τ satisfy

$$\zeta = h\tau + \omega, \qquad h = \text{const}, \tag{2.1.2}$$

where τ and ω are independent. In this case,

$$\xi = \zeta - a\tau = \omega - \frac{a_\omega}{a_\tau}\tau, \qquad a = \frac{a_\omega}{a_\tau} + h, \qquad \sigma^2 = \frac{a_\tau^2\sigma_\omega^2 + a_\omega^2\sigma_\tau^2}{a_\tau^3},$$

where $a_\omega := \mathbf{E}\omega, \sigma_\omega^2 = \mathbf{D}\omega$. If $h = 0$, then (2.1.2) means that ζ and τ are independent.

Let

$$\Delta[x) = [x, x + \Delta)$$

be the half-open interval of length $\Delta > 0$, and $\chi = \chi_\infty$ be the overshoot of the random walk $\{T_k\}_{k=1}^\infty$ over the infinitely remote barrier, so that

$$\mathbf{P}(\chi \geq t) = \frac{1}{a_\tau}\int_t^\infty \mathbf{P}(\tau \geq u)\,du.$$

We note again that, unless otherwise stated, the sequence $\{\tau_j, \zeta_j\}_{j=1}^\infty$ in §§2.1–2.3 will be assumed to be *homogeneous*. Our aim in this section is to prove the following integro-local theorem for homogeneous CRPs.

Theorem 2.1.2 I. *Let* (2.1.2) *hold, let* $a_\omega \neq 0$, $\sigma_\tau < \infty$, $\sigma_\omega < \infty$, *and let random variables* τ *and* ω *be nonlattice. Then, as* $t \to \infty$,

$$\frac{1}{\Delta}\mathbf{P}\Big(Y(t) - at \in \Delta[x),\ \gamma(t) \geq u,\ \chi(t) \geq v\Big) = \frac{\mathbf{P}(\chi \geq u + v)}{\sqrt{t}}\,\phi_\sigma\Big(\frac{x}{\sqrt{t}}\Big) + o\Big(\frac{1}{\sqrt{t}}\Big), \tag{2.1.3}$$

where the remainder term $o(1/\sqrt{t})$ *is uniform in all* x *and in* Δ *from any fixed compact set separated from zero.*

If $a_\omega = 0$, $\sigma_\omega < \infty$ *and if the distribution of* τ *for some* $r \in (3/2, 2)$ *and all* $t > 1$ *satisfies the condition*

$$\mathbf{P}(\tau \geq t) \leq c_1 t^{-r}, \qquad c_1 = \text{const}, \tag{2.1.4}$$

then $\xi = \omega$ *and equality* (2.1.3) *remains valid for* $\sigma^2 = \frac{\sigma_\omega^2}{a_\tau}$.

II. *If the random variable* τ *is arithmetic and if* ω *is nonlattice, then assertion* I *remains valid, but with integer-valued* t, $\gamma(t)$, $\chi(t)$, u, v *and with*

$$\mathbf{P}(\chi \geq k) = \frac{1}{a_\tau} \sum_{j=k+1}^{\infty} \mathbf{P}(\tau \geq j).$$

III. *If the random variable* ω *is arithmetic, then assertions* I, II *remain valid with integer* $\Delta \geq 1$.

IV. *Assertions* I–III *remain valid if the process* $Y(t)$ *is replaced by the process* $Z(t)$.

The case when, instead of $Y(t)$, one deals with the process $Y(t) + qt$ does not cause additional difficulties, because the centered process

$$Y(t) + qt - (a + q)t = Y(t) - at$$

has the same form as $Y(t) - at$.

It is seen from Theorem 2.1.2 that the random variables $Y(t) - at$ (on the one hand) and the variables $\gamma(t)$ and $\chi(t)$ (on the other hand) are asymptotically independent.

In the case of independent τ and ζ, the result of Theorem 2.1.2 with $u = v = 0$ was obtained in [21, Theorem 10.6.3]. In the case (2.1.2), it was established in [30].

In the proof of Theorem 2.1.2 that follows, we shall use the one-dimensional versions of Theorem 2.1.1 either with $\boldsymbol{\xi} = \tau$ (and hence $\boldsymbol{a} = a_\tau$, $\boldsymbol{\sigma}^2 = \sigma_\tau^2$) or with $\boldsymbol{\xi} = \omega$ (and hence $\boldsymbol{a} = a_\omega$, $\boldsymbol{\sigma}^2 = \sigma_\omega^2$).

Proof of Theorem 2.1.2 I. Let ω and τ be nonlattice. We set

$$B_t = B_t(u, v) := \{\gamma(t) \geq u, \chi(t) \geq v\}. \tag{2.1.5}$$

Using the relations $t = T_{\eta(t)} - \chi(t)$ and

$$Y(t) - at = S_{\eta(t)} + a\chi(t),$$

where

$$S_n = \sum_{j=1}^{n} \xi_j, \qquad \xi_j = \zeta_j - a\tau_j,$$

we find the asymptotic representation for the probability

$$P_t := \mathbf{P}(Y(t) - at \in \Delta[x), B_t^\delta) = \mathbf{P}(S_{\eta(t)} + a\chi(t) \in \Delta[x), B_t^\delta) \tag{2.1.6}$$

where

$$B_t^\delta := B_t^\delta(u, v) = \{\gamma(t) \geq u, \chi(t) \in \delta[v)\}, \qquad \delta[v) = [v, v + \delta).$$

We set

$$n_{\pm} = \left[\frac{t}{a_\tau} \pm N t^{3/4}\right], \tag{2.1.7}$$

where $N = N_t \to \infty$ (slowly enough) as $t \to \infty$. We have

$$\begin{aligned}\mathbf{P}(\eta(t) \le n_-) &= \mathbf{P}(T_{n_-} > t)\\ &= \mathbf{P}(T_{n_-} - a_\tau n_- > a_\tau N t^{3/4}(1 + o(1))\\ &\le \frac{\sigma_\tau^2 n_-}{a_\tau^2 N^2 t^{3/2}} = o\left(\frac{1}{\sqrt{t}}\right).\end{aligned} \tag{2.1.8}$$

A similar argument shows that

$$\mathbf{P}(\eta(t) > n_+) = o\left(\frac{1}{\sqrt{t}}\right). \tag{2.1.9}$$

Hence

$$\begin{aligned}P_t &= o\left(\frac{1}{\sqrt{t}}\right) + \sum_{n=n_-}^{n_+} \mathbf{P}(S_n \in \Delta[x), B_t^\delta, \eta(t) = n)\\ &= o\left(\frac{1}{\sqrt{t}}\right) + \sum_{n=n_-}^{n_+} \mathbf{P}(\eta(t) = n, B_t^\delta)\mathbf{P}(S_n \in \Delta[x) \mid \eta(t) = n, B_t^\delta).\end{aligned} \tag{2.1.10}$$

The further argument we will split into four stages.

(1) First, we will find the asymptotics of the first factor in the sum on the right of (2.1.10) (with n replaced by $n+1$ for simplicity, which does not affect the asymptotics),

$$\mathbf{P}(\eta(t) = n + 1, B_t^\delta) = \int_0^{t-u} \mathbf{P}(T_n \in dy)\mathbf{P}(\tau \in t - y + \delta[v)). \tag{2.1.11}$$

Let $\sigma_\tau < \infty$. Then, using the integro-local theorem (Theorem 2.1.1) for $\mathbf{S}_n = T_n$, and arguing as in the proof of Theorem 10.5.8 in [21], we find by (2.1.11) that

$$\mathbf{P}(\eta(t) = n + 1, B_t^\delta) = \frac{1}{\sqrt{n}}\,\phi_{\sigma_\tau}\left(\frac{t - a_\tau n}{\sqrt{n}}\right) I_{u+v}^{u+v+\delta} + o\left(\frac{1}{\sqrt{n}}\right), \tag{2.1.12}$$

where

$$I_{u+v}^{u+v+\delta} = \int_{u+v}^{u+v+\delta} \mathbf{P}(\tau \ge z)\, dz.$$

(In Theorem 10.5.8 in [21], relation (2.1.12) was proved in the absence of the event B_t^δ under the probability sign in (2.1.12); the presence of this event does not change the argument.)

It is easily checked that exactly the same asymptotic representation holds for $\mathbf{P}(\eta(t) = n, B_t^\delta)$. Setting

$$n_t = \frac{t}{a_\tau}, \qquad k = n - n_t, \tag{2.1.13}$$

and noting that $n \sim n_t$ for $n \in [n_-, n_+]$, $t \to \infty$, we get

$$\mathbf{P}(\eta(t) = n, B_t^\delta) = \frac{1}{\sqrt{n_t}}\,\phi_{\sigma_\tau}\Big(\frac{ka_\tau}{\sqrt{n_t}}\Big) I_{u+v}^{u+v+\delta} + o\Big(\frac{1}{\sqrt{n_t}}\Big). \qquad (2.1.14)$$

(2) Now we will find the asymptotics of the second factor in the sum on the right of (2.1.10). Setting

$$\Omega_n = \sum_{j=1}^{n} \omega_j,$$

we have

$$S_n = \Omega_n + bT_n, \quad \text{where} \quad b = h - a, \quad a_\omega + ba_\tau = 0,$$

and moreover, on the set $\{\eta(t) = n\} \cap B_t^\delta$, one has $T_n \in t + \delta[v)$, and

$$\{S_n \in \Delta[x)\} = \{\Omega_n - a_\omega n \in \Delta[x) - a_\omega(n_t + k) - b(a_\tau n_t + \chi(t))\}.$$

We have $ba_\tau = -a_\omega$, and hence on the set $\{\eta(t) = n\} \cap B_t^\delta$ for sufficiently small δ (such that $|b|\delta < \Delta$) we have

$$\{S_n \in \Delta[x)\} \subset \{\Omega_n - a_\omega n \in [x - b\delta, x + \Delta) - bv - a_\omega k\}, \qquad (2.1.15)$$

and similarly,

$$\{S_n \in \Delta[x)\} \supset \{\Omega_n - a_\omega n \in [x, x + \Delta - b\delta) - bv - a_\omega k\}. \qquad (2.1.16)$$

Assume for definiteness that $b \ge 0$. The lengths of the half-open intervals $[x - b\delta, x + \Delta)$ and $[x, x+\Delta-b\delta)$ in (2.1.15), (2.1.16) are equal to $\Delta + b\delta$ and $\Delta - b\delta$, respectively. Next, ω are nonlattice and Ω_n does not depend on $\{T_k\}_1^\infty$, hence by (2.1.15), (2.1.16), from the integro-local theorem for Ω_n, and since $n \sim n_t$ for $n \in [n_-, n_+]$, we have

$$\mathbf{P}\Big(S_n \in \Delta[x) \mid \eta(t) = n, B_t^\delta\Big) = \frac{\Delta + \rho_\delta}{\sqrt{n_t}}\,\phi_{\sigma_\omega}\Big(\frac{x - a_\omega k}{\sqrt{n_t}}\Big) + o\Big(\frac{1}{\sqrt{n_t}}\Big), \qquad (2.1.17)$$

where $|\rho_\delta| \le b\delta$, the remainder term $o(\frac{1}{\sqrt{n_t}})$ is uniform in x (the additions of the terms $-bv$, $-b\delta$ to x in the right-hand sides of (2.1.15), (2.1.16) do not change the asymptotics (2.1.17), and so one can neglect such changes of x in (2.1.17)). For a similar reason, the quantities n_t and the numbers k in (2.1.17) can be assumed to be integer if one puts $n_t = [\frac{t}{a_\tau}]$ in (2.1.13).

The case $b \le 0$ is dealt with in the same way when $|\rho_\delta| < |b|\delta$.

(3) Now consider the sum on the right of (2.1.10). This sum controls the main part of the quantity P_t defined in (2.1.6). We let $Q_{n,1}$, $Q_{n,2}$ denote the principal parts of the asymptotics in (2.1.14) and (2.1.17), respectively. Hence by (2.1.10) and since

$$\sum_n \mathbf{P}(\eta(t) = n, B_t^\delta) \le 1,$$

we find that

$$P_t = o\Big(\frac{1}{\sqrt{t}}\Big) + \sum_{n=n_-}^{n_+} \mathbf{P}(\eta(t) = n, B_t^\delta) \Big(Q_{n,2} + o\Big(\frac{1}{\sqrt{n_t}}\Big)\Big)$$
$$= o\Big(\frac{1}{\sqrt{t}}\Big) + \sum_{n=n_-}^{n_+} \Big(Q_{n,1} + o\Big(\frac{1}{\sqrt{n_t}}\Big)\Big) Q_{n,2} = o\Big(\frac{1}{\sqrt{t}}\Big) + \sum_{n=n_-}^{n_+} Q_{n,1}Q_{n,2}.$$

Here the last equality holds, because the sum $\sum_{n=n_-}^{n_+} Q_{n,2}$ is bounded, and so

$$\sum_{n=n_-}^{n_+} o\Big(\frac{1}{\sqrt{n_t}}\Big)Q_{n,2} = o\Big(\frac{1}{\sqrt{t}}\Big).$$

Therefore it remains to consider the sum (which we denote by Σ) of the products of the main parts in (2.1.14) and (2.1.17):

$$\Sigma := \sum_{k=-[Nt^{3/4}]}^{[Nt^{3/4}]} \frac{(\Delta + \rho_\delta)}{\sigma_\tau \sigma_\omega 2\pi n_t} I_{u+v}^{u+v+\delta} \exp\Big\{-\frac{k^2 a_\tau^2}{2\sigma_\tau^2 n_t} - \frac{(x - a_\omega k)^2}{2\sigma_\omega^2 n_t}\Big\}. \tag{2.1.18}$$

There are two cases to consider: $a_\omega \neq 0$ and $a_\omega = 0$.

If $a_\omega \neq 0$, then σ_τ is finite and Σ is the integral sum approximating (up to a constant factor) the convolution at the point $\frac{x}{a_\omega}$ of two normal densities with parameters $(0, \frac{n_t\sigma_\tau^2}{a_\tau^2})$ and $(0, \frac{n_t\sigma_\omega^2}{a_\omega^2})$, respectively. The variance of the random variable corresponding to this convolution is equal to

$$n_t \left(\frac{a_\omega^2\sigma_\tau^2 + a_\tau^2\sigma_\omega^2}{a_\tau^2 a_\omega^2}\right), \tag{2.1.19}$$

and so we can find the asymptotics of Σ. More precisely, we put

$$m_\tau^2 = \frac{\sigma_\tau^2 n_t}{a_\tau^2}, \qquad m_\omega^2 = \frac{\sigma_\omega^2 n_t}{a_\omega^2}.$$

Hence, as $t \to \infty$, the sum in (2.1.18) becomes

$$\Sigma = \frac{(\Delta + \rho_\delta) I_{u+v}^{u+v+\delta}}{a_\tau a_\omega} \int_{-\infty}^{\infty} \phi_{m_\tau}(y)\phi_{m_\omega}\Big(\frac{x}{a_\omega} - y\Big)\, dy(1 + o(1)), \tag{2.1.20}$$

where the integral on the right is

$$\phi_m\Big(\frac{x}{a_\omega}\Big) = a_\omega \phi_{a_\omega m}(x), \tag{2.1.21}$$

where

$$a_\omega^2 m^2 = a_\omega^2(m_\tau^2 + m_\omega^2) = \frac{(a_\omega^2\sigma_\tau^2 + a_\tau^2\sigma_\omega^2)n_t}{a_\tau^2}.$$

In our case, $\xi = \zeta - a\tau = \omega + b\tau$, $b = -\frac{a_\omega}{a_\tau}$,

$$\sigma_\xi^2 = \mathbf{E}(\omega + b\tau)^2 = \sigma_\omega^2 + \frac{a_\omega^2\sigma_\tau^2}{a_\tau^2} = \frac{a_\tau^2\sigma_\omega^2 + a_\omega^2\sigma_\tau^2}{a_\tau^2} = \frac{a_\omega^2 m^2}{n_t},$$

and in (2.1.21) we have

$$\phi_m\left(\frac{x}{a_\omega}\right) = a_\omega \phi_{\sigma_\xi\sqrt{n_t}}(x).$$

Since

$$n_t = \frac{t}{a_\tau}, \qquad \frac{\sigma_\xi^2}{a_\tau} = \sigma^2,$$

we finally obtain from (2.1.20) the following value for the sum Σ in (2.1.18):

$$\Sigma = \frac{(\Delta + \rho_\delta) I_{u+v}^{u+v+\delta}}{a_\tau} \phi_{\sigma\sqrt{t}}(x)(1 + o(1)). \tag{2.1.22}$$

In view of the above, we arrive at the following intermediate assertion.

If (2.1.2) *holds, then as* $t \to \infty$, *for any fixed* $u \geq 0$, $v \geq 0$,

$$\begin{aligned}\mathbf{P}(Y(t) - at \in \Delta[x), \gamma(t) \geq u,\ \chi(t) \in \delta[v))& \\ = \frac{(\Delta + \rho_\delta) I_{u+v}^{u+v+\delta}}{a_\tau \sqrt{t}} & \phi_\sigma\left(\frac{x}{\sqrt{t}}\right) + o\left(\frac{1}{\sqrt{t}}\right)\end{aligned} \tag{2.1.23}$$

uniformly in x *and in* Δ *from any fixed compact set bounded away from zero. The value of* $|\rho_\delta|$ *in* (2.1.23) *does not exceed* $|b|\delta$.

Note that a slightly different approach could also be used for the proof of (2.1.23): after the proof of the existence of the asymptotically normal density in (2.1.20), (2.1.21), the central limit theorem should be employed (see §1.4), using which the parameters of the density thus obtained could be found.

If $a_\omega = 0$, then one cannot in general use relation (2.1.18) under the conditions of the theorem (see (2.1.4)), because in this relation it is assumed that σ_τ is finite and T_n is asymptotically normal. For $a_\omega = 0$, instead of (2.1.7), we set

$$n_\pm = \left[\frac{t}{a_\tau} \pm t^s\right], \qquad s \in \left(\frac{3}{2r}, 1\right),$$

where r is defined in (2.1.4). Hence by Corollary 3.1.1 in [38], instead of (2.1.8) we have $1 - rs < -1/2$,

$$\begin{aligned}\mathbf{P}(\eta(t) \leq n_-) &= \mathbf{P}(T_{n_-} - a_\tau n_- > a_\tau t^s + O(1)) \\ &\leq n_-(a_\tau t^s)^{-r} c_1(1 + o(1)) \sim c_1 a_\tau^{-1-r} t^{1-rs} = o\left(\frac{1}{\sqrt{t}}\right).\end{aligned}$$

A similar analysis establishes (2.1.9).

Next, in the case $a_\omega = 0$, one just has to observe at Stage 1) of the proof that

$$\sum_{k=-[t^s]}^{[t^s]} \mathbf{P}\left(\eta(t) = \left[\frac{t}{a_\tau}\right] + k, B_t^\delta\right) = a_\tau^{-1} I_{u+v}^{u+v+\delta} + o(1). \tag{2.1.24}$$

Stage (2) of the proof remains unchanged. At Step (3) of the proof we come to the estimate of the value

$$\left[\frac{\Delta+\rho_\delta}{\sigma_\omega\sqrt{2\pi n_t}}\exp\left\{-\frac{x^2}{2\sigma_\omega^2 n_t}\right\}+o\left(\frac{1}{\sqrt{n_t}}\right)\right]\sum_{k=-[t^s]}^{[t^s]}\mathbf{P}\left(\eta(t)=\left[\frac{t}{a_\tau}\right]+k,B_t^\delta\right)$$

(cf. (2.1.17)), which by (2.1.24) gives

$$\begin{aligned}&\mathbf{P}(Y(t)-at\in\Delta[x),\gamma(t)\geq u,\ \chi(t)\in\delta[v))\\&=\frac{(\Delta+\rho_\delta)I_{u+v}^{u+v+\delta}}{a_\tau\sqrt{n_t}}\phi_{\sigma_\omega}\left(\frac{x}{\sqrt{n_t}}\right)+o\left(\frac{1}{\sqrt{t}}\right)=\frac{(\Delta+\rho_\delta)I_{u+v}^{u+v+\delta}}{a_\tau\sqrt{t}}\phi_{\sigma}\left(\frac{x}{\sqrt{t}}\right)+o\left(\frac{1}{\sqrt{t}}\right)\end{aligned}$$

(cf. (2.1.23)). This gives us the same asymptotic representation as (2.1.23) with $\sigma^2=\frac{\sigma_\omega^2}{a_\tau}$.

(4) Now we will establish (2.1.3). Given a fixed integer $K>0$, we split the half-open interval $[v,v+\delta)$ into small half-open intervals $[v_k,v_k+\delta/K)$ of length δ/K, where $v_k=k\delta/K$, $k=0,1,\ldots,K$. We let $\rho_\delta^{(k)}$ denote the "correction term" for Δ in (2.1.23) that corresponds to the half-open interval $[v_k,v_k+\delta/K)$. Note that $|\rho_\delta^{(k)}|<|b|\delta/K$ for all k. Summing up over $k=0,\ldots,K$ both sides of (2.1.23) for the new half-open intervals, we get, uniformly in x (and in Δ in the corresponding domain),

$$\begin{aligned}&\mathbf{P}(Y(t)-at\in\Delta[x),\gamma(t)\geq u,\chi(t)\in[v,v+\delta))\\&\qquad=\frac{(\Delta+\rho)I_{u+v}^{u+v+\delta}}{a_\tau\sqrt{t}}\phi_\sigma\left(\frac{x}{\sqrt{t}}\right)+o\left(\frac{1}{\sqrt{t}}\right),\end{aligned}\tag{2.1.25}$$

where $|\rho|<|b|\delta/K$. Since K in (2.1.25) is arbitrary, for the left-hand side of (2.1.25), which is independent of K, we obtain the representation like the right-hand side of (2.1.25), in which the factor $(\Delta+\rho)$ is replaced by $(1+o(1))\Delta$ (as $t\to\infty$).

Next, since in (2.1.25)

$$\{\chi(t)\in[v,v+\delta_1+\delta_2)\}=\{\chi(t)\in[v,v+\delta_1)\}\bigcup\{\chi(t)\in[v+\delta_1,v+\delta_1+\delta_2)\},$$

the events on the right of this relation are disjoint and since

$$I_{u+v}^{u+v+\delta_1}+I_{u+v+\delta_1}^{u+v+\delta_1+\delta_2}=I_{u+v}^{u+v+\delta_1+\delta_2},$$

relation (2.1.25) is also preserved for an arbitrary $\delta>0$. We have

$$\frac{1}{a_\tau}I_{u+v}^{u+v+\delta}\to\frac{1}{a_\tau}\int_{u+v}^{\infty}\mathbf{P}(\tau\geq y)\,dy=\mathbf{P}(\chi\geq u+v)\quad\text{as}\quad\delta\to\infty,$$

and now (2.1.3) follows from (2.1.25).

II. If the variable τ is arithmetic, the proof of the theorem becomes simpler. Instead of the events B_t^δ in (2.1.6) one should use, for integer u and v, the event

$$B_t^1=B_t^1(u,v)=\{\gamma(t)\geq u,\chi(t)=v\}$$

(corresponding to the value $\delta=1$). In this case, there is no need to use inclusions (2.1.15), (2.1.16), while the function I_{u+v}^{u+v+1} turns into the probability $\mathbf{P}(\tau\geq u+v)$.

There is also no need in some other parts of the argument; for example, one does not need the considerations at Stage I, 4) related to the use of arbitrary large K.

III. If the random variable ω is arithmetic, then in (2.1.15), (2.1.16) one should use the local theorem for Ω_n in the arithmetic case.

IV. Now consider the process $Z(t) = Z_{\nu(t)}$. We have

$$t = T_{\nu(t)} + \gamma(t),$$

and so

$$Z(t) - at = Z_{\nu(t)} - aT_{\nu(t)} - a\gamma(t) = S_{\nu(t)} - a\gamma(t). \tag{2.1.26}$$

Similarly to (2.1.6), one should introduce the events

$${}^{\delta}B_t := \{\gamma(t) \in \delta[u), \chi(t) \geq v\} \tag{2.1.27}$$

and look for a representation for

$$P^t = \mathbf{P}(Z(t) - at \in \Delta[x), {}^{\delta}B_t) = \mathbf{P}(S_{\nu(t)} - a\gamma(t) \in \Delta[x), {}^{\delta}B_t). \tag{2.1.28}$$

The remaining steps repeat, with minor modifications, those given at Stage I of the proof. As in (2.1.11), we have

$$\mathbf{P}(\nu(t) = n, {}^{\delta}B_t) = \int_{t-u-\delta}^{t-u} \mathbf{P}(T_n \in dy)\mathbf{P}(\tau \geq t - y + v). \tag{2.1.29}$$

If $a_\omega \neq 0$, then, after using the integro-local theorem for T_n to approximate the integral in (2.1.29) (for small lengths of half-open intervals), we obtain for the left-hand side of (2.1.29) the same asymptotic representation as on the right of (2.1.12).

Next, on the set ${}^{\delta}B_t$ the term $-bv$ in the right-hand sides of (2.1.15), (2.1.16) should be replaced by bu. As a result, for the probability

$$\mathbf{P}(Z(t) - at \in \Delta[x), {}^{\delta}B_t) \tag{2.1.30}$$

we get the same intermediate assertion as in (2.1.23): *the probability* (2.1.30) *has the same asymptotic representation as appears on the right of* (2.1.23).

If $a_\omega = 0$, then the similar arguments give (2.1.23).

The remaining part of the proof proceeds with obvious changes as the final part of Stage I of the proof. Theorem 2.1.2 is proved. □

2.2 Refinement of Stone's Integro-Local Theorem for Random Walks

The following refinement of the integro-local theorem (Theorem 2.1.1) will be needed in the proofs in §2.3 of the integro-local theorems for CRPs for arbitrarily dependent τ and ζ.

Let us get back to the presentation in §2.1 and consider the following additional conditions.

$[\mathbf{M}_{2+r}]$ *For some* $r \in (0,1]$, *there exists* $\mathbf{E}|\boldsymbol{\xi}|^{2+r} < \infty$.

$[\mathbf{C}_\varphi]$ *The characteristic function* $\varphi(\boldsymbol{\lambda})$ *in the subspace* $\mathbb{R}^l$ *of variables* $\lambda_1, \ldots, \lambda_l$ $(\lambda_{l+1} = \cdots = \lambda_d = 0)$ *satisfies the Cramér condition*

$$\overline{\lim_{\substack{|\boldsymbol{\lambda}|\to\infty \\ \boldsymbol{\lambda}\in\mathbb{R}^l}}} |\varphi(\boldsymbol{\lambda})| < 1.$$

Recall that condition $[\mathbf{C}_\varphi]$ is always satisfied if the distribution of the subvector $(\xi_{(1)}, \ldots, \xi_{(l)})$ has an absolutely continuous spatial component in $\mathbb{R}^l$. (In this case, each of $\xi_{(j)}$, $j \le l$, has an absolutely continuous component but not vice versa.) We also recall that, in the context of CRP-problems, which will be considered in §2.3, the vector $\boldsymbol{\xi}$ is (τ, ζ) and $\mathbf{S}_n = (T_n, Z_n)$.

We will prove the following result, which refines the remainder term $o(n^{-d/2})$ in (2.1.1) and extends the range of $\boldsymbol{\Delta}[\boldsymbol{x}]$.

In this section, we denote by $\boldsymbol{\Delta}$ the set of positive numbers $\boldsymbol{\Delta} = (\Delta_1, \ldots, \Delta_d)$, where $\Delta_j \ge 1$ are integer for $j > l$, and define

$$\Pi = \prod_{j=1}^{d} \Delta_j, \qquad \underline{\Delta} = \min_{1\le j\le d} \Delta_j, \qquad \overline{\Delta} = \max_{1\le j\le d} \Delta_j. \tag{2.2.1}$$

Next, consider the half-open parallelepiped

$$\boldsymbol{\Delta}[\boldsymbol{x}) = \prod_{j=1}^{d} [x_j, x_j + \Delta_j). \tag{2.2.2}$$

Theorem 2.2.1 *Let the conditions of Theorem* 2.1.1 *be satisfied and let the additional conditions* $[\mathbf{M}_{2+r}]$, $[\mathbf{C}_\varphi]$ *be met. Then*

$$\frac{\mathbf{P}(\mathbf{S}_n \in \boldsymbol{\Delta}[\boldsymbol{x}))}{\Pi} = \frac{1}{n^{d/2}} \phi_\sigma\left(\frac{\boldsymbol{x} - \boldsymbol{a}n}{\sqrt{n}}\right) + O(n^{-\frac{d+r}{2}}), \tag{2.2.3}$$

where the remainder term $O(n^{-\frac{d+r}{2}})$ *is uniform in* x *and in* $\boldsymbol{\Delta}$ *from the domain*

$$q^n \le \underline{\Delta} \le \overline{\Delta} \le cn^{\frac{1-r}{2}} \tag{2.2.4}$$

for some $q < 1$ *and any* $c = \text{const}$.

From the proof of the theorem it will follow that the remainder term $O(n^{-\frac{d+r}{2}})$ is uniform also in all distributions for which

$$\mathbf{E}|\xi|^{2+r} < c_1 < \infty, \qquad \sup_{|\boldsymbol{\lambda}|>1} |\varphi(\boldsymbol{\lambda})| < c_2 < 1,$$

and $\boldsymbol{\sigma}$ lies in a compact set bounded away from degenerate values.

Theorem 2.2.1 will be used, for example, in proving the integro-local theorem for CRPs in §2.3.

Under the conditions of Theorem 2.2.1, in the nonlattice case ($l = d$) we have even more reason to call the left-hand side in (2.2.3) the "predensity" of $\mathbf{S}_n$, since Δ_j may decay exponentially fast with increasing n (see (2.2.4)).

We also note that the exponential lower decay bound for $\boldsymbol{\Delta}$ in (2.2.4) is "qualitatively" the best possible. This is indicated by the following example. Assume that $d = 1$ and that $\boldsymbol{\xi} = \xi$ is nonlattice, satisfies the conditions of Theorem 2.2.1, and has an atom at zero, $\mathbf{P}(\xi = 0) = p > 0$, $a = \mathbf{E}\xi = 0$. Then $\mathbf{S}_n$ also has an atom at zero, and $\mathbf{P}(\mathbf{S}_n = 0) \geq p^n$. This means that the left-hand side of (2.2.3) for $\boldsymbol{x} = 0$, $\boldsymbol{\Delta} = \Delta = p^n$ will be not smaller than 1, and the right-hand side is $O(n^{-1/2})$, that is, it converges to 0 as $n \to \infty$, so that (2.2.3) ceases to hold for $\Delta \leq p^n$. Therefore, the bound p^n for Δ (which is also exponentially decreasing like q^n in Theorem 2.2.1, $q > p$) controls the range of those Δ for which (2.2.3) is not true.

Proof of Theorem 2.2.1 We first consider the case $l = d$ in which all the coordinates $\xi_{(j)}$ are nonlattice. We can assume without loss of generality that $\boldsymbol{a} = 0$. In order to use a convenient inversion formula, we shall use the "smoothing method" and, in addition to $\mathbf{S}_n$, consider the sums

$$\widetilde{\mathbf{S}}_n = \mathbf{S}_n + \boldsymbol{\theta}_{\boldsymbol{\delta}},$$

where $\boldsymbol{\theta}_{\boldsymbol{\delta}}$ does not depend on $\{\boldsymbol{\xi}_1, \boldsymbol{\xi}_2, \dots\}$ and has uniform distribution in the parallelepiped

$$[\boldsymbol{\delta}] = \prod_{j=1}^{d} [-\delta_j, 0],$$

where the coordinates $\delta_j \leq \Delta_j$ of the vector $\boldsymbol{\delta} = (\delta_1, \dots, \delta_d)$ will be chosen later. The characteristic function $\varphi_{[\boldsymbol{\delta}]}(\boldsymbol{\lambda})$ of the variable $\boldsymbol{\theta}_{\boldsymbol{\delta}}$ has the form

$$\varphi_{[\boldsymbol{\delta}]}(\boldsymbol{\lambda}) := \prod_{j=1}^{d} \frac{(1 - e^{-i\lambda_j \delta_j})}{i\lambda_j \delta_j}, \tag{2.2.5}$$

and the characteristic function $\widetilde{\mathbf{S}}_n$ is equal to $\varphi^n(\boldsymbol{\lambda})\varphi_{[\boldsymbol{\delta}]}(\boldsymbol{\lambda})$.

The probability

$$\mathbf{P}(\widetilde{\mathbf{S}}_n \in \boldsymbol{\Delta}[\boldsymbol{x})) = \int_{\boldsymbol{v} \in [\boldsymbol{\Delta}]} \mathbf{P}(\widetilde{\mathbf{S}}_n \in \boldsymbol{x} + d\boldsymbol{v})$$

is equal, up to the factor Π, to the value of the *density* at the point $\boldsymbol{x}$ of the convolution of the uniform distribution on $[\boldsymbol{\Delta}]$ and the distribution of $\widetilde{\mathbf{S}}_n$. The characteristic function of this convolution equals the product

$$\varphi^n(\boldsymbol{\lambda})\varphi_{[\boldsymbol{\delta}]}(\boldsymbol{\lambda})\varphi_{[\boldsymbol{\Delta}]}(\boldsymbol{\lambda})$$

and is integrable on $\mathbb{R}^d$. Therefore, we have the inversion formula

$$\mathbf{P}(\widetilde{\mathbf{S}}_n \in \boldsymbol{\Delta}[\boldsymbol{x})) = \frac{\Pi}{(2\pi)^d} \int_{\mathbb{R}^d} e^{-i\langle \boldsymbol{\lambda}, \boldsymbol{x} \rangle} \varphi^n(\boldsymbol{\lambda})\varphi_{[\boldsymbol{\delta}]}(\boldsymbol{\lambda})\varphi_{[\boldsymbol{\Delta}]}(\boldsymbol{\lambda})\, d\lambda_1 \dots d\lambda_d. \tag{2.2.6}$$

We split the integral on the right of (2.2.6) into three parts $I(B_1), I(B_2), I(B_3)$, which are the integrals over the domains

$$B_1 = \{|\lambda| < \gamma_1 n^{-\rho}\}, \quad B_2 = \{\gamma_1 n^{-\rho} \le |\lambda| < \gamma\}, \quad B_3 = \{|\lambda| \ge \gamma\}, \tag{2.2.7}$$

respectively, for fixed $\gamma_1 > 0$, $\gamma > 0$, where $\rho \in (0, 1/2)$ will be chosen later.

In the domain $B_1 \cup B_2$, for sufficiently small γ, we shall use the expansion (recall that without loss of generality we assume that $\boldsymbol{a} = 0$)

$$1 - \varphi(\boldsymbol{\lambda}) = \frac{\boldsymbol{\lambda}\sigma^2\boldsymbol{\lambda}^*}{2} + O(|\boldsymbol{\lambda}|^{2+r}) \quad \text{as} \quad |\boldsymbol{\lambda}| \to 0, \tag{2.2.8}$$

where the remainder term depends only on $\mathbf{E}|\boldsymbol{\xi}|^{2+r}$ (see, e.g., [81], §12.4).

Consider the integral $I(B_1)$, in which we put

$$\varphi_{[\delta]}(\boldsymbol{\lambda})\varphi_{[\Delta]}(\boldsymbol{\lambda}) =: \widehat{\varphi}(\boldsymbol{\lambda}),$$

and change the variables

$$\boldsymbol{x} = \boldsymbol{v}\sqrt{n}, \qquad \boldsymbol{\lambda} = \frac{\boldsymbol{u}}{\sqrt{n}}.$$

We get

$$I(B_1) = n^{-d/2} \int_{|\boldsymbol{u}|<\gamma_1 n^{1/2-\rho}} e^{-i\langle \boldsymbol{u},\boldsymbol{v}\rangle} \varphi^n\Big(\frac{\boldsymbol{u}}{\sqrt{n}}\Big)\widehat{\varphi}\Big(\frac{\boldsymbol{u}}{\sqrt{n}}\Big)\, du_1 \dots du_d. \tag{2.2.9}$$

By (2.2.8),

$$n \ln \varphi\Big(\frac{\boldsymbol{u}}{\sqrt{n}}\Big) = n \ln\Big[1 - \Big(1 - \varphi\Big(\frac{\boldsymbol{u}}{\sqrt{n}}\Big)\Big)\Big] = -\frac{\boldsymbol{u}\sigma^2\boldsymbol{u}^*}{2} + n^{-r/2} O(|\boldsymbol{u}|^{2+r}). \tag{2.2.10}$$

If $\rho = \frac{1}{2+r}$ and $|\boldsymbol{u}| < \gamma_1 n^{1/2-\rho}$, then $-r/2 + (1/2 - \rho)(2 + r) = 0$, and

$$n^{-r/2}|\boldsymbol{u}|^{2+r} \le \gamma_1^{2+r},$$

and the remainder term in (2.2.10) can be made arbitrarily small by choosing an appropriate γ_1. Therefore,

$$\varphi^n\Big(\frac{\boldsymbol{u}}{\sqrt{n}}\Big) = \big[1 + n^{-r/2} O(|\boldsymbol{u}|^{2+r})\big] \exp\Big\{-\frac{\boldsymbol{u}\sigma^2\boldsymbol{u}^*}{2}\Big\} \tag{2.2.11}$$

and, moreover, for $\delta_j \le \Delta_j$,

$$\widehat{\varphi}\Big(\frac{\boldsymbol{u}}{\sqrt{n}}\Big) = \prod_{j=1}^{d}\Big(1 + O\Big(\frac{|u_j|\Delta_j}{\sqrt{n}}\Big)\Big)\Big(1 + O\Big(\frac{|u_j|\delta_j}{\sqrt{n}}\Big)\Big) = 1 + \frac{|\boldsymbol{u}|\overline{\Delta}}{\sqrt{n}} O(1) \tag{2.2.12}$$

with the uniform remainder term $O(1)$. We have

$$\int_{\mathbb{R}^d} |\boldsymbol{u}|^b e^{-\frac{\boldsymbol{u}\sigma^2\boldsymbol{u}^*}{2}} du_1 \dots du_d < \infty \quad \text{for any} \quad b > 0,$$

and hence, by (2.2.9), (2.2.11), (2.2.12),

$$I(B_1) = n^{-d/2} \int_{|u|<\gamma_1 n^{1/2-\rho}} e^{-i\langle u,v\rangle - \frac{u\sigma^2 u^*}{2}} du_1 \dots du_d$$
$$+ n^{-\frac{d+r}{2}} O(1) + n^{-d/2} \frac{\overline{\Delta}}{\sqrt{n}} O(1). \qquad (2.2.13)$$

Therefore, if $\delta_j \le \Delta_j$, $j = 1, \dots, d$, and

$$\overline{\Delta} \le n^{\frac{1-r}{2}} \qquad (2.2.14)$$

then both uniform remainder terms in (2.2.13) can be combined into one in the form $O(n^{-\frac{d+r}{2}})$.

The integral on the right of (2.2.13) over the complementary domain $\{|u| \ge \gamma n^{1/2-\rho}\}$ is

$$O(\exp\{-c_1 n^{1-2\rho}\}) \qquad (2.2.15)$$

with some $c_1 > 0$. Therefore, representation (2.2.13) for $I(B_1)$ with the remainder term of order $O(n^{-\frac{d+r}{2}})$ remains valid if we replace the integral in (2.2.13) by the integral over the entire space. By the inversion formula for the normal distribution, this integral equals

$$\int_{\mathbb{R}^d} e^{-i\langle u,v\rangle - \frac{u\sigma^2 u^*}{2}} = \frac{(2\pi)^{d/2}}{|\sigma|} \exp\left\{-\frac{v\sigma^{-2}v^*}{2}\right\}.$$

As a result, for $\delta_j \le \Delta_j$ and $\overline{\Delta} \le n^{\frac{1-r}{2}}$, $n \to \infty$, we obtain

$$I(B_1) = \frac{(2\pi)^{d/2} n^{-d/2}}{|\sigma|} e^{-\frac{v\sigma^{-2}v^*}{2}} + O(n^{-\frac{d+r}{2}}) \qquad (2.2.16)$$

uniformly in x.

Now we consider the integral $I(B_2)$. By choosing an appropriate γ (see (2.2.7)), the value $-\mathrm{Re}(1-\varphi(\lambda))$ in (2.2.8) can be made less than $-\frac{\lambda\sigma^2\lambda^*}{3}$ for $-\gamma < |\lambda| < \gamma$, and the value $\mathrm{Re}(n \ln \varphi(\lambda))$ (cf. (2.2.10)) can be made less than $-\frac{u\sigma^2 u^*}{4}$ (with the same change $\lambda = \frac{u}{\sqrt{n}}$ as in (2.2.10)). Therefore,

$$I(B_2) < n^{-d/2} \int_{|u|\ge\gamma_1 n^{1/2-\rho}} e^{-\frac{u\sigma^2 u^*}{4}} du_1 \dots du_d,$$

and so, by the above (see (2.2.15)),

$$I(B_2) = O(\exp\{-c_2 n^{1-2\rho}\})$$

with some $c_2 \in (0, \infty)$ and all sufficiently large n. This means that uniformly in x

$$I(B_2) = o(n^{-\frac{d+r}{2}}). \qquad (2.2.17)$$

It remains to estimate $I(B_3)$. Since ξ is nonlattice, we have, by condition $[\mathbf{C}_\varphi]$,

$$q_\gamma := \sup_{|\lambda|>\gamma} |\varphi(\lambda)| < 1.$$

Hence

$$I(B_3) \le q_\gamma^n \int_{|\lambda|\ge\gamma} |\widehat{\varphi}(\lambda)| d\lambda_1 \ldots d\lambda_d, \tag{2.2.18}$$

where

$$|\widehat{\varphi}(\lambda)| \le \frac{2^{2d}}{\prod_{j=1}^{d} |\lambda_j|^2 \Delta_j \delta_j}, \tag{2.2.19}$$

the function $\prod_{j=1}^{d} |\lambda_j|^{-2}$ being integrable over the domain $|\lambda| \ge \gamma$. Hence, for $\delta_j = n^{-r/2}\Delta_j$,

$$I(B_3) \le \frac{c_3 q_\gamma^n n^{\frac{rd}{2}}}{\underline{\Delta}^{2d}}, \qquad c_3 = \text{const},$$

and so, for

$$\underline{\Delta}^{2d} \ge q_\gamma^n n^{\frac{rd+d+r}{2}}, \tag{2.2.20}$$

we have

$$I(B_3) = O(n^{-\frac{d+r}{2}}). \tag{2.2.21}$$

Clearly, there always exists $q \in (q_\gamma^{\frac{1}{2d}}, 1)$ such that (2.2.20) and (2.2.21) are fulfilled for $\underline{\Delta} \ge q^n$ and all sufficiently large n.

Now by comparing (2.2.6), (2.2.16), (2.2.17), and (2.2.21) we get

$$\begin{aligned} \mathbf{P}(\widetilde{\mathbf{S}}_n \in \mathbf{\Delta}[\boldsymbol{x}]) &= \frac{\Pi}{(2\pi n)^{d/2}|\sigma|} e^{-\frac{v\sigma^{-2}v^*}{2}} + \Pi O(n^{-\frac{d+r}{2}}) \\ &= \Pi\left[n^{-d/2}\phi_\sigma(\boldsymbol{v}) + O(n^{-\frac{d+r}{2}})\right]. \end{aligned} \tag{2.2.22}$$

But

$$\{\widetilde{\mathbf{S}}_n \in \mathbf{\Delta}_\delta[\boldsymbol{x}]\} \subset \{\mathbf{S}_n \in \mathbf{\Delta}[\boldsymbol{x}]\}, \tag{2.2.23}$$

where $\mathbf{\Delta}_\delta[\boldsymbol{x}]$ is the parallelepiped

$$\mathbf{\Delta}_\delta[\boldsymbol{x}) := \prod_{j=1}^{d} [x_j, x_j + \Delta_j - \delta_j).$$

Hence, for $\delta_j = n^{-r/2}\Delta_j$,

$$\mathbf{P}(\mathbf{S}_n \in \mathbf{\Delta}[\boldsymbol{x})) \ge \Pi\left[n^{-d/2}(1 - n^{-r/2})^d \phi_\sigma(\boldsymbol{v}) + O(n^{-\frac{d+r}{2}})\right]. \tag{2.2.24}$$

Similarly, putting

$$\widetilde{\mathbf{S}}_n = \mathbf{S}_n - \boldsymbol{\theta}_\delta,$$

we have

$$\{\mathbf{S}_n \in \mathbf{\Delta}[\boldsymbol{x})\} = \{\widetilde{\mathbf{S}}_n \in \mathbf{\Delta}[\boldsymbol{x}) - \theta_\delta\} \subset \{\widetilde{\mathbf{S}}_n \in \mathbf{\Delta}^\delta[\boldsymbol{x})\},$$

where

$$\boldsymbol{\Delta}^{\delta}[\boldsymbol{x}) := \prod_{j=1}^{d}[x_j, x_j + \Delta_j + \delta_j),$$

and therefore,

$$\mathbf{P}(\mathbf{S}_n \in \boldsymbol{\Delta}[\boldsymbol{x})) \leq \Pi\left[n^{-d/2}(1 + n^{-r/2})^d \phi_\sigma(\boldsymbol{v})\right] + O(n^{-\frac{d+r}{2}}). \tag{2.2.25}$$

This together with (2.2.24) proves (2.2.3) with $l = d$.

Now let $l = 0$, that is, we suppose that all coordinates $\xi_{(k)}$ are arithmetic. In this case, for $\Delta_j \equiv 1$ the inversion formula for $\mathbf{S}_n$ for integer x_k has the form

$$\mathbf{P}(\mathbf{S}_n = \boldsymbol{x}) = \frac{1}{(2\pi)^d} \int_{\mathbf{Z}^d} e^{-i\langle \boldsymbol{\lambda}, \boldsymbol{x}\rangle} \varphi^n(\boldsymbol{\lambda}) d\lambda_1 \dots d\lambda_d,$$

where $\mathbf{Z}^d$ is the cube in $\mathbb{R}^d$ defined as the direct product of the intervals

$$\mathbf{Z}^d = \prod_{j=1}^{d}[-\pi \leq \lambda_j \leq \pi].$$

Here, no smoothing operation is needed, and all considerations are simplified. For simplicity, we assume first that $\boldsymbol{a} = 0$, although this is a constraint for arithmetic $\boldsymbol{\xi}$. Then all the calculations in the first part of the proof of the theorem remain valid for the same partition of the integration domain $\mathbf{Z}^d$ into the subdomains B_1, B_2, and B_3; here by B_3 we mean the intersection

$$\{\boldsymbol{\lambda} : |\boldsymbol{\lambda}| \geq \gamma\} \bigcap \mathbf{Z}^d,$$

for which, as before,

$$\sup_{\boldsymbol{\lambda} \in B_3} |\varphi(\boldsymbol{\lambda})| =: q_\gamma < 1.$$

If $\boldsymbol{a} \neq 0$, then, after the change

$$\boldsymbol{x} - \boldsymbol{a}n = \boldsymbol{v}\sqrt{n}, \qquad \boldsymbol{\lambda} = \frac{\boldsymbol{u}}{\sqrt{n}} \tag{2.2.26}$$

we have, instead of (2.2.9),

$$I(B_1) = n^{-d/2} \int_{|\boldsymbol{u}| < \gamma_1 n^{1/2-\rho}} e^{-i\langle \boldsymbol{u}, \boldsymbol{v}\rangle + i\sqrt{n}\langle \boldsymbol{u}, \boldsymbol{a}\rangle} \varphi^n\left(\frac{\boldsymbol{u}}{\sqrt{n}}\right) du_1 \dots du_d, \tag{2.2.27}$$

and, instead of (2.2.8), (2.2.10), and (2.2.11), respectively, we have

$$\begin{aligned}
1 - \varphi(\boldsymbol{\lambda}) &= -i\langle \boldsymbol{a}, \boldsymbol{\lambda}\rangle - \frac{\boldsymbol{\lambda}\sigma^2\boldsymbol{\lambda}^*}{2} + O(|\boldsymbol{\lambda}|^{2+r}),\\
n \ln \varphi(\boldsymbol{\lambda}) &= -i\sqrt{n}\langle \boldsymbol{a}, \boldsymbol{u}\rangle - \frac{\boldsymbol{u}\sigma^2\boldsymbol{u}^*}{2} + n^{-r/2} O(|\boldsymbol{u}|^{2+r}), \qquad (2.2.28)\\
\varphi^n\left(\frac{\boldsymbol{u}}{\sqrt{n}}\right) &= \left[1 + n^{-r/2} O(|\boldsymbol{u}|^{2+r})\right] \exp\left\{-i\sqrt{n}\langle \boldsymbol{a}, \boldsymbol{u}\rangle - \frac{\boldsymbol{u}\sigma^2\boldsymbol{u}^*}{2}\right\}.
\end{aligned}$$

From (2.2.27) and (2.2.28) it follows that representation (2.2.13) for $I(B_1)$ is preserved for $\boldsymbol{v} = \frac{\boldsymbol{x}-\boldsymbol{a}n}{\sqrt{n}}$. The rest of the calculations in the proof of the theorem remain valid with obvious adjustments due to the change of variables (2.2.26).

If $\Delta_j \geq 1$ are integer, then the easiest way to obtain the theorem is by summing up the probabilities $\mathbf{P}(\mathbf{S}_n = \boldsymbol{x} + \boldsymbol{y})$, where $\boldsymbol{y} = (y_1, \ldots, y_d)$, y_j are integer, $0 \leq y_j < \Delta_j$, $j = 1, \ldots, d$.

It remains to consider the intermediate case $0 < l < d$. Here, as before, the distribution of $\mathbf{S}_n$ should be smoothed, but only with respect to the first l coordinates and for the smoothed sum $\widetilde{\mathbf{S}}_n$ the inversion formula

$$\mathbf{P}(\widetilde{\mathbf{S}}_n \in \Delta[\boldsymbol{x})) = \frac{\Delta^l}{(2\pi)^d} \int_{\mathbb{R}^l} \int_{Z^{d-l}} e^{-i\langle \boldsymbol{\lambda}, \boldsymbol{x} \rangle} \varphi^n(\boldsymbol{\lambda}) \varphi^{(l)}_{[\Delta]}(\boldsymbol{\lambda}) \varphi^{(l)}_{[\delta]}(\boldsymbol{\lambda}) d\lambda_1 \ldots d\lambda_d \qquad (2.2.29)$$

should be applied, where $\varphi^{(l)}_{[\delta]}(\boldsymbol{\lambda})$ is the characteristic function of the uniform distribution on the parallelepiped $\prod_{k=1}^{l} [-\delta_j \leq \lambda_j \leq 0]$.

The remaining arguments in deriving (2.2.3), related to the asymptotic analysis of the integrands in (2.2.29) in B_1–B_3, are completely preserved except for obvious and insignificant modifications. In particular, inequality (2.2.19) remains valid if d is replaced by l. The function $\prod_{j=1}^{l} |\lambda_j|^{-2}$ is integrable at infinity in $\mathbb{R}^l$, and so, for $\delta_j = n^{-r/2}\Delta_j$, $1 \leq j \leq l$,

$$I(B_3) \leq \frac{c_3 q_\gamma^n n^{rl/2}}{\underline{\Delta}^{2l}}$$

and (2.2.21) holds for

$$\underline{\Delta}^{2l} \geq q_\gamma^n n^{\frac{lr+d+r}{2}}.$$

Theorem 2.2.1 is proved. □

2.3 Integro-Local Theorems for Compound Renewal Processes in the General Case

Let us return to integro-local theorems for CRPs. We now consider the general case when the joint distribution of τ and ζ is arbitrary. However, in this case, one can obtain integro-local theorems only under additional moment and structural conditions. In that case, one can refine the remainder term and establish more complete results.

In this section, as in §2.1, the processes $Z(t) = Z_{\nu(t)}$ and $Y(t) = Z_{\eta(t)}$, are assumed to be *homogeneous*.

We first consider the process $Z(t)$. By

$$\varphi(\boldsymbol{\lambda}) := \mathbf{E}e^{i(\lambda_1 \tau + \lambda_2 \zeta)}, \quad \text{where} \quad \boldsymbol{\lambda} = (\lambda_1, \lambda_2), \qquad \operatorname{Im} \lambda_j = 0, \quad j = 1, 2,$$

we denote the characteristic function of a vector $\boldsymbol{\xi} = (\tau, \zeta)$. We also set

$$\mathbf{B}_t(u, v, \mathbf{w}) := \{\gamma(t) \ge u,\ \chi(t) \ge v,\ \zeta(t) \in \mathbf{w}\},$$

where $\zeta(t) = \zeta_{\eta(t)}$, $\mathbf{w}$ is an arbitrary measurable subset of $\mathbb{R}$. Let, as before, $a = a_\zeta / a_\tau$, $\sigma^2 = \sigma_\xi^2 / a_\tau$, and $\xi = \zeta - a\tau$.

Theorem 2.3.1 I. *Let a vector (τ, ζ) be nonlattice and satisfy condition $[\mathbf{M}_r]$ for $r \in (2, 3]$ and the Cramér condition $[\mathbf{C}_\varphi]$ for the characteristic function:*

$$\overline{\lim_{|\boldsymbol{\lambda}| \to \infty}} |\varphi(\boldsymbol{\lambda})| < 1. \tag{2.3.1}$$

Also let

$$\mathbf{E}\tau^r < \infty, \qquad \mathbf{E}|\zeta|^r < \infty, \qquad r \in (2, 3] \tag{2.3.2}$$

(the parameter r in Theorem 2.2.1 and in (2.3.2) is different). Then

$$\begin{aligned} \mathbf{P}(Z(t) - at \in \Delta[x), \mathbf{B}_t(u, v, w)) \\ = \frac{\Delta\, I_{u+v}(w)}{a_\tau \sqrt{t}}\, \phi_\sigma\left(\frac{x}{\sqrt{t}}\right) + (p_t(w) + \Delta I_{u+v}(w)) O(t^{-\beta}), \end{aligned} \tag{2.3.3}$$

where $p_t(w)$ is a distribution, which depends, in general, on t,

$$I_u(w) := \int_u^\infty \mathbf{P}(\tau \ge y, \zeta \in w)\, dy,$$

$\beta = \frac{r(r-1)}{2(1+r)} > \frac{1}{2}$ for $r > 1 + \sqrt{2}$, the remainder $O(t^{-\beta})$ is uniform with respect to all x, $u \le \overline{u}$ (for any fixed $\overline{u} > 0$), in v, w, and also in Δ from the interval $(t^{1/2-\beta}, ct^{\frac{3-r}{2}})$ for any $c = \mathrm{const} > 0$.

II. *If the random variable τ is arithmetic and if condition (2.3.2) and condition (2.3.1) for the characteristic function of ζ are met, then representation (2.3.3) holds true for integer t, $\gamma(t)$, $\chi(t)$, u, v (cf. assertion II of Theorem 2.1.2).*

III. *If the random variable ζ is arithmetic, then assertions I, II hold true for integer $\Delta \ge 1$.*

The question of whether an integro-local representation of the form (2.1.3) remains valid in the general case without the additional conditions $[\mathbf{M}_r]$, $[\mathbf{C}_\varphi]$ remains open.

In §2.5, we shall need the following result, which follows from Theorem 2.3.1.

Corollary 2.3.2 *Let the conditions of Theorem 2.3.1 hold with $r \in (1 + \sqrt{2}, 3)$, let τ be an arithmetic random variable, and let t assume integer values. Then, for any fixed l,*

$$\mathbf{P}(Z(t) - at \in \Delta[x) \mid \gamma(t) = l) = \frac{\Delta}{\sqrt{t}}\, \phi_\sigma\left(\frac{x}{\sqrt{t}}\right) + o(t^{-1/2})$$

uniformly in x and in Δ from the interval $(t^{-\varepsilon}, t^{\varepsilon})$ with some $\varepsilon > 0$.

Proof of Theorem 2.3.1 I. We set, for brevity $(u, v, \mathbf{w}) = W$, so that

$$\mathbf{B}_t(W) := \{\gamma(t) \ge u,\ \chi(t) \ge v,\ \zeta(t) \in \mathbf{w}\}$$

and consider the probability

$$P(t, \Delta, W) := \mathbf{P}(Z(t) - at \in \Delta[x), \mathbf{B}_t(W)) = \sum_{n=1}^{\infty} P_n(t, \Delta, W), \quad (2.3.4)$$

where

$$P_n(t, \Delta, W) := \mathbf{P}(\nu(t) = n,\ S_n + a\gamma(t) \in \Delta[x), \mathbf{B}_t(W)) \quad (2.3.5)$$

(see (2.1.26)).

We first estimate the part of the sum in (2.3.4) when n varies outside the interval $[n_-, n_+]$, where

$$n_\pm := \left[\frac{t}{a_\tau} \pm t^h\right],$$

the number $h \in (1/2, 1)$ will be chosen later. The probability

$$P_n(t, \Delta, W) = \mathbf{P}(\eta(t) = n + 1, S_n + a\gamma(t) \in \Delta[x),\ \mathbf{B}_t(W), \zeta(t) \in \mathbf{w}),$$

for each t and Δ is a measure with respect to $\mathbf{w}$. Hence

$$H_t(\mathbf{w}) := \sum_{n \notin [n_-, n_+]} P_n(t, \Delta, W)$$

is also a measure. Let us estimate $H_t(\mathbb{R})$. Since $\mathbf{E}\tau^r < \infty$, we have

$$\mathbf{E}|T_n - a_\tau n|^r \le c n^{r/2}, \qquad c = \text{const},$$

$$\mathbf{P}(\eta(t) \le n_-) = \mathbf{P}(T_{n_-} > t) \le \mathbf{P}(T_{n_-} - a_\tau n_- > a_\tau t^h) \le \frac{c n_-^{r/2}}{(a_\tau t^h)^r} = O(t^{r\left(\frac{1}{2} - h\right)}).$$

In a similar way we establish that

$$\mathbf{P}(\eta(t) > n_+) = O(t^{r\left(\frac{1}{2} - h\right)}).$$

As a corollary, we get, uniformly in $x, \Delta, u, v, \mathbf{w}$

$$H_t(\mathbb{R}) = O(t^{\frac{r}{2} - hr}),$$

$$P(t, \Delta, W) = p_t(\mathbf{w}) O(t^{\frac{r}{2} - hr}) + \sum_{n=n_-}^{n_+} P_n(t, \Delta, W) \quad (2.3.6)$$

where $p_t(\mathbf{w}) := H_t(\mathbf{w})/H_t(\mathbb{R})$ is a distribution, $n \sim t/a_\tau$ for all $n \in [n_-, n_+]$ and $t \to \infty$.

The main part of the proof of the theorem consists in finding the asymptotic behavior of the terms $P_n(t, \Delta, W)$ for $n \in [n_-, n_+]$. We will split the analysis of this asymptotics into several steps.

(1) We first estimate $P_n(t, \Delta, W)$ from above and below. Setting

$$\delta[u) = [u, u+\delta), \quad {}^\delta\mathbf{B}_t = {}^\delta\mathbf{B}_t(W) = \{\gamma(t) \in \delta[u),\ \chi(t) \ge v, \zeta(t) \in \mathbf{w}\},$$
$$u_k = u + \delta k,\ k = 0, 1, \ldots, K = \left[\frac{t-u}{\delta}\right],$$

we have

$$P_n(t, \Delta, W) = \sum_{k=0}^{K} P_n(t, \Delta, W_k, \delta), \tag{2.3.7}$$

where $W_k = (u_k, v, \mathbf{w})$,

$$\begin{aligned} P_n(t, \Delta, W_k, \delta) &= \mathbf{P}(\nu(t) = n,\ Z(t) - at \in \Delta[x),\ {}^\delta\mathbf{B}_t(W_k)) \\ &= \mathbf{P}(\nu(t) = n,\ S_n - a\gamma(t) \in \Delta[x),\ {}^\delta\mathbf{B}_t(W_k)) \end{aligned}$$

(see (2.1.28)). Assume for definiteness that $a \ge 0$. Then we have the following inclusions of events:

$$\begin{aligned} \{S_n - a\gamma(t) \in \Delta[x), {}^\delta\mathbf{B}_t\} &\subset \{S_n \in [x + au,\ x + \Delta + au + a\delta),\ {}^\delta\mathbf{B}_t\} \\ &= \{S_n \in \Delta_{+\delta}[x + au), {}^\delta\mathbf{B}_t\}; \\ \{S_n - a\gamma(t) \in \Delta[x), {}^\delta\mathbf{B}_t\} &\supset \{S_n \in \Delta_{-\delta}[x + au + a\delta), {}^\delta\mathbf{B}_t\}, \end{aligned}$$

where $\Delta_{\pm\delta} = \Delta \pm a\delta$. Therefore,

$$\begin{aligned} P_n(t, \Delta, W_k, \delta) &\le \mathbf{P}(\nu(t) = n,\ S_n \in \Delta_{+\delta}[x + au_k), {}^\delta\mathbf{B}_t(W_k)) \\ &= \int_{t-u_k-\delta}^{t-u_k} \mathbf{P}(T_n \in dy,\ S_n \in \Delta_{+\delta}[x + au_k))\mathbf{P}(\tau_{n+1} \ge t - y + v, \zeta_{n+1} \in \mathbf{w}) \\ &\le \mathbf{P}(T_n \in \delta[t - u_k - \delta),\ S_n \in \Delta_{+\delta}[x + au_k))\mathbf{P}(\tau \ge u_k + v,\ \zeta \in \mathbf{w}); \end{aligned} \tag{2.3.8}$$

and similarly,

$$\begin{aligned} &P_n(t, \Delta, W_k, \delta) \\ &\ge \int_{t-u_k-\delta}^{t-u_k} \mathbf{P}(T_n \in dy,\ S_n \in \Delta_{-\delta}[x + au_k + a\delta))\mathbf{P}(\tau_{n+1} \ge t - y + v,\ \zeta_{n+1} \in \mathbf{w}) \\ &\ge \mathbf{P}(T_n \in \delta[t - u_k - \delta),\ S_n \in \Delta_{-\delta}[x + au_k + a\delta))\mathbf{P}(\tau \ge u_k + v + \delta,\ \zeta \in \mathbf{w}). \end{aligned} \tag{2.3.9}$$

When considering the integrals in (2.3.8), (2.3.9), for definiteness, we included the left-hand end points of the intervals $[t - u_k - \delta, t - u_k)$ into the domain of integration and excluded the right-hand end points. Since in what follows we shall deal with sums of these integrals over k (see (2.3.7)) and since the half-open intervals $\delta[t - u_k - \delta)$ thus obtained are disjoint, this assumption does not impair the generality.

(2) We now use (2.3.8) and (2.3.9) to obtain an asymptotic representation for $P_n(t, \Delta, W)$. We set

$$s = t - a_\tau n, \quad Q(s, x) = \frac{s^2}{\sigma_\tau^2} - 2\rho\frac{sx}{\sigma_\tau\sigma_\xi} + \frac{x^2}{\sigma_\xi^2}, \quad C = \frac{1}{2\pi\sigma_\tau\sigma_\xi\sqrt{1-\rho^2}},$$

where

$$\rho = \frac{\mathbf{E}(\tau - a_\tau)\xi}{\sigma_\tau \sigma_\xi}$$

is the correlation coefficient between τ and ξ, $|\rho| < 1$.

Lemma 2.3.3 *Let the conditions of Theorem* 2.3.1 *be satisfied and let* $n \in [n_-, n_+]$. *Then, as* $t \to \infty$,

$$P_n(t, \Delta, W) = \Delta I_{u+v}(\mathbf{w}) O(t^{-r/2}) + \frac{C \Delta I_{u+v}(\mathbf{w})}{n} \exp\Big\{-\frac{1}{2(1-\rho^2)n} Q(s, x)\Big\} \quad (2.3.10)$$

uniformly in x, $u \le \overline{u} = \text{const} > 0$, v, $\mathbf{w}$, *and also in* Δ *from the interval* $(t^{1-\frac{r}{2}}, ct^{\frac{3-r}{2}})$, $c = \text{const}$.

Proof In this range of n, we have

$$|s| < a_\tau t^h, \qquad h \in (1/2, 1), \qquad n \sim \frac{t}{a_\tau} \quad \text{as} \quad t \to \infty.$$

We will make use of the following refinement of the integro-local theorem, which follows from Theorem 2.2.1.

Let the conditions of Theorem 2.3.1 *be met. Then, as* $t \to \infty$,

$$\mathbf{P}(T_n \in \delta[t), S_n \in \Delta[x)) = \Delta\delta O(n^{-r/2}) + \frac{C\Delta\delta}{n} \exp\Big\{-\frac{1}{2(1-\rho^2)n} Q(s, x)\Big\} \quad (2.3.11)$$

uniformly in x *and in* Δ, δ *from the interval* $(q^n, cn^{\frac{3-r}{2}})$, $q < 1$, $c = \text{const}$.

By virtue of (2.3.8) we have

$$P_n(t, \Delta, W, \delta) \le \sum_{k=0}^{K} \mathbf{P}(T_k \in \delta[t - u_k - \delta), S_n \in \Delta_{+\delta}[x + au_k)) \times \mathbf{P}(\tau \ge u_k + v, \zeta \in \mathbf{w}).$$

Using representation (2.3.10) with

$$s_k = s - u_k - \delta = t - a_\tau n - u_k - \delta, \qquad x_k = x + au_k, \qquad \Delta = \Delta_{+\delta} = \Delta + a\delta,$$

we get

$$P_n(t, \Delta, W, \delta) \le (\Delta + a\delta)\delta O(n^{-r/2}) \sum_{k\ge 0} \mathbf{P}(\tau \ge u_k + v, \zeta \in \mathbf{w}) + \frac{C(\Delta + a\delta)\delta}{n} \sum_{k\ge 0} E(s_k, x_k) \mathbf{P}(\tau \ge u_k + v, \ \zeta \in \mathbf{w}), \quad (2.3.12)$$

where

$$E(s, x) = \exp\Big\{-\frac{Q(s, x)}{2(1-\rho^2)n}\Big\}.$$

Note that

$$\begin{aligned} \frac{\partial E(s,x)}{\partial s} &= E(s,x)\left(\frac{c_1 x}{n} + \frac{c_2 s}{n}\right) = O\Big(\frac{1}{\sqrt{t}}\Big), \\ \frac{\partial E(s,x)}{\partial x} &= E(s,x)\left(\frac{c_3 x}{n} + \frac{c_4 s}{n}\right) = O\Big(\frac{1}{\sqrt{t}}\Big) \end{aligned} \tag{2.3.13}$$

as $t \to \infty$, where c_1–c_4 are independent of s, x, n and all the terms $O(\cdot)$ are uniform in x, $|s| \le a_\tau t^h$, $n \in [n_-, n_+]$. Hence, for $\delta = \frac{1}{n}\min(1,\Delta)$, $a \ge 0$, we have

$$\begin{aligned} &\sum_{k\ge 0} \delta E(s_k, x_k)\mathbf{P}(\tau \ge u_k + v, \zeta \in \mathbf{w}) \\ &\quad = \sum_{k\ge 0} \delta \left[E(s,x) + u_k O\Big(\frac{1}{\sqrt{t}}\Big)\right] \mathbf{P}(\tau \ge u_k + v,\ \zeta \in \mathbf{w}) \\ &\quad = E(s,x) I_{u+v}(\mathbf{w})(1 + O(1/t)) + O\Big(\frac{1}{\sqrt{t}}\Big) \end{aligned} \tag{2.3.14}$$

(here we used the fact that $\sum_{k\ge 0} u_k \delta \mathbf{P}(\tau \ge u_k + v) \le \mathbf{E}\tau^2$). The estimates $O(\frac{1}{t})$, $O(\frac{1}{\sqrt{t}})$ in (2.3.14) are uniform in x, $u \le \overline{u} = \text{const}$, v, $\mathbf{w}$. Since, moreover, in (2.3.12),

$$\sum_{k\ge 0} \delta \mathbf{P}(\tau \ge u_k + v, \zeta \in \mathbf{w}) \to I_{u+v}(\mathbf{w}) \quad \text{as} \quad t \to \infty,$$

it follows from (2.3.7), (2.3.12), and (2.3.14) with $\delta = \frac{1}{n}\min(1,\Delta)$ that $(\Delta + \delta) = \Delta(1 + O(1/n))$,

$$P_n(t,\Delta,W) \le \frac{C\Delta}{n} E(s,x) I_{u+v}(\mathbf{w}) + \Delta I_{u+v}(\mathbf{w}) O(n^{-r/2}),\ \ \Delta \ge t^{1-\frac{r}{2}} \tag{2.3.15}$$

(here we used the fact that $\frac{r}{2} \le \frac{3}{2}$, and so one can ignore the last term on the right of (2.3.14); it is seen from (2.3.15) that it makes no sense to consider the values $\Delta < t^{1-r/2}$).

Similarly, from (2.3.9) we get the lower estimate. The case $a \le 0$ is dealt with in the same way. Lemma 2.3.3 is proved. □

(3) In the final part of the proof of Theorem 2.3.1 we will use Lemma 2.3.3 and the initial representation (2.3.6) for $P(t,\Delta,W)$. Setting

$$h = \frac{r}{1+r}$$

in (2.3.6), we have

$$\beta = rh - \frac{r}{2} = \frac{r(r-1)}{2(1+r)}.$$

Summing up over $n \in [n_-, n_+]$ the right- and left-hand sides in (2.3.10), we get

$$\begin{aligned} P(t,\Delta,W) &= p_t(\mathbf{w}) O(t^{-\beta}) + C\Delta I_{u+v}(\mathbf{w}) \sum_{n=n_-}^{n_+} \frac{1}{n} \exp\Big\{-\frac{1}{2(1-\rho^2)n} Q(s,x)\Big\} \\ &\quad + (n_+ - n_-)\Delta I_{u+v}(\mathbf{w}) O(t^{-r/2}), \end{aligned} \tag{2.3.16}$$

the estimates being uniform in x, Δ, n, $u \le \overline{u}$, v, $\mathbf{w}$ in the regions stated in Lemma 2.3.3. Putting here $h = \frac{r}{1+r}$, we find that $h - \frac{r}{2} = \frac{r(1-r)}{2(1+r)} = -\beta$,

$$(n_+ - n_-)O(t^{-r/2}) = O(t^{h-r/2}) = O(t^{-\beta}).$$

It is easily seen, for $r > 1 + \sqrt{2}$,

$$-\beta + \frac{1}{2} = \frac{r(1-r)}{2(1+r)} + \frac{1}{2} = -\frac{(r-1)^2 - 2}{2(1+r)} < 0,$$

the common remainder term in (2.3.16) being $(p_t(\mathbf{w}) + \Delta I_{u+v}(\mathbf{w}))O(t^{-\beta})$.

Now consider the sum on the right of (2.3.16), in which the terms are summed over (noninteger) values $s = t - a_\tau n$ in the range $|s| \le a_\tau t^h$. Since

$$n = \frac{t}{a_\tau} - \frac{s}{a_\tau}, \qquad \frac{1}{n} = \frac{a_\tau}{t} + O\Big(\frac{s}{t^2}\Big) \quad \text{as} \quad t \to \infty,$$

we have

$$\sum_{n=n_-}^{n_+} \frac{1}{n} \exp\Big\{-\frac{1}{2(1-\rho^2)n} Q(s,x)\Big\}$$
$$= \frac{a_\tau}{t} \sum_{|s| \le a_\tau t^h} \exp\Big\{-\frac{1}{2(1-\rho^2)}\Big(\frac{a_\tau}{t} + O\Big(\frac{s}{t^2}\Big)\Big) Q(s,x)\Big\}$$
$$+ O\Big(\frac{1}{t^2}\Big) \sum_{|s| \le a_\tau t^h} s \exp\Big\{-\frac{a_\tau}{(1-\rho^2)t} Q(s,x)\Big\}. \tag{2.3.17}$$

Now we will make a general comment on sums of the form

$$\sum_{|s| \le a_\tau t^h} \exp\Big\{-\frac{cQ(s,x)}{t}\Big\}, \qquad c > 0, \tag{2.3.18}$$

which appear in (2.3.17) and which determine the asymptotics of $P(t, \Delta, u, v)$ in (2.3.16). The quadratic form $Q(s,x)$ attains its minimum over s at the point $s = \frac{\rho x \sigma_\tau}{\sigma_\xi}$, the minimum value being

$$\frac{x^2(1-\rho^2)}{\sigma_\xi^2}.$$

Therefore, for $|x| \ge \sqrt{t}\ln t$, the quadratic form $\frac{1}{t}Q(s,x)$ in the argument of the exponential function in (2.3.18) exceeds the value $c_1 \ln^2 t$, $c_1 = \text{const}$, and hence the exponent itself on the right of (2.3.18) is $o(t^{-l})$ for any $l > 0$ and all s, $|s| < a_\tau t^h$. The same applies to the sums in (2.3.17).

We will obtain a similar result if, in the above argument, we swap x and s: hence the sum of the terms in (2.3.18) over s with $|s| > \sqrt{t}\ln t$ is also $O(t^{-l})$ for any $l > 0$. Thus, up to an additive term $O(t^{-l})$, it suffices to consider in (2.3.17) the values $|x| \le \sqrt{t}\ln t$ and compute the sum over s satisfying the inequality $|s| < \sqrt{t}\ln t$.

Now let us return to (2.3.17) and change the variables as follows:

$$s_1 = \frac{s}{\sqrt{t}}, \qquad x_1 = \frac{x}{\sqrt{t}}.$$

Hence

$$\frac{1}{t}\, Q(s,x) = Q(s_1, x_1)$$

and the first sum on the right-hand side of (2.3.17) is

$$O(t^{-l}) + \frac{a_\tau}{\sqrt{t}} \sum_{|s_1| \le \ln t} \frac{1}{\sqrt{t}} \exp\Big\{-\frac{a_\tau}{2(1-\rho^2)}\, Q(s_1, x_1)\big[1 + O(s_1 t^{-1/2})\big]\Big\}. \qquad (2.3.19)$$

Since for the argument of the exponential function for $|x_1| < \ln t$ we have

$$Q(s_1, x_1) s_1 t^{-1/2} = (\ln t)^3 O(t^{-1/2}) = o(1)$$

as $t \to \infty$, the exponent in (2.3.19) equals

$$[1 + s_1 O(t^{-1/2} \ln^2 t)] \exp\Big\{-\frac{a_\tau Q(s_1, x_1)}{2(1-\rho^2)}\Big\}.$$

As the function

$$\exp\left\{-\frac{a_\tau}{2(1-\rho^2)}\, Q(s_1, x_1)\right\}$$

is smooth and decreasing, the sum in (2.3.19) is equal to

$$O(t^{-l}) + \frac{a_\tau}{\sqrt{t}}\Big[\Big[\int_{-\infty}^{\infty} \exp\Big\{-\frac{a_\tau}{2(1-\rho^2)}\, Q(s_1, x_1)\Big\} ds_1\Big] + O\Big(\frac{1}{\sqrt{t}}\Big)\Big] + O(t^{-1} \ln^2 t). \qquad (2.3.20)$$

It is clear that the main remainder term in (2.3.20) is $O(t^{-1} \ln^2 t)$.

The second sum in (2.3.17) is estimated in a similar way: it is $O(t^{-3/2})$ and hence can be neglected.

Returning to (2.3.16) and observing that $\beta < 1$, we get

$$P(t, \Delta, W) = (p_t(\mathbf{w}) + \Delta I_{u+v}(\mathbf{w}))O(t^{-\beta}) + \frac{C \Delta a_\tau I_{u+v}(\mathbf{w})}{\sqrt{t}}\, J(x_1, \rho), \qquad (2.3.21)$$

where

$$J(x_1, \rho) := \int_{-\infty}^{\infty} \exp\Big\{-\frac{a_\tau}{2(1-\rho^2)}\, Q(s_1, x_1)\Big\} ds_1. \qquad (2.3.22)$$

The integral $J(x_1, \rho)$ in (2.3.22), (2.3.21) can be computed directly or found in tables of integrals. Note that the following considerations can be also applied.

For a fixed a_τ, the function $J(x_1, \rho)$ in (2.3.22) depends only on x_1 and ρ. We apply (2.3.21) in the case of linearly dependent variables $\zeta = \omega + c\tau$, which was discussed in §1.8. For fixed a_τ, σ_τ, the distribution of ω can always be chosen so that the correlation coefficient between $\xi = \omega + (c - a)\tau = \omega - \frac{a_\omega \tau}{a_\tau}$ and τ would be equal to a given ρ. Therefore, applying (2.3.21) with $\mathbf{w} = \mathbb{R}$ and Theorem 2.1.2 to such (τ, ζ), we see that the main parts of the asymptotics in (2.3.21) and (2.1.3) must coincide. This fact and relation (2.3.21) establish (2.3.3). The bounds for possible values of Δ are given in Lemma 2.3.3. However, relation (2.3.3) shows that it makes sense to consider the case when the range of Δ is narrowed down to the interval $(t^{\frac{1}{2}-\beta}, ct^{\frac{3-r}{2}})$, where $\frac{1}{2} - \beta > 1 - \frac{r}{2}$.

II, III. Extending the assertions of the theorem to the case of arithmetic τ and arithmetic ζ can be done as in Stages II, III of the proof of Theorem 2.1.2. Theorem 2.3.1 is proved. □

Now we will obtain an analog of Theorem 2.3.1 for the processes $Y(t)$. Let

$$I_u = \int_u^\infty \mathbf{P}(\tau > v)dv$$

and let, as before,

$$B_t = B_t(u, v) = \{\gamma(t) \geq u, \chi(t) \geq v\}.$$

Theorem 2.3.4 *Let the conditions of assertion* I *of Theorem* 2.3.1 *be satisfied. Then, as* $t \to \infty$,

$$\frac{1}{\Delta}\mathbf{P}(Y(t) - at \in \Delta[x), B_t) = \frac{I_{u+v}}{a_\tau \sqrt{t}}\, \phi_\sigma\left(\frac{x}{\sqrt{t}}\right)(1 + o(1)) + o(t^{-\beta'}), \tag{2.3.23}$$

for any $\beta' < \beta$, *where* β *was defined in Theorem* 2.3.1, *the remainder terms* $o(1)$, $o(t^{-\beta'})$ *are uniform in all* x, $u \leq \overline{u} = \text{const}$, v, *and also in* Δ *from any compact set bounded away from* 0.

For simplicity, in Theorem 2.3.4 we only consider the case when the values of Δ lie in a compact set. Similarly to Theorem 2.3.1, one can also consider the values of Δ from broader ranges, including the cases of decaying or growing Δ as $t \to \infty$. This, however, will make both the formulations and proofs more complicated.

The proof of Theorem 2.3.4 is based on Theorem 2.3.1. Note that, for $\mathbf{w} = \Delta[w)$,

$$\mathbf{P}(Y(t) - at \in \Delta[x);\ B_t(u, v)) = \int \mathbf{P}(Z(T) \in x - dw;\ \mathbf{B}_t(u, v, \Delta[w)). \tag{2.3.24}$$

For small δ, the integral on the right-hand side of (2.3.24) can be approximated by the sum $\sum_k P_k$, where

$$P_k := \mathbf{P}(Z(T) - at \in \delta[x - w_k);\ \mathbf{B}_t(u, v, \Delta[w_k)),\ w_k = k\delta), \tag{2.3.25}$$

$k = \cdots - 1, 0, 1, \ldots$ Let us use Theorem 2.3.1, but first we note the following fact. Assume, for simplicity, that δ exactly divides Δ. It is easily seen that

$$\delta \sum_k I_u(\Delta[w_k)) = \Delta I_u(\mathbb{R}) = \Delta I_u, \tag{2.3.26}$$

where the measure $I_u(\mathbf{w})$ is defined in Theorem 2.3.1,

$$\sum_k p_t(\Delta[w_k)) = \frac{\Delta}{\delta}, \qquad \frac{1}{\delta}O(t^{-\beta}) = o(t^{-\beta'})$$

for any $\beta' < \beta$ if $\delta \to 0$ sufficiently slowly as $t \to \infty$. Hence by Theorem 2.3.1 the sum of the remainder terms for the terms in (2.3.25) is

$$\Big[\sum_k p_t(\Delta[w_k)) + \delta \sum_k I_{u+v}(\Delta[w_k))\Big]O(t^{-\beta}) = \Delta O(t^{-\beta'}) \tag{2.3.27}$$

for any $\beta' < \beta$.

Let us now consider the sum P_k' (see (2.3.3)) of the principal parts of the terms P_k in (2.3.25). We split it into two subsums $\sum_1$ and $\sum_2$ over the domains

$$\mathcal{N}_1 := \Big\{k : |w_k| < \frac{\sqrt{t}}{\ln t}\Big\} \quad \text{and} \quad \mathcal{N}_2 := \Big\{k : |w_k| \geq \frac{\sqrt{t}}{\ln t}\Big\},$$

respectively. For the second sum. we have

$$\sum\nolimits_2 = \sum_{k \in \mathcal{N}_2} P_k' \leq \frac{\Delta}{\sigma\sqrt{2\pi t}} I_{u+v}(R_t) \tag{2.3.28}$$

(cf. (2.3.26)), where

$$R_t := \Big\{w \in \mathbb{R} : |w| \geq \frac{\sqrt{t}}{\ln t}\Big\}.$$

We have the following estimate:

$$\begin{aligned}\int_0^\infty \mathbf{P}\Big(\tau \geq u,\, \zeta \geq \frac{\sqrt{t}}{\ln t}\Big)\, du &\leq \frac{\sqrt{t}}{\ln t}\mathbf{P}\Big(\zeta \geq \frac{\sqrt{t}}{\ln t}\Big) + \int_{\sqrt{t}/\ln t}^\infty \mathbf{P}(\tau \geq u)\, du \\ &\leq \mathbf{E}|\zeta|^r\Big(\frac{\sqrt{t}}{\ln t}\Big)^{-r+1} + \frac{\mathbf{E}\tau^r}{r-1}\Big(\frac{\sqrt{t}}{\ln t}\Big)^{-r+1}.\end{aligned}$$

The integral $\int_0^\infty \mathbf{P}(\tau \geq u,\, \zeta \leq -\frac{\sqrt{t}}{\ln t})\, du$ is estimated similarly. Taking into account the factor $\frac{\Delta}{\sqrt{t}}$ in (2.3.28), this gives

$$\sum\nolimits_2 = \Delta O\Big(\frac{t^{-r/2}}{(\ln t)^{r-1}}\Big) = \Delta o(t^{-\beta}). \tag{2.3.29}$$

We now consider the first sum $\sum_1$. There are two cases to consider:

(a) $|x| \leq c\sqrt{t}\ln t$,

(b) $|x| \geq c\sqrt{t}\ln t$,

where the number c will be chosen later.

In case (a), $k \in \mathcal{N}_1$, we have

$$\exp\Big\{-\frac{(x-w_k)^2}{2\sigma^2 t}\Big\} = \exp\Big\{-\frac{x^2}{2\sigma^2 t} + O\Big(\frac{1}{\sqrt{\ln t}}\Big)\Big\} = \exp\Big\{-\frac{x^2}{2\sigma^2 t}\Big\}\Big(1 + O\Big(\frac{1}{\sqrt{\ln t}}\Big)\Big),$$

$$\sum\nolimits_1 = \frac{\Delta I_{u+v}}{a_\tau \sigma\sqrt{2\pi t}} e^{-\frac{x^2}{2\sigma^2 t}}(1 + o(1)).$$

In case (b), $k \in \mathcal{N}_1$, we have $x - w_k \sim x$ as $t \to \infty$, and

$$\exp\Big\{-\frac{(x-w_k)^2}{2\sigma^2 t}\Big\} \leq \exp\Big\{-\frac{c^2 \ln t(1+o(1))}{2\sigma^2}\Big\} = O(t^{-1/2})$$

for $c > \sigma\sqrt{2}$. Taking into account the factor $\frac{\delta}{\sqrt{t}}$ on the right of (2.3.3), we get for $\sum_1$ the estimate

$$\sum\nolimits_1 = \Delta O(t^{-1}) = \Delta o(t^{-\beta}). \tag{2.3.30}$$

Now (2.3.23) follows by combining (2.3.27), (2.3.29), and (2.3.30). Theorem 2.3.4 is proved.

2.4 Extension of Results to the Inhomogeneous Case

In this section, we will prove an auxiliary assertion enabling one to extend the results of Theorems 2.1.2, 2.3.1, and 2.3.4 to inhomogeneous CRPs.

In what follows, it will be convenient for us to denote by $(\tau_{(1)}, \zeta_{(1)})$ the inhomogeneous element (τ_1, ζ_1) of the governing random sequence, retaining the notation $Z(t)$, $Y(t)$ for the homogeneous processes that are governed by the sequence $\{\tau_j, \zeta_j\}_{j=1}^{\infty}$ of independent identically distributed random vectors. Correspondingly, the inhomogeneous CRPs will be denoted, as before, by $\widetilde{Z}(t)$, $\widetilde{Y}(t)$, so that, for $\tau_{(1)} > 0$,

$$\widetilde{Z}(t) = \begin{cases} 0 & \text{for} \quad t \in [0, \tau_{(1)}), \\ \zeta_{(1)} + Z(t - \tau_{(1)}) & \text{for} \quad t \geq \tau_{(1)}; \end{cases}$$

$$\widetilde{Y}(t) = \begin{cases} \zeta_{(1)} & \text{for} \quad t \in [0, \tau_{(1)}), \\ \zeta_{(1)} + Y(t - \tau_{(1)}) & \text{for} \quad t \geq \tau_{(1)}, \end{cases}$$

where $Z(t)$, $Y(t)$ are homogeneous CRPs that do not depend on $(\tau_{(1)}, \zeta_{(1)})$ (the case $\tau_{(1)} = 0$ is simpler and hence will not be considered). Further, let $\widetilde{\gamma}(t)$, $\widetilde{\chi}(t)$ be random variables of the form $\gamma(t)$, $\chi(t)$ (see (1.1.1), (1.1.2))), but defined for the inhomogeneous governing sequence $\widetilde{T}_k = \tau_{(1)} + T_{k-1}$, $k = 1, 2, \ldots$, so that, for instance,

$$\widetilde{\chi}(t) \underset{p}{=} \begin{cases} \tau_{(1)} - t, & \text{if} \quad \tau_{(1)} > t, \\ \chi(t - \tau_{(1)}) & \text{if} \quad \tau_{(1)} \leq t, \end{cases}$$

where $\chi(t)$ corresponds to the homogeneous process $\{T_k\}_{k=1}^{\infty}$. We set

$$\widetilde{B}_t = \{\widetilde{\gamma}(t) \geq u, \ \widetilde{\chi}(t) \geq v\}.$$

We will consider a "partial triangular array scheme," meaning that the distributions of the random variables $\tau_{(1)} = \tau_{(1)}(t)$, $\zeta_{(1)} = \zeta_{(1)}(t)$ can depend on the parameter t.

We will say that the random variable $\tau_{(1)} \geq 0$, depending on the parameter t, is *uniformly integrable* if

$$\sup_t \mathbf{E}(\tau_{(1)};\ \tau_{(1)} > N) \to 0 \quad \text{as} \quad N \to \infty.$$

It is easily seen that $\tau_{(1)}$ is uniformly integrable if there exists an integrable function $q(u)$ such that $\mathbf{P}(\tau_{(1)} > u) \leq q(u)$ for all t and u.

Thus, let $\tau_{(1)} \geq 0$, $\zeta_{(1)}$ be given random variables that in general depend on the parameter t.

Theorem 2.4.1 I. *Assume that τ and ζ are independent or linearly dependent, a homogeneous CRP $Y(t)$ satisfies the conditions of Theorem* 2.1.2 *so that the integro-local representation* (2.1.3) *holds true and the random variables $\tau_{(1)} \geq 0$, $|\zeta_{(1)}|$ are uniformly integrable. Then the inhomogeneous process $\widetilde{Y}_1(t)$ also satisfies the integro-local representation* (2.1.3)*, that is, the probability $\mathbf{P}(\widetilde{Y}(t) - at \in \Delta[x);\ \widetilde{B}_t)$ is equal to the right-hand side of* (2.1.3).

II. *Assume that a homogeneous CRP $Z(t)$ satisfies the conditions of Theorem* 2.3.1 *so that the integro-local representation* (2.3.3) *holds true and the random variables $(\tau_{(1)})^{2\beta}$, $|\zeta_{(1)}|^{2\beta}$ are uniformly integrable, where $\beta \in (1/2, 1)$ is the same as in Theorem* 2.3.1. *Then the inhomogeneous process $\widetilde{Z}(t)$ also satisfies the integro-local representation* (2.3.3).

Proof I. By the inequality

$$\mathbf{P}(|\zeta_{(1)}| \geq v) \leq \frac{\mathbf{E}(|\zeta_{(1)}|;\ |\zeta_{(1)}| \geq v)}{v} =: \frac{\delta_v}{v},$$

where $\delta_v \to 0$ as $v \to \infty$, there exists a sequence $\varepsilon = \varepsilon_t \to 0$ as $t \to \infty$ such that

$$\mathbf{P}(|\zeta_{(1)}| > \varepsilon_t \sqrt{t}) = o\Big(\frac{1}{\sqrt{t}}\Big).$$

Similarly, there exists a sequence $\varepsilon_t \to 0$ as $t \to \infty$ such that

$$\mathbf{P}(\tau_{(1)} > \varepsilon_t \sqrt{t}) = o\Big(\frac{1}{\sqrt{t}}\Big).$$

Putting

$$\begin{aligned}
&\tau' = \min(\tau_{(1)}, \varepsilon_t \sqrt{t}\,),\\
&\zeta' = \max\big[\min(\zeta_{(1)}, \varepsilon_t \sqrt{t}\,), -\varepsilon_t \sqrt{t}\,\big],\\
&A = \{\tau_{(1)} < \varepsilon_t \sqrt{t},\ |\zeta_{(1)}| < \varepsilon_t \sqrt{t}\,\}, \qquad B_t = \{\gamma(t) \geq u,\ \chi(t) \geq v\},
\end{aligned}$$

we have $\mathbf{P}(\overline{A}) = o(1/\sqrt{t})$ and

$$\begin{aligned}
&\mathbf{P}(\widetilde{Y}(t) - at \in \Delta[x),\ \widetilde{B}_t)\\
&\quad = \mathbf{P}(\zeta_{(1)} - a\tau_1 + Y(t - \tau_{(1)}) - a(t - \tau_{(1)}) \in \Delta[x), \widetilde{B}_t)\\
&\quad = o\Big(\frac{1}{\sqrt{t}}\Big) + \mathbf{P}(\zeta' - a\tau' + Y(t - \tau') - a(t - \tau') \in \Delta[x), B_{t-\tau'} A)\\
&\quad = o\Big(\frac{1}{\sqrt{t}}\Big) + \int_0^{\varepsilon_t t} \int_{-\varepsilon_t \sqrt{t}}^{\varepsilon_t \sqrt{t}} \mathbf{P}(Y(t-\theta) - a(t-\theta) \in \Delta[x - y + a\theta), B_{t-\theta})\\
&\qquad \times \mathbf{P}(\tau' \in d\theta, \zeta' \in dy). \qquad (2.4.1)
\end{aligned}$$

But for any θ and y from the domain of integration in (2.4.1), as $t \to \infty$, one has $t - \theta \sim t$,

$$\frac{(x-y+a\theta)^2}{2\sigma^2(t-\theta)} \quad \begin{cases} = o(1) & \text{if} \quad |x| = o(\sqrt{t}), \\ = \frac{x^2}{2\sigma^2 t} + o(1) & \text{if} \quad |x| \text{ is comparable to } \sqrt{t}, \\ \to \infty & \text{if} \quad |x| \gg \sqrt{t}. \end{cases}$$

Hence the integrand on the right of (2.4.1) has the same form as the right-hand side in (2.1.3). Therefore, after integrating with respect to the distribution (τ', ζ') in (2.4.1), we will again obtain a representation of the form (2.1.3).

II. The proof of the second part of the assertion of the theorem requires a few more precise considerations. Since $(\tau_{(1)})^{2\beta}$, $|\zeta_{(1)}|^{2\beta}$ are integrable, for some sequence $\varepsilon_t \to 0$ as $t \to \infty$ we have

$$\mathbf{P}(\tau_{(1)} > \varepsilon_t \sqrt{t}\,) = o(t^{-\beta}),$$
$$\mathbf{P}(|\zeta_{(1)}| > \varepsilon_t \sqrt{t}) = o(t^{-\beta}),$$

so that $\mathbf{P}(\overline{A}) = o(t^{-\beta})$. In view of (2.3.3), the first factor in the integrand in (2.4.1) has the form

$$\frac{c}{\sqrt{t-\theta}}\, e^{-\frac{(x-y+a\theta)^2}{2\sigma^2(t-\theta)}} + O((t-\theta)^{-\beta}), \tag{2.4.2}$$

where c does not depend on x, y, t and θ, and $\theta = o(\sqrt{t}\,)$, $y = o(\sqrt{t})$ (for simplicity, we will only consider the case when $\Delta \le \text{const}$). We set

$$e(x,t) = e^{-\frac{x^2}{2\sigma^2 t}}.$$

Hence as in (2.3.13) we have, as $t \to \infty$,

$$\frac{\partial e(x,t)}{\partial x} = c_1 e(x,t)\,\frac{x}{t} = O\Big(\frac{1}{\sqrt{t}}\Big),$$
$$\frac{\partial e(x,t)}{\partial t} = c_2 e(x,t)\,\frac{x^2}{t^2} = O\Big(\frac{1}{t}\Big),$$

where c_1, c_2 do not depend on x, t, and the terms $O(\cdot)$ are uniform in x. Besides,

$$\frac{1}{\sqrt{t-\theta}} = \frac{1}{\sqrt{t}} + O\Big(\frac{\theta}{t^{3/2}}\Big) \quad \text{as} \quad t \to \infty.$$

Hence the expression in (2.4.2) is equal to

$$\frac{c}{\sqrt{t}}\, e^{-\frac{x^2}{2\sigma^2 t}} + O(t^{-\beta}) + |y| O\Big(\frac{1}{t}\Big) + \theta O\Big(\frac{1}{t}\Big).$$

Since the expectations $\mathbf{E}\tau'$ and $\mathbf{E}\zeta'$ are finite, the integral in (2.4.1) is equal to

$$\frac{c}{\sqrt{t}}\, e^{-\frac{x^2}{2\sigma^2 t}} + O(t^{-\beta}) + O\Big(\frac{1}{t}\Big) \quad \text{as} \quad t \to \infty.$$

This establishes (2.3.3) for the process $\widetilde{Z}$. Theorem 2.4.1 is proved. □

Using (2.4.1) from the proof of Theorem 2.4.1 one can easily check that in the case $a = 0$ the conclusion of Theorem 2.4.1 remains valid if the conditions of uniform integrability of $\tau_{(1)}$ in assertion I and of $\tau_{(1)}^{2\beta}$ in assertion II are replaced by the conditions of uniform integrability of $\sqrt{\tau_{(1)}}$ and $\tau_{(1)}^{\beta}$, respectively.

As was already pointed out, the "partial triangular array scheme" for inhomogeneous CRPs, when the distributions of $\tau_1 = \tau_{(1)}, \zeta_1 = \zeta_{(1)}$ depend on t, appears when, for example, we are studying the behavior of the increments $Y_u(t) = Y(u+t) - Y(u)$ as $t \to \infty$, $u = u(t) \to \infty$, which can be looked upon as an inhomogeneous process with initial jumps distributed as $\chi(u)$, $\zeta(u)$. In this case, according to Theorem 2.4.1, under the conditions of, say, assertion I, one has to assume the uniform integrability of $\sqrt{\chi(u)}$ and $|\zeta(u)|$, which always takes place provided that $\mathbf{E}\tau^{3/2} < \infty$, $\mathbf{E}\zeta^2 < \infty$.

If the distribution of (τ, ζ) satisfies the Cramér moment condition, then integro-local theorems can be derived in a wider deviation zone, which involves in particular the large deviation zone. Chapter 5 is devoted to this topic.

2.5 Integro-Local Theorems for Markov Additive Processes

Markov additive process X_n (MAPs) were defined in §1.8, in which the principal "integral" limit laws for MAPs in the normal deviation zone were established. Let us now derive integro-local theorems.

In this section, keeping all the notation of §1.8, we consider only "homogeneous" processes with $\mathrm{x}_0 = \mathrm{e}_0$ and cycles generated by returns to the state e_0. The passage to inhomogeneous processes is carried out in the same way as for inhomogeneous CRPs.

Theorem 2.5.1 *Let a random variable ζ either satisfy the Cramér condition* (2.3.1) *on the characteristic function or be arithmetic, and let*

$$\mathbf{E}\tau^3 < \infty, \quad \mathbf{E}|\zeta|^3 < \infty, \quad \mathbf{E}\widehat{\zeta}^3 < \infty. \tag{2.5.1}$$

Then

I. *The asymptotics*

$$\mathbf{P}(\mathrm{X}_n - an \in \Delta[x)) = \frac{\Delta}{\sqrt{n}}\,\phi_\sigma\left(\frac{x}{\sqrt{n}}\right) + o(n^{-1/2}) \tag{2.5.2}$$

holds uniformly in x and in Δ from the interval $(n^{-\varepsilon}, n^{-\varepsilon})$ with some $\varepsilon > 0$. If ζ is arithmetic, then in (2.5.2) $\Delta = 1$ *and x is integer.*

II. *If $\{\mathrm{x}_n\}$ is finite irreducible aperiodic Markov chain, then $\mathbf{P}(\tau > n) < cq^n$ with some $q < 1$ and all n. If, in addition,*

$$\max_{j \le m} \mathbf{E}|\xi(\mathrm{e}_j)|^3 < \infty, \tag{2.5.3}$$

where $\mathrm{e}_0, \ldots, \mathrm{e}_m$ are the states of the chain, then (2.5.1) *and* (2.5.2) *hold.*

Condition (2.5.1) can be relaxed as follows: the exponent 3 can be replaced by some value $r \in (1+\sqrt{2}, 3)$ (cf. Theorem 2.3.1).

Proof of Theorem 2.5.1 We have

$$\mathrm{X}_n = Z(n) + \rho_n,$$

where ρ_n is defined in (1.8.9). Consider the probability

$$P_{n,x} := \mathbf{P}(\mathrm{X}_n - an \in \Delta[x)) = \mathbf{P}(Z(n) - an + \rho_n \in \Delta[x)).$$

For any $b > 1/a_\tau$,

$$\mathbf{P}(\eta(n) > bn) = \mathbf{P}(T_{bn} < n) < \frac{bn\sigma_\tau^2}{((a_\tau b - 1)n)^2} = o\Big(\frac{1}{n}\Big)$$

as $n \to \infty$. Next, for $\varepsilon_n = \overline{o}(1)$, $n \to \infty$, we get

$$\begin{aligned}
\mathbf{P}(|\rho_n| > \varepsilon_n\sqrt{n}) &= o\Big(\frac{1}{n}\Big) + \mathbf{P}(|\rho_n| > \varepsilon_n\sqrt{n};\ \eta(n) \le bn) \\
&\le o\Big(\frac{1}{n}\Big) + \mathbf{P}(\max_{j \le bn} \widehat{\zeta}_j > \varepsilon_n\sqrt{n}) \le o\Big(\frac{1}{n}\Big) + bn\mathbf{P}(\widehat{\zeta} > \varepsilon_n\sqrt{n}) \\
&\le bno\Big(\frac{1}{(\varepsilon_n\sqrt{n})^3}\Big) = o(n^{-1/2}), \qquad (2.5.4)
\end{aligned}$$

where $\widehat{\zeta}_j$ are independent copies of $\widehat{\zeta}$. Hence, as $n \to \infty$,

$$\begin{aligned}
P_{n,x} &= \mathbf{P}(Z(n) - an \in \Delta[x - \rho_n);\ |\rho_n| \le \varepsilon_n\sqrt{n}) + o(n^{-1/2}) \\
&= \mathbf{P}(Z(n) - an \in \Delta[x - r_n);\ |\rho_n| \le \varepsilon_n\sqrt{n}) + o(n^{-1/2}),
\end{aligned}$$

where

$$r_n = \begin{cases} \rho_n & \text{if} \quad |\rho_n| < \varepsilon_n\sqrt{n}, \\ -\varepsilon_n\sqrt{n} & \text{if} \quad \rho_n \le -\varepsilon_n\sqrt{n}, \\ \varepsilon_n\sqrt{n} & \text{if} \quad \rho_n \ge \varepsilon_n\sqrt{n}. \end{cases}$$

Hence from (2.5.4) it follows that, as $n \to \infty$,

$$P_{n,x} = \mathbf{P}(Z(n) - an \in \Delta[x - r_n)) + o(n^{-1/2}).$$

We now note that, for a fixed l, all the above arguments can be applied also to the probability

$$\begin{aligned}
P_{n,x,l} :&= \mathbf{P}(\mathrm{X}_n - an \in \Delta[x), \gamma(n) = l) \\
&= \mathbf{P}(Z(n) - an \in \Delta[x), \gamma(n) = l) + o(n^{-1/2}).
\end{aligned}$$

Moreover, the random variables ρ_n, r_n are conditionally independent of $Z(n)$ provided $\gamma(n) = l$. Hence by Corollary 2.3.2,

$$\mathbf{P}(Z(n) - an \in \Delta[x - r_n) \,|\, \gamma(n) = l, r_n) = \frac{\Delta}{\sqrt{n}}\phi_\sigma\Big(\frac{x - r_n}{\sqrt{n}}\Big) + o(n^{-1/2}).$$

Here

$$\phi_\sigma\left(\frac{x - r_n}{\sqrt{n}}\right) = \begin{cases} \phi_\sigma\left(\frac{x}{\sqrt{n}}\right)(1 + o(1)) & \text{for } |x| < \sqrt{n}\ln n, \\ o(1) & \text{for } |x| \geq \sqrt{n}\ln n, \end{cases} \tag{2.5.5}$$

where the asymptotics given by the right-hand side of (2.5.5) does not depend on r_n. Therefore,

$$P_{n,x,l} = \frac{\Delta}{\sqrt{n}}\,\phi_\sigma\left(\frac{x}{\sqrt{n}}\right) + o(n^{-2/3}).$$

Multiplying $P_{n,x,l}$ by $\mathbf{P}(\gamma(n) = l)$ and summing with respect to l from $l = 0$ to $l = N$, we get

$$\mathbf{P}(\mathrm{X}_n - an \in \Delta[x), \gamma(n) \leq N) = \mathbf{P}(\gamma \leq N)\left[\frac{\Delta}{\sqrt{n}}\,\phi_\sigma\left(\frac{x}{\sqrt{n}}\right) + o(n^{-1/2})\right],$$

where

$$\mathbf{P}(\gamma \leq N) = \lim_{n\to\infty} \mathbf{P}(\gamma(n) \leq N) = a_\tau^{-1} \sum_{l=0}^{N} \mathbf{P}(\tau > l) \to 1$$

as $N \to \infty$. As a result, we get (2.5.2).

II. The inequality

$$\mathbf{P}(\tau > n) < cq^n, \quad q < 1, \quad c = \text{const}, \tag{2.5.6}$$

for the regeneration cycle length for any fixed state of an ergodic finite Markov chain is well known.

Next consider the random variable $\overline{\xi}$ with the distribution

$$\mathbf{P}(\overline{\xi} \geq t) = \max_{j\leq m} \mathbf{P}\big(|\xi(\mathrm{e}_j)| \geq t\big).$$

We have $\mathbf{P}(\overline{\xi} \geq t) \leq \sum_{j\leq m} \mathbf{P}\big(|\xi(\mathrm{e}_j)| \geq t\big)$, and hence by (2.5.3), it is clear that $\mathbf{E}\overline{\xi}^3 < \infty$. Next, on the event $\{\tau = l\}$ we have

$$\widehat{\zeta} \underset{p}{\leq} \sum_{k=1}^{l} \overline{\xi}_k,$$

where $\overline{\xi}_k$ are independent copies of $\overline{\xi}$,

$$\mathbf{E}(\widehat{\zeta}^3 \mid \tau = l) \leq l^3 \mathbf{E}\overline{\xi}^3.$$

Now (2.5.1) and (2.5.2) follow from (2.5.6). Theorem 2.5.1 is proved. □

3

Large Deviation Principles for Compound Renewal Processes

3.1 Introduction

We have seen in §§1.2–1.7 that distributions of CRPs behave on large time intervals in the *normal* deviation zone in much the same way as those of classical random walks. For example, in the case $\sigma < \infty$, after normalization $Z(T) - aT$ has the same limit distribution as the sums $S_n = \sum_{j=1}^{n} \xi_j$ of the random variables $\xi_j \underset{d}{=} \xi = \zeta - a\tau$ as $n = [T/a_\tau] \to \infty$.

Things get much more complicated in the zone of *large* deviations. The reason is not only that the CRP is governed by two sequences of random variables $\{\tau_j\}$ and $\{\zeta_j\}$ (the coordinates of the vector (τ, ζ) can be dependent), but also that in many cases these coordinates make the CRP deviate "in different directions." For example, for independent τ and ζ, in the case $a = 0$ large values of ζ_j promote large deviations of $Z(T)$, while large values of τ_j promote small deviations (for $\tau_1 > T$, there are no deviations from the line $z = at = 0$ at all). At the same time, in the moderately large deviation zone, the well-known analogy with the behavior of classical random walks (of the sequence $\{S_n\}$; see §4.8 and Chapter 7) is still present.

In this chapter, it will be assumed that the Cramér's moment condition is satisfied:

[C] $\mathbf{E}e^{\lambda\tau+\mu\zeta} < \infty$ *in some neighborhood of the origin* $|\lambda| + |\mu| < v, v > 0$. *The same condition holds for* (τ_1, ζ_1).

Problems of large deviations for random processes have been extensively studied. For basic approaches to the derivation of the large deviation principles (LDPs) for random processes of general form and the survey of the available results, see, e.g., [64] (and the references given there). In our case, in studying CRPs, more effective and more constructive are direct approaches based on the knowledge of the specific nature of the processes (as defined in (3)). They enable us to obtain much more advanced results including the explicit form of the deviation rate function in the form of the Legendre transform of the fundamental function (see §3.5).

For Markov additive processes (sums of random variables defined on states or transitions of a Markov chain) related results were obtained in [95, 96]. If a chain is a Harris chain (a φ-irreducible chain), then the trajectories of such processes can be split into independent cycles generated by returns of the chain to a positive atom (possibly artificial; see §1.8). These cycles form a CRP, which in many cases has the principal contribution into the sums under consideration. However, it should be noted that the cycle-generated CRPs form only a certain special subclass of general CRPs (not all possible joint distributions of τ and ζ can be realized via cycles). Note that some results in the above sources involve conditions whose meaning is sometimes unclear.

If a Markov chain is finite, then LDPs for cycle-generated CRPs are studied in [82] for a slightly more general object (the vectors (τ_j, ζ_j) are defined on transitions of a Markov chain) with the help of the results from [95, 96].

Let us return to CRPs of general form. In what follows, we shall frequently use the phrase "$\Delta = \Delta_T$ *converges sufficiently slowly to* 0 *as* $T \to \infty$." To shorten this phrase, we will write

$$\Delta = \Delta_T = \overline{o}(1) \quad \text{as} \quad T \to \infty. \tag{3.1.1}$$

The phrase "there exists a $\Delta = \Delta_T = \overline{o}(1)$ such that a relation A (depending on Δ) holds" means that:

(1) there exists a sequence $\Delta = \Delta_T \to 0$ as $T \to \infty$ such that relation A holds;
(2) the relation A remains valid for any sequence $\Delta = \Delta'_T \geq \Delta_T$, $\Delta'_T \to 0$;
(3) if $\Delta'_T < \Delta_T$, then the relation A may fail to hold.

For example, by the law of large numbers $|\frac{Z(T)}{T} - a| \underset{p}{\to} 0$ as $T \to \infty$. It follows that there exists an $\varepsilon = \varepsilon_T = \overline{o}(1)$ such that

$$\mathbf{P}\left(\left|\frac{Z(T)}{T} - a\right| > \varepsilon_T\right) \to 0 \quad \text{as} \quad T \to \infty.$$

It is clear that this relation holds for all $\varepsilon_T \to 0$, $\varepsilon \gg T^{-1/2}$ if $\sigma < \infty$ and fails to hold if the sequence ε_T converges to 0 faster than $T^{-1/2}$.

Recall that under the Cramér's moment condition on the distribution of ξ, in accordance with the local LDP for S_n, for $x = \alpha n$ and $n \to \infty$, there exists a $\Delta = \Delta_n = \overline{o}(1)$ such that

$$\frac{1}{n} \ln \mathbf{P}\left(\frac{S_n}{n} \in \Delta[\alpha)\right) = -\Lambda^{(\xi)}(\alpha) + o(1), \tag{3.1.2}$$

where $\Delta[\alpha) = [\alpha, \alpha + \Delta)$, $\Lambda^{(\xi)}(\alpha)$ is the deviation function (rate function) corresponding to the random variable ξ, that is, the Legendre transform

$$\Lambda^{(\xi)}(\alpha) = \sup_{\mu} \{\alpha\mu - A^{(\xi)}(\mu)\} \tag{3.1.3}$$

of the function $A^{(\xi)}(\mu) = \ln \psi^{(\xi)}(\mu)$, $\psi^{(\xi)}(\mu) = \mathbf{E}e^{\mu\xi}$.

It turns out that CRPs also satisfy the analog of (3.1.2) (that is, a local LDP), according to which the limit

$$\lim_{T\to\infty} \frac{1}{T} \ln \mathbf{P}\Big(\frac{Z(T)}{T} \in \Delta[\alpha)\Big)$$

exists, where $\Delta = \Delta_T = \overline{o}(1)$ as $T \to \infty$. This limit, in accordance with our statements that follow, we denote by $\widehat{D}(\alpha)$. This limit is much more complicated than the deviation function $\Lambda^{(\xi)}(\alpha)$. A remarkable fact is that the function $\widehat{D}(\alpha)$ is the Legendre transform

$$\widehat{D}(\alpha) = \sup_{\mu} \{\mu\alpha - \widehat{A}(\mu)\}$$

(cf. (3.1.3)) of the so-called fundamental function $\widehat{A}(\mu)$, which shares many (but not all) properties of the logarithm of the Laplace transform of the distribution of some random variable, and besides,

$$\widehat{A}(\mu) = \lim_{T\to\infty} \frac{1}{T} \ln \mathbf{E}^{\mu Z(T)}. \tag{3.1.4}$$

For example, the function $\widehat{A}(\mu)$ is convex, $\widehat{A}(0) = 0$,

$$\begin{aligned} \widehat{A}'(0) &= a \qquad \Big(\sim \frac{\mathbf{E}Z(T)}{T} \quad \text{as} \quad T \to \infty\Big), \\ \widehat{A}''(0) &= \sigma^2 \qquad \Big(\sim \frac{\mathbf{D}Z(T)}{T} \quad \text{as} \quad T \to \infty\Big), \end{aligned} \tag{3.1.5}$$

$\widehat{A}(\mu) \to \infty$ as $|\mu| \to \infty$ if the random variable ζ takes values of both signs (that is, $\mathbf{P}(\zeta \gtrless 0) > 0$) and so on. Moreover, the function $\widehat{A}(\mu)$ can be found "explicitly" in terms of the Laplace transform of the distribution of (τ, ζ). Namely, if we set

$$\mathbf{A}(\lambda, \mu) = \ln \mathbf{E} e^{\lambda\tau + \mu\zeta}$$

and consider the "generalized" solution $A(\mu)$ of the equation

$$A(-\lambda, \mu) = 0$$

for λ defining $A(\mu)$ as

$$A(\mu) = \inf \{\lambda : \mathbf{A}(-\lambda, \mu) \le 0\},$$

then

$$\widehat{A}(\mu) = \max\,(A(\mu), -\lambda_+), \qquad \lambda_+ = \sup\{\lambda : \psi^{(\tau)}(\lambda) = \mathbf{E}e^{\lambda\tau} < \infty\}$$

and hence one can single out the domain of analyticity of the function $\widehat{A}(\mu)$ in which $\widehat{A}(\mu) = A(\mu)$ is the "exact" solution for λ of the equation

$$\mathbf{A}(-\lambda, \mu) = 0.$$

Nevertheless, "despite" properties (3.1.4), (3.1.5), there do not generally exist a distribution of probabilities for which the function $\widehat{A}(\mu)$ would be the logarithm of the Laplace transform. In other words, roughly speaking, in the large deviation zone the CRPs $Z(T)$ behave as $T \to \infty$ as sums of some "pseudorandom" variables (see properties (3.1.4), (3.1.5) and §3.5 below) corresponding to the "transform" $\widehat{A}(\mu)$.

In the present chapter, we study the nature of the distributions of CRPs in the large deviation zone and establish the corresponding LDPs. In §3.2, we clarify the relation of the distribution of a CRP with the renewal measure for the sequence $\{T_n, Z_n\}$. Section 3.3 is devoted to the investigation of some properties of the deviation functions of the renewal measures that appear in this problem. In §3.4, we prove the LDPs for the CRPs $Z(t)$. The definition of the fundamental function is given in §3.5, where we also study its properties and relations to the deviation functions. In §3.6, we present several results on LDPs for the process $Y(t)$ and for Markov additive processes. Assertion (3.1.4) will be proved in §3.7. Moderately large deviation principles will be established later, in §4.8.

3.2 Relationship between Compound Renewal Processes and the Renewal Measure. Deviation Function for the Renewal Measure

3.2.1 Renewal Measures and CRPs

In §1.1, we introduced the renewal function $H(t)$ corresponding to the sequence $\{T_k\}$,

$$H(t) = \mathbf{E}\eta(t) = \sum_{k=0}^{\infty} \mathbf{P}(T_k \le t).$$

The natural analog of this concept in the multidimensional case is the *renewal measure*. As in §§2.1–2.4, by

$$\mathbf{S}_n = (T_n, Z_n) = \sum_{j=1}^{n} \boldsymbol{\xi}_j \tag{3.2.1}$$

we denote sums of vectors $\boldsymbol{\xi}_j = (\tau_j, \zeta_j)$, $j = 1, 2, \ldots$. The renewal measure $H(B)$ corresponding to a *homogeneous* sequence $\{\mathbf{S}_n\}$ is defined by

$$H(B) = \sum_{n=0}^{\infty} \mathbf{P}(\mathbf{S}_n \in B), \tag{3.2.2}$$

where B is a measurable set from $\mathbb{R}^2$.

If a sequence $\{\boldsymbol{\xi}_j\}$ is *inhomogeneous* ($\boldsymbol{\xi}_1 \underset{d}{\neq} \boldsymbol{\xi}$), then the corresponding renewal measure will be denoted by $\widetilde{H}(B)$. As measurable sets B we will frequently consider the rectangles $B_{t,x} = \delta[t) \times \Delta[x)$, $t \ge 0$, $x \in \mathbb{R}$; in this case, we will write $H(B_{t,x}) = H(\delta[t), \Delta[x))$.

Since in parallel with the distribution of $Z(t)$ in the state space we will be also interested in the distribution of the *trajectories* of the CRP $Z(t)$, it will be convenient, as in §§1.6, 1.7, to consider the process $Z(\cdot)$ on an increasing time interval $[0, T]$,

$T \to \infty$, and study the distribution of the normalized processes $\frac{Z(tT)}{T}$, but now for $t \in [0, 1]$.

We shall start with rough integro-local theorems for the values of $\frac{Z(T)}{T}$ or, equivalently, with the *local* LDP for CRPs in the state space.

By its very nature, a CRP is closely related to the renewal measure, because by the total probability formula,

$$\mathbf{P}(Z(T) \in T\Delta[\alpha)) = \int_0^T \sum_{n=0}^{\infty} \mathbf{P}(T_n \in du, Z_n \in T\Delta[\alpha), \ \tau_{n+1} > T - u). \qquad (3.2.3)$$

If $\alpha \notin (-\Delta, 0]$, then $\mathbf{P}(Z_0 \in T\Delta[\alpha)) = 0$ and

$$\mathbf{P}(Z(T) \in T\Delta[\alpha)) = \int_0^T H_1(du, T\Delta[\alpha))\mathbf{P}(\tau > T - u), \qquad (3.2.4)$$

where

$$H_1(B) = \sum_{n=1}^{\infty} \mathbf{P}(\mathbf{S}_n \in B).$$

If $\alpha \in (-\Delta, 0]$, then on the right-hand side of (3.2.3) there appears the first term $\mathbf{P}(\tau_1 > T)$ corresponding to $n = 0$, so that, for all α,

$$\mathbf{P}(Z(T) \in T\Delta[\alpha))$$
$$= \mathbf{1}_{\{\alpha\in(-\Delta,0]\}}\mathbf{P}(\tau_1 > T) + \int_0^T H_1(du, T\Delta[\alpha))\mathbf{P}(\tau > T - u). \quad (3.2.5)$$

If a sequence $\{\xi_k\}$ is *inhomogeneous* ($\xi_1 \underset{d}{\neq} \xi$), then the measure $H_1(B)$ (which we also call the renewal measure, like the measure $H(B)$ in (3.2.2)) will be denoted by $\widetilde{H}_1(B)$.

We first consider the *homogeneous* case.

To describe the asymptotic behavior of the integral in the right-hand side of (3.2.4), (3.2.5), we will approximate this integral using sums involving the values of the renewal measure H_1 for large rectangles: $H_1(T\delta[u_k), T\Delta[\alpha))$ for small δ and Δ, $u_k = k\delta$, $\alpha = x/T$, so that $T\Delta[\alpha) = [x, x + T\Delta)$. More precisely, we employ the following inequalities, which estimate the required integral from above and below:

$$\sum_{k=0}^{N-1} H_1(T\delta[u_k), T\Delta[\alpha))\mathbf{P}(\tau > T - u_k)$$
$$\leq \int_0^T H_1(du, T\Delta[\alpha))\mathbf{P}(\tau > T - u)$$
$$\leq \sum_{k=0}^{N-1} H_1(T\delta[u_k), T\Delta[\alpha))\mathbf{P}(\tau > T - u_{k+1}); \qquad (3.2.6)$$

here we assume for simplicity that the number $N = 1/\delta$ is integer. We will also need the rough asymptotics of the products

$$H_1(T\delta[t), T\Delta[\alpha))\mathbf{P}(\tau > T(1-t)) \quad \text{for} \quad t \in [0,1), \tag{3.2.7}$$

where, as before,

$$H_1(T\delta[t), T\Delta[\alpha)) = \sum_{n=1}^{\infty} \mathbf{P}(T_n \in T\delta[t), Z_n \in T\Delta[\alpha)), \tag{3.2.8}$$

and hence, we shall also need the rough asymptotics of the renewal measure H_1 for the sets

$$T(\delta[t) \times \Delta[\alpha)) = [tT, tT + \delta T) \times [\alpha T, \alpha T + \Delta T). \tag{3.2.9}$$

3.2.2 Asymptotics for the Renewal Measure and the Corresponding Deviation Function

As was already pointed out, we assume everywhere in this chapter that the Cramér condition [**C**] is met. Together with condition [**C**], we shall sometimes use the following stronger condition for τ (for ζ):

$$[\mathbf{C}_\infty] \qquad \begin{aligned} \mathbf{E}e^{\lambda\tau} &< \infty \quad \text{for all } \lambda, \\ (\mathbf{E}e^{\mu\zeta} &< \infty \quad \text{for all } \mu) \end{aligned}$$

and write, respectively, $\tau \in [\mathbf{C}_\infty]$ ($\zeta \in [\mathbf{C}_\infty]$).

We set

$$\begin{aligned} \mathbf{A}(\lambda,\mu) &:= \ln \mathbf{E}e^{\lambda\tau+\mu\zeta}, \\ \mathbf{A}_1(\lambda,\mu) &:= \ln \mathbf{E}e^{\lambda\tau_1+\mu\zeta_1}, \\ \mathcal{A} &:= \{(\lambda,\mu) : \mathbf{A}(\lambda,\mu) < \infty\}, \\ \mathcal{A}_1 &:= \{(\lambda,\mu) : \mathbf{A}_1(\lambda,\mu) < \infty\}. \end{aligned} \tag{3.2.10}$$

In accordance with condition [**C**], the interiors $(\mathcal{A})$, $(\mathcal{A}_1)$ of the sets $\mathcal{A}$, $\mathcal{A}_1$ contain the point $(\lambda, \mu) = (0, 0)$. We shall need the known integro-local theorems and the local LDP for the sums $\mathbf{S}_n$ (see [23], §2.9). In the study of LDPs for the sums $\mathbf{S}_n$, an important role is played by the *deviation function*

$$\mathbf{\Lambda}(v,\alpha) := \sup_{(\lambda,\mu)} \{\lambda v + \mu\alpha - \mathbf{A}(\lambda,\mu)\}$$

corresponding to a random vector $\boldsymbol{\xi} = (\tau, \zeta)$. This is the Legendre transform of a convex lower-continuous function $\mathbf{A}(\lambda,\mu)$, and hence the function $\mathbf{\Lambda}(v,\alpha)$ is also convex and lower-continuous. The deviation function is studied as a probabilistic characteristic, for example, in §1.1 of [23] and in §9.1 of [21].

Regarding notation (3.2.1), (3.2.10), we note the following. Some symbols, for example, S_n, ξ_j and some other ones will be used to denote both scalar and vector variables. To make them easier to distinguish, vector variables will be denoted by bold letters. Similarly, A, Λ and some other symbols will be used to denote functions of one or two variables. For convenience of notation, functions of two variables will be denoted by bold letters.

Let us return to the problem of asymptotic behavior of the renewal measure (3.2.8). Rough asymptotics of the measure (3.2.8) has been studied very thoroughly in [23], §2.9 (see also the references given there). Here we will briefly reproduce the arguments that lead naturally to the appearance of the corresponding deviation function (the so-called second deviation function), which describes the asymptotics of the logarithm of the renewal measure. The logarithmic asymptotics of the terms in the sum on the right of (3.2.8) is known from the local LDP for $\mathbf{S}_n$. Namely, in the *homogeneous* case,

$$\ln \mathbf{P}(T_n \in T\delta[t), Z_n \in T\Delta[\alpha)) = -n\mathbf{\Lambda}\Big(\frac{Tt}{n}, \frac{T\alpha}{n}\Big) + o(n) = -T\theta\mathbf{\Lambda}\Big(\frac{t}{\theta}, \frac{\alpha}{\theta}\Big) + o(\theta T),$$

where $\theta = \frac{n}{T}$, $\delta = \bar{o}(1)$, $\Delta = \bar{o}(1)$ as $n \to \infty$ (see, e.g., §2.8 in [23]). Hence it is natural to expect that the exponential part of the renewal measure $H_1(T\delta[t), T\Delta[\alpha))$ as $T \to \infty$ has the form

$$e^{-T\mathbf{D}_\Lambda(t,\alpha)+o(T)}, \tag{3.2.11}$$

where

$$\mathbf{D}_\Lambda(t,\alpha) := \inf_{\theta>0} \theta\mathbf{\Lambda}\Big(\frac{t}{\theta}, \frac{\alpha}{\theta}\Big), \tag{3.2.12}$$

so that

$$\mathbf{D}_\Lambda(t,\alpha) \le \mathbf{\Lambda}(t,\alpha). \tag{3.2.13}$$

This is the deviation function for the renewal measure (the *second deviation function*), which plays the crucial role in describing the asymptotics of the renewal measure. The following properties of the functions $\mathbf{D}_\Lambda(t,\alpha)$ were established in [23], §2.9: this function is *nonnegative* and *is linear along any ray*, that is, for $v > 0$ and any t, α,

$$\mathbf{D}_\Lambda(vt, v\alpha) = v\mathbf{D}_\Lambda(t,\alpha) \tag{3.2.14}$$

(the ruled property; this property clearly follows from definition (3.2.12)). We have $\mathbf{\Lambda}(t,\alpha) = 0$ for $(t,\alpha) = (a_\tau, a_\zeta) = (\mathbf{E}\tau, \mathbf{E}\zeta)$, and hence, for all $v > 0$,

$$\mathbf{D}_\Lambda(va_\tau, va_\zeta) = v\mathbf{D}_\Lambda(a_\tau, a_\zeta) = 0. \tag{3.2.15}$$

The function $\mathbf{D}_\Lambda(t,\alpha)$ is *convex*, that is, for any $p \ge 0, q \ge 0, p+q = 1, (t,\alpha), (u,\beta) \in \mathbb{R}^2$,

$$\mathbf{D}_\Lambda(pt + qu, p\alpha + q\beta) \le p\mathbf{D}_\Lambda(t,\alpha) + q\mathbf{D}_\Lambda(u,\beta). \tag{3.2.16}$$

From the convexity (3.2.16) and the ruled property (3.2.14) it follows that $\mathbf{D}_\Lambda(t,\alpha)$ is *subadditive*,

$$\mathbf{D}_\Lambda(t+u, \alpha+\beta) \le \mathbf{D}_\Lambda(t,\alpha) + \mathbf{D}_\Lambda(u,\beta). \tag{3.2.17}$$

We denote by $\mathcal{D}^{<\infty} := \{(t,\alpha) : \mathbf{D}_\Lambda(t,\alpha) < \infty\}$ the effective domain of the function $\mathbf{D}_\Lambda(t,\alpha)$. Being convex, the function $\mathbf{D}_\Lambda(t,\alpha)$ is continuous inside $\mathcal{D}^{<\infty}$.

Outside $\mathfrak{D}^{<\infty}$, this function is ∞, and on the boundary $\partial\mathfrak{D}^{<\infty}$ of the set $\mathfrak{D}^{<\infty}$ it may have points of discontinuity.

As was already pointed out, it is natural to expect that the function $\mathbf{D}_\Lambda(t,\alpha)$ will describe the asymptotics of the function

$$-\frac{1}{T}\ln H_1(T\delta[t], T\Delta[\alpha]) \quad \text{as } T \to \infty.$$

However, this is not entirely so, because the following function of t and α,

$$-\lim_{T\to\infty}\frac{1}{T}\ln H_1(T\delta[t], T\Delta[\alpha]),$$

is lower-continuous (see also §4.1 in [23]), while the function $\mathbf{D}_\Lambda(t,\alpha)$, as defined in (3.2.12), is not in general lower-continuous. Indeed, if, for example, $\mathbf{P}(\zeta > 0) = 1$, then $\Lambda(0,0) = \infty$, and hence, $\mathbf{D}_\Lambda(0,0) = \infty$. However, by (3.2.15),

$$\lim_{v\to 0}\mathbf{D}_\Lambda(va_\tau, va_\zeta) = 0 < \mathbf{D}_\Lambda(0,0) = \infty,$$

that is, the function $\mathbf{D}_\Lambda$ is not lower-continuous at the point $(t,\alpha) = (0,0)$. Now if we "adjust" the function $\mathbf{D}_\Lambda(t,\alpha)$ at the points of the boundary $\partial\mathfrak{D}^{<\infty}$ (that is, at the points of discontinuity of the function $\mathbf{D}_\Lambda$ if such points exist), replacing it by the lower-continuous function

$$\mathbf{D}(t,\alpha) := \lim_{\varepsilon\to 0}\mathbf{D}_\Lambda((t,\alpha)_\varepsilon), \text{ where } \mathbf{D}_\Lambda(B) := \inf_{(u,\beta)\in B}\mathbf{D}_\Lambda(u,\beta) \tag{3.2.18}$$

(here $(t,\alpha)_\varepsilon$ is the ε-neighborhood of the point $(t,\alpha) \in \mathbb{R}^2$) and putting

$$\mathbf{D}(t,\alpha) := \mathbf{D}_\Lambda(t,\alpha)$$

at the remaining points (t,α) (that is, at the points $(t,\alpha) \notin \partial\mathfrak{D}^{<\infty}$), then we get the lower-continuous function $\mathbf{D}(t,\alpha)$ (representation (3.2.18) holds for all points $(t,\alpha) \in \mathbb{R}^2$), all the properties (3.2.13)–(3.2.17) of the function $\mathbf{D}_\Lambda(t,\alpha)$ being preserved. In what follows, the function $\mathbf{D}(t,\alpha)$, as well as the function $\mathbf{D}_\Lambda(t,\alpha)$, will also be called the *deviation function* (the *second deviation function*) corresponding to the renewal measure of the random vector $\boldsymbol{\xi} = (\tau,\zeta)$.

So, from the above and using (3.2.11) we get that in the homogeneous case, as $T \to \infty$,

$$\frac{1}{T}\ln H_1(T\delta[t], T\Delta[\alpha]) \sim -\mathbf{D}(t,\alpha). \tag{3.2.19}$$

This clarifies the "probabilistic" meaning of the function $\mathbf{D}(t,\alpha)$.

We now give a precise formulation of the theorem on the asymptotical behavior of the renewal measure $\widetilde{H}_1$ in the general *inhomogeneous* case. For this purpose, we use the "admissible inhomogeneity" condition, under which an inhomogeneous jump $\boldsymbol{\xi}_1$ will not "aggravate" the asymptotic formula (3.2.19).

Theorem 3.2.1 (see Theorem 2.9.6 in [12][1]) *Let ξ_1 satisfy the "admissible inhomogeneity" condition*

$$\mathcal{A}^{\le 0} := \{(\lambda, \mu) : \mathbf{A}(\lambda, \mu) \le 0\} \subset [\mathcal{A}_1]. \tag{3.2.20}$$

Assume that $\delta = \delta_T = \overline{o}(1)$, $\Delta = \Delta_T = \overline{o}(1)$ as $T \to \infty$. Then

$$\frac{1}{T} \ln \widetilde{H}_1(T\delta[t), T\Delta[\alpha)) = -\mathbf{D}(t, \alpha) + o(1). \tag{3.2.21}$$

It is clear that if $t > 0$ or $\alpha \notin (-\Delta, 0]$, then the "zero" term $\mathbf{P}(T_0 \in T\delta[t), Z_0 \in T\Delta[\alpha))$ in the definition of the renewal function is zero and so $\widetilde{H}_1$ in (3.2.21) can be replaced by $\widetilde{H}$.

3.2.3 Preliminary Version of the Local LDP for CRPs

Using the above considerations we can obtain the following "preliminary version" of the local LDP for CRPs.

We set

$$\alpha = \frac{x}{T}, \qquad l(T, y) := -\frac{1}{T} \ln \mathbf{P}(\tau > Ty),$$
$$\lambda_+ := \sup\{\lambda : \psi^{(\tau)}(\lambda) = \mathbf{E}e^{\lambda\tau} < \infty\}. \tag{3.2.22}$$

Theorem 3.2.2 *Let the admissible inhomogeneity condition* (3.2.20) *hold. Then*

$$\mathbf{P}(Z(T) \in T\Delta[\alpha)) = \mathbb{1}_{\{\alpha \in (-\Delta, 0]\}} \mathbf{P}(\tau_1 \ge T) + P_T(T\Delta(\alpha)), \tag{3.2.23}$$

where for $\Delta = \overline{o}(1)$, as $T \to \infty$,

$$P_T(T\Delta[\alpha)) = \exp\{-T\widehat{D}_T(\alpha) + o(T)\},$$
$$\widehat{D}_T(\alpha) := \inf_{0 \le t \le 1} [\mathbf{D}(t, \alpha) + l(T, 1-t)].$$

Proof Relation (3.2.23) follows from (3.2.5) (in the inhomogeneous case in (3.2.5) one should replace H_1 by $\widetilde{H}_1$), in which the integral in the right-hand side is denoted by $P_T(T\Delta[\alpha))$. In view of (3.2.6) for $u_k = k\delta$, $N = 1/\delta$, we have

$$\begin{aligned} P_T(T\Delta[\alpha)) &= \int_0^T \widetilde{H}_1(du, T\Delta[\alpha))\mathbf{P}(\tau > T - u) \\ &\ge \sum_{k=0}^{N-1} \widetilde{H}_1(T\delta[u_k), T\Delta[\alpha))\mathbf{P}(\tau > T - Tu_k) \qquad (3.2.24) \\ &\ge \max_{k<N} \widetilde{H}_1(T\delta[u_k), T\Delta[\alpha))\mathbf{P}(\tau > T - Tu_k). \qquad (3.2.25) \end{aligned}$$

Next, an appeal to Theorem 3.2.1 for $\delta = \overline{o}(1)$, $\Delta = \overline{o}(1)$ shows that, as $T \to \infty$,

$$\frac{1}{T} \ln P_T(T\Delta[\alpha)) \ge -\min_{k<N} [\mathbf{D}(u_k, \alpha) + l(T, 1 - u_k)] + o(1) = -\widehat{D}_T(\alpha) + o(1).$$

[1] In the Russian version [23] of the monograph [12], Theorem 2.9.6 was proved under the more stringent condition $\mathcal{A}^{\le 0} \subset \mathcal{A}_1$.

Let us now obtain the converse inequality. As in (3.2.25), using (3.2.6) we get

$$P_T(T\Delta[\alpha)) \le \sum_{k=0}^{N-1} \widetilde{H}_1(T\delta[u_k), T\Delta[\alpha))\mathbf{P}(\tau > T - Tu_{k+1})$$
$$\le N \max_{k<N} \widetilde{H}_1(T\delta[u_k), T\Delta[\alpha))\mathbf{P}(\tau > T - Tu_{k+1}).$$

Next, since $\delta = \overline{o}(1)$ as $T \to \infty$, $N = 1/\delta$, and since the function $\mathbf{D}(t, \alpha)$ is continuous on its effective domain, we have $\frac{\ln N}{T} \to 0$ as $T \to \infty$, and

$$\frac{1}{T} \ln P_T(T\Delta[\alpha)) \le \frac{\ln N}{T} - \widehat{D}_T(\alpha) + o(1) = -\widehat{D}_T(\alpha) + o(1).$$

Theorem 3.2.2 is proved. □

To obtain explicit representations for the function $\widehat{D}_T(\alpha)$, we shall need some properties of the second deviation function $\mathbf{D}(t, \alpha)$.

3.3 Deviation Functions for the Renewal Measure and for Compound Renewal Processes

3.3.1 Properties of the Function $\mathbf{D}(t, \alpha)$ and of the Deviation Functions for CRPs

The preliminary version of the LDP (Theorem 3.2.2) shows that the functions $\widehat{D}_T(\alpha)$ and $l(T, y)$ (see (3.2.2), (3.2.2)) turn out to be extremely important in describing the asymptotic behavior of $\ln \mathbf{P}(Z(T) \in T\Delta[\alpha))$. To ascertain the asymptotics of the function $\widehat{D}_T(\alpha)$, we shall need some properties of the functions $\mathbf{D}(t, \alpha)$ and $\ln \mathbf{P}(\tau > Ty)$.

As was already pointed out, the function $\mathbf{D}(t, \alpha)$, as defined in (3.2.18), shares the properties (3.2.14)–(3.2.17) of the function $\mathbf{D}_\Lambda(t, \alpha)$.

Namely, *the function* $\mathbf{D}(t, \alpha)$ *is convex and ruled*:

$$\mathbf{D}(vt, v\alpha) = v\mathbf{D}(t, \alpha).$$

Besides, it is lower-continuous and semiadditive:

$$\mathbf{D}(t + u, \alpha + \beta) \le \mathbf{D}(t, \alpha) + \mathbf{D}(u, \beta).$$

The following representation for the function $\mathbf{D}(t, \alpha)$ underlies many constructions that follow below (see, e.g., the definition of the fundamental function in §3.5.1, relation (3.5.3), Theorem 3.5.1, etc.). For a proof, see [23] (Theorem 2.9.1).

Theorem 3.3.1 *For all* (t, α), *the following representation holds*

$$\mathbf{D}(t, \alpha) = \sup_{(\lambda,\mu)\in\mathcal{A}^{\le 0}} (\lambda t + \mu\alpha) = \sup_{(\lambda,\mu)\in\partial\mathcal{A}^{\le 0}} (\lambda t + \mu\alpha), \qquad (3.3.1)$$

where $\partial\mathcal{A}^{\le 0}$ *is the boundary of the set* $\mathcal{A}^{\le 0} := \{(\lambda, \mu) : \mathbf{A}(\lambda, \mu) \le 0\}$.

This result gives, in parallel with (3.2.18), another characteristic of the second deviation function $\mathbf{D}(t, \alpha)$, which plays an important role in what follows.

Note that since the function $\mathbf{D}$ is ruled for $t > 0$, we have

$$\mathbf{D}(t, \alpha) = t\mathbf{D}(1, \alpha/t) \tag{3.3.2}$$

and hence, the function $\mathbf{D}(t, \alpha)$ of two variables is completely determined by the values of the univariate function

$$D(\alpha) := \mathbf{D}(1, \alpha). \tag{3.3.3}$$

It turns out that under broad assumptions the function $D(\alpha)$ is the deviation function for a CRP. In view of (3.3.1), we have

$$D(\alpha) = \sup_{(\lambda,\mu)\in\partial\mathcal{A}^{\le 0}} \{\lambda + \mu\alpha\}. \tag{3.3.4}$$

Next, let $(\lambda(\alpha), \mu(\alpha))$ be a point from $\partial\mathcal{A}^{\le 0}$ at which the supremum in (3.3.4) for the function $D(\alpha)$ is attained. Hence

$$D(\alpha) = \lambda(\alpha) + \mu(\alpha)\alpha. \tag{3.3.5}$$

To describe the deviation function for CRPs in the general case, we introduce, in addition to the function $D(\alpha)$, the function

$$\widehat{D}(\alpha) = \begin{cases} \inf_{t\in[0,1]} [\mathbf{D}(t, \alpha) + (1-t)\lambda_+] & \text{if } \lambda_+ < \infty, \\ D(\alpha) & \text{if } \lambda_+ = \infty, \end{cases} \tag{3.3.6}$$

where, as before,

$$\lambda_+ := \sup\{\lambda : \mathbf{E}e^{\lambda\tau} < \infty\}.$$

In what follows, an important role will be played by the value

$$D(0) = \sup_{(\lambda,\mu)\in\mathcal{A}^{\le 0}} \lambda = \sup\{\lambda : \inf_{\mu} \mathbf{A}(\lambda, \mu) \le 0\}. \tag{3.3.7}$$

This value is finite if the random variable ζ takes values of both signs, that is, if

$$\mathbf{P}(\zeta < 0) > 0, \qquad \mathbf{P}(\zeta > 0) > 0 \tag{3.3.8}$$

or if

$$\mathbf{P}(\zeta = 0) > 0. \tag{3.3.9}$$

If (3.3.8) or (3.3.9) holds, then $\mathbf{\Lambda}(t, 0) < \infty$ for some $t > 0$,

$$D(0) = \mathbf{D}(1, 0) \le \frac{1}{t}\mathbf{\Lambda}(t, 0) < \infty.$$

The possibility $D(0) = \infty$ appears in special alternatives to (3.3.8), (3.3.9):

$$\mathbf{P}(\zeta > 0) = 1 \quad \text{or} \quad \mathbf{P}(\zeta < 0) = 1. \tag{3.3.10}$$

In the first case, $\mathbf{\Lambda}(t,\alpha) \uparrow \infty$ as $\alpha \downarrow 0$. At the same time,

$$D(\alpha) \to D(0) \quad \text{as } \alpha \downarrow 0,$$

where the value $D(0)$, as in (3.3.8), (3.3.9), can be finite (see Example 3.5.2 below).

Note that the number $D(0) = D(1,0) < \infty$ is the value at which the line $\lambda = D(0)$ in the plane (λ, μ) touches from the right the boundary $\partial\mathcal{A}^{\le 0}$ of the convex set $\mathcal{A}^{\le 0}$ (see (3.3.7)).

Some properties of the functions $D(\alpha)$ and $\widehat{D}(\alpha)$ are described in the following result. The study of these properties will be continued in §3.5.

Theorem 3.3.2 (i) *The functions $\widehat{D}(\alpha) \le D(\alpha)$ are convex and lower-continuous (inside their effective domains). They are nonnegative, vanish at the unique point $\alpha = a = \frac{a_\zeta}{a_\tau}$, and*

$$\widehat{D}(\alpha) = D(\alpha) \tag{3.3.11}$$

near the point $\alpha = a$. We always have,

$$\widehat{D}(0) = \min\{D(0), \lambda_+\}. \tag{3.3.12}$$

(ii) *If the condition*

$$[\boldsymbol{\lambda}_+] \qquad \lambda_+ \ge D(0) \tag{3.3.13}$$

is satisfied, or, equivalently,

$$\widehat{D}(0) = D(0),$$

then (3.3.11) *holds for all α.*

If, for a given α, the point $(\lambda(\alpha), \mu(\alpha)) \in \partial\mathcal{A}^{\le 0}$ (see (3.3.5)*) is such that $\lambda(\alpha) \le \lambda_+$, then equality* (3.3.11) *also holds.*

(iii) *If at least one of the following four conditions*

$$\begin{cases} (1) \text{ the region } (\mathcal{A}) \text{ lies in the half-plane } \lambda < \lambda_+, \\ \qquad (\text{as in the case of independent } \tau \text{ and } \zeta), \\ (2)\ \mathbf{E}\zeta = 0, \\ (3)\ \tau \in [\mathbf{C}_\infty] \text{ or } z \in [\mathbf{C}_\infty] \end{cases} \tag{3.3.14}$$

is satisfied, then condition $[\boldsymbol{\lambda}_+]$ is met.

(iv) *If at least one of the following three conditions*

$$\mathcal{A}^{\le 0} \subset \{\lambda < \lambda_+\}, \qquad \mathbf{E}\zeta = 0, \qquad \lambda_+ = \infty$$

is satisfied, then $\lambda_+ > D(0)$.

(v) *The inequality*

$$\varliminf_{\alpha\to\infty} \frac{1}{\alpha} D(\pm\alpha) > 0 \tag{3.3.15}$$

always holds. If, in addition, $\zeta \in [\mathbf{C}_\infty]$, then

$$\lim_{\alpha\to\infty} \frac{1}{\alpha} D(\pm\alpha) = \infty. \tag{3.3.16}$$

Remark 3.3.3 (1) Below in Example 3.5.14 we show that condition $[\lambda_+]$ is not always met; that is, the inequality $\lambda_+ < D(0)$ is possible.

(2) It is easily seen that condition (1) in (3.3.14) is always satisfied if, for all $v > 0$, $w > 0$,

$$\mathbf{P}(\tau \geq v, \zeta \gtrless \pm w) \leq \mathbf{P}(\tau \geq v)\mathbf{P}(\zeta \gtrless \pm w) r(v, w), \quad w > 0, \tag{3.3.17}$$

where

$$\ln r(v, w) \leq \varepsilon_v v + \varepsilon_{|w|}|w|, \qquad \varepsilon_t \to 0 \quad \text{as} \quad t \to \infty. \tag{3.3.18}$$

In this case, $(\mathcal{A})$ is a subset of the rectangle $(-\infty, \lambda_+) \times (\mu_-^{(\zeta)}, \mu_+^{(\zeta)})$ (as in the case of independent τ and ζ). For example, if for some $\beta \in (0, 1)$, the inequality

$$\ln r(v, w) \leq c(v + |w|)^\beta, \qquad c = \text{const}$$

is satisfied outside the unit ball, then condition (3.3.17) is met, because $(v + |w|)^\beta \leq v^\beta + |w|^\beta$.

Proof of Theorem 3.3.2 (i) If $\lambda_+ = \infty$, then $\widehat{D}(\alpha) = D(\alpha)$ by definition (3.3.6), where the function $D(\alpha)$ is convex and lower-continuous. This follows from the fact that the function $\mathbf{D}(t, \alpha)$ is convex and lower-continuous (see the beginning of §3.3.1).

If $\lambda_+ < \infty$, then, for any α, there exists a $t_\alpha \in [0, 1]$ such that

$$\widehat{D}(\alpha) = \mathbf{D}(t_\alpha, \alpha) + (1 - t_\alpha)\lambda_+. \tag{3.3.19}$$

Hence, since the function $\mathbf{D}(t, \alpha)$ is convex, for any $p \geq 0, q \geq 0, p+q = 1, \alpha, \beta \in \mathbb{R}$, we have

$$\begin{aligned} p\widehat{D}(\alpha) + q\widehat{D}(\beta) &= p\mathbf{D}(t_\alpha, \alpha) + p(1 - t_\alpha)\lambda_+ + q\mathbf{D}(t_\beta, \beta) + q(1 - t_\beta)\lambda_+ \\ &\geq \mathbf{D}(pt_\alpha + qt_\beta, p\alpha + q\beta) + (1 - (pt_\alpha + qt_\beta))\lambda_+ \geq \widehat{D}(p\alpha + q\beta). \end{aligned}$$

This shows that the function $\widehat{D}(\alpha)$ is convex.

Let us verify that the function $\widehat{D}(\alpha)$ is lower-continuous. Let $\alpha_n \to \alpha$ as $n \to \infty$. Assuming, without loss of generality that the sequence t_{α_n} converges to $t \in [0, 1]$ as $n \to \infty$, we get

$$\varliminf_{n\to\infty} \widehat{D}(\alpha_n) = \varliminf_{n\to\infty} \mathbf{D}(t_{\alpha_n}, \alpha_n) + \lim_{n\to\infty} (1 - t_{\alpha_n})\lambda_+ \geq \mathbf{D}(t, \alpha) + (1 - t)\lambda_+ \geq \widehat{D}(\alpha).$$

Hence the function $\widehat{D}(\alpha)$ is lower-continuous. The inequality $\widehat{D}(\alpha) \leq D(\alpha)$ is a direct consequence of definition (3.3.6).

According to Theorem 4.10.1 in [23], $\widehat{D}(\alpha) = D(\alpha)$ near the point $a = \frac{a_\zeta}{a_\tau}$, and as $\alpha \to a$,

$$D(\alpha) = \frac{1}{2}(\alpha - a)^2 D_a''[1 + o(1)],$$

where the coefficient

$$D''_a := \frac{\partial^2}{\partial \alpha^2} \mathbf{D}(1,\alpha)\big|_{\alpha=a} \in (0,\infty)$$

can be found explicitly (see [23], Theorem 2.9.2). Hence the convex functions $D(\alpha)$ and $\widehat{D}(\alpha)$ vanish only at the point $\alpha = a$.

Let us prove (3.3.12). To this end, we estimate $\widehat{D}(\alpha)$ from below. By (3.2.14), we have

$$\mathbf{D}(t,0) = tD(0),$$
$$\widehat{D}(0) = \inf_{t\in[0,1]} \{tD(0) + (1-t)\lambda_+\} = \min\{D(0), \lambda_+\}.$$

(ii) Let $\lambda_+ \geq D(0)$. By Theorem 3.3.1, for any $(\lambda,\mu) \in \mathcal{A}^{\leq 0}$ and $t \geq 0$,

$$\mathbf{D}(t,\alpha) \geq t\lambda + \alpha\mu = \lambda + \alpha\mu - \lambda(1-t). \tag{3.3.20}$$

Assume first that $D(\alpha) < \infty$. Setting $\lambda = \lambda(\alpha)$, $\mu = \mu(\alpha)$, it follows by (3.3.20) that

$$\begin{aligned}\mathbf{D}(t,\alpha) + \lambda_+(1-t) &\geq \lambda(\alpha) + \alpha\mu(\alpha) + (1-t)(\lambda_+ - \lambda(\alpha))\\ &= D(\alpha) + (1-t)(\lambda_+ - \lambda(\alpha)).\end{aligned}$$

We have $\lambda_+ \geq D(0)$, and hence by (3.3.7) $\lambda(\alpha) \leq D(0) \leq \lambda_+$ and for all $t \in [0,1]$,

$$(1-t)(\lambda_+ - \lambda(\alpha)) \geq 0. \tag{3.3.21}$$

Therefore,

$$\mathbf{D}(t,\alpha) + \lambda_+(1-t) \geq D(\alpha) \quad \text{and} \quad \widehat{D}(\alpha) \geq D(\alpha).$$

If $D(\alpha) = \infty$, then for a given $N > 0$, instead of $(\lambda(\alpha),\mu(\alpha))$ one should choose a point $(\widehat{\lambda},\widehat{\mu}) \in \mathcal{A}^{\leq 0}$ such that $\widehat{\lambda} + \alpha\widehat{\mu} \geq N$ and then repeat the above arguments. As a result, we get $\widehat{D}(\alpha) \geq N$, which, since N is arbitrary, means that $\widehat{D}(\alpha) = \infty$.

The converse inequality $\widehat{D}(\alpha) \leq D(\alpha)$ is clear. This proves (3.3.11) for all α.

If α is such that $\lambda_+ \geq \lambda(\alpha)$, then inequality (3.3.21) (for which the condition $\lambda_+ \geq D(0)$ was required) clearly holds. The remaining arguments leading to (3.3.11) are the same.

(iii) (1) Let $(\mathcal{A})$ lie in the half-plane $\lambda < \lambda_+$. In view of (3.3.7) we have

$$D(0) = \sup\{\lambda : \inf_\mu \mathbf{A}(\lambda,\mu) \leq 0\},$$

and hence this supremum cannot exceed λ_+, because by the condition we have $\mathbf{A}(\lambda,\mu) = \infty$ for $\lambda > \lambda_+$ and any μ. Hence $D(0) \leq \lambda_+$, $\widehat{D}(\alpha) = D(\alpha)$.

(2) If $\mathbf{E}\zeta = 0$, then $\mathbf{A}(0,\mu) > 0$ for $\mu \neq 0$, $\mathbf{A}(\lambda,0) < 0$ for $\lambda < 0$, so that $D(0) = 0$ by (3.3.7) and condition $[\boldsymbol{\lambda}_+]$ is met (see (3.3.13)).

(3) If $\tau \in [\mathbf{C}_\infty]$, then $\lambda_+ = \infty$; in this case (3.3.13) clearly holds.

Now let $\zeta \in [\mathbf{C}_\infty]$, $\lambda_+ < \infty$. We shall first show that, for any $\lambda > \lambda_+$ and μ,

$$\mathbf{A}(\lambda,\mu) = \infty. \tag{3.3.22}$$

Assume on the contrary that $\mathbf{A}(\widetilde{\lambda},\widetilde{\mu}) < \infty$ for some $\widetilde{\lambda} > \lambda_+$ and $\widetilde{\mu}$. Since $A^{(\zeta)}(-\mu) = \ln \mathbf{E}e^{-\mu\zeta} < \infty$ for any μ, we have

$$e^{p\mathbf{A}(\widetilde{\lambda},\widetilde{\mu})} = (\mathbf{E}e^{\widetilde{\lambda}\tau+\widetilde{\mu}\zeta})^p < \infty$$

for

$$\lambda \in (\lambda_+, \widetilde{\lambda}), \quad p := \frac{\lambda}{\widetilde{\lambda}}, \quad q := 1-p, \quad \mu := \widetilde{\mu}p, \quad \mu' := \frac{\mu}{q}.$$

Hence, by Hölder's inequality,

$$\infty > e^{p\mathbf{A}(\widetilde{\lambda},\widetilde{\mu})+qA^{(\zeta)}(-\mu')} = (\mathbf{E}e^{\widetilde{\lambda}\tau+\widetilde{\mu}\zeta})^p(\mathbf{E}e^{-\mu'\zeta})^q \geq \mathbf{E}e^{\lambda\tau+\mu\zeta-\mu\zeta} = \mathbf{E}e^{\lambda\tau} = \infty.$$

This contradiction establishes (3.3.22). From (3.3.22), (3.3.7) it follows that the set $\mathcal{A}$ lies in the half-plane

$$\{(\lambda,\mu) : \ \lambda \leq \lambda_+\},$$

and therefore, the first condition in (3.3.14) is met.

(iv) If $\mathbf{E}\zeta = 0$ or $\lambda_+ = \infty$, then the inequality $\lambda_+ > D(0)$ is already established. The inclusion $\mathcal{A}^{\leq 0} \subset \{\lambda < \lambda_+\}$ means that λ_+ is greater than the first coordinate of any point (λ,μ) from $\mathcal{A}^{\leq 0}$, i.e. (see (3.3.7)),

$$\lambda_+ > \sup_{(\lambda,\mu)\in\mathcal{A}^{\leq 0}} \lambda = D(0).$$

The condition $\mathcal{A}^{\leq 0} \subset \{\lambda < \lambda_+\}$ can be clearly replaced by the condition $\partial\mathcal{A}^{\leq 0} \subset \{\lambda < \lambda_+\}$.

(v) Relation (3.3.15) follows from the convexity of the function $D(\alpha) \geq 0$, which vanishes only at the point $a = \frac{a_\zeta}{a_\tau}$.

Let us prove (3.3.16) in the case $\zeta \in [\mathbf{C}_\infty]$. Assume that, for $\lambda \leq 0$, the function $h_+(\lambda)$ (the function $h_-(\lambda)$) defines the upper (lower) part of the boundary $\partial\mathcal{A}^{\leq 0}$ of the convex set $\mathcal{A}^{\leq 0}$. The ray $\{(\lambda,\mu) : \ \lambda \leq 0, \mu = 0\}$ wholly lies in the set $\mathcal{A}^{\leq 0}$, and hence the function $h_+(\lambda)$ is nonnegative, concave, and nondecreasing as $|\lambda| \to \infty$ in the range $\lambda \leq 0$. Similarly, the function $h_-(\lambda)$ is nonpositive, convex, and nonincreasing as $|\lambda| \to \infty$ in the range $\lambda \leq 0$. Since, by Cauchy's inequality, for any $N < \infty$,

$$\sup_{|\mu|\leq N} \mathbf{E}e^{\lambda\tau+\mu\zeta} \leq \sup_{|\mu|\leq N} \sqrt{\mathbf{E}e^{2\mu\zeta}}\sqrt{\mathbf{E}e^{2\lambda\tau}} \to 0$$

as $\lambda \to -\infty$, it follows that $h_-(\lambda) \leq -N < N \leq h_+(\lambda)$ for all sufficiently large R for $\lambda \leq -R$. This means that

$$h_\pm(\lambda) \to \pm\infty \quad \text{as} \quad \lambda \to -\infty. \tag{3.3.23}$$

Next, we have $(-\alpha, \ h_\pm(-\alpha)) \in \partial\mathcal{A}^{\leq 0}$ for all $\alpha > 0$, and hence by (3.3.4)

$$D(\alpha) = \sup_{(\lambda,\mu)\in\partial\mathcal{A}^{\leq 0}} \{\lambda + \mu\alpha\} \geq -\alpha + h_+(-\alpha)\alpha = \alpha(h_+(-\alpha) - 1);$$

$$D(-\alpha) = \sup_{(\lambda,\mu)\in\partial\mathcal{A}^{\leq 0}} \{\lambda - \mu\alpha\} \geq -\alpha - h_-(-\alpha)\alpha = \alpha(-h_-(-\alpha) - 1).$$

From these inequalities and from (3.3.23) it follows that the convex lower-continuous function $D(\alpha)$ grows as $|\alpha| \to \infty$ faster than any linear function, and hence (3.3.16) holds. This proves assertion (iv) of Theorem 3.3.2, and therefore, Theorem 3.3.2. □

In §3.5, it will be shown that in the case $\lambda_+ < D(0)$ there exists a neighborhood (β_-, β_+) of 0 in which $\widehat{D}(\alpha) < D(\alpha)$, while $\widehat{D}(\alpha) = D(\alpha)$ outside this neighborhood. We shall find the explicit form of the boundaries $\beta_\pm$ of this neighborhood and clarify the probabilistic sense of the condition $\lambda_+ < D(0)$.

In Chapter 5 we shall need more details about the deviation functions of CRPs. In this regard, in §3.5 we shall continue the study of properties of the functions D, $\widehat{D}$ and of some other characteristics of CRPs.

3.4 Large Deviation Principles for $Z(T)$

From Theorem 3.3.2 we have the following alternative:

$$\begin{aligned} &\text{either } \widehat{D}(0) = D(0) \le \lambda_+ \ \ (\text{condition } [\boldsymbol{\lambda}_+]) \text{ is met,} \\ &\text{or } \widehat{D}(0) = \lambda_+ < D(0). \end{aligned} \tag{3.4.1}$$

In the first case,

$$\widehat{D}(\alpha) = D(\alpha) \quad \text{for all } \alpha.$$

In the second case, we will require the additional condition (including the inequality $\lambda_+ < D(0)$).

$[\overline{\boldsymbol{\lambda}}_+]$ $\lambda_+ < D(0)$ *and the "rough smoothness" condition for the distribution of* τ (the LDP for τ) is met:

$$\ln \mathbf{P}(\tau \ge T) \sim -\lambda_+ T \quad \text{as} \quad T \to \infty. \tag{3.4.2}$$

3.4.1 The General Case

The following theorem is the main result of §3.4.1.

Theorem 3.4.1 *Assume that the admissible inhomogeneity condition*

$$\mathcal{A}^{\le 0} \subset [\mathcal{A}_1] \tag{3.4.3}$$

and the condition $[\boldsymbol{\lambda}_+] \bigcup [\overline{\boldsymbol{\lambda}}_+]$ *are met. If* $\alpha = 0$, *then, in addition to* (3.4.3), *we assume that*

$$\lambda_+^{(\tau_1)} := \sup\{\lambda : \mathbf{E}e^{\lambda \tau_1} < \infty\} \ge \min\{D(0), \lambda_+\} = \widehat{D}(0). \tag{3.4.4}$$

Then, under the above conditions,

(i) *the local LDP holds, that is, for any* $\alpha \in \mathbb{R}$,

$$\lim_{T \to \infty} \frac{1}{T} \ln \mathbf{P}\Big(\frac{1}{T} Z(T) \in \Delta[\alpha)\Big) = -\widehat{D}(\alpha), \tag{3.4.5}$$

where $\Delta = \Delta_T = \overline{o}(1)$ *as* $T \to \infty$ (see (3.1.1))*;*

(ii) *the "integral" LDP holds, that is, for any Borel set* $B \subset \mathbb{R}$,

$$\overline{\lim_{T\to\infty}} \frac{1}{T} \ln \mathbf{P}\Big(\frac{1}{T} Z(T) \in B\Big) \le -\widehat{D}([B]), \tag{3.4.6}$$

$$\underline{\lim}_{T\to\infty} \frac{1}{T} \ln \mathbf{P}(\frac{1}{T} Z(T) \in B) \ge -\widehat{D}((B)), \tag{3.4.7}$$

where $[B]$ *and* (B) *are the closure and interior of a set* B*, respectively,*

$$\widehat{D}(B) := \inf_{\alpha\in B} \widehat{D}(\alpha).$$

The function $\widehat{D}(\alpha)$ will be called the (the *deviation function for CRPs in its general form*, that is, for CRPs without condition (3.3.13)). If $\lambda_+ \ge D(0)$, then $\widehat{D}(\alpha)$ becomes $D(\alpha)$, which will be simply called the *deviation function for CRPs*.

Assertion (i) of Theorem 3.4.1 under some simplified assumption was proved in [23] (see Theorem 4.10.1).

From Theorem 3.4.1 we can derive analogous results for the compound renewal process

$$Z^{(q)}(t) = Z(t) + qt, \quad t \in [0, T],$$

with linear drift qt.

Corollary 3.4.2 *Let all the conditions of Theorem* 3.4.1 *be satisfied. Then, for any* $q \in \mathbb{R}$*, all the assertions of Theorem* 3.4.1 *hold for the process* $Z^{(q)}(T)$ *with the deviation function*

$$D^{Z^{(q)}}(\alpha) = \widehat{D}(\alpha - q).$$

Proof of Theorem 3.4.1 (i) We will use the preliminary version of the LDP in Theorem 3.2.2. According to this theorem (see (3.2.2)–(3.2.25)), for the proof of the local LDP (3.4.5) we need to verify that

$$\lim_{T\to\infty} \widehat{D}_T(\alpha) = \widehat{D}(\alpha), \tag{3.4.8}$$

and that in the case $\alpha = 0$ we have

$$\ln \mathbf{P}(\tau_1 \ge T) \le -T\widehat{D}(0) + o(T) \quad \text{as} \quad T \to \infty. \tag{3.4.9}$$

Let us prove (3.4.8). We set

$$\widehat{D}_+(\alpha) := \overline{\lim_{T\to\infty}} \widehat{D}_T(\alpha), \qquad \widehat{D}_-(\alpha) := \underline{\lim}_{T\to\infty} \widehat{D}_T(\alpha).$$

(a) *The upper estimate*. We claim that

$$\widehat{D}_+(\alpha) \le \widehat{D}(\alpha). \tag{3.4.10}$$

It is clear that

$$\widehat{D}_T(\alpha) \le \mathbf{D}(t_\alpha, \alpha) + l(T, 1 - t_\alpha),$$

where the function $l(T, y)$ was defined in (3.2.22) and t_α is from (3.3.19).

If $t_\alpha = 1$ (this is always the case if $\lambda_+ \geq D(0)$), then by $l(T, 1) = 0$ we have

$$\widehat{D}_T(\alpha) \leq \mathbf{D}(1, \alpha) = \widehat{D}(\alpha),$$

which implies (3.4.10).

If $t_\alpha < 1$, then $\lambda_+ = \widehat{D}(0) < D(0)$ and by condition $[\overline{\boldsymbol{\lambda}}_+]$, the limit

$$\lim_{T\to\infty} \frac{1}{T} \ln \mathbf{P}(\tau > Ty) = -\lambda_+ y$$

exists. Hence

$$\widehat{D}_+(\alpha) \leq \mathbf{D}(t_\alpha, \alpha) + \lim_{T\to\infty} l(T, 1 - t_\alpha) = \mathbf{D}(t_\alpha, \alpha) + (1 - t_\alpha)\lambda_+ = \widehat{D}(\alpha)$$

(see (3.3.19), (3.2.2)). This establishes (3.4.10).

(b) *The lower estimate*. We claim that

$$\widehat{D}_-(\alpha) \geq \widehat{D}(\alpha) \tag{3.4.11}$$

(here the condition $[\boldsymbol{\lambda}_+] \cup [\overline{\boldsymbol{\lambda}}_+]$ is not required).

We first consider the case $\lambda_+ < \infty$. By Chebyshev's inequality, for any $\varepsilon > 0$, there exists a constant $c = c(\varepsilon)$ such that

$$\mathbf{P}(\tau > y) \leq c e^{-\lambda_+(1-\varepsilon)y}. \tag{3.4.12}$$

Hence (see (3.2.2))

$$\begin{aligned}
\widehat{D}_T(\alpha) &\geq \inf_{0\leq t\leq 1} \left[\mathbf{D}(t, \alpha) + \lambda_+(1 - \varepsilon)(1 - t)\right] + O(1/T) \\
&\geq (1 - \varepsilon)\widehat{D}(\alpha) + O(1/T), \\
\widehat{D}_-(\alpha) &= \varliminf_{T\to\infty} \widehat{D}_T(\alpha) \geq (1 - \varepsilon)\widehat{D}(\alpha).
\end{aligned} \tag{3.4.13}$$

This proves (3.4.11) in the case $\lambda_+ < \infty$, since $\varepsilon > 0$ is arbitrary.

Consider now the case $\lambda_+ = \infty$. Repeating the previous arguments with λ_+ replaced in (3.4.12) by an arbitrary large N, and putting $\varepsilon = 0$, we get as in (3.4.13) the inequality

$$\widehat{D}_-(\alpha) \geq \min_{0\leq t\leq 1} \{\mathbf{D}(t, \alpha) + (1 - t)N\}.$$

The function $\mathbf{D}(t, \alpha)$ is convex with respect to t, and hence there exists a constant d such that $\mathbf{D}(t, \alpha) \geq \mathbf{D}(1, \alpha) + (t - 1)d$. Therefore,

$$\mathbf{D}(t, \alpha) + (1 - t)N \geq \mathbf{D}(1, \alpha) + (1 - t)(N - d) \geq \mathbf{D}(1, \alpha) \ \text{ for } \ N > d,$$
$$\widehat{D}_-(\alpha) \geq \mathbf{D}(1, \alpha) = \widehat{D}(\alpha).$$

This establishes (3.4.8).

If $\alpha \neq 0$, then the first term on the right of (3.2.5) is zero, which proves (3.4.5) in Theorem 3.4.1. If $\alpha = 0$, then in (3.2.5) we have the equality

$$\mathbf{P}\Big(\frac{1}{T} Z(T) \in \Delta[0)\Big) = \mathbf{P}(\tau_1 \geq T) + \int_0^T \widetilde{H}_1(dt, T\Delta[0))\mathbf{P}(\tau > T - t). \tag{3.4.14}$$

We have already established that the second term on the right of (3.4.14) has the form $e^{-T\widehat{D}(0)(1+o(1))}$. So it remains to verify (3.4.9). By the exponential Chebyshev inequality and assumption (3.4.4), we have

$$\mathbf{P}(\tau_1 \geq T) \leq e^{-T\widehat{D}(0)(1+o(1))}.$$

Therefore,

$$\overline{\lim_{T\to\infty}} \frac{1}{T} \ln \mathbf{P}(\tau_1 \geq T) \leq -\widehat{D}(0). \tag{3.4.15}$$

This inequality is equivalent to (3.4.9). Assertion (i) of Theorem 3.4.1 is proved.

(ii) The local LDP for $\frac{1}{T}Z(T)$ is now established, and so to prove the "integral" LDP it suffices to verify (see, e.g., [23], §4.1) that, for any $N < \infty$, there exists $R = R_N < \infty$ such that

$$\overline{\lim_{T\to\infty}} \frac{1}{T} \ln \mathbf{P}\Big(\frac{1}{T}|Z(T)| \geq R\Big) \leq -N. \tag{3.4.16}$$

Note that, for any arbitrary $M < \infty$,

$$\mathbf{P}\left(\frac{1}{T}|Z(T)| \geq R\right) \leq \mathbf{P}(\nu(T) \geq TM) + \mathbf{P}(|\zeta_1| + \cdots + |\zeta_{TM}| \geq RT), \tag{3.4.17}$$

where

$$\mathbf{P}(\nu(T) \geq TM) \leq \mathbf{P}(T_{MT} \leq T)$$

(we assume, for simplicity, that MT is an integer number). If $\frac{1}{M} < \mathbf{E}\tau$, then by the exponential Chebyshev inequality we get

$$\mathbf{P}(\nu(T) \geq TM) \leq e^{-MT\Lambda^{(\tau)}(1/M)}.$$

Since $\Lambda^{(\tau)}(v) \to \infty$ as $v \to 0$, we can always choose a sufficiently large $M = N$ so as to have

$$\mathbf{P}(\nu(T) \geq TM) \leq e^{-NT}. \tag{3.4.18}$$

The random vectors $\boldsymbol{\xi}_1 = (\tau_1, \zeta_1)$, $\boldsymbol{\xi} = (\tau, \zeta)$ that determine the process $Z(T)$ satisfy condition [**C**], and hence using the exponential Chebyshev inequality we can similarly estimate the second term on the right of (3.4.17), thereby verifying (3.4.16). Theorem 3.4.1 is proved. □

Remark 3.4.3 In the case $\widehat{D}(0) = D(0)$, the admissible inhomogeneity conditions (3.4.3), (3.4.4) are *minimal* conditions for the local LDP (3.4.5) to hold. This follows from the fact that these conditions are necessary conditions for (3.4.5). Indeed, let $\widehat{D}(0) = D(0)$ and let the local LDP (3.4.5) hold. Let us prove inequality (3.4.4). We have

$$\mathbf{P}(\tau_1 > T) \leq \mathbf{P}\Big(\frac{1}{T} Z(T) \in \Delta[0)\Big),$$

and hence,

$$\overline{\lim_{T\to\infty}} \frac{1}{T} \ln \mathbf{P}(\tau_1 > T) \le -D(0).$$

It follows that, for any $\lambda < D(0)$,

$$\int_0^\infty e^{\lambda t}\mathbf{P}(\tau_1 > t)\, dt < \infty, \qquad \mathbf{E}e^{\lambda\tau_1} < \infty,$$

which means that $\lambda_+^{(\tau_1)} \ge D(0)$, thereby proving inequality (3.4.4).

A more complicated argument in [45] shows that condition (3.4.3) is also a necessary condition.

Remark 3.4.4 If the process $Z(t)$ satisfies the conditions of Theorem 3.4.1, then, clearly, the process $-Z(t)$ also satisfies these conditions. Hence the local LDP (3.4.5) can be equivalently stated as follows:

$$\lim_{T\to\infty} \frac{1}{T} \ln \mathbf{P}\Big(\frac{Z(T)}{T} \in (\alpha)_\varepsilon\Big) = -\widehat{D}(\alpha);$$

here $(\alpha)_\varepsilon$ is the ε-neighborhood of the point α, $\varepsilon = \overline{o}(1)$ as $T \to \infty$ ($\varepsilon_T \to 0$ sufficiently slowly as $T \to \infty$). This in turn implies that, for any sequence $T' \sim T$, as $T \to \infty$,

$$\lim_{T\to\infty} \frac{1}{T} \ln \mathbf{P}\Big(\frac{Z(T')}{T} \in (\alpha)_\varepsilon\Big) = \lim_{T\to\infty} \frac{1}{T} \ln \mathbf{P}\Big(\frac{Z(T)}{T} \in (\alpha)_\varepsilon\Big) = -\widehat{D}(\alpha).$$

Remark 3.4.5 From the proof of Theorem 3.4.1 (see also Remark 2 in [44]) it follows that the conclusions of this theorem remain valid also in the case when the distribution of the first jump (we denote it by $\boldsymbol{\xi}_{1,T} = (\tau_{1,T}, \zeta_{1,T})$) depends on T so that, for all (λ, μ) and all sufficiently large T,

$$\psi_{1,T}(\lambda,\mu) := \mathbf{E}e^{\lambda\tau_{1,T}+\mu\zeta_{1,T}} \le c\psi_1(\lambda,\mu), \quad c = \text{const}, \tag{3.4.19}$$

where $\psi_1(\lambda,\mu)$ is the Laplace transform of the distribution of some *fixed* random vector $\boldsymbol{\xi}_1 = (\tau_1, \zeta_1)$. In this case, the conditions of Theorem 3.4.1 should be applied to this fixed vector $\boldsymbol{\xi}_1$.

3.4.2 Homogeneous Processes and Processes with Stationary Increments

It is clear that relations (3.4.3) and (3.4.4) are satisfied for *homogeneous* renewal processes $Z(t)$, for which the random vector $\boldsymbol{\xi}_1 = (\tau_1, \zeta_1)$ has the same distribution as the vector $\boldsymbol{\xi} = (\tau, \zeta)$. Hence, *for the homogeneous renewal processes in Theorem* 3.4.1 *and in Corollary* 3.4.2, *conditions* (3.4.3), (3.4.4) *are superfluous* (that is, they are satisfied automatically).

Let us now consider the case, important for applications, when the process $Z(t)$ has *stationary increments*, that is, the distribution of the differences

$$Z(t+h) - Z(t)$$

does not depend on $t \geq 0$. As is known, $Z(t)$ is a process with stationary increments if the “initial” vector $\boldsymbol{\xi}_1 = (\tau_1, \zeta_1)$ is chosen in a special way so that

$$\psi_1(\lambda, \mu) := \mathbf{E}e^{\lambda\tau_1+\mu\zeta_1} = \frac{1}{\lambda\mathbf{E}\tau}(\psi(\lambda, \mu) - \psi(0, \mu)), \tag{3.4.20}$$

where $\psi(\lambda, \mu) := \mathbf{E}e^{\lambda\tau+\mu\zeta}$ (see Lemma 1.1.5, and also [23], §4.10).

The following natural question arises: in which cases will the admissible inhomogeneity conditions (3.4.3), (3.4.4) of Theorem 3.4.1 be satisfied for processes with stationary increments? The next result gives a partial answer to this question.

Everywhere in what follows, the initial jumps $\boldsymbol{\xi}_1$ for a CRP with stationary increments will be denoted by $\boldsymbol{\xi}^{(\mathrm{st})}$ in order to differ them from the initial jumps in the general case. Correspondingly, all the entities corresponding to $\boldsymbol{\xi}^{(\mathrm{st})}$ will be given the superscript (st). For example, the function $\psi^{(\mathrm{st})}(\lambda, \mu)$ is equal to the right-hand side of (3.4.20),

$$\mathcal{A}^{(\mathrm{st})} := \{(\lambda, \mu) : \psi^{(\mathrm{st})}(\lambda, \mu) < \infty\},$$

and so on.

Corollary 3.4.6 *Let $Z(t) = Z^{(\mathrm{st})}(t)$ be a CRP with stationary increments, (that is, $\xi_1 = \xi^{(\mathrm{st})}$) and let at least one of the two following conditions be satisfied:*

(a) *the set $(\mathcal{A})$ is rectangular (as in the case of independent τ and ζ), that is,*

$$(\mathcal{A}) = (-\infty, \lambda_+) \times (\mu_-^{(\zeta)}, \mu_+^{(\zeta)});$$

(b) $\zeta \in [\mathbf{C}_\infty]$.

Then $\widehat{D}(\alpha) = D(\alpha)$ for all α, and assertions (3.4.5)–(3.4.7) *of Theorem* 3.4.1 *are valid.*

Proof The fact that $\widehat{D}(\alpha) = D(\alpha)$ follows from conditions (a), (b) and Theorem 3.3.2 (see (3.3.14)). It remains to verify the admissible inhomogeneity conditions (3.4.3), (3.4.4). In view of (3.4.20) we have $\lambda_+^{(\tau_1)} = \lambda_+$ for the processes under consideration, and hence condition (3.4.4) is met.

Consider condition (3.4.3). If (a) is satisfied, then $(0, \mu) \in \mathcal{A}$ for $\mu \in (\mu_-^{(\zeta)}, \mu_+^{(\zeta)})$, and therefore, $(\mathcal{A}^{(\mathrm{st})}) = (\mathcal{A})$ by (3.4.20), which implies (3.4.3). If (b) is satisfied, then $\psi(0, \mu) < \infty$ for any μ, and again we have $(\mathcal{A}^{(\mathrm{st})}) = (\mathcal{A})$ and (3.4.3) is satisfied. Corollary 3.4.6 is proved. □

For dependent τ and ζ it may happen that there exists a point $(\lambda, \mu) \in \mathcal{A}^{\leq 0}$ with $\lambda < 0$ such that $\psi^{(\mathrm{st})}(\lambda, \mu) = \infty$, and condition (3.4.3) is not met, as the following example shows.

Example 3.4.7 Let $\zeta = \tau + \omega$, where ω does not depend on τ. Then

$$\psi(\lambda, \mu) := \mathbf{E}e^{\lambda\tau+\mu(\tau+\omega)} = \psi^{(\tau)}(\lambda + \mu)\psi^{(\omega)}(\mu),$$

and by (3.4.20)

$$\psi^{(\mathrm{st})}(\lambda, \mu) = \frac{1}{\lambda\mathbf{E}\tau}\psi^{(\omega)}(\mu)[\psi^{(\tau)}(\lambda + \mu) - \psi^{(\tau)}(\mu)].$$

We choose τ and ω so that, for some $\delta \in (0, 1/2)$,

$$\psi^{(\tau)}(1-\delta) = \infty, \quad \psi^{(\omega)}(1) \le 1, \quad \psi^{(\omega)}(1+\delta) < \infty.$$

Hence, for $(\lambda, \mu) = (-1+\delta_1, 1+\delta_2)$ and all δ_1, δ_2 such that $|\delta_1|+|\delta_2| < \delta$, we have

$$\boldsymbol{\psi}^{(\mathrm{st})}(\lambda, \mu) = \frac{1}{(-1+\delta_1)\mathbf{E}\tau}\psi^{(\omega)}(1+\delta_2)\,[\psi^{(\tau)}(\delta_1+\delta_2) - \psi^{(\tau)}(1+\delta_2)] = \infty.$$

This means that the point $(-1, 1)$ together with some neighborhood lies outside the set $\mathcal{A}^{(\mathrm{st})}$, that is, $(-1, 1) \notin [\mathcal{A}^{(\mathrm{st})}]$. At the same time, $\psi(-1, 1) = \psi^{(\omega)}(1) \le 1$, that is, $(-1, 1) \in \mathcal{A}^{\le 0}$. This means that condition (3.4.3) for $\mathcal{A}_1 = \mathcal{A}^{(\mathrm{st})}$ is not met.

So, the fulfillment of the LDP for the homogeneous CRPs does not mean in general that it holds for CRPs with stationary increments. If conditions (a), (b) of Corollary 3.4.2 are not met, then the admissible inhomogeneity condition (3.4.3) should be tested in addition.

3.4.3 LDP for the Process $(Z(t), \gamma(t))$ and Its Consequences

If in the case $\lambda_+ < \infty$ the distribution of τ satisfies the "rough smoothness" condition, then one can establish an LDP for the vector $(Z(t), \gamma(t))$. This result has several useful corollaries. For simplicity, we consider only the homogeneous case.

Theorem 3.4.8 (i) *Let $\lambda_+ < \infty$ and let condition* (3.4.2) *be met. Then, for $u < 1$, $\Delta = \overline{o}(1)$, $\delta = \overline{o}(1)$, as $T \to \infty$,*

$$\frac{1}{T} \ln \mathbf{P}(Z(T) \in T\Delta[\alpha), \gamma(T) \in T\delta[u)) = -D(1-u, \alpha) - \lambda_+ u + o(1). \tag{3.4.21}$$

(ii) *If $\lambda_+ = \infty$, then, as $T \to \infty$,*

$$\begin{aligned}\frac{1}{T} \ln \mathbf{P}(Z(T) \in T\Delta[\alpha)) &= \frac{1}{T} \ln \mathbf{P}(Z(T) \in T\Delta[\alpha), \gamma(T) < \delta T) + o(1)\\ &= -D(\alpha) + o(1).\end{aligned} \tag{3.4.22}$$

If in (3.4.21) we put $u = u_k = k\delta$, $k = 0, 1, \ldots, N-1$, $N = 1/\delta$ (we assume that N is an integer number), then $\mathbf{P}(Z(T) \in T\Delta[\alpha))$ is the sum of the probabilities under the logarithm sign in (3.4.21) with $u = u_k$, $k = 0, \ldots, N-1$. Now the result of Theorem 3.4.1 follows from Theorem 3.4.8.

The proof of assertion (i) of Theorem 3.4.8 repeats the arguments given in the proofs of Theorems 3.2.2 and 3.4.1. However, now the analysis is simpler, because in our setting here the right-hand side of the analog of representation (3.2.25) (now for $\ln \mathbf{P}(Z(T) \in T\Delta[\alpha), \gamma(T) \in T\delta[u))$) involves the integral $\int_{T(1-u-\delta)}^{T(1-u)}$ in place of $\int_0^T$ (cf. (3.2.25)). The former integral is estimated from above and below by the quantities

$$\begin{aligned}&H_1(T\delta[1-u-\delta), T\Delta[\alpha))\mathbf{P}(\tau > uT),\\ &H_1(T\delta[1-u-\delta), T\Delta[\alpha))\mathbf{P}(\tau > T(u+\delta)).\end{aligned} \tag{3.4.23}$$

We have

$$\frac{1}{T}\ln \mathbf{P}(\tau > uT) \sim -\lambda_+ u + o(1) \tag{3.4.24}$$

for all $\lambda_+ < \infty$, and hence the remaining arguments remain unchanged.

In case (ii), in the representation for $\ln \mathbf{P}(Z(T) \in T\Delta[\alpha),\ \gamma(T) \in T\delta[u))$ of the form (3.4.21), for $u \geq \delta T$, a term will appear on the right of (3.4.21) in place of the term $\lambda_+ u$, the former term unboundedly increasing as $T \to \infty$. This yields (3.4.22). Theorem 3.4.8 is proved.

From Theorem 3.4.8 we get:

Corollary 3.4.9 *Under the conditions of Theorem* 3.4.8,

$$\frac{1}{T}\ln \mathbf{P}(Z(T) \in T\Delta[\alpha), \gamma(T) < \delta T) = -D(\alpha) + o(1). \tag{3.4.25}$$

Note that the conditions sufficient for (3.4.25) to hold can be slightly relaxed.

Corollary 3.4.10 *If condition* $[\boldsymbol{\lambda}_+]\bigcup[\overline{\boldsymbol{\lambda}}_+]$ *is met, then* (3.4.25) *is satisfied.*

Proof If condition $[\overline{\boldsymbol{\lambda}}_+]$ is met, then (3.4.25) follows from Corollary 3.4.9.

If $\lambda_+ \geq D(0)$, then the upper estimate

$$\ln \mathbf{P}(Z(T) \in T\Delta[\alpha), \gamma(T) < \delta T) < \ln \mathbf{P}(Z(T) \in T\Delta[\alpha)) = -TD(\alpha) + o(T)$$

follows from Theorem 3.4.1. To obtain the lower estimate, we shall employ the estimate in (3.4.23) with $u = 0$, which gives

$$\mathbf{P}(Z(T) \in T\Delta[\alpha), \gamma(T) > \delta T) \geq H_1(T\delta[1-\delta), T\Delta[\alpha))\mathbf{P}(\tau > \delta T),$$

and use the fact that by the proof of Theorem 3.4.1 the infimum in (3.2.2) is attained in the case $\lambda_+ \geq D(0)$ at the point $t = 1$. Corollary 3.4.10 is proved. □

So, if by adding the high-probability event $\{\gamma(T) < \delta T\}$ (the probability $\mathbf{P}(\gamma(T) \geq \delta T)$ is exponential with respect to δT) to the event $\{Z(T) \in T\Delta[\alpha)\}$ we slightly narrow the latter event, then, under condition $[\boldsymbol{\lambda}_+] \cup [\overline{\boldsymbol{\lambda}}_+]$, the LDP with the deviation function $D(\alpha)$ always holds for this event.

3.5 Fundamental Functions and Their Properties. Further Properties of the Deviation Function $D(\alpha)$, $\widehat{D}(\alpha)$. On the Condition $\lambda_+ < D(0)$

3.5.1 Fundamental Functions and Their Properties. Further Properties of the Function $D(\alpha)$

Recall that by Theorem 3.3.1 (see (3.3.4)),

$$D(\alpha) = \mathbf{D}(1, \alpha) = \sup_{(\lambda,\mu)\in\partial\mathcal{A}^{\leq 0}} (\lambda + \alpha\mu). \tag{3.5.1}$$

Hence, in order to study the function $D(\alpha)$, we need to describe the behavior of the boundary $\partial\mathcal{A}^{\le 0}$.

First of all, we note that the set $\mathcal{A}^{\le 0}$ is a body (has nonempty interior), because $\mathbf{E}\tau > 0$, $\mathbf{A}(0,0) = 0$, and $\frac{\partial}{\partial\lambda}\mathbf{A}(\lambda, 0) > 0$. To describe the boundary $\partial\mathcal{A}^{\le 0}$, consider the section $\mathcal{A}_{(\mu)}$ of the set $\mathcal{A}$ at the level μ,

$$\mathcal{A}_{(\mu)} := \{\lambda : \mathbf{A}(\lambda, \mu) < \infty\},$$

and define

$$\mu^+ := \sup\{\mu : \mathcal{A}_{(\mu)} \ne \varnothing\}, \qquad \mu^- := \inf\{\mu : \mathcal{A}_{(\mu)} \ne \varnothing\}.$$

Let $\mathcal{M}$ be the set of all values of μ such that $\mathbf{A}(\lambda, \mu) < \infty$ for some λ,

$$(\mu^-, \mu^+) \subset \mathcal{M} := \{\mu : \min_{\lambda} \mathbf{A}(\lambda, \mu) < \infty\} \subset [\mu^-, \mu^+].$$

If $\mu \in \mathcal{M}$, then the function $\mathbf{A}(\lambda, \mu)$ is strictly increasing with respect to λ from $-\infty$ to ∞. Hence the values

$$\begin{aligned} A(\mu) &:= \inf\{\lambda : \mathbf{A}(-\lambda, \mu) \le 0\} = -\sup\{\lambda : \mathbf{A}(\lambda, \mu) \le 0\}, \\ A^{\infty}(\mu) &:= \inf\{\lambda : \mathbf{A}(-\lambda, \mu) < \infty\} = -\sup\{\lambda : \mathbf{A}(\lambda, \mu) < \infty\} \end{aligned} \tag{3.5.2}$$

are well defined, and so $A(\mu)$ is a generalized solution for λ of the equation

$$\mathbf{A}(-\lambda, \mu) = 0.$$

It is clear that

$$(-A(\mu), \mu) \in \partial\mathcal{A}^{\le 0}, \qquad (-A^{\infty}(\mu), \mu) \in \partial\mathcal{A},$$

that is, the curve $\lambda = -A(\mu)$ in $\mathbb{R}^2$ parametrizes the boundary $\partial\mathcal{A}^{\le 0}$, and the curve $\lambda = -A^{\infty}(\mu)$ parametrizes the boundary of the set $\mathcal{A}$ for $\mu \in \mathcal{M}$. If, for $\mu \in \mathbb{R}$, there does not exist a point λ such that $\mathbf{A}(\lambda, \mu) < \infty$, then we set $A(\mu) = \infty$, $A^{\infty}(\mu) = \infty$. This extends the function $A(\mu)$ to the entire real line; moreover, $A(\mu) = \infty$ for $\mu \notin \mathcal{M}$ ($\mu^{\pm}$ are points of the discontinuity; $A(\mu^{\pm})$ can be finite), and so (3.5.1) can be written as

$$D(\alpha) = \sup_{\mu}\{\mu\alpha - A(\mu)\}. \tag{3.5.3}$$

This means that $D(\alpha)$ is the *Legendre transform* of $A(\mu)$.

The function $A(\mu)$ defined in (3.5.2) will play an important role in what follows. We will call this function the *fundamental function* for CRPs.

If $\tau_1 \equiv \tau \equiv 1$ (that is, $Z(k) = Z_k$ is an ordinary random walk), then

$$\mathbf{A}(-\lambda, \mu) = -\lambda + A^{(\zeta)}(\mu) \quad \text{and} \quad A(\mu) = A^{(\zeta)}(\mu).$$

If τ and ζ are independent and if τ is exponentially or geometrically distributed, then the homogeneous CRP $Z(t)$ becomes a process with independent increments; the function $A^{(\tau)}(\lambda)$ can be written down explicitly, as well as the solution for λ of the equation

$$A^{(\tau)}(-\lambda) + A^{(\zeta)}(\mu) = 0$$

(in terms of the function $A^{(\zeta)}(\mu)$). If τ is exponentially distributed with parameter β, then the CRP $Z(t)$ becomes a compound Poisson process,

$$\mathbf{E}e^{\mu Z(t)} = \exp\{t\beta(\psi^{(\zeta)}(\mu) - 1)\}, \quad A^{(\tau)}(\lambda) = \ln\frac{\beta}{\beta - \lambda},$$
$$A(\mu) = \beta(\psi^{(\zeta)}(\mu) - 1) = \ln \mathbf{E}e^{\mu Z(1)}.$$

However, as we will see below, in the general case the function $A(\mu)$ is not necessarily the logarithm of the Laplace transform.

Let us formulate the main properties of the fundamental function.

Theorem 3.5.1 *The fundamental function $A(\mu)$ is convex, lower-continuous (inside its effective domain), and is the Legendre transform of the deviation function $D(\alpha)$, so that,* (3.5.3) *holds, and moreover,*

$$A(\mu) = \sup_{\alpha}\{\mu\alpha - D(\alpha)\}. \tag{3.5.4}$$

So, $A(\mu)$ and $D(\alpha)$ form a pair of functions related by the dual Legendre transforms.

Proof We first observe that

$$\mathbf{A}(-A(\mu), \mu) \le 0 \quad \text{whenever} \quad A(\mu) < \infty. \tag{3.5.5}$$

Indeed, from definition (3.5.2) it follows that in the case $A(\mu) < \infty$ there exists a sequence $\lambda_n < -A(\mu)$ such that $\lambda_n \to -A(\mu)$ as $n \to \infty$ and $\mathbf{A}(\lambda_n, \mu) \le 0$, that is, $(\lambda_n, \mu) \in (\mathcal{A}^{\le 0})$. However, the function $\mathbf{A}(\lambda, \mu)$ is lower-continuous (inside its effective domain). Hence

$$0 \ge \mathbf{A}(\lambda_n, \mu) \to \mathbf{A}(-A(\mu), \mu) \quad \text{as} \quad n \to \infty.$$

This proves (3.5.5).

Let us show that $A(\mu)$ is convex, that is, we need to verify that

$$A(p\mu_1 + q\mu_2) \le pA(\mu_1) + qA(\mu_2) \tag{3.5.6}$$

for any $\mu_1, \mu_2 \in \mathbb{R}$, $p \ge 0$, $q \ge 0$, $p + q = 1$.

If $A(\mu_1) = \infty$ or $A(\mu_2) = \infty$, then inequality (3.5.6) holds. If $A(\mu_1) < \infty$ and $A(\mu_2) < \infty$, then by (3.5.5),

$$\mathbf{A}(-A(\mu_1), \mu_1) \le 0, \qquad \mathbf{A}(-A(\mu_2), \mu_2) \le 0.$$

Since the function $\mathbf{A}(\lambda, \mu)$ is convex, we have

$$\begin{aligned}&\mathbf{A}(-pA(\mu_1) - qA(\mu_2), p\mu_1 + q\mu_2)\\ &\qquad\qquad \le p\mathbf{A}(-A(\mu_1), \mu_1) + q\mathbf{A}(-A(\mu_2), \mu_2) \le 0.\end{aligned} \tag{3.5.7}$$

The left-hand side in (3.5.7) is nonpositive, and hence by definition (3.5.2) the value of the first argument of the function $\mathbf{A}(\cdot, \cdot)$ in the left-hand side of (3.5.7) is majorized by the value of the function $-A(\cdot)$ of the second argument of the function $\mathbf{A}(\cdot,\cdot)$, that is,

$$-pA(\mu_1) - qA(\mu_2) \le -A(p\mu_1 + q\mu_2).$$

This proves (3.5.6).

Let us now verify that the function $A(\mu)$ is lower-continuous. For the convex function $A(\mu)$, it suffices to test its lower-continuity only at points $\mu^{\pm}$ of discontinuity. Let

$$A_+ := \lim_{\mu\uparrow\mu^+} A(\mu).$$

If $A_+ = \infty$, then since $A(\mu)$ is convex, we have $A(\mu^+) = \infty = A_+$. Now let $A_+ < \infty$. Since the function $\mathbf{A}(\lambda,\mu)$, which defines the set $\mathcal{A}^{\le 0}$, is convex and lower-continuous, the set $\mathcal{A}^{\le 0}$ is convex and closed. Hence, the "horizontal ray"

$$L_+ := \{(\lambda,\mu^+) : \lambda \le -A_+\}$$

is contained in $\mathcal{A}^{\le 0}$. From definition (3.5.2) it follows that the ray L_+ is a part of the boundary $\partial\mathcal{A}^{\le 0}$ of the set $\mathcal{A}^{\le 0}$. Hence $A(\mu^+) = A_+$, which means that the function $A(\mu)$ is lower-continuous at the right-hand point of discontinuity μ^+. The case of the left-hand point of discontinuity μ^- is considered similarly.

So, the function $A(\mu)$ is convex and lower-continuous (as well as the function $D(\alpha)$), and hence by (3.5.3) we have (3.5.4) (see, e.g., [106, §12]). Theorem 3.5.1 is proved. □

Let us continue with the study of the fundamental function $A(\lambda)$.

It is clear that

$$A^\infty(\mu) \le A(\mu).$$

Since $(0,0) \in (\mathcal{A})$ and $A(0) = 0$, we always have $A^\infty(\mu) < A(\mu)$ near the point $\mu = 0$. We set

$$\begin{aligned} \mu_+ &:= \min\{\mu > 0 : A(\mu) = A^\infty(\mu)\},\\ \mu_- &:= \max\{\mu < 0 : A(\mu) = A^\infty(\mu)\}. \end{aligned} \tag{3.5.8}$$

For $\mu \in (\mu_-,\mu_+)$, we have

$$A^\infty(\mu) < A(\mu), \qquad (-A(\mu),\mu) \in (\mathcal{A}),$$

and $\lambda = -A(\mu)$ is the unique solution of the equation

$$\mathbf{A}(\lambda,\mu) = 0. \tag{3.5.9}$$

By the implicit function theorem, the function $A(\mu)$ is analytic on the interval (μ_-,μ_+). This interval will be called the *main domain of analyticity* of the function $A(\mu)$.

On the boundary $\partial\mathcal{A}$ (for $\lambda = -A^\infty(\mu)$), the function $\mathbf{A}(\lambda,\mu)$ may have finite or infinite values; moreover, they may oscillate (decrease or increase). If these oscillations happen around the zero value, then for $\mu \in (\mu_+,\mu^+)$, in parallel with the main domain of analyticity (μ_-, μ_+), there may appear new domains on which $A(\mu) > A^\infty(\mu)$, and therefore, the function $A(\mu)$ again becomes analytic, because it is the solution of (3.5.9) which lies inside $\mathcal{A}$. We set

$$M = \{\mu : A(\mu) > A^\infty(\mu)\}, \quad M^\infty = \{\mu : A(\mu) = A^\infty(\mu)\}. \tag{3.5.10}$$

The function $A(\mu)$ is analytic in the domain $M \supset (\mu_-, \mu_+)$ (which need not be simply connected; the function $A(\mu)$ can also have some domains of analyticity inside M^∞). In the domain

$$(\mu^-, \mu^+) = M \cup M^\infty$$

the function $A(\mu)$ is convex, continuous, and hence, is differentiable almost everywhere, but the derivative $A'(\mu)$ in the domain M^∞ may be discontinuous.

Given a function $F(u, v)$ of two variables u and v, the subscripts (1) and (2) will denotes its derivatives with respect to the first and second argument, respectively. For example,

$$F'_{(1)}(u_1, v_1) = \frac{\partial}{\partial u} F(u, v)|_{(u,v)=(u_1,v_1)},$$
$$F''_{(1,2)}(u_1, v_1) = \frac{\partial}{\partial u \partial v} F(u, v)|_{(u,v)=(u_1,v_1)}.$$

Consider some analytic properties of the function $A(\mu)$ at the point 0. We have

$$\mathbf{A}(0,0) = 0, \qquad A(0) = 0,$$
$$\mathbf{A}'_{(1)}(0,0) = a_\tau = \mathbf{E}\tau, \qquad \mathbf{A}'_{(2)}(0,0) = a_\zeta = \mathbf{E}\zeta, \qquad \mathbf{A}''_{(1,1)}(0,0) = \mathbf{D}\tau,$$
$$\mathbf{A}''_{(1,2)}(0,0) = \mathbf{E}\tau\zeta - a_\tau a_\zeta, \qquad \mathbf{A}''_{(2,2)}(0,0) = \mathbf{D}\zeta.$$

Differentiating the identity $\mathbf{A}(-A(\mu), \mu) = 0$ with respect to μ at the point $\mu = 0$, we get

$$A'(0) = \frac{a_\zeta}{a_\tau} = a, \qquad A''(0) = \frac{1}{a_\tau}\mathbf{E}(\zeta - a\tau)^2 \tag{3.5.11}$$

and therefore (see Theorem 1.2.1),

$$A'(0) = \lim_{t\to\infty} \frac{\mathbf{E}Z(t)}{t}, \qquad A''(0) = \lim_{t\to\infty} \frac{\mathbf{D}Z(t)}{t}, \tag{3.5.12}$$

so that $A'(0)$ and $A''(0)$ are, respectively, the average expectation and variance of $Z(t)$ per unit time. This shows that the function $A(\mu)$ shares many important properties with the logarithm of the Laplace transform of a distribution. Below (see Theorem 3.7.1) we shall show that, in addition to (3.5.12), we have

$$A(\mu) = \lim_{t\to\infty} \frac{1}{t} \ln \mathbf{E}e^{\mu Z(t)}$$

provided that $\lambda_+ \geq D(0)$. An analog of this relation will be also obtained in the general case.

Using the Cramér transform one can also easily check that, for $\mu \in (\mu_-, \mu_+)$, in addition to (3.5.11), we have the following strict inequality:

$$A''(\mu) > 0.$$

The possible discrepancy of the domains M and (μ^-, μ^+) is the property that distinguishes the function $A(\mu)$ from the logarithm of the Laplace transform of some distribution, say, of the function $A^{(\zeta)}(\mu)$. Note that all the distributions frequently encountered in applications (e.g., exponential (for τ), normal, and bounded) satisfy $\mathbf{A}(\lambda, \mu) = \infty$ for $(\lambda, \mu) \in \partial\mathcal{A}$ (for $\lambda = -A^{\infty}(\mu)$) and $A(\mu) > A^{\infty}(\mu)$ for $\mu \in (\mu^-, \mu^+) = (\mu_-, \mu_+)$, so that the domains M and (μ^-, μ^+) coincide, and the above discrepancy is absent.

Let us return to the function $D(\alpha) = \mathbf{D}(1, \alpha)$. As was already pointed out, representation (3.5.1) can be written as

$$D(\alpha) = \sup_{(\lambda,\mu)\in\partial\mathcal{A}^{\le 0}} \{\lambda + \alpha\mu\} = \sup_{\mu} (\alpha\mu - A(\mu)), \tag{3.5.13}$$

so that $D(\alpha)$ *is the Legendre transform at the point* α *of the convex lower-continuous function* $A(\mu)$. In other words, the function $D(\alpha)$ is defined from the function $A(\mu)$ in the same way as the deviation function $\Lambda(\alpha)$ for random walks is defined from the logarithm of the Laplace transform of the distribution of a jump. By the above properties of the function $A(\mu)$, the second deviation function $D(\alpha)$ shares many analytic properties with the deviation function $\Lambda(\alpha)$. Most of these properties have been already established; they are also immediate from (3.5.13). We augment these properties with some other properties of the functions $D(\alpha)$ and $A(\mu)$ and summarize them in the theorem.

We let $\mu(\alpha)$ denote the value at which the supremum on the right of (3.5.13) is attained, so that

$$D(\alpha) = \alpha\mu(\alpha) - A(\mu(\alpha)). \tag{3.5.14}$$

Note that since the function $A(\mu)$ is convex, the derivative $A'(\mu)$, $A(\mu) = \int_0^{\mu} A'(v)dv$, is monotone increasing almost everywhere. In what follows, we shall have to deal with inverse functions of increasing (not necessarily strictly monotone increasing) and not necessarily continuous functions (like $A'(\mu)$). If $F(u)$ is such a function, then one can set, for example,

$$F^{(-1)}(v) = \sup\{u : \ F(u) < v\} \tag{3.5.15}$$

(the generalized inverse function). This function is left- (lower-)continuous. If the inequality in (3.5.15) is replaced by $F(u) \le v$, then the function $F^{(-1)}(v)$ is right (upper) continuous. If the original function $F(u)$ is left-continuous and if the inverse function is defined as in (3.5.15), then the inverse function of $F^{(-1)}(v)$ coincides with F, and hence

$$(F^{(-1)})^{(-1)}(u) = F(u). \tag{3.5.16}$$

In a number of cases, it will be more convenient to use definition (3.5.15) on one half-axis $v > v_0$, while on the other half-axis $v \le v_0$ it would be better to use the inequality $F(u) \le v$ in place of the $F(u) < v$ in (3.5.15). When we are free to extend the original function $F(u)$ by continuity, which is always so for the increasing function $F(u) = A'(u)$), we can always satisfy (3.5.16) by taking F to be left-continuous for $u > F^{(-1)}(v_0)$ and right-continuous for $u \le F^{(-1)}(v_0)$.

In what follows, in order not to overburden the text with the above refinements, which mostly concern the function $A'(\mu)$, one may keep in mind the most common case of a strictly monotone continuous function $A'(\mu)$ on $\{\mu : A(\mu) \le 0\}$ (in this case such refinements are unnecessary). In order to pass to the general case, we should define $A'(\mu)$ as a right-continuous function on $(\mu^-, \mu(0)]$ and a left-continuous one on $(\mu(0), \mu^+)$. Similarly, we define the function $\mu(\alpha) = (A')^{(-1)}(\alpha)$; for this function the direction of continuity changes at the point $\alpha = 0$. Now (3.5.16) holds with $F = A'$. Thus, we can assume that the function $A'(\mu)$ (and therefore, the function $\mu(\alpha) = (A')^{(-1)}(\alpha)$) is defined everywhere as a function continuous "inside" the interval $[\mu^-, \mu^+]$ in both directions from the point $\mu(0)$ (from the point 0 in $[\alpha^-, \alpha^+]$, $\alpha^\pm := A'(\mu^\pm \mp 0)$, for the function $(A')^{(-1)}(\alpha)$). The required relations (3.5.43)–(3.5.46) are then satisfied (see below).

Note that the equality of functions (3.5.16) does not generally imply the equality of their arguments, that is, the relation $F^{(-1)}(F(u)) = u$. The latter equality is valid whenever the point u lies outside the intervals where F is constant.

Theorem 3.5.2 (i) *Let d_+, d_- be the boundary points of the effective domain $\mathcal{D}_1^{<\infty}$ of the function $D(\alpha)$,*

$$d_+ := \sup\{\alpha : D(\alpha) < \infty\}, \qquad d_- = \inf\{\alpha : D(\alpha) < \infty\}.$$

Then

$$d_- = \operatorname{ess\,inf} \frac{\zeta}{\tau}, \qquad d_+ = \operatorname{ess\,sup} \frac{\zeta}{\tau}. \tag{3.5.17}$$

(ii) *For $\mu \in (\mu_-, \mu_+)$ (see (3.5.8)), the function*

$$A(\mu) = -\sup\{\lambda : \mathbf{A}(\lambda, \mu) \le 0\}$$

is analytic and is the unique solution for λ of the equation $\mathbf{A}(-\lambda, \mu) = 0$.

(iii) *The function $\mu(\alpha)$ is the "generalized" inverse function of the derivative $A'(\mu)$:*

$$\mu(\alpha) = (A')^{(-1)}(\alpha) \tag{3.5.18}$$

(the required refinement of the continuity properties of the function $A'(\mu)$ was discussed above, together with the refinement of the inverse function $(A')^{(-1)}$ so as to satisfy (3.5.16)*), so that*

$$\mu(\alpha) = \mu^+ \quad \textit{for} \quad \alpha > \alpha^+ := A'(\mu^+ - 0),$$
$$\mu(\alpha) = \mu^- \quad \textit{for} \quad \alpha < \alpha^- := A'(\mu^- + 0).$$

(iv) *For all* α,

$$D'(\alpha) = (A')^{(-1)}(\alpha) = \mu(\alpha) \tag{3.5.19}$$

or, equivalently,

$$D(\alpha) = \int_a^\alpha \mu(v)dv, \quad A'(\mu) = (D')^{(-1)}(\mu); \\ \lim_{\alpha\to\infty} \frac{D(\pm\alpha)}{\alpha} = \pm\mu^{\pm}. \tag{3.5.20}$$

For $\alpha \geq a$, *the function* $D(\alpha)$ *is glued (with the preservation of the continuity of the derivatives) from the three pieces in general*:

(1) *an analytic function on* $[a, \alpha_+)$, $\alpha_+ := A'(\mu_+ - 0)$ (μ_+ *is defined in* (3.5.8)). *The function* $\mu(\alpha)$ *is also analytic on* $[a, \alpha_+)$;

(2) *an "intermediate" part on* (α_+, α^+), $\alpha^+ = A'(\mu^+ - 0)$, *in which some intervals of analyticity are possible; this piece is absent if* $\mu_+ = \mu^+$, *or, equivalently, if*

$$\mathbf{A}(\lambda, \mu) > 0 \quad for \quad (\lambda, \mu) \in \partial\mathcal{A}, \quad \mu \in (\mu_-, \mu^+); \tag{3.5.21}$$

(3) *the linear part*

$$D(\alpha) = D(\alpha^+) + (\alpha - \alpha^+)\mu^+ \quad for \quad \alpha \geq \alpha^+$$

(*this part is absent if* $\mu^+ = \infty$ *or* $D(\alpha^+) = \infty$).

A similar gluing takes place for $\alpha \leq a$ *and for* α_-, α^- *defined by* $\alpha_- := A'(\mu_- + 0)$, $\alpha^- := A'(\mu^- + 0)$.

(v) *The following equalities hold*:

$$D(a) = D'(a) = 0, \qquad D''(a) = \frac{1}{A''(0)} = \frac{a_\tau}{\mathbf{E}(\zeta - a\tau)^2}. \tag{3.5.22}$$

Note that, for independent τ and ζ,

$$d_- = \frac{\text{ess inf}\,\zeta}{\text{ess sup}\,\tau}, \qquad d_+ = \frac{\text{ess sup}\,\zeta}{\text{ess inf}\,\tau}.$$

In view of assertion (ii), the principal "analytic" difference of the second deviation function $D(\alpha)$ from the "ordinary" deviation function consists in the possible appearance of the "intermediate zones" (α^-, α_-), (α_+, α^+) between the domains of analyticity and linearity of the function $D(\alpha)$. The "ordinary" deviation function does not have these zones (see, e.g., [21, §9.1], [23, §1.1.1]). Recall that, similarly, the difference of the function $A(\mu)$ from the logarithm of the Laplace transform of some distribution is exhibited in the possible appearance of the intermediate zones (μ^-, μ_-), (μ_+, μ^+).

Proof of Theorem 3.5.2 (i) Let us verify the second of (3.5.17). By definition of $b := \text{ess sup}\,\frac{\zeta}{\tau}$, for any $\varepsilon > 0$, there exists $p > 0$ such that

$$\mathbf{P}\left(\frac{\zeta}{\tau} > b - \varepsilon\right) \geq 2p.$$

Moreover, there always exists $t_0 > 0$ such that

$$\mathbf{P}\Big(\frac{\zeta}{\tau} > b - \varepsilon, \tau > t_0\Big) \geq p.$$

Assume, for definiteness, that $b > 0$. We set

$$B_j = \Big\{\frac{\zeta_j}{\tau_j} > b - \varepsilon, \tau_j > t_0\Big\}, \quad B(n) = \prod_{j=1}^{n} B_j,$$

where $(\tau_j, \zeta_j) \underset{p}{=} (\tau, \zeta)$ (the process is homogeneous) and define $n_t = t/t_0$. On the set $B(n_t)$, we have

$$\eta(t) < \frac{t}{t_0} = n_t. \tag{3.5.23}$$

Assume, for simplicity, that t is such that n_t is an integer number. If we set

$$A_t = \Big\{\frac{Z(t)}{t} > b - \varepsilon\Big\},$$

then, clearly,

$$\mathbf{P}(A_t) \geq \mathbf{P}(A_t B(n_t)) = p^{n_t} \mathbf{P}(A_t \mid B(n_t)).$$

All the jumps of the process $Z(u)$ with numbers $j \leq n_t$ have the same "slope" $\frac{\zeta_j}{\tau_j} > b - \varepsilon$ on the set $B(n_t)$. Therefore, the trajectory of $Z(u)$ on the first $\eta(t) \leq n_t$ jumps has the average slope

$$\frac{Z(t)}{t} > b - \varepsilon.$$

This means that

$$\mathbf{P}(A_t \mid B(n_t)) = 1, \qquad \mathbf{P}(A_t) \geq p^{n_t}$$

and by (3.5.23),

$$\frac{1}{t} \ln \mathbf{P}(A_t) \geq \frac{n_t}{t} \ln p = \frac{\ln p}{t_0} > -\infty.$$

By the LDP,

$$\begin{aligned} -\infty < \varliminf_{t\to\infty} \frac{1}{t} \ln \mathbf{P}\Big(\frac{Z(t)}{t} > b - \varepsilon\Big) &\leq \varlimsup \frac{1}{t} \ln \mathbf{P}\Big(\frac{Z(t)}{t} > b - \varepsilon\Big) \\ &\leq -\widehat{D}((b - \varepsilon, \infty)). \end{aligned} \tag{3.5.24}$$

On the other hand, the slope of any jump (τ_j, ζ_j) of the process $Z(u)$ cannot exceed $b + \varepsilon$, and hence the average slope $\frac{Z(t)}{t}$ also has this property. By the LDP,

$$-\infty = \varliminf \frac{1}{t} \ln \mathbf{P}\Big(\frac{Z(t)}{t} > b + \varepsilon\Big) \geq -\widehat{D}((b + \varepsilon, \infty)). \tag{3.5.25}$$

From (3.5.24), (3.5.25) it follows that $\widehat{D}((b - \varepsilon, \infty)) < \infty$ and $\widehat{D}((b + \varepsilon, \infty)) = \infty$. The functions D and $\widehat{D}$ have the same effective domains (see Theorem 3.5.9 below), and hence, since $\varepsilon > 0$ is arbitrary, we get $d_+ = b = \operatorname{ess\,sup} \frac{\zeta}{\tau}$.

The proof of the first relation in (3.5.17) is the same (the modifications required in the case $b < 0$ are not essential). Assertion (i) of Theorem 3.5.2 is proved.[2]

(ii) The derivative $A'(\mu)$ is monotone increasing, and hence for $\alpha \in (\alpha_-, \alpha_+)$ the equation

$$A'(\mu) = \alpha \tag{3.5.26}$$

for the point $\mu = \mu(\alpha)$ at which the supremum in (3.5.13) is attained, is uniquely solvable, and moreover, $\mu(\alpha)$, which is the inverse function of $A'(\mu)$, lies in the analyticity domain of the function $A(\mu)$. Hence by the implicit function theorem, the functions $\mu(\alpha)$, $A(\mu(\alpha))$ and $D(\alpha) = \alpha\mu(\alpha) + A(\mu(\alpha))$ are analytic on (α_-, α_+).

(iii) To make the presentation more transparent, we *adopt the convention* that the set M^∞ in (3.5.10) consists of an at most *finite* number of closed intervals or half-open intervals (at terminal points of the domain (μ^-, μ^+)), and hence, the *derivative* $A'(\mu)$ *has a finite number of points of discontinuity*. Then the function $A(\mu)$ is everywhere differentiable, except possibly at a finite number of points. Assume, for definiteness, that $\alpha > a$ and $\mu_j > 0$, $j = 1, \ldots, k$ are the positive points of discontinuity. If the derivative $A'(\mu) = \alpha$ exists at the point μ, then by (3.5.26) $\mu = \mu(\alpha)$ is the value of the inverse function of $A'(\mu)$. Consider now the points of discontinuity. Let $\alpha_1 = A'(\mu_1 - 0)$. Then, for $\alpha > \alpha_1$, by definition the quantity $\mu(\alpha)$ has the constant value μ_1 in the domain $(\alpha_1, \alpha_1^+]$, where $\alpha_1^+ = A'(\mu_1 + 0)$. For $\alpha > \alpha_1^+$, relation (3.5.26) is again effective up to the point $\alpha_2 = A'(\mu_2 - 0)$, and so on. But the above construction of the function $\mu(\alpha)$ is nothing other than the definition (given in (3.5.18)) of the generalized inverse function.

Next, differentiating the identity

$$D(\alpha) = \alpha\mu(\alpha) - A(\mu(\alpha))$$

with respect to $\alpha \in (\alpha_-, \alpha_+)$ and using (3.5.26), we find that

$$D'(\alpha) = \mu(\alpha) + \alpha\mu'(\alpha) - A'(\mu(\alpha))\mu'(\alpha) = \mu(\alpha). \tag{3.5.27}$$

Hence, for $\alpha_0 \in (\alpha_-, \alpha_+)$ and $\alpha \in (\alpha_-, \alpha_+)$,

$$D(\alpha) = D(\alpha_0) + \int_{\alpha_0}^{\alpha} \mu(\alpha) d\alpha. \tag{3.5.28}$$

Let us now ascertain the behavior of the function $D(\alpha)$ for $\alpha > \alpha_+$ and $\alpha < \alpha_-$. Let $\alpha > \alpha_+$. Outside the intervals $[\alpha_j, \alpha_j^+]$, where $\alpha_j = A'(\mu_j - 0)$, $\alpha_j^+ = A'(\mu_j + 0)$, $j = 1, \ldots, k$, the above relation (3.5.27) is preserved, and $\mu(\alpha) = D'(\alpha)$. For $\alpha \in [\alpha_j, \alpha_j^+]$,

$$D(\alpha) = \alpha\mu(\alpha) - A(\mu(\alpha)) = \alpha\mu_j - A(\mu_j) = D(\alpha_j) + (\alpha - \alpha_j)\mu_j$$

[2] For independent τ and ζ, assertion (i) of Theorem 3.5.2 was proved in assertion (ii) of Theorem 2.2 in [26], but in [26] the independence condition in assertion (ii) of Theorem 2.2 is missing.

and the function D is linear, remaining as a whole convex and continuous. This means that the integral relation (3.5.28) is preserved for $\mu(\alpha) = \mu_j$ in the domains $\alpha \in [\alpha_j, \alpha_j^+]$, $j = 1, \dots, k$.

Similarly, for $\mu^+ < \infty$, $\alpha > \alpha^+ := A'(\mu^+ - 0)$,

$$D(\alpha) = \alpha\mu^+ - A(\mu^+) = D(\alpha^+) + (\alpha - \alpha^+)\mu^+, \tag{3.5.29}$$

so that $\mu(\alpha) = \mu^+$ for $\alpha \geq \alpha^+$. This proves (3.5.20), from which it follows that

$$\frac{D(\alpha)}{\alpha} \to \mu^+ \quad \text{as} \quad \alpha \to \infty. \tag{3.5.30}$$

The cases when $\alpha < \alpha_-$, the points of jumps are negative, and the case $\alpha < a$ are dealt with similarly.

Now the properties of the function D claimed in assertions (1)–(3), (iv) of the theorem follow from (3.5.20), (3.5.18), (3.5.29) and the above.

In the general case (without the simplifying assumption that $A'(\mu)$ has a finite number of discontinuities), relations (3.5.18), (3.5.20) were proved in [89]. It remains unclear to what extent this assumption reduces the generality. The author is unaware of any examples of distributions of (τ, ζ) for which the function $A'(\mu)$ has an infinite number of points of discontinuity. The construction of such an example (if it exists) requires considerable effort. Possible points of discontinuity are the tacnodes (points of contact) of the curves $\partial\mathcal{A}^{\leq 0}$ and $\partial\mathcal{A}$. Hence, if $\mathcal{A} = \mathbb{R}^2$ or if $\mathbf{A}(\lambda, \mu) > 0$ for $(\lambda, \mu) \in \partial\mathcal{A}$, then the section of the surface $z = \mathbf{A}(\lambda, \mu)$ at $z = 0$ is analytic, and moreover, $A'(\mu)$ has no points of discontinuity at all. In the case $(\mathcal{A}) = \{\lambda < \lambda_+\}$, the function $A'(\mu)$ has at most two points of discontinuity. If the domain $(\mathcal{A})$ is rectangular (as in the case of independent τ and ζ), the number of discontinuities is at most four.

(v) Differentiating with respect to $\alpha \in (\alpha_-, \alpha_+)$ the identity

$$A'(\mu(\alpha)) = \alpha,$$

we find that

$$\mu'(\alpha) = \frac{1}{A''(\mu(\alpha))}.$$

We have $\mu(a) = 0$, $D'(\alpha) = \mu(\alpha)$, and hence at the point $\alpha = a$ we get

$$D(a) = D'(a) = 0, \qquad D''(a) = \mu'(a) = \frac{1}{A''(0)}.$$

Now (3.5.22) follows from (3.5.11).

Theorem 3.5.2 is proved. □

Corollary 3.5.3 *From assertion* (v) *of Theorem* 3.5.2 *it follows that, for* α *from the neighborhood of* $a \in (\alpha_-, \alpha_+)$,

$$D(\alpha) = \frac{a_\tau(\alpha - a)^2}{2\mathbf{E}(\zeta - a\tau)^2} + O(|\alpha - a|^3). \tag{3.5.31}$$

Remark 3.5.4 We note the following property of the function $A(\mu)$ clarifying to some extent its nature, which is determined by analytic properties of the Laplace transform $\mathcal{H}(\lambda,\mu)$ of the renewal measure $H(dt,d\alpha)$ in the homogeneous case. Indeed,

$$\mathcal{H}(\lambda,\mu) := \int e^{\lambda t+\mu\alpha} H(dt,d\alpha) = \sum_{n=0}^{\infty} \int e^{\lambda t+\mu\alpha} \mathbf{P}(T_n \in dt, Z_n \in d\alpha)$$
$$= \sum_{n=0}^{\infty} \psi^n(\lambda,\mu) = \frac{1}{1-\boldsymbol{\psi}(\lambda,\mu)} \tag{3.5.32}$$

and the convex curve $\lambda = -A(\mu)$ in $\mathbb{R}^2$ consists of singular points of the function $\frac{1}{1-\psi(\lambda,\mu)}$; the domain of analyticity of the function $\mathcal{H}(\lambda,\mu)$ lies inside this curve. If, for example, $\mathbf{A}(\lambda,\mu) > 0$ for $(\lambda,\mu) \in \partial\mathcal{A}$, then all singularities on the curve $\lambda = -A(\mu)$ form the continuum of poles (zeros) of the function $1-\psi(\lambda,\mu)$.

The LDP for $H(T\delta[t], T\Delta[\alpha])$ in Theorem 3.2.1 (see (3.2.21)) together with formula (3.3.1) for the function $\mathbf{D}(t,\alpha)$ arise from the "rough" inversion of representation (3.5.32) for $\mathcal{H}(\lambda,\mu)$ based on the location of singularities of this function (that is, on the curve $\lambda = -A(\mu)$).

In the one-dimensional case, such an inversion is simpler. For the renewal measure $H^{(\zeta)}(B) = \sum_{n=0}^{\infty} \mathbf{P}(Z_n \in B)$, the Laplace transform has the form $\mathcal{H}^{(\zeta)}(\mu) = (1-\psi^{(\zeta)}(\mu))^{-1}$, and for $\alpha > 0$,

$$\frac{1}{T} \ln H^{(\zeta)}(T\Delta[\alpha]) \sim -\mu^{(\zeta)} \quad \text{as} \quad T \to \infty, \ \Delta = \overline{o}(1),$$

where

$$\mu^{(\zeta)} = \max\left(\mu : A^{(\zeta)}(\mu) \le 0\right) = \max\left(\mu : \psi^{(\zeta)}(\mu) \le 1\right).$$

If $\mu^{(\zeta)} > 0$, $A^{(\zeta)}(\mu^{(\zeta)} + \varepsilon) < \infty$ with some $\varepsilon > 0$, then the point $\mu^{(\zeta)}$ is an isolated pole of the function $\mathcal{H}^{(\zeta)}(\mu)$. Moreover, one can easily derive the more accurate asymptotic representation

$$H^{(\zeta)}(T\Delta[\alpha]) \sim c e^{-\mu^{(\zeta)}\alpha T} \quad \text{as} \quad T \to \infty.$$

Remark 3.5.5 We also note that the functions $A(\mu)$, $D(\alpha)$ are invariant with respect to the "aggregations" of the jumps of $\boldsymbol{\xi}$. Namely, consider the CRP $Z^*(t)$ controlled by the vector $\boldsymbol{\xi}^* = \boldsymbol{\xi}_1 + \boldsymbol{\xi}_2$, where $\boldsymbol{\xi}_1$ and $\boldsymbol{\xi}_2$ are independent. Hence, with obvious changes in notation, for the logarithm of the Laplace transform $\mathbf{A}^*(\lambda,\mu)$ of the distribution of $\boldsymbol{\xi}^*$, we have

$$\mathbf{A}^*(\lambda,\mu) = 2\mathbf{A}(\lambda,\mu), \quad A^*(\mu) = A(\mu), \quad D^*(\alpha) = D(\alpha),$$

even though the distributions of $Z^*(t)$ and $Z(t)$ are different (cf. Remark 3.5.17).

3.5.2 Properties of the Functions $\mu(\alpha)$ and $\mu(\alpha)$

Let us continue the study of the functions $\mathbf{D}(\theta, \alpha)$, $D(\alpha)$ and describe the set of values α for which $\widehat{D}(\alpha) < D(\alpha)$. To this end, we shall need the properties of the functions $\mu(\alpha)$, $\lambda(\alpha)$ and of the value $D(0)$, which together with the value λ_+ determine the appearance of the inequalities $\widehat{D}(\alpha) < D(\alpha)$.

Some properties of the parameter $D(0)$ are given in the following lemma.

Lemma 3.5.6 (i) *If*

$$\mathcal{A}^{\leq 0} \subset \{\lambda \leq \lambda_+\}, \tag{3.5.33}$$

then $\lambda_+ \geq D(0)$.

(ii) *If* ζ *takes values of both signs (that is,* $\mathbf{P}(\zeta \gtrless 0) > 0$*), then* $D(0) < \infty$.

(iii) *Let* $\mathbf{P}(\zeta \geq 0) = 1$, $\mathbf{P}(\zeta = 0) > 0$, $\psi_0^{(\tau)}(\lambda) := \mathbf{E}(e^{\lambda\tau}; \zeta = 0)$, *and let*

$$\lambda_{0,+} := \sup\{\lambda : \psi_0^{(\tau)}(\lambda) < \infty\}.$$

Then $D(0) = \lambda_0 \leq \lambda_{0,+}$, *where* λ_0 *is the "generalized" solution of the equation* $\psi_0^{(\tau)}(\lambda) = 1$, *that is,*

$$\lambda_0 = \begin{cases} \text{is the solution of the equation } \psi_0^{(\tau)}(\lambda) = 1 \text{ if } \psi_0^{(\tau)}(\lambda_{0,+}) \geq 1 \\ \lambda_{0,+} \text{ if } \psi_0^{(\tau)}(\lambda_{0,+}) < 1. \end{cases}$$

The inequality $D(0) < \infty$ *always holds.*

If τ *and* ζ *are independent, then* $\lambda_{0,+} = \lambda_+$.

(iv) *If*

$$\mathbf{P}(\zeta > 0) = 1, \qquad (\mathcal{A}) = (-\infty, \lambda_+) \times (-\infty, \mu_+^{(\zeta)}) \tag{3.5.34}$$

(like for independent τ *and* ζ*), then* $D(0) = \lambda_+$.

The first assertion of the lemma generalizes assertion (iii), (1) of Theorem 3.3.2. The analogs of assertions (iii), (iv) hold in the case $\mathbf{P}(\zeta \leq 0) = 1$.

It is worth augmenting Lemma 3.5.6 with Examples 3.5.1 (see below), in which the possibilities $\lambda_+ < D(0)$ and $D(0) = \infty$ are realized.

Proof of Lemma 3.5.6 (i) By (3.5.1),

$$D(0) = \sup_{(\lambda,\mu)\in\mathcal{A}^{\leq 0}} \lambda.$$

In view of condition (3.5.33), this means that $D(0) \leq \lambda_+$.

(ii) If ζ takes values of both signs, then $A(\mu) \to \infty$ as $|\mu| \to \infty$, and therefore, $\inf A(\mu) = -D(0) > -\infty$.

(iii) If $\mathbf{P}(\zeta \geq 0) = 1$, $\mathbf{P}(\zeta = 0) > 0$, then

$$\mathbf{A}(\lambda, \mu) \to \ln \mathbf{E}(e^{\lambda\tau}; \zeta = 0) = \psi_0^{(\tau)}(\lambda) \quad \text{as} \quad \mu \to -\infty.$$

As a result, $A(\mu) \to -\lambda_0$ as $\mu \to -\infty$, and $D(0) = -\inf A(\mu) = \lambda_0$. Since λ_0 is always finite, $D(0) < \infty$.

If τ and ζ are independent, then $\psi_0^{(\tau)}(\lambda) = \psi^{(\tau)}(\lambda)\mathbf{P}(\zeta = 0)$, $\lambda_{0,+} = \lambda_+$, and $D(0) = \lambda_0 \le \lambda_+$.

(iv) Since $\zeta > 0$, we have, for any $\varepsilon > 0$,

$$\mathbf{A}(\lambda_+ - \varepsilon, \mu) \to \infty \quad \text{as} \quad \mu \to -\infty.$$

On the other hand, for all μ,

$$\mathbf{A}(\lambda_+ + 0, \mu) = \infty. \tag{3.5.35}$$

Therefore,

$$A(\mu) \to -\lambda_+ \text{ as } \mu \to -\infty, \quad D(0) = -\inf A(\mu) = \lambda_+. \tag{3.5.36}$$

Lemma 3.5.6 is proved. □

Recall that the condition $\lambda_+ \ge D(0)$ is quite broad. It is satisfied if at least one of the conditions among (3.3.14) of Theorem 3.3.2 and (i), (iv) of Lemma 3.5.6 is met.

Condition $\lambda_+ < D(0)$ is a "pathological" condition – under this condition both the "regularity" of the behavior of the function $\widehat{D}(\alpha)$ (see below) and the summarizing character of the LDP are lost, because in this case in accordance with the LDP

$$\ln \mathbf{P}(Z(T) \in T\Delta[0)) \sim -T\widehat{D}(0) = -T\lambda_+ \quad \text{as} \quad T \to \infty$$

and the principal contribution into the probability of the event $\{Z(T) \in T\Delta[v)\}$ comes from one great jump $\tau_1 > T$ (recall that $\{\tau_1 > T\} \subset \{Z(T) = 0\}$, and moreover, $\ln \mathbf{P}(\tau_1 > T) \sim -T\lambda_+$ as $T \to \infty$ in the homogeneous case by condition $[\overline{\lambda_+}]$).

To find representations for the function $\widehat{D}(\alpha)$, we shall need to know the form of the derivatives of the function $\mathbf{D}(\theta, \alpha)$.

Lemma 3.5.7 *For all $\theta > 0$ and α,*

$$\mathbf{D}'_{(1)}(\theta, \alpha) = \lambda(\alpha/\theta), \qquad \mathbf{D}'_{(2)}(\theta, \alpha) = \mu(\alpha/\theta).$$

Proof By the above, the function $\mathbf{D}(\theta, \alpha) = \theta D(\alpha/\theta)$ is finite in the cone

$$\mathcal{D}^{<\infty} = \left\{(\theta, \alpha) : D\left(\frac{\alpha}{\theta}\right) < \infty\right\}$$

and is analytic in the cone

$$\mathcal{D} = \left\{(\theta, \alpha) : \frac{\alpha}{\theta} \in (\alpha_-, \alpha_+)\right\}. \tag{3.5.37}$$

If we set

$$\lambda(\alpha) := -A(\mu(\alpha)),$$

then from (3.5.14) for $\alpha \in (\alpha_-, \alpha_+)$ we have

$$\begin{gathered} D(\alpha) = \lambda(\alpha) + \alpha\mu(\alpha), \quad \mathbf{A}(\lambda(\alpha), \mu(\alpha)) = 0, \\ (\lambda(\alpha), \mu(\alpha)) \in (\mathcal{A}), \end{gathered} \tag{3.5.38}$$

that is, with increasing $\alpha \in (\alpha_-, \alpha_+)$ the point $(\lambda(\alpha), \mu(\alpha))$ moves along the boundary $\partial\mathcal{A}^{\le 0}$ and passes through the point $(0,0)$ for $\alpha = a$. The pair $(\lambda(\alpha), \mu(\alpha))$ determines, in parallel with $(-A(\mu), \mu)$, another parameterization of the smooth part of the boundary of the set $\mathcal{A}^{\le 0}$.

In view of $\mathbf{D}(\theta, \alpha) = \theta D(\alpha/\theta)$ (see (3.3.2)) and (3.5.20), for $(\theta, \alpha) \in \mathcal{D}$ we have

$$\mathbf{D}'_{(1)}(\theta, \alpha) = D\left(\frac{\alpha}{\theta}\right) - \frac{\alpha}{\theta} D'\left(\frac{\alpha}{\theta}\right) = \frac{\alpha}{\theta}\mu\left(\frac{\alpha}{\theta}\right) - A\left(\mu\left(\frac{\alpha}{\theta}\right)\right) - \frac{\alpha}{\theta}\mu\left(\frac{\alpha}{\theta}\right)$$
$$= -A\left(\mu\left(\frac{\alpha}{\theta}\right)\right) = \lambda\left(\frac{\alpha}{\theta}\right), \tag{3.5.39}$$
$$\mathbf{D}'_{(2)}(\theta, \alpha) = D'\left(\frac{\alpha}{\theta}\right) = \mu\left(\frac{\alpha}{\theta}\right). \tag{3.5.40}$$

But by Theorem 3.5.2 the function $D(\alpha)$ is differentiable everywhere, $D'(\alpha) = \mu(\alpha)$, so that these relations remain valid everywhere. Lemma 3.5.7 is proved. □

Consider now the properties of the function $\lambda(\alpha)$. In order to formulate the next lemma, we define, in the case $\lambda_+ < D(0)$, the values

$$m_- = \begin{cases} \min\{\mu : A(\mu) = -\lambda_+\} & \text{if} \quad A(\mu^-) \ge -\lambda_+, \\ \mu^- & \text{if} \quad A(\mu^-) < -\lambda_+; \end{cases}$$
$$m_+ = \begin{cases} \max\{\mu : A(\mu) = -\lambda_+\} & \text{if} \quad A(\mu^+) \ge -\lambda_+, \\ \mu^+ & \text{if} \quad A(\mu^+) < -\lambda_+. \end{cases} \tag{3.5.41}$$

Note that the condition $A(\mu^\pm) < -\lambda_+$ *is equivalent to the inequality*

$$\mathbf{A}(\lambda_+, \mu^\pm) < 0.$$

Indeed, by definition

$$A(\mu^-) = -\sup\{\lambda : \mathbf{A}(\lambda, \mu^-) \le 0\}.$$

Assume, for simplicity, that $\mathbf{A}(-A(\mu^-), \mu^-) = 0$. Hence if $\lambda_+ < -A(\mu^-)$, then $\mathbf{A}(\lambda_+, \mu^-) < 0$, and vice versa. The case $\lambda_+ < -A(\mu^+)$ is treated similarly.

Lemma 3.5.8 (i) *The function $\lambda(\alpha) = -A(\mu(\alpha))$ attains its maximum at the point $\alpha = 0$, $\lambda(0) = D(0)$. It increases (with possible intervals of constancy) for $\alpha < 0$ and it decreases for $\alpha > 0$; $\lambda(a) = 0$.*

(ii) *Let $\lambda_+ < D(0)$ and let (β_-, β_+) be the maximal interval on which $\lambda(\alpha) > \lambda_+$, so that*

$$\lambda(\alpha) \le \lambda_+ \quad \textit{for} \quad \alpha \notin (\beta_-, \beta_+), \quad 0 \in (\beta_-, \beta_+).$$

Next, let $A(\mu^-) \ge -\lambda_+$ (this is possible only if $a > 0$ and if ζ takes values of both signs, that is, $\mathbf{P}(\zeta \gtrless 0) > 0$). Then there exist two finite solutions $m_- < m_+ < 0$ of the equation $A(\mu) = -\lambda_+$, and moreover,

$$\beta_- = A'(m_-) := A'(m_- + 0),$$
$$\beta_+ = A'(m_+) := A'(m_+ - 0) \in (0, a). \tag{3.5.42}$$

The equality $\beta_- = -\infty$ *is possible only in the case* $A(\mu^-) = -\lambda_+$ $(m_- = \mu^-)$, $A'(\mu^-) = -\infty$.

If $A(\mu^-) < -\lambda_+$, *then* $A(m_+) = -\lambda_+$, *the second relation in* (3.5.42) *holds, and* $\beta_- = -\infty$.

The analogous results also hold in the case $a < 0$.

(iii) *Let* $\lambda_+ < D(0)$ *and let* $\mathbf{P}(\zeta \geq 0) = 1$. *Then* $\mu^- = -\infty$, $A(\mu_-) = -D(0) < -\lambda_+$, $\beta_- = -\infty$. *The analogous result also holds in the case* $\mathbf{P}(\zeta \leq 0) = 1$.

Proof (i) The first assertion of the lemma follows from the equality $\lambda(\alpha) = -A(\mu(\alpha))$, since the function $\mu(\alpha)$ is increasing, and from the relations

$$\inf_{\alpha} A(\mu(\alpha)) = \inf_{\mu} A(\mu) = -D(0) = A(\mu(0)).$$

The equality $\lambda(a) = 0$ is clear, because $\lambda(a) = -A(\mu(a)) = -A(0) = 0$.

(ii) Let us prove the second assertion. Since $a > 0$, $A(0) = 0$, the function $A(\mu)$ is convex and continuous from inside of the effective domain, we have

$$\inf A(\mu) = \inf_{\mu \leq 0} A(\mu) = -D(0) < -\lambda_+.$$

Therefore, in the case $A(\mu^-) > -\lambda_+$ there exist two solutions $m_- < m_+ < 0$, $m_- > \mu^-$ of the equation $A(\mu) = -\lambda_+$. If $A(\mu^-) = -\lambda_+$, then $m_- = \mu^-$, but the case $\mu^- = -\infty$ is impossible, since $A(\mu)$ is convex, so that $m_- = \mu^- > -\infty$.

Consider the values $\beta_\pm$. We have

$$\beta_+ = \sup\{\alpha > 0 : \lambda(\alpha) > \lambda_+\} = \sup\{\alpha > 0 : \mu(\alpha) < m_+\}, \tag{3.5.43}$$

where on the right we have nothing else as the definition of the left-continuous inverse function of $\mu(\alpha) = (A')^{(-1)}(\alpha)$ at the point m_+. Since $A'(\mu)$ is also left-continuous on $(\mu(0), \mu^+)$, $m_+ < 0$, we have by (3.5.16)

$$\beta_+ = \mu^{(-1)}(m_+) = A'(m_+) \in (0, a). \tag{3.5.44}$$

Next,

$$\beta_- = \inf\{\alpha < 0 : \ \lambda(\alpha) = -A(\mu(\alpha)) > \lambda_+\}.$$

Since the function $-A(\mu)$ in increasing in the domain $(\mu^-, \mu(0))$, we have

$$\beta_- = \inf\{\alpha < 0 : \ \mu(\alpha) > m_-\} = \sup\{\alpha < 0 : \ \mu(\alpha) \leq m_-\}. \tag{3.5.45}$$

Here on the right we have the relation determining the value of the right-continuous inverse function of $(A')^{(-1)}(\mu)$ at the point m_-. The function $A'(\mu)$ is also right-continuous on $(\mu^-, \mu(0))$, and hence by (3.5.16),

$$\beta_- = \mu^{(-1)}(m_-) = A'(m_-). \tag{3.5.46}$$

If

$$A(\mu^-) = -\lambda_+ \ (m_- = \mu^-) \text{ and } \alpha^- = A'(\mu^- + 0) = A'(\mu^-) = A'(m_-) = -\infty,$$

then $\beta_- = -\infty$.

If

$$A(\mu^-) > -\lambda_+ \text{ or } A(\mu^-) = -\lambda_+ \ (m_- = \mu^-),\ A'(\mu^-) = A'(\mu^- + 0) > -\infty,$$

then $\beta_- > -\infty$.

The last claim of assertion (ii) is clear. Assertion (iii) is straightforward.

Lemma 3.5.8 is proved. □

To conclude this section, we summarize in Figure 3.1 the relations between the principal characteristics of CRPs (without devitalization of the conditions).

Figure 3.1. *Main characteristics of CRPs*

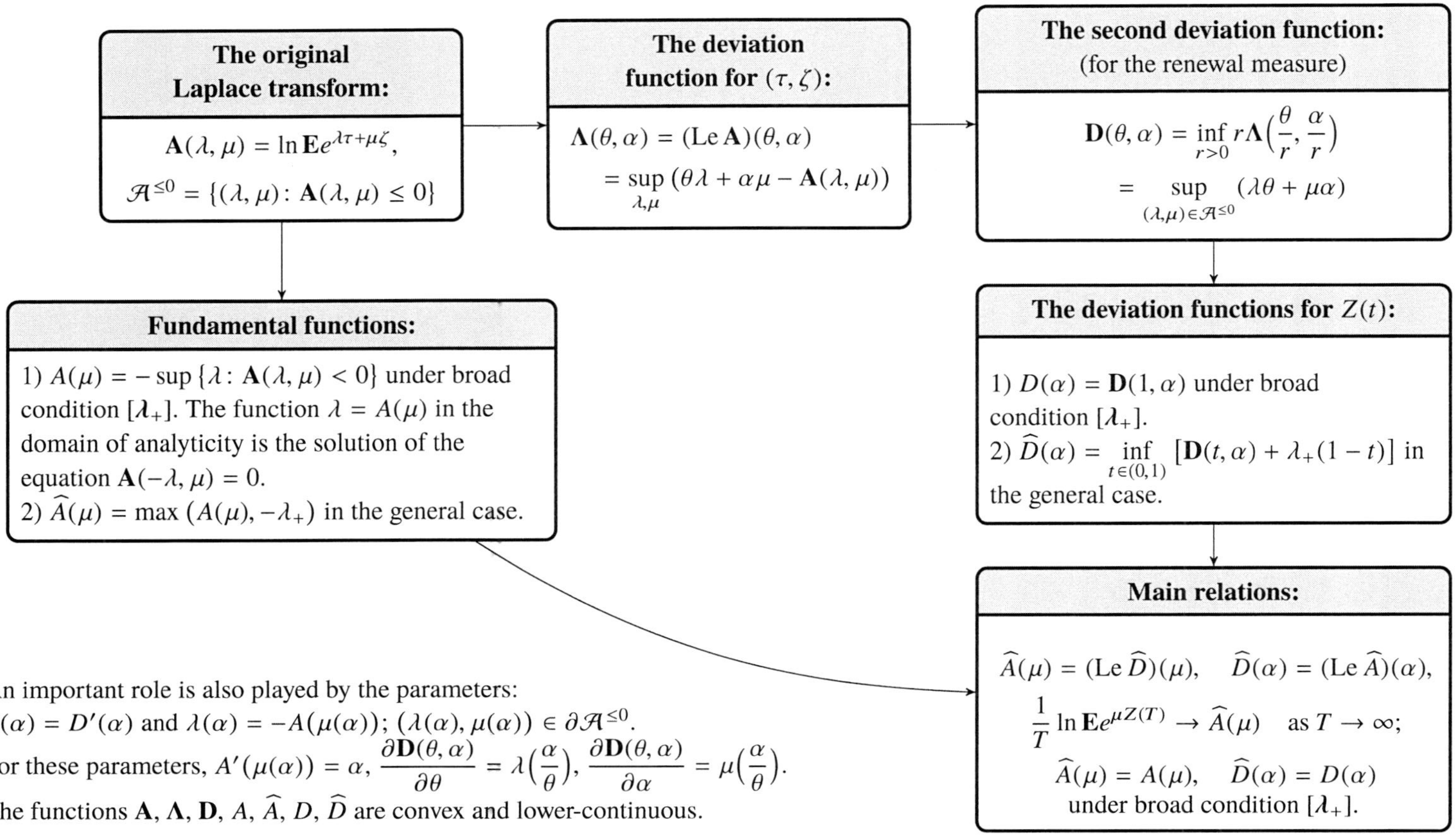

An important role is also played by the parameters:
$\mu(\alpha) = D'(\alpha)$ and $\lambda(\alpha) = -A(\mu(\alpha))$; $(\lambda(\alpha), \mu(\alpha)) \in \partial\mathcal{A}^{\le 0}$.
For these parameters, $A'(\mu(\alpha)) = \alpha$, $\dfrac{\partial \mathbf{D}(\theta, \alpha)}{\partial \theta} = \lambda\left(\dfrac{\alpha}{\theta}\right)$, $\dfrac{\partial \mathbf{D}(\theta, \alpha)}{\partial \alpha} = \mu\left(\dfrac{\alpha}{\theta}\right)$.
The functions $\mathbf{A}$, $\mathbf{\Lambda}$, $\mathbf{D}$, A, $\widehat{A}$, D, $\widehat{D}$ are convex and lower-continuous.

3.5.3 Properties of the Deviation Function in Its General Form and of the Corresponding Fundamental Function

We can now describe the properties of the deviation function $\widehat{D}(\alpha)$ in its general form (3.3.6). As was already pointed out, $\widehat{D}(\alpha) = D(\alpha)$ for all α if $\lambda_+ \geq D(0)$. For $\lambda_+ > D(0)$, the interval $[\beta_-, \beta_+]$ is empty. For $\lambda_+ = D(0)$, in a broad class of cases the interval $[\beta_-, \beta_+]$ degenerates to the point $\beta_\pm = 0$, for example, in the case when the point 0 lies in the interval $(A'(\mu_- + 0), A'(\mu_+ - 0))$, where $(\mu_-, \mu_+) \ni 0$ is the analyticity domain of the function $A(\mu)$.

Consider the case $\lambda_+ < D(0)$. We have the following theorem, in which for completeness of presentation, some results repeat those from Lemma 3.5.8.

Theorem 3.5.9 *Let $\lambda_+ < D(0)$ and let, for definiteness, $a > 0$ (in this case $\beta_+ \in (0, a)$).*

(i) *If $A(\mu^-) \geq -\lambda_+$ (this is possible only if ζ takes values of both signs, that is, $\mathbf{P}(\pm\zeta > 0) > 0$), then there exist solutions $m_- < m_+ < 0$ of the equation $A(\mu) = -\lambda_+$,*

$$\beta_\pm = A'(m_\pm), \tag{3.5.47}$$

and $\beta_- = -\infty$ only if $A(\mu^-) = -\lambda_+$, $A'(\mu^-) = -\infty$. Furthermore,

$$\widehat{D}(\alpha) = \begin{cases} D(\alpha) & \text{for } \alpha \notin (\beta_-, \beta_+), \\ D(\beta_-) + (\alpha - \beta_-)m_- = \alpha m_- + \lambda_+ & \text{for } \alpha \in (\beta_-, 0], \\ D(\beta_+) + (\alpha - \beta_+)m_+ = \alpha m_+ + \lambda_+ & \text{for } \alpha \in [0, \beta_+). \end{cases} \tag{3.5.48}$$

If the point $m_\pm$ does not lie in an interval of constancy of the function $A'(\mu)$ (for example, if $A''(m_\pm) > 0$), then $m_\pm = \mu(\beta_\pm)$ holds, which is the "inverse" equality of (3.5.47).

(ii) *If $A(\mu^-) < -\lambda_+$ and ζ takes values of both signs, then there exists a unique solution $m_+ < 0$ of the equation $A(\mu) = -\lambda_+$. Moreover, $\mu^- > -\infty$, and (3.5.48) holds, where the middle row should be replaced by*

$$\widehat{D}(\alpha) = \alpha\mu^- + \lambda_+ \quad \text{for} \quad \alpha < 0. \tag{3.5.49}$$

If $A(\mu^-) < -\lambda_+$ and $\mathbf{P}(\zeta \geq 0) = 1$, then $\mu^- = -\infty$, and the middle row in (3.5.48) should be replaced by

$$\widehat{D}(\alpha) = \infty \quad \text{for} \quad \alpha < 0. \tag{3.5.50}$$

The analogous ("symmetric") *relations hold in the case $a < 0$.*

If the definition (3.5.41) of the values $m_\pm$ is used, then all assertions (3.5.48)–(3.5.50) of Theorem 3.5.9 can be unified for any a as follows.

Corollary 3.5.10 *If $\lambda_+ < D(0)$, then*

$$\widehat{D}(\alpha) = \begin{cases} D(\alpha) & \text{for} \quad \alpha \notin (\beta_-, \beta_+), \\ \alpha m_- + \lambda_+ & \text{for} \quad \alpha \in (\beta_-, 0], \\ \alpha m_+ + \lambda_+ & \text{for} \quad \alpha \in [0, \beta_+). \end{cases}$$

From Theorem 3.5.9 it also follows that if $a > 0$, $A''(m_+) > 0$, then the deviation function $\widehat{D}(\alpha)$ is differentiable at the point $\alpha = \beta_+$,

$$\widehat{D}'(\beta_+) = m_+ = \mu(\beta_+) = D'(\beta_+),$$

that is, the straight line $y = D(\beta_+) + (\alpha - \beta_+)m_+$ (see (3.5.48)) is a tangent line to the curve $y = D(\alpha)$ at the point $\alpha = \beta_+$.

A similar assertion also holds for the point $\alpha = \beta_-$ for $a < 0$.

Proof of Theorem 3.5.9 (i) Let $A(\mu^-) \geq -\lambda_+$. Then, by Lemma 3.5.8, the values $m_- < m_+ < 0$ are finite, $\beta_\pm = A'(m_\pm)$. From the definition of the function $\mathbf{D}(\theta, \alpha)$ it follows that it is ruled, that is, $\mathbf{D}(t\theta, t\alpha) = t\mathbf{D}(\theta, \alpha)$ for $t > 0$, so that

$$\mathbf{D}(t, \alpha) = tD\left(\frac{\alpha}{t}\right). \tag{3.5.51}$$

Next, by Lemma 3.5.7,

$$\frac{\partial}{\partial t}\left[\mathbf{D}(t, \alpha) + \lambda_+(1 - t)\right] = \lambda\left(\frac{\alpha}{t}\right) - \lambda_+. \tag{3.5.52}$$

For $t = 1$ and $\alpha \in (\beta_-, \beta_+)$, the right-hand side of (3.5.52) is equal to $\lambda(\alpha) - \lambda_+ > 0$. In other words, the function inside the derivative sign in (3.5.52) is convex and decreases with decreasing t from $t = 1$. Hence

$$\widehat{D}(\alpha) < \mathbf{D}(1, \alpha) = D(\alpha).$$

For $t = 1$ and $\alpha \notin (\beta_-, \beta_+)$, the right-hand side in (3.5.52) is $\lambda(\alpha) - \lambda_+ \leq 0$, and the convex function inside the derivative sign in (3.5.52) increases with decreasing t from $t = 1$. Hence $\widehat{D}(\alpha) = \mathbf{D}(1, \alpha) = D(\alpha)$.

Next, for $\alpha \in [0, \beta_+)$, we let t_α denote the point from $[0, 1]$ at which the infimum in the definition of the function $\widehat{D}(\alpha)$ is attained (see (3.3.6)).

From the definition of the number β_+ (see Lemma 3.5.8) we have

$$\lambda(\beta_+ - 0) > \lambda_+, \qquad \lambda(\beta_+ + 0) \leq \lambda_+. \tag{3.5.53}$$

The function $D(t, \alpha)$ is convex, and hence by (3.5.52) inequalities (3.5.53) mean that

$$\frac{\alpha}{t_\alpha} = \beta_+, \qquad t_\alpha = \frac{\alpha}{\beta_+}. \tag{3.5.54}$$

Hence, for $\alpha \in [0, \beta_+)$, from (3.3.6) we have

$$\widehat{D}(\alpha) = \mathbf{D}(t_\alpha, \alpha) + \lambda_+(1 - t_\alpha) = \frac{\alpha}{\beta_+}D(\beta_+) + \lambda_+\left(1 - \frac{\alpha}{\beta_+}\right).$$

As a result, we get

$$\widehat{D}(\alpha) = \begin{cases} D(\alpha) & \text{for } \alpha \notin (\beta_-, \beta_+), \\ D(\beta_+) + \frac{\alpha - \beta_+}{\beta_+}(D(\beta_+) - \lambda_+) & \text{for } \alpha \in [0, \beta_+). \end{cases} \tag{3.5.55}$$

We claim that

$$D(\beta_+) = m_+\beta_+ + \lambda_+. \tag{3.5.56}$$

There are two cases to consider:

$$\text{either} \quad \mu(\beta_+) = m_+ \qquad \text{or} \quad \mu(\beta_+) \neq m_+. \tag{3.5.57}$$

In the first case (this case always holds if $A''(m_+) > 0$),

$$D(\beta_+) = \mu(\beta_+)\beta_+ - A(\mu(\beta_+)) = \mu(\beta_+)\beta_+ + \lambda(\beta_+) = m_+\beta_+ + \lambda_+.$$

In the second case, the function $\mu(\alpha)$ is necessarily discontinuous at the point $\alpha = \beta_+$, and the left-continuous (for $\mu > \mu(0)$) generalized inverse function $A'(\mu)$ of $\mu(\alpha)$ has the half-open interval of constancy

$$\mathcal{M} := \big[\mu(\beta_+),\ \mu(\beta_+ + 0)\big).$$

Moreover, the point m_+ lies in this half-open interval (so that the second possibility in (3.5.57) can be written in the form $m_+ \in \mathcal{M}$, which does not exclude the first possibility). Next,

$$D(\beta_+) = \sup_{\mu} \{\mu\beta_+ - A(\mu)\} = \sup_{\mu \in \mathcal{M}} \{\mu\beta_+ - A(\mu)\}, \tag{3.5.58}$$

where, for $\mu \in \mathcal{M}$,

$$A(\mu) = A(m_+) + (\mu - m_+)A'(m_+) = -\lambda_+ + (\mu - m_+)\beta_+,$$

so that the difference under the supremum sign in (3.5.58)

$$\mu\beta_+ - A(\mu) = \lambda_+ + m_+\beta_+$$

does not depend on $\mu \in \mathcal{M}$,

$$D(\beta_+) = \lambda_+ + m_+\beta_+.$$

We again arrive at relation (3.5.56).

Substituting the value $D(\beta_+) - \lambda_+ = m_+\beta_+$ in the second row on the right of (3.5.55), we get the first equality in the third row in (3.5.48). Inserting there again the value $D(\beta_+)$, we get from (3.5.56) the second equality. This proves assertion (3.5.48) for $\alpha \geq 0$.

The case when α lies in the half-open interval $(\beta_-, 0]$ is treated similarly. This completes the proof of (3.5.48).

The result relating to the case $\beta_- = -\infty$ follows from (3.5.42).

(ii) Let ζ take values of both signs, $A(\mu^-) < -\lambda_+$. Since ζ takes values of both signs, we have $A(\mu) \to \infty$ as $\mu \to -\infty$. However, $A(\mu^-) < -\lambda_+$, and therefore, $\mu^- > -\infty$.

(a) We first consider the case $\alpha^- = A'(\mu^- + 0) \geq 0$. As before, for all $\alpha \in [0, \beta_+)$ we have $t_\alpha = \frac{\alpha}{\beta_+}$ and $\widehat{D}(\alpha)$ is defined by the last row in (3.5.55). Moreover, $\widehat{D}(\alpha) = D(\alpha)$ for $\alpha \geq \beta_+$.

Consider the domain $\alpha < 0$. We have

$$\inf A(\mu) = A(\mu^-) = -D(0), \quad \mu^- = \mu(0),$$

$$\mu(\alpha) = \mu^-, \quad \lambda(\alpha) = -A(\mu^-) > \lambda_+ \quad \text{for all} \quad \alpha < 0; \quad \beta_- = -\infty.$$

Hence, for $\alpha < 0$,

$$\frac{\partial}{\partial t}\mathbf{D}(t,\alpha) = \lambda\left(\frac{\alpha}{t}\right) = -A(\mu^-) > \lambda_+, \tag{3.5.59}$$

and so the derivative in (3.5.52) for all t is constant and positive. This means that

$$\widehat{D}(\alpha) = \mathbf{D}(0,\alpha) + \lambda_+,$$

and by (3.5.59) we have

$$\mathbf{D}(1,\alpha) = \mathbf{D}(0,\alpha) - A(\mu^-),$$

$$\widehat{D}(\alpha) = D(\alpha) + A(\mu^-) + \lambda_+ = D(\alpha) + \lambda_+ - D(0).$$

However, for $\alpha < 0$,

$$D(\alpha) = \alpha\mu(\alpha) - A(\mu(\alpha)) = \alpha\mu^- - A(\mu^-) = \alpha\mu^- + D(0),$$

so that

$$\widehat{D}(\alpha) = \alpha\mu^- + \lambda_+.$$

(b) Now let us consider the case $\alpha^- = A(\mu^- + 0) < 0$. Here

$$A(\mu^-) > -D(0), \qquad \mu(\alpha) = \mu^-, \qquad \lambda(\alpha) = -A(\mu^-) > 0$$

for all $\alpha \le \alpha^-$. Therefore,

$$\frac{\partial}{\partial t}\mathbf{D}(t,\alpha) = -A\left(\mu\left(\frac{\alpha}{t}\right)\right) > 0 \quad \text{for all} \quad \frac{\alpha}{t} \le \alpha^-$$

and the derivative $\frac{\partial}{\partial t}\mathbf{D}(t,\alpha)$ is constant and is equal to $-A(\mu^-)$ for $\frac{\alpha}{t} \le \alpha^-$, that is, for $t \le \frac{\alpha}{\alpha^-}$ ($\frac{\alpha}{\alpha^-} \ge 1$ for $\alpha \le \alpha^- < 0$). Hence $\mathbf{D}(1,\alpha) = \mathbf{D}(0,\alpha) - A(\mu^-)$ for $\alpha \le \alpha^-$. Therefore, as before, we get

$$\widehat{D}(\alpha) = \alpha\mu^- + \lambda_+ \quad \text{for} \quad \alpha \le \alpha^-. \tag{3.5.60}$$

If $\alpha \in (\alpha^-, 0)$, then the derivative $\frac{\partial}{\partial t}\mathbf{D}(t,\alpha)$ is constant and is equal to $-A(\mu^-)$ only for $t \le \frac{\alpha}{\alpha^-} < 1$. Hence in this case

$$\mathbf{D}\left(\frac{\alpha}{\alpha^-},\alpha\right) = \mathbf{D}(0,\alpha) - \frac{\alpha}{\alpha^-}A(\mu^-), \ \mathbf{D}(0,\alpha) = \frac{\alpha}{\alpha^-}D(\alpha^-) + \frac{\alpha}{\alpha^-}A(\mu^-),$$

$$\widehat{D}(\alpha) = \mathbf{D}(0,\alpha) + \lambda_+ = \frac{\alpha}{\alpha^-}(D(\alpha^-) + A(\mu^-)) + \lambda_+ = \alpha\mu^- + \lambda_+$$

for $\alpha \in (\alpha^-, 0)$. Thus, equality (3.5.60) holds for all $\alpha < 0$. Assertion (3.5.49) is proved.

If $\mathbf{P}(\zeta \ge 0) = 1$, $A(\mu^-) < -\lambda_+$, then $\widehat{D}(\alpha) = \infty = D(\alpha)$ for $\alpha < 0$. This follows, for example, from the LDP and because $\mathbf{P}(Z(T) < 0) = 0$. This fact can be also derived from the original definition of $\widehat{D}(\alpha)$ (see [47]), because $\Lambda(t,\alpha) = \mathbf{D}(t,\alpha) = \infty$ for $\alpha < 0$. Equality (3.5.50) is proved.

The case $a < 0$ is dealt with similarly. The proof of Theorem 3.5.9 is complete. □

Consider the function

$$\widehat{A}(\mu) = \max\left(A(\mu), -\lambda_+\right). \tag{3.5.61}$$

This function (like the function $A(\mu)$) is convex and lower-continuous. It becomes $A(\mu)$ if $\lambda_+ \geq D(0)$ $(= -\inf A(\mu))$.

Theorem 3.5.11 *The functions $\widehat{D}(\alpha)$ and $\widehat{A}(\mu)$ are related via the dual Legendre transforms,*

$$\widehat{A}(\mu) = \sup_{\alpha}\left(\alpha\mu - \widehat{D}(\alpha)\right), \qquad \widehat{D}(\alpha) = \sup_{\mu}\left(\alpha\mu - \widehat{A}(\mu)\right). \tag{3.5.62}$$

In analogy with the above, the function $\widehat{A}(\mu)$ can be called a *general fundamental function* for the CRP $Z(t)$. It coincides with $A(\mu)$ in the case $\lambda_+ \geq D(0)$ and is always equal to $A(\mu)$ in the neighborhood of the origin. Moreover, it has the property

$$\ln \mathbf{E}e^{\mu Z(T)} \sim T\widehat{A}(\mu)$$

as $T \to \infty$ (see Theorem 3.7.1 below).

Proof The functions $\widehat{A}$ and $\widehat{D}$ are convex and lower-continuous, and hence it suffices to prove only one of relations (3.5.62) and only in the case $\lambda_+ < D(0)$, but for $\lambda_+ \geq D(0)$ the conclusion of Theorem 3.5.11 has been already verified; see Theorem 3.5.1.

We first consider the case when ζ takes values of both signs, $|\beta_\pm| < \infty$, and prove the second of (3.5.62); that is, setting

$$\widetilde{D}(\alpha) = \sup_{\mu}\left(\alpha\mu - \widehat{A}(\mu)\right),$$

we shall show that $\widetilde{D}(\alpha) = \widehat{D}(\alpha)$, where the function $\widehat{D}(\alpha)$ is defined in (3.3.6). The function $\widehat{A}(\mu)$ is almost everywhere differentiable in its effective domain, the point $\mu = \widehat{\mu}(\alpha)$ at which the supremum in (3.5.62) is attained satisfies, for almost all α, the equation

$$\widehat{A}'(\mu) = \alpha$$

so that, as before (cf. Theorem 3.5.2), $\widehat{\mu}(\alpha) = (\widehat{A}')^{(-1)}(\alpha)$ is the generalized inverse function of $\widehat{A}'$. Since $\widehat{A}$ is convex, the function $\widehat{\mu}(\alpha)$ is nondecreasing. Since $\widehat{A}(\mu) = A(\mu)$ for $\mu \leq \mu(\beta_-) = m_-$, we get, for $\alpha \leq \beta_-$, the equality $\widehat{\mu}(\alpha) = \mu(\alpha)$, and besides,

$$\widetilde{D}(\alpha) = \alpha\widehat{\mu}(\alpha) - \widehat{A}(\widehat{\mu}(\alpha)) = \alpha\mu(\alpha) - A(\mu(\alpha)) = D(\alpha).$$

A similar relation holds $\alpha \geq \beta_+$.

If $\alpha \in (\beta_-, 0)$, then the supremum in (3.5.62) is attained at the minimal possible value $\mu \geq \mu(\beta_-)$, that is, at $\mu = \mu(\beta_-) = m_- = \widehat{\mu}(\alpha)$, so that

$$\widetilde{D}(\alpha) = \alpha\mu(\beta_-) - \widehat{A}(\mu(\beta_-)) = \alpha\mu(\beta_-) + \lambda_+ = D(\beta_-) + (\alpha - \beta_-)\mu(\beta_-).$$

For $\alpha \in (0, \beta_+)$, a similar argument can be applied to show that

$$\widetilde{D}(\alpha) = D(\beta_+) + (\alpha - \beta_+)\mu(\beta_+).$$

So, $\widetilde{D}(\alpha) = \widehat{D}(\alpha)$.

Consider now the case $a > 0$, $\beta_- = -\infty$, $\mu^- > -\infty$ and prove the first relation in (3.5.62). By (3.5.49), $\sup_\alpha (\alpha\mu - \widehat{D}(\alpha))$ for $\mu < \mu^-$ is attained at $\alpha = -\infty$ and equals ∞. The last relation on the right of (3.5.48) implies that the supremum under consideration for $\mu \in [\mu^-, \mu(\beta_+)]$ is attained at $\alpha = 0$ and equals $-\lambda_+$. Finally, for $\mu > \mu(\beta_+)$ the supremum is attained at the same point $\alpha(\mu)$ as when considering $\sup_\alpha (\alpha\mu - D(\alpha))$, and therefore it equals $A(\mu)$. This description of $\sup_\alpha (\alpha\mu - \widehat{D}(\alpha))$ coincides with the definition of $\widehat{A}(\mu)$ in (3.5.61).

The case when $a > 0$, $\beta_- = -\infty$, $\mu^- = -\infty$ (that is, the case when $\mathbf{P}(\zeta \geq 0) = 1$) is treated similarly. In the same "symmetric" fashion, we also consider the case when $a < 0$ and $\beta_+ = \infty$.

Theorem 3.5.11 is proved. □

3.5.4 Condition $\lambda_+ < D(\alpha)$ and Strong Dependence between τ and ζ in the Large Deviation Zone

From Theorem 3.3.2 (see (3.3.14)) it follows that the condition $\lambda_+ < D(0)$ implies the dependence of τ and ζ; moreover, it implies that $\mathbf{E}\zeta \neq 0$ and λ_+ and $\max\{|\mu_-^{(\zeta)}|, \mu_+^{(\zeta)}\}$ are both finite. The fact that the possibility $\lambda_+ < D(0)$ is implemented (on a dependence of the form $\zeta = c\tau + \omega$, where τ and ω are independent) will be illustrated in Example 3.5.14 (see below).

As was already pointed out, the probabilistic sense of the condition $\lambda_+ < D(0)$ is that in this case the probability that the homogeneous process $Z(t)$ hits at time T the neighborhood of the origin with its first jump $\tau_1 > T$ is much higher than the probability for the random walk $\{\mathbf{S}_n\}$ to hit somewhere the Δ-neighborhood of the point $(1, 0)$ (see Theorem 3.5.1). This fact, which is related to the "excessive" influence of the first jump $\tau_1 > T$ on $Z(T)$ (here the coordinate ζ_1 has no effect on $Z(T)$, but has an influence on $Y(T)$), "aggravates" the analytic properties of the CRP deviation function near the origin (thereby generating the "irregularity domain" (β_-, β_+)) and the properties of the CRP trajectories.

Next, the condition $[\overline{\lambda}_+]$ for $D(0) < \infty$ means a strong dependence of τ and ζ in the large deviation zone. Namely, we will show that, in parallel with (3.4.2), for some $\delta > 0$ and $T \to \infty$,

$$\ln \mathbf{P}(\tau > T,\ \zeta = o(T)) \leq -(\lambda_+ + \delta)T, \tag{3.5.63}$$

so that the conditional probability $\mathbf{P}(\zeta = o(T)|\ \tau \geq T)$ is exponentially (with respect to T) small and that deviations of τ of order T imply with high probability deviations of $|\zeta|$ of the same order (see (3.5.66) below). Inequality (3.5.63) takes place, for example, for dependences of the form

$$\zeta = c\tau + \omega, \qquad c \neq 0, \tag{3.5.64}$$

where ω does not depend on τ, $\omega \in [\mathbf{C}_0]$. Relation (3.5.63) follows from the next lemma.

Lemma 3.5.12 *Let condition $[\overline{\lambda}_+]$ hold. Then, for any sequence $\varepsilon_T \to 0$ as $T \to \infty$,*

$$\overline{\lim_{T\to\infty}} \frac{1}{T} \ln \mathbf{P}(\tau > T,\ |\zeta| < \varepsilon_T T) \le -D(0). \tag{3.5.65}$$

Since by $[\overline{\lambda}_+]$ we have (3.4.2) and the inequality $D(0) > \lambda_+ + \delta$ holds for some $\delta > 0$, using Lemma 3.5.12 we conclude that, for all sufficiently large T,

$$\ln \mathbf{P}(|\zeta| < \varepsilon_T T \mid \tau \ge T) < -\delta T. \tag{3.5.66}$$

Proof of Lemma 3.5.12 Consider the set

$$B := \{(t,\alpha) : t > 1,\ |\alpha| < \varepsilon\}$$

(this set depends on T via $\varepsilon = \varepsilon_T$). Then the event under the probability sign in (3.5.65) can be written in the form $\{(\tau,\zeta) \in TB\}$. The set TB is open and convex, and hence by the exponential Chebyshev inequality (see §4.4 in [23]),

$$\mathbf{P}((\tau,\zeta) \in TB) \le e^{-\mathbf{\Lambda}(TB)}, \qquad \mathbf{\Lambda}(TB) = \inf_{(t,\alpha)\in TB} \mathbf{\Lambda}(t,\alpha).$$

In view of (3.2.12), (3.2.18) we have

$$\mathbf{\Lambda}(t,\alpha) \ge \mathbf{D}_\Lambda(t,\alpha) \ge \mathbf{D}(t,\alpha),$$

and hence

$$\mathbf{P}((\tau,\zeta) \in TB) \le e^{-T\mathbf{D}(B)}, \text{ where } \mathbf{D}(B) := \inf_{(t,\alpha)\in B} \mathbf{D}(t,\alpha). \tag{3.5.67}$$

So to prove (3.5.65) it suffices to verify that

$$\underline{\lim_{T\to\infty}}\, \mathbf{D}(B) \ge D(0).$$

We have

$$\begin{aligned}
\mathbf{D}(B) = \inf_{t>1,\ |\alpha|<\varepsilon} \mathbf{D}(t,\alpha) &= \inf_{t>1} t \inf_{|\alpha|<\varepsilon} \mathbf{D}(1,\alpha/t) \\
&= \inf_{t>1} t \inf_{|\beta|<\varepsilon/t} \mathbf{D}(1,\beta) \ge \inf_{t>1} t \inf_{|\beta|<\varepsilon} \mathbf{D}(1,\beta) \\
&= \inf_{|\beta|<\varepsilon} \mathbf{D}(1,\beta) \inf_{t>1} t = \inf_{|\beta|<\varepsilon} \mathbf{D}(1,\beta).
\end{aligned}$$

The function $\mathbf{D}(1,\beta)$ is lower-continuous, and hence

$$\underline{\lim_{T\to\infty}}\, \mathbf{D}(B) \ge \underline{\lim_{T\to\infty}} \inf_{|\beta|<\varepsilon} \mathbf{D}(1,\beta) = \mathbf{D}(1,0) = D(0).$$

Now (3.5.65) follows from (3.5.67). Lemma 3.5.12 is proved. □

The above means that the condition $[\overline{\lambda}_+]$ is very special and "less inclusive" than $[\lambda_+]$. Note also that relations (3.5.63) and (3.5.66) cannot characterize the possibility $\lambda_+ < D(0)$ in (3.4.1), because, for example, for dependences (3.5.64) for $\mathbf{E}\omega = -c\mathbf{E}\tau$ both the first alternative in (3.4.1) takes place and inequality (3.5.63) holds.

Remark 3.5.13 Using the lower continuity of the function $D(\alpha)$ and arguing as in the proof of Lemma 3.5.12, we can prove that, for a *fixed* $\varepsilon > 0$,

$$\overline{\lim_{T\to\infty}} \frac{1}{T} \ln \mathbf{P}(\tau > T,\ |\zeta| < \varepsilon T) \le -D(0) + \delta(\varepsilon),$$

where $\delta(\varepsilon) > 0$ can be made arbitrarily small by choosing a suitable ε.

3.5.5 Examples

Example 3.5.14 Let $\zeta = c\tau + \omega$, where τ and ω are independent. Then, as was already pointed out, with obvious notational conventions,

$$\mathbf{A}(\lambda,\mu) = \ln \mathbf{E} e^{\lambda\tau+\mu\zeta} = \ln \mathbf{E} e^{\lambda\tau+\mu(c\tau+\omega)} = A^{(\tau)}(\lambda + c\mu) + A^{(\omega)}(\mu). \tag{3.5.68}$$

Hence the interior of the effective domain $\mathcal{A}$ of the function $\mathbf{A}(\lambda,\mu)$ has the form

$$(\mathcal{A}) = \{(\lambda,\mu) :\ \lambda + c\mu < \lambda_+,\quad \mu_-^{(\omega)} < \mu < \mu_+^{(\omega)}\},$$

where $(\mu_-^{(\omega)},\ \mu_+^{(\omega)})$ is the effective domain (the domain of analyticity) of the function $A^{(\omega)}(\mu)$.

The function $A^{(\tau)}(\lambda)$ is monotone increasing, and hence there exists the (generalized) inverse function

$$(A^{(\tau)})^{(-1)}(v) = \sup\{\lambda :\ A^{(\tau)}(\lambda) \le v\}.$$

It is clear that

$$(A^{(\tau)})^{(-1)}(v) = \lambda_+ \ \text{ for }\ v \ge A^{(\tau)}(\lambda_+). \tag{3.5.69}$$

By (3.5.68),

$$\begin{aligned} A(\mu) = -\sup\{\lambda :\ \mathbf{A}(\lambda,\mu) \le 0\} &= -\sup\{\lambda :\ A^{(\tau)}(\lambda + c\mu) \le -A^{(\omega)}(\mu)\} \\ &= -(-c\mu + (A^{(\tau)})^{(-1)}(-A^{(\omega)}(\mu)) \\ &= c\mu - (A^{(\tau)})^{(-1)}(-A^{(\omega)}(\mu)). \end{aligned} \tag{3.5.70}$$

It is clear that if $\omega > 0$, then

$$-A^{(\omega)}(\mu) \to \infty, \quad (A^{(\tau)})^{(-1)}(-A^{(\omega)}(\mu)) \to \lambda_+ \quad \text{as} \quad \mu \to -\infty$$

and hence, $A(\mu) \sim c\mu \to -\infty$ as $\mu \to -\infty$, $\lambda_+ < \infty$, $c > 0$, so that

$$D(0) = -\inf A(\mu) = \infty, \qquad \lambda_+ < D(0).$$

Let us now show that the alternative $\lambda_+ < D(0)$ can also be realized for ζ that take values of both signs, that is, for $D(0) < \infty$.

If $\mathbf{P}(\tau > t) = e^{-t}$, then

$$A^{(\tau)}(\lambda) = -\ln(1-\lambda), \qquad (A^{(\tau)})^{(-1)}(v) = 1 - e^{-v},$$

$$A(\mu) = c\mu + e^{A^{(\omega)}(\mu)} - 1,$$

$$D(0) = \sup_\mu\{-A(\mu)\} = 1 + \sup_\mu[-c\mu - (e^{A^{(\omega)}(\mu)} - 1)].$$

This means that $D(0) - 1$ is the Legendre transform at the point $-c$ of the convex function $e^{A^{(\omega)}(\mu)} - 1$. If $\mathbf{P}(\omega \gtrless 0) > 0$, then $A^{(\omega)}(\mu) \to \infty$ as $|\mu| \to \infty$ and hence, $D(0) \to \infty$ as $|c| \to \infty$. So, by choosing c we can again satisfy the inequality $\lambda_+ < D(0)$, so that the forbidden set (β_-, β_+) is nonempty for sufficiently large $|c|$.

Let us return to (3.5.70). Let

$$\mu \in (\mu_-^{(\omega)}, \mu_+^{(\omega)}), \qquad v = -A^{(\omega)}(\mu) < A^{(\tau)}(\lambda_+).$$

For such μ, the function $(A^{(\tau)})^{(-1)}(v)$ becomes the "ordinary" inverse function of $A^{(\tau)}(\lambda)$ (i.e., $A^{(\tau)}((A^{(\tau)})^{(-1)}(v)) \equiv v$), while the function $(A^{(\tau)})^{(-1)}(- A^{(\omega)}(\mu))$ is an analytic function of μ.

But if

$$-A^{(\omega)}(\mu) \geq A^{(\tau)}(\lambda_+)$$

(this is possible for $\mathbf{E}\omega \neq 0$, $|\mu_\pm^{(\omega)}| > 0$, because in this case $\min_\mu A^{(\omega)}(\mu) < 0$), then, for such μ,

$$A(\mu) = c\mu - \lambda_+ = A^\infty(\mu)$$

(see (3.5.69), (3.5.70)). If, for example, $\mathbf{E}\omega < 0$, then $\min_{\mu>0} A^{(\omega)}(\mu) < 0$,

$$\mu_- = \mu^- = \mu_-^{(\omega)}, \quad \mu_+ < \mu^+ = \mu_+^{(\omega)}, \quad \mu_+ > 0.$$

If $c = 0$, $\tau \equiv 1$, then

$$A^{(\tau)}(\lambda) = \lambda, \qquad \lambda_+ = \infty, \qquad A(\mu) = A^{(\omega)}(\mu) = A^{(\zeta)}(\mu).$$

In the important particular case when τ and ζ are independent ($c = 0$), we have the following result (which we have already practically proved).

Lemma 3.5.15 *If τ and ζ are independent, then*

(i)
$$\mu^- = \mu_-^{(\zeta)} := \inf\{\mu : \ A^{(\zeta)}(\mu) < \infty\},$$

$$\mu^+ = \mu_+^{(\zeta)} := \sup\{\mu : \ A^{(\zeta)}(\mu) < \infty\},$$

$$A^\infty(\mu) = -\lambda_+ \quad \textit{for} \quad \mu \in (\mu^-, \mu^+).$$

(ii) *If*

$$-\inf_\mu A^{(\zeta)}(\mu) < A^{(\tau)}(\lambda_+) \tag{3.5.71}$$

(this is always the case if $\mathbf{E}\zeta = 0$), then $\lambda_+ > D(0)$, the interval $(\mu^-, \mu^+) = (\mu_-, \mu_+)$ is the main domain of analyticity of the function $A(\mu)$,

$$A(\mu) = -(A^{(\tau)})^{(-1)}(-A^{(\zeta)}(\mu)). \tag{3.5.72}$$

If

$$-\inf_\mu A^{(\zeta)}(\mu) \geq A^{(\tau)}(\lambda_+), \tag{3.5.73}$$

then the solutions $\mu'_- \leq \mu'_+$ of the equation

$$A^{(\zeta)}(\mu) = -A^{(\tau)}(\lambda_+)$$

are well defined.

For $\mathbf{E}\zeta > 0$,

$$\mu'_{\pm} < 0, \quad \mu_- = \mu'_+, \quad \mu_+ = \mu^+,$$
$$A(\mu) = A^{\infty}(\mu) \quad \textit{for} \quad \mu \in [\mu'_-, \mu'_+]. \tag{3.5.74}$$

For $\mathbf{E}\zeta < 0$,

$$\mu'_{\pm} > 0, \quad \mu_- = \mu^-, \quad \mu_+ = \mu'_-,$$
$$A(\mu) = A^{\infty}(\mu) \quad \textit{for} \quad \mu \in [\mu'_-, \mu'_+]. \tag{3.5.75}$$

Proof of Lemma 3.5.15 Assertion (i) is obvious.

(ii) If $-A^{(\zeta)}(\mu) \geq A^{(\tau)}(\lambda_+)$, then

$$A(\mu) = -\sup\{\lambda : A^{(\tau)}(\mu) \leq -A^{(\zeta)}(\mu)\} = -\lambda_+ = A^{\infty}(\mu). \tag{3.5.76}$$

If $-A^{(\zeta)}(\mu) < A^{(\tau)}(\lambda_+)$ (see condition (3.5.71)), then since the inverse function $(A^{(\tau)})^{(-1)}(y)$ of $A^{(\tau)}(\lambda)$ is monotone, we have

$$A(\mu) = -\sup\{\lambda : \lambda \leq (A^{(\tau)})^{(-1)}(-A^{(\zeta)}(\mu))\} = -(A^{(\tau)})^{(-1)}(-A^{(\zeta)}(\mu)),$$

where the right-hand side is an analytic function. This proves (3.5.72).

If (3.5.73) holds, then (3.5.76), (3.5.15), (3.5.75) are satisfied for $\mu \in [\mu'_-, \mu'_+]$. Lemma 3.5.15 is proved. □

From the proof it also follows that *conditions* (3.5.71) *and* $D(0) < \lambda_+$ *are equivalent*, and the domains (μ^-, μ'_-) for $\mathbf{E}\zeta < 0$ and (μ'_+, μ_+) for $\mathbf{E}\zeta > 0$ are domains of "nondegenerate" analyticity ($A(\mu) \not\equiv -\lambda_+$).

Note that there are only few examples known in which the functions $\mathbf{D}(\theta, \alpha)$, $A(\mu)$ can be found explicitly. Let us present one of them.

Example 3.5.16 Let τ and ζ be independent Γ-distributed with parameters (λ_+, γ) and (μ_+, γ). Let the left superscript γ indicate the dependence of the characteristics in question on γ, so that, in our case,

$${}^{\gamma}\mathbf{A}(\lambda, \mu) = -\gamma \ln\left(1 - \frac{\lambda}{\lambda_+}\right) - \gamma \ln\left(1 - \frac{\mu}{\mu_+}\right). \tag{3.5.77}$$

We also set

$$\mathbf{A}(\lambda, \mu) := {}^{1}\mathbf{A}(\lambda, \mu) = -\ln\left(1 - \frac{\lambda}{\lambda_+}\right)\left(1 - \frac{\mu}{\mu_+}\right),$$
$$\mathbf{\Lambda}(\theta, \alpha) := {}^{1}\mathbf{\Lambda}(\theta, \alpha), \quad \mathbf{D}_{\Lambda}(\theta, \alpha) := {}^{1}\mathbf{D}_{\Lambda}(\theta, \alpha).$$

Then

$$
\begin{aligned}
{}^{\gamma}\mathbf{A}(\lambda,\mu) &= \gamma\mathbf{A}(\lambda,\mu),\\
{}^{\gamma}\mathbf{\Lambda}(\theta,\alpha) &= \sup_{(\lambda,\mu)}\{\lambda v+\mu\alpha-\gamma\mathbf{A}(\lambda,\mu)\} = \gamma\mathbf{\Lambda}\Big(\frac{\theta}{\gamma},\frac{\alpha}{\gamma}\Big),\\
{}^{\gamma}\mathbf{D}_{\Lambda}(\theta,\alpha) &= \inf_{r>0} r\gamma\mathbf{\Lambda}\Big(\frac{\theta}{r\gamma},\frac{\alpha}{r\gamma}\Big) = \mathbf{D}_{\Lambda}(\theta,\alpha),
\end{aligned}
\tag{3.5.78}
$$

that is, ${}^{\gamma}\mathbf{D}_{\Lambda}(\theta,\alpha)$ does not depend on γ.

Remark 3.5.17 The above setting corresponds to the property that the second deviation function $\mathbf{D}(\theta,\alpha)$ is invariant with respect to aggregation of jumps – here, by aggregation of jumps we mean that instead of $\boldsymbol{\xi}$ we consider the jumps of ${}^{k}\boldsymbol{\xi} \overset{d}{=} \mathbf{S}_k$ with fixed k. In this setting it is natural to expect that the number of hits of the aggregated random walk $\{{}^{k}\mathbf{S}_n\}$, $n=0,1,\dots$, in the remote set TB is about k times less than for the original walk $\{\mathbf{S}_n\}$. This means that the corresponding renewal measure ${}^{k}H$ satisfies the relation

$$
{}^{k}H(TB) \approx \frac{1}{k}H(TB).
$$

In turn, as a result, $\ln {}^{k}H(TB)$ and $\ln H(TB)$ have the same asymptotic behavior as $T\to\infty$. This clarifies the invariance of the second deviation function $\mathbf{D}(\theta,\alpha)$ under “aggregations” (cf. Remarks 3.5.4 and 3.5.5).

However, if by an “aggregation” we mean the change $\boldsymbol{\xi}$ to ${}^{(b)}\boldsymbol{\xi} := b\boldsymbol{\xi}$ (the variation of the scale), then with obvious notational conventions we have

$$
\begin{aligned}
{}^{(b)}\mathbf{A}(\lambda,\mu) &= \mathbf{A}(b\lambda,b\mu),\\
{}^{(b)}\Lambda(\theta,\alpha) &= \sup_{(\lambda,\mu)}\{\lambda\theta+\mu\alpha-\mathbf{A}(b\lambda,b\mu)\} = \Lambda\Big(\frac{\theta}{b},\frac{\alpha}{b}\Big),\\
{}^{(b)}\mathbf{D}_{\Lambda}(\theta,\alpha) &= \inf_{r>0} r\Lambda\Big(\frac{\theta}{rb},\frac{\alpha}{rb}\Big) = \frac{1}{b}\mathbf{D}_{\Lambda}(\theta,\alpha) = \mathbf{D}_{\Lambda}\Big(\frac{\theta}{b},\frac{\alpha}{b}\Big),
\end{aligned}
$$

where the right-hand side obviously depends on b.

Let us return to Example 3.5.16. Relation (3.5.78) means that in order to find ${}^{\gamma}\mathbf{D}(\theta,\alpha)=\mathbf{D}(\theta,\alpha)$ it suffices to consider the case $\gamma=1$, that is, when τ and ζ are exponentially distributed and $Z(t)$ is a compound Poisson process. In this case,

$$
\begin{aligned}
\mathbf{\Lambda}(\theta,\alpha) &= \Lambda^{(\tau)}(\theta)+\Lambda^{(\zeta)}(\alpha) = \lambda_+\theta+\mu_+\alpha-2-\ln\theta\alpha\lambda_+\mu_+,\\
\mathbf{D}_{\Lambda}(\theta,\alpha) &= \lambda_+\theta+\mu_+\alpha+\inf_{r>0} r[2\ln r-2-\ln\theta\alpha\lambda_+\mu_+].
\end{aligned}
$$

Differentiating with respect to r the function under the infimum, we get

$$
2\ln r-2-\ln\theta\alpha\lambda_+\mu_+ +2 = 0
$$

for the point

$$
r = r(\theta,\alpha) = \sqrt{\theta\alpha\lambda_+\mu_+},
$$

at which the infimum is attained. This implies that

$$\mathbf{D}(\theta,\alpha) = \mathbf{D}_\Lambda(\theta,\alpha) = \lambda_+\theta + \mu_+\alpha - 2\sqrt{\theta\alpha\lambda_+\mu_+} \tag{3.5.79}$$

for positive θ and α.

Next, if points (θ,α) are assumed to lie on the boundary of the positive quadrant, then, for instance, for $\theta = 0$ we have $\Lambda(0,\alpha) = \infty$ and $\mathbf{D}_\Lambda(0,\alpha) = \infty$. Similarly, $\mathbf{D}_\Lambda(\theta,0) = \infty$. The function $\mathbf{D}(\theta,\alpha)$ removes this discontinuity, and moreover, $\mathbf{D}(\theta,\alpha)$ for all $\theta \geq 0$, $\alpha \geq 0$, is equal to the right-hand side of (3.5.79). In particular,

$$D(\alpha) = \mathbf{D}(1,\alpha) = \mu_+\alpha - 2\sqrt{\alpha\lambda_+\mu_+} + \lambda_+, \tag{3.5.80}$$

$$D(0) = \lambda_+, \quad \mu(\alpha) = D'(\alpha) = \mu_+ - \sqrt{\frac{\lambda_+\mu_+}{\alpha}}. \tag{3.5.81}$$

Next, the fundamental function ${}^\gamma A(\mu)$ satisfies the equation ${}^\gamma\mathbf{A}(-\lambda,\mu) = 0$ or, equivalently (see (3.5.77)), the equation

$$\left(1 - \frac{\lambda}{\lambda_+}\right)\left(1 - \frac{\mu}{\mu_+}\right) = 1.$$

Moreover, the function ${}^\gamma A(\mu)$ also does not depend on γ,

$$\begin{gathered} {}^\gamma A(\mu) = A(\mu) = \frac{\lambda_+\mu}{\mu_+ - \mu}, \\ \lambda(\alpha) = -A(\mu(\alpha)) = \lambda_+ - \sqrt{\alpha\lambda_+\mu_+}. \end{gathered} \tag{3.5.82}$$

It can be verified that from (3.5.80)–(3.5.82) we get that

$$D(\alpha) = \lambda(\alpha) + \mu(\alpha)\alpha,$$

and that $A(\mu)$ is the Legendre transform of $D(\alpha)$.

3.6 On Large Deviation Principles for the Process $Y(t)$ and for Markov Additive Processes

If we try an LDP for $Y(t)$ using the same approach as in §3.4.1, then in the homogeneous case we get, for $x = \alpha T$,

$$\begin{aligned} \mathbf{P}(Y(T) \in T\Delta[\alpha)) &= \mathbf{P}(\tau > T,\ \zeta \in T\Delta[\alpha)) \\ &+ \int_0^T \int_{-\infty}^{\infty} H_1(du,dy)\,\mathbf{P}(\tau > T-u, \zeta \in T\Delta[\alpha - y)). \end{aligned} \tag{3.6.1}$$

However, the study of the rough asymptotics of the two-dimensional integral in (3.6.1) involves quite laborious calculations. At the same time, if as in §3.4.3 the event $\{Y(T) \in T\Delta[\alpha)\}$ is narrowed down to the event $\{Y(T) \in T\Delta[\alpha), \chi(T) < T\delta\}$ as $\delta \to 0$, then we can easily establish an LDP (with the deviation function $D(\alpha)$).

Another way to simplify the problem is to assume that τ and ζ are independent. Consider these two possibilities.

3.6.1 LDP for the Process $Y(t)$ on the Narrowing of the Set $\{Y(T) \in T\Delta[\alpha)\}$

Theorem 3.6.1 (i) *For all* $\alpha \in \mathbb{R}$, $\Delta = \overline{o}(1)$, $\delta = \overline{o}(1)$ *and* $T \to \infty$,

$$\frac{1}{T} \ln \mathbf{P}(Y(T) \in T(\alpha)_\Delta,\ \chi(T) < \delta T) \le -D(\alpha) + o(1). \tag{3.6.2}$$

(ii) *Assume that at least one of the following conditions is satisfied for some* $h > 0$ *and all sufficiently large* t:

(a) $\lambda_+ < \infty$, (3.4.2) *is met, and*

$$\mathbf{P}(\tau \in (t, 2t), |\zeta| \ge ht) < \frac{1}{2}\mathbf{P}(\tau \in (t, 2t)); \tag{3.6.3}$$

(b) *the condition* $[\boldsymbol{\lambda}_+] \bigcup [\overline{\boldsymbol{\lambda}}_+]$ *is met and, for some* $h > 0$, p, q, $p + q < 1$, *and all* $v < t$, $t \to \infty$,

$$\mathbf{P}(\tau \ge v + t) < p\,\mathbf{P}(\tau \ge v), \quad \mathbf{P}(\tau \in (v, v + t), |\zeta| \ge ht) < q\,\mathbf{P}(\tau \ge v).$$

Then

$$\frac{1}{T} \ln \mathbf{P}(Y(T) \in T(\alpha)_\Delta,\ \chi(T) < \delta T) = -D(\alpha) + o(1). \tag{3.6.4}$$

Condition (b) is quite broad; for example, it is satisfied in the absence of growing lacunas in the distribution for τ, and if ζ, τ are independent, weakly dependent, or linearly dependent.

Proof of Theorem 3.6.1 The first assertion follows from Theorem 3.2.1 and the clear inequality

$$\mathbf{P}(Y(T) \in T(\alpha)_\Delta,\ \chi(T) < \delta T) \le H_1(T\delta[1), T(\alpha)_\Delta).$$

Let us verify the second assertion. Assume that condition (a) is met. Setting

$$B(\delta, \Delta) = \{Y(T) \in T(\alpha)_\Delta,\ \chi(T) < \delta T\},$$

we note that the event

$$\bigcup_{k=1}^{\infty} \{T_k \in T\delta[1 - \delta), Z_k \in T(\alpha)_\Delta, \gamma(T) < \delta T,\ \tau_{k+1} \in (\delta T, 2\delta T), |\zeta_{k+1}| < h\delta T\}$$

implies the event $B(2\delta, \Delta + h\delta)$. Hence

$$\mathbf{P}(B(2\delta, \Delta + h\delta)) \ge H_1(T\delta[1 - \delta),\ T(\alpha)_\Delta)\,\mathbf{P}(\tau \in (\delta T, 2\delta T),\ |\zeta| < h\delta T). \tag{3.6.5}$$

Here by Theorem 3.2.1 the logarithm of the first factor on the right is asymptotically equivalent to

$$-T\mathbf{D}(1 - \delta, \alpha) + o(T) = -T\mathbf{D}(1, \alpha) + o(T). \tag{3.6.6}$$

For the second factor, by condition (a) we have, as $t \to \infty$,

$$\mathbf{P}(\tau \in (t, 2t), |\zeta| < ht) = \mathbf{P}(\tau \in (t, 2t)) - \mathbf{P}(\tau \in (t, 2t), |\zeta| \ge ht) \ge \frac{1}{2}\mathbf{P}(\tau \in (t, 2t)).$$

Hence, for $t = \delta T$, by (3.4.2) the logarithm of the second factor on the right of (3.6.5) is not smaller than $-\lambda_+ t + o(t) = -\lambda_+ \delta T + o(\delta T) = o(T)$. Now using (3.6.2), (3.6.5) and (3.6.6), we get

$$\frac{1}{T} \ln \mathbf{P}(B(2\delta, \Delta + h\delta)) = -D(\alpha) + o(1) \quad \text{as} \quad T \to \infty,$$

which proves (3.6.4) (for $\Delta > h\delta$).

Now let condition (b) be met. Then, for $\Delta_1 = \Delta + h\delta$, we have

$$\{Z(T) \in T(\alpha)_\Delta, \gamma(T) < \delta T,\ \chi(T) < \delta T,\ |\zeta_{\eta(T)}| < h\delta T\} \subset B(\delta, \Delta_1).$$

Hence

$$\begin{aligned}
&\mathbf{P}(B(\delta, \Delta_1)) \\
&\quad \geq \int_0^{\delta T} \mathbf{P}(Z(T) \in T(\alpha)_\Delta,\ \gamma(T) \in dv,\ \chi(T) < \delta T,\ |\zeta_{\eta(T)}| < h\delta T) \\
&= \int_0^{\delta T} \mathbf{P}(Z(T) \in T(\alpha)_\Delta, \gamma(T) \in dv)\, \mathbf{P}(\chi(T) < \delta T,\ |\zeta_{\eta(T)}| < h\delta T \mid \gamma(T) = v) \\
&= \int_0^{\delta T} \mathbf{P}(Z(T) \in T(\alpha)_\Delta,\ \gamma(T) \in dv)\, \mathbf{P}(\tau \in (v, v + \delta T),\ |\zeta| < h\delta T \mid \tau > v).
\end{aligned} \tag{3.6.7}$$

Here for $t = \delta T \to \infty$ we have

$$\begin{aligned}
&\mathbf{P}(\tau \in (v, v+t),\ |\zeta| < ht \mid \tau > v) \\
&\qquad = \frac{\mathbf{P}(\tau \in (v, v+t))}{\mathbf{P}(\tau > v)} - \frac{\mathbf{P}(\tau \in (v, v+\delta t),\ |\zeta| \geq ht)}{\mathbf{P}(\tau > v)}.
\end{aligned}$$

By condition (b) this quantity is greater than $1 - p - q = r > 0$. Hence by (3.6.7),

$$\mathbf{P}(B(\delta, \Delta_1)) \geq r\, \mathbf{P}(Z(T) \in T(\alpha)_\Delta, \gamma(T) < \delta T),$$

where (3.4.25) is satisfied for the right-hand side in view of condition $[\lambda_+] \bigcup [\overline{\lambda}_+]$ and Corollary 3.4.10. This proves the required lower estimate for $\mathbf{P}(B(\delta, \Delta_1))$. Now (3.6.4) follows from (3.6.2). Theorem 3.6.1 is proved. □

The following result is also worth pointing out.

Theorem 3.6.2 *If* $|\zeta| < c = \text{const}$ *or* $\tau < c$ *and*

$$\sup_{v \in (0,c)} \mathbf{P}(|\zeta| > t \mid \tau > v) < q < 1 \quad \textit{as} \quad t \to \infty, \tag{3.6.8}$$

then

$$\frac{1}{T} \ln \mathbf{P}(Y(T) \in T\Delta[\alpha]) = -D(\alpha) + o(1) \quad \textit{as} \quad T \to \infty. \tag{3.6.9}$$

Proof If $|\zeta| < c$, then $\lambda_+ \geq D(0)$, $|Y(T) - Z(T)| < c$ and (3.6.9) is clear. If $\tau < c$ and if (3.6.8) is satisfied, then condition (b) of Theorem 3.6.1 is met, and hence

$$\ln \mathbf{P}(Y(T) \in T\Delta[\alpha]) = \ln \mathbf{P}(Y(T) \in T\Delta[\alpha], \chi(T) < \delta T) = -TD(\alpha) + o(T).$$

Theorem 3.6.2 is proved. □

3.6.2 LDP for $Y(t)$ when τ and ζ Are Independent

Theorem 3.6.3 *Let τ and ζ be independent. Then*

$$\frac{1}{T}\ln \mathbf{P}(Y(T) \in T\Delta[\alpha)) = -D(\alpha) + o(1), \tag{3.6.10}$$

where $\Delta = \overline{o}(1)$ as $T \to \infty$.

As in Theorem 3.4.1, from the local LDP (3.6.10) we get the integral LDP of the form (3.4.6), (3.4.7). So, in the case of independent τ and ζ, the LDPs for the processes $Z(t)$ and $Y(t)$ coincide.

Proof of Theorem 3.6.3 If τ and ζ are independent, then

$$P_{T,\alpha} := \mathbf{P}(Y(T) \in T\Delta[\alpha)) = \int_{-\infty}^{\infty} \mathbf{P}(\zeta \in dy)\mathbf{P}(Z(T) \in T\Delta[\alpha - y/T)).$$

Setting $y_k = k\Delta$, $k = \dots, -1, 0, 1, \dots$, we get the rough approximation of $P_{t,\alpha}$ by sums

$$P_{T,\alpha} \approx \sum_k \mathbf{P}(\zeta \in T\Delta[y_k))\mathbf{P}(Z(T) \in T\Delta[\alpha - y_k)). \tag{3.6.11}$$

As in the proof of Theorem 3.4.1, this approximation is justified via inequalities.

We denote by $P_{T,\alpha,k}$ the term in the sum in (3.6.11) corresponding to the summation index k and consider the terms with $k \geq 0$. These terms are nonzero if $\mathbf{P}(\zeta > 0) > 0$. Assume first that $\mu_+^{(\zeta)} < \infty$. Then by the Chebyshev inequality, for $k \geq 1$, we get

$$\mathbf{P}(\zeta \in T\Delta[y_k)) \leq \mathbf{P}(\zeta \geq Ty_k) \leq e^{-Ty_k\mu_+^{(\zeta)}(1+o(1))}. \tag{3.6.12}$$

Next, by Theorem 3.4.1,

$$P_{T,\alpha,k} = \mathbf{P}(\zeta \in T\Delta[y_k))e^{-TD(\alpha-y_k)+o(T)},$$

where, since $D(\alpha)$ is convex,

$$TD(\alpha - y_k) \geq TD(\alpha) - D'(\alpha)y_kT. \tag{3.6.13}$$

Returning back to the terms $P_{T,\alpha,k}$, we find that

$$P_{T,\alpha,0} = e^{-TD(\alpha)+o(T)},$$

and further, for $k \geq 1$,

$$P_{T,\alpha,k} \leq e^{-TD(\alpha)+D'(\alpha)Ty_k-Ty_k\mu_+^{(\zeta)}(1+o(1))}.$$

In Theorem 3.5.2 it was shown that $D(\beta)/\beta \to \mu^+$ as $\beta \to \infty$, where for independent τ and ζ we have $\mu^+ = \mu_+^{(\zeta)}$. Hence, since $D(\alpha)$ is convex,

$$D'(\alpha) - \mu_+^{(\zeta)} \leq 0, \tag{3.6.14}$$

as $\Delta \to 0$, $\Delta T \to \infty$, $T \to \infty$, and moreover, for any $N < \infty$, by (3.6.12) and (3.6.13),

$$\sum_{k=1}^{N} P_{T,\alpha,k} \leq N e^{-TD(\alpha)+o(T)}.$$

We choose N so as to have $\alpha - N\Delta < a$. Then the sum in (3.6.11) with respect to $k \geq N$ satisfies the inequality

$$\sum_{k\geq N} P_{T,\alpha,k} \leq \mathbf{P}(Z(T) < T(\alpha - N\Delta))\mathbf{P}(\zeta > TN\Delta). \tag{3.6.15}$$

In accordance with (3.6.12)–(3.6.14) and the integral LDP for $Z(T)$, the right-hand side in (3.6.15) is majorized by

$$e^{-TD(\alpha-N\Delta)-TN\Delta\mu_+^{(\zeta)}+o(T)} \leq e^{-TD(\alpha)+o(T)}.$$

As a result, we get

$$\sum_{k=0}^{\infty} P_{T,\alpha,k} = e^{-TD(\alpha)+o(T)}.$$

The sums $\sum_{k<0} P_{T,\alpha,k}$ in the case $\mathbf{P}(\zeta < 0) > 0$, $\mu_-^{(\zeta)} > -\infty$ are considered similarly.

If $\mu_+^{(\zeta)} = \infty$ or $\mu_-^{(\zeta)} = -\infty$, then the above estimates become even stronger. Theorem 3.6.3 is proved. □

3.6.3 On Large Deviation Principles for Markov Additive Processes

Let $\{\mathrm{x}_n, \mathrm{X}_n\}$ be an MAP defined in §1.8 on a Harris chain $\{\mathrm{x}_n\}$ in a measurable space $(E, \mathcal{E})$ and let τ be the cycle length for returning of the chain $\{\mathrm{x}_n\}$ to a positive atom (possibly artificial), and let ζ be the increment of the sums of X_n on this cycle.

In order to simplify the statements, we shall consider only countable ergodic Markov chains with states $0, 1, \ldots$, transition probabilities $P_i(j)$, $i, j = 0, 1, \ldots$, and the atom $\mathrm{e}_0 = 0$. It is known that in this case, the "rough" asymptotics of the distribution of X_n is determined by the Legendre transform of the logarithm of the maximal eigenvalue of the matrix $M(\mu) = \|P_i(j)\mathbf{E}e^{\mu\xi(j)}\|$ (see, e.g., [98], [95], [96] and the notation in §1.8).

To be more precise, let, as before, $\mathbf{A}(\lambda, \mu) = \mathbf{E}e^{\lambda\tau+\mu\zeta}$ and let $\mathcal{A}$ be the effective domain of the function $\mathbf{A}(\lambda, \mu)$. Next, let $A_M(\mu)$ be the logarithm of the maximal eigenvalue of the matrix $M(\mu)$. The following fact follows from the results of [95].

If the set $\mathcal{A}$ is open, then, for any fixed $\mathrm{e} \in E$ *and any set $E_1 \in \mathcal{E}$ satisfying some (poorly formalizable) conditions,*

$$\lim_{n\to\infty} \frac{1}{n} \ln \mathbf{P}(\mathrm{X}_n \in n\Delta[\alpha]; \mathrm{x}_n \in E_1 \mid \mathrm{x}_0 = \mathrm{e}) = -D(\alpha), \tag{3.6.16}$$

where $D(\alpha) = \sup_\mu (\alpha\mu - A_M(\mu))$ is the Legendre transform of $A_M(\mu)$, $\Delta = \overline{o}(1)$

as $n \to \infty$. *Moreover, the function* $A_M(\mu)$ *satisfies the equation (for* λ*)*

$$\mathbf{A}(-\lambda, \mu) = 0.$$

So, in the context of MAPs and the cycles (τ, ζ) generated by these processes, the *fundamental function* $A(\mu)$ *appears as the function* $A_M(\mu)$. If $\mathcal{A}$ is open, then $A(\mu)$ is an analytic function and, as in the case of CRPs, the right-hand side in (3.6.16) is the Legendre transform of $A(\mu)$:

$$-D(\alpha) = -(\mathrm{Le}A)(\alpha).$$

From the results of [96] we can extend (3.6.16) to the general case as follows.

Under some conditions on e *and* E_1,

$$\lim_{n\to\infty} \frac{1}{n} \ln \mathbf{P}(\mathrm{X}_n \in n\Delta[\alpha]; \mathrm{x}_n \in E_1 \mid \mathrm{x}_0 = \mathrm{e}) = -D(\alpha) = -(\mathrm{Le}A)(\alpha), \qquad (3.6.17)$$

where $A(\mu) = \sup\{\lambda : \mathbf{A}(\lambda, \mu) \le 0\}$, $\Delta = \overline{o}(1)$ *as* $n \to \infty$.

Remark 3.6.4 In (3.6.16), (3.6.17), the conditions on the set E_1 are essential, because if we remove these conditions and we put $E_1 = E$, then the above results cease to be true in general. This is supported by the example of the MAP $(\mathrm{x}_n^*, \mathrm{X}_n^*)$ constructed in Remark 1.8.8, for which the characteristics (τ^*, ζ^*) of the cycle for the chain $(\mathrm{x}_n^*, \mathrm{X}_n^*)$ are equal to $(\tau^*, \zeta^*) = (\tau + 2, \zeta)$. Let, in the notation of Remark 1.8.8, $\gamma^*(n) := \min(j \ge 0 : \mathrm{x}_{n-j} = \mathrm{e}_0) = l$ (that is, $n - l$ is the moment of last hit of the chain $\{\mathrm{x}_k\}$ in the atom $\mathrm{e}_0 = 0$ prior to the moment n). Let, in this example, l be chosen so that $p := P_{0,l-1}(N) > 0$. Then

$$\mathbf{P}(\mathrm{x}_n^* = N + k) = ck^{-2} \quad \text{for} \quad k = 1, 2, \ldots; \quad \xi^*(N + k) = k^s.$$

Hence, setting $k = n^{2/s}$, we see that the increment ξ_n^* of the sums X_n^* at the last moment with probability

$$\mathbf{P}(\gamma^*(n) = l)pck^{-2} = \frac{\mathbf{P}(\tau > l)}{a_\tau} pcn^{-4/s}(1 + o(1)) \quad \text{as } n \to \infty$$

is n^2. It is clear that in this case the LDP cannot hold.

Remark 3.6.5 From (3.6.16), (3.6.17) it follows that the deviation function for X_n for all α is $D(\alpha)$. However, this is not so for the CRP $Z(n)$ in the case $\lambda_+ < D(0)$. As was already pointed out, this can be explained by the fact that under condition $[\overline{\lambda}_+]$ we have $\ln \mathbf{P}(\tau_1 > n) \sim -\lambda_+ n$ as $n \to \infty$ and the inequality $\ln \mathbf{P}(Z(n) = 0) \ge -\lambda_+ n(1 + o(1)) \gg -nD(0)$ is satisfied. For MAPs, there are no relations of this kind, apparently because the relation $\lambda_+ < D(0)$ for MAP-generated cycles is not realized. This is supported by the case of finite Markov chains, in which a simple proof of the LDP for an MAP with the deviation function $D(\alpha)$ can be obtained without utilizing CRPs; see §4.11 in [23].

Remark 3.6.6 The method of the proof of the LDP for MAPs in [95, 96] is quite complicated and depends on several other studies. At the same time, we have seen in §1.8 that all basic limit laws for MAPs in the normal deviation zone can be established quite easily using the same laws for CRPs. One can proceed similarly in the large deviation zone.

Let $Z(n)$ be a CRP constructed from cycles with parameters (τ, ζ). Then, under condition $\lambda_+ \geq D(0)$, from Theorem 3.4.1 we have

$$\lim_{n\to\infty} \frac{1}{n} \ln \mathbf{P}(Z(n) \in n\Delta[\alpha)) = -D(\alpha) \tag{3.6.18}$$

for $\Delta = \overline{o}(1)$, $n \to \infty$. Moreover, we have the representation

$$\mathrm{X}_n = Z(n) + \rho_n, \quad \text{where} \quad \rho_n = \sum_{k=n-\gamma(n)+1}^{n} \xi(\mathrm{x}_k) \tag{3.6.19}$$

(see (1.8.9)). The above results (3.6.16), (3.6.17) mean that the asymptotics of $\ln \mathbf{P}(Z(n) \in n\Delta[\alpha))$ remains the same if $Z(n)$ is augmented with the term ρ_n. This indicates that the LDP for MAPs can be established directly using (3.6.18), (3.6.19) and by estimating the term ρ_n. In view of Remark 3.6.4, this requires, as in §1.8, conditions on the "oscillations" $\widehat{\zeta} = \max_{k\leq\tau} \mathrm{X}_k - \min_{k\leq\tau} \mathrm{X}_k$ (for $\mathrm{x}_0 = \mathrm{e}_0$) on the cycle.

So, the alternative approach to establishing the LDP for MAPs consists of the following steps: (a) for the process $\{Z(n)\}$ induced by the process $\{\mathrm{X}_k\}$, to check condition $[\mathbf{\lambda}_+]\bigcap[\overline{\mathbf{\lambda}}_+]$ in order to be able to apply Theorem 3.4.1; it seems that the case $[\overline{\mathbf{\lambda}}_+]$ is not realized (checking this condition for finite Harris Markov chains does not cause any difficulties); and (b) to show that the term ρ_n in (3.6.19) can be neglected.

Possible approaches to estimation of the term ρ_n will be illustrated in §5.6 when proving much more precise integro-local limit theorems for MAPs.

3.7 Rough Asymptotics for the Laplace Transform of the Distribution of a Compound Renewal Process

Recall that in general case by $\mu^\pm$ we denote the left- and right-hand boundary points of the effective domain for the function $A(\mu)$:

$$\mu^- := \inf\{\mu : A(\mu) < \infty\}, \qquad \mu^+ := \sup\{\mu : A(\mu) < \infty\}.$$

Theorem 3.7.1 *Let condition* $[\mathbf{\lambda}_+]\bigcup[\overline{\mathbf{\lambda}}_+]$ *be met and the admissible inhomogeneity condition*

$$\mathcal{A}^{\leq 0} \subset [\mathcal{A}_1], \qquad \lambda_+^{(\tau_1)} := \sup\{\lambda : \mathbf{E}e^{\lambda\tau_1} < \infty\} \geq D(0)$$

be satisfied. Then, for any $\mu \neq \mu^\pm$,

$$\lim_{T\to\infty} \frac{1}{T} \ln \mathbf{E}e^{\mu Z(T)} = \widehat{A}(\mu). \tag{3.7.1}$$

Apparently the assertion (3.7.1) also holds for $\mu = \mu^{\pm}$.

This assertion in the case $\lambda_+ \geq D(0)$ $(\widehat{A} = A)$ was announced in assertion (v) of Theorem 2.2. in [26], but its proof was given only under simplifying assumptions (for transparency of presentation) and hence cannot be considered to be complete. Let us give the complete proof of Theorem 3.7.1.

Proof of Theorem 3.7.1 If $\mu = 0$, then the left- and right-hand sides of equality (3.7.1) vanish and in this case (3.7.1) is clear.

Consider the case $\mu \in (0, \mu^+)$. We set

$$E(B) := \mathbf{E}\Big(e^{\mu Z(T)};\ \frac{Z(T)}{T} \in B\Big),$$

$$\overline{L}(B) := \overline{\lim_{T\to\infty}} \frac{1}{T} \ln E(B), \qquad \underline{L}(B) := \underline{\lim}_{T\to\infty} \frac{1}{T} \ln E(B).$$

Our purpose here is to estimate the values $\overline{L}(\mathbb{R})$, $\underline{L}(\mathbb{R})$ and to prove the inequality

$$\overline{L}(\mathbb{R}) \leq \widehat{A}(\mu) \leq \underline{L}(\mathbb{R}). \tag{3.7.2}$$

Let $d_- < d_+$ be the boundaries of the set $\mathcal{D}_1^{<\infty} = \{\alpha : D(\alpha) < \infty\} = \{\alpha : \widehat{D}(\alpha) < \infty\}$. By Theorems 3.5.2 and 3.5.11,

$$\widehat{A}(\mu) = \sup_{\alpha} \{\alpha\mu - \widehat{D}(\alpha)\} = \sup_{\alpha\in\mathcal{D}_1^{<\infty}} \{\alpha\mu - \widehat{D}(\alpha)\} = \sup_{(d_-,d_+)} \{\alpha\mu - \widehat{D}(\alpha)\}.$$

There are four cases to consider (depending on whether d_-, d_+ are finite or infinite). The proof of (3.7.2) will be divided into several stages, which we formulate as lemmas.

Lemma 3.7.2 *Let $\mu \in (0, \mu^+)$, $|d_-| < \infty$, $d_+ = \infty$. Then*

$$\overline{L}(\mathbb{R}) \leq \widehat{A}(\mu). \tag{3.7.3}$$

Proof Given a small $\varepsilon > 0$ and a large $N > d_-$, we set

$$B_1 = (-\infty, d_- - \varepsilon],\ \ B_2 = (d_- - \varepsilon, d_- + \varepsilon),\ \ B_3 = [d_- + \varepsilon, N],\ \ B_4 = (N, \infty).$$

Then

$$\mathbf{E}e^{\mu Z(T)} = E(\mathbb{R}) = \sum_{i=1}^{4} E(B_i).$$

Under the conditions of Theorem 3.7.1, the large deviation principle for the CRP $Z(t)$, as given in Theorem 3.4.1, can be written in the form: *for any* $\alpha \in \mathbb{R}$,

$$\lim_{\varepsilon\to0} \overline{\lim_{T\to\infty}} \frac{1}{T} \ln \mathbf{P}\Big(\frac{Z(T)}{T} \in (\alpha)_\varepsilon\Big) \leq -\widehat{D}(\alpha), \tag{3.7.4}$$

$$\lim_{\varepsilon\to0} \underline{\lim}_{T\to\infty} \frac{1}{T} \ln \mathbf{P}\Big(\frac{Z(T)}{T} \in (\alpha)_\varepsilon\Big) \geq -\widehat{D}(\alpha). \tag{3.7.5}$$

As a result, we have

$$D(B_1) = \infty, \qquad \overline{L}(B_1) = -\infty \tag{3.7.6}$$

and besides, for any $G > 0$, there exists a small $\varepsilon > 0$ such that

$$\overline{L}(B_2) \le \begin{cases} -G & \text{if } \widehat{D}(d_-) = \infty, \\ \mu d_- - \widehat{D}(d_-) + \delta_\varepsilon \le \sup_{[d_-,d_+]} \{\mu\alpha - \widehat{D}(\alpha)\} + \delta_\varepsilon & \\ \qquad\qquad \widehat{A}(\mu) + \delta_\varepsilon & \text{if } \widehat{D}(d_-) < \infty, \end{cases}$$

where $\delta_\varepsilon \to 0$ as $\varepsilon \to 0$. So,

$$\overline{L}(B_2) \le \widehat{A}(\mu) + \delta_\varepsilon. \tag{3.7.7}$$

Let us now estimate $\overline{L}(B_3)$. By (3.7.4), for any $\alpha \in B_3$ and given $\varepsilon > 0$, $h > 0$, there exists $\varepsilon(\alpha) \le \varepsilon$ such that

$$\overline{\lim_{T\to\infty}} \frac{1}{T} \ln \mathbf{P}\Big(\frac{Z(T)}{T} \in (\alpha)_{\varepsilon(\alpha)}\Big) \le -\widehat{D}(\alpha) + h. \tag{3.7.8}$$

The neighborhoods $\{(\alpha)_{\varepsilon(\alpha)};\ \alpha \in B_3\}$ in (3.7.8) cover the compact set B_3. This cover contains a finite subcover

$$\{(\alpha_j, \varepsilon_j);\ j = 1, \ldots, M\}, \tag{3.7.9}$$

where $\varepsilon_j := \varepsilon(\alpha_j) \le \varepsilon$ for a given $\varepsilon > 0$. Hence by (3.7.8),

$$\begin{aligned} E(B_3) = \mathbf{E}\Big(e^{\mu Z(T)};\ \frac{Z(T)}{T} \in B_3\Big) &\le \sum_{j=1}^{M} E((\alpha_j)_{\varepsilon_j}) \\ &\le M \max_{1\le j\le M} E((\alpha_j)_{\varepsilon_j}). \end{aligned} \tag{3.7.10}$$

From (3.7.8), (3.7.9) it follows that

$$\begin{aligned} \overline{L}(B_3) \le \max_{1\le j\le M} \{\mu(\alpha_j + \varepsilon_j) - \widehat{D}(\alpha_j) + h\} & \\ \le \sup_{\alpha\in B_3} \{\mu\alpha - \widehat{D}(\alpha)\} + \mu\varepsilon + h. & \end{aligned} \tag{3.7.11}$$

We have

$$\sup_{\alpha\in B_3} \{\mu\alpha - \widehat{D}(\alpha)\} \le \widehat{A}(\mu),$$

and since $\varepsilon > 0$ and $h > 0$ are arbitrary small, we get

$$\overline{L}(B_3) \le \widehat{A}(\mu). \tag{3.7.12}$$

To obtain the required estimate for $L(\mathbb{R})$, it remains to estimate $\overline{L}(B_4)$.

Let us show that, for any $G < \infty$, there exists $N = N_G < \infty$ such that

$$\overline{L}(B_4) \le -G. \tag{3.7.13}$$

Assuming, for simplicity, that N and T are integer numbers, we split the domains $t \le T$, $z \ge TN$ into half-open intervals

$$\delta[j) := [j, j+1), \qquad \Delta[k) := [k, k+1),$$

which correspond to the values $\delta = \Delta = 1$. Then

$$E(B_4) = \sum_{k \geq TN} \mathbf{E}(e^{\mu Z(T)};\ Z(T) \in \Delta[k)) = \sum_{k \geq TN} E\Big(\frac{1}{T}\,\Delta[k)\Big).$$

Next, to estimate from above, we have

$$E\Big(\frac{1}{T}\,\Delta[k)\Big) = \sum_{n \geq 1} \int_0^T \int_{z \in \Delta[k)} e^{\mu z}\mathbf{P}(T_n \in dt, Z_n \in dz)\mathbf{P}(\tau > T - t)$$
$$\leq e^{\mu(k+1)} \sum_{n \geq 1} \sum_{j \leq T} P_n(j,k)\mathbf{P}(\tau > T - j - 1), \tag{3.7.14}$$

where

$$P_n(j,k) := \mathbf{P}(T_n \in \delta[j), T_n \leq T,\ Z_n \in \Delta[k)).$$

In the sum $E(B_4)$, consider the subsum

$$\Sigma_1 := \sum_{k \geq NT} e^{\mu(k+1)} \sum_{n \geq k+1} \sum_{j \leq T} P_n(j,k)\mathbf{P}(\tau > T - j - 1), \tag{3.7.15}$$

which is taken over all n such that

$$n - 1 \geq k \geq NT > 2\frac{T}{a_\tau} \quad \text{for} \quad N > \frac{2}{a_\tau}.$$

From the exponential Chebyshev inequality and taking into account that the distribution of τ_1 differs in general from that of τ, we get

$$P_n(j,k) \leq \mathbf{P}(T_n \leq T) \leq \mathbf{P}(T_n - \tau_1 \leq T) \leq e^{-(n-1)\Lambda^{(\tau)}\left(\frac{T}{n-1}\right)} \leq e^{-(n-1)\Lambda^{(\tau)}(1/N)},$$

where $\Lambda^{(\tau)}$ is the deviation function for $\tau > 0$ such that $\Lambda^{(\tau)}(1/N) \to \infty$ as $N \to \infty$. Hence

$$\sum_{n \geq k+1} P_n(j,k)\mathbf{P}(\tau > T - j - 1) \leq [1 - e^{-\Lambda^{(\tau)}(1/N)}]^{-1}\, e^{-k\Lambda^{(\tau)}(1/N)}\mathbf{P}(\tau > T - j - 1),$$

and for sufficiently large N, we have

$$[1 - e^{-\Lambda^{(\tau)}(1/N)}]^{-1} \leq 2,$$
$$\mu(k+1) - k\Lambda^{(\tau)}\Big(\frac{1}{N}\Big) \leq -k, \qquad \sum_{j \leq T} \mathbf{P}(\tau > T - j - 1) \leq \mathbf{E}\tau + 1,$$
$$\Sigma_1 \leq 2\sum_{k \geq NT} (1 + \mathbf{E}\tau)e^{-k} \leq c_1 e^{-NT}, \qquad c_1 = \text{const.}$$

Let us now estimate the remaining subsum Σ_2 of the sum $E(B_4)$ (over $n \leq k$ in (3.7.14), cf. (3.7.15)). To this end, we choose $h > 0$ such that $\mu_h := \mu + 2h < \mu^+$. We set $\lambda_h := -A(\mu_h)$, so that the point (λ_h, μ_h) lies on the boundary $\partial\mathcal{A}^{\leq 0}$ of the set $\mathcal{A}^{\leq 0}$, $\psi(\lambda_h, \mu_h) \leq 1$, $\psi_1(\lambda_h, \mu_h) < \infty$. We have

$$\begin{aligned}
\mathbf{P}(T_n \in \delta[j), T_n \le T, Z_n \in \Delta[k)) \\
&= \mathbf{E}(e^{\pm\lambda_h T_n + \pm\mu_h Z_n}; T_n \in \delta[j), T_n \le T, Z_n \in \Delta[k)) \\
&\le e^{-\lambda_h j - \mu_h k + |\lambda_h| + \mu_h}\mathbf{E}(e^{\lambda_h T_n + \mu_h Z_n}) \le e^{|\lambda_h|T + |\lambda_h| - \mu_h k + \mu_h}\psi_1(\lambda_h, \mu_h) \\
&\le c_2 e^{|\lambda_h|T - \mu_h k}, \ c_2 = \text{const.}
\end{aligned}$$

Hence, uniformly in $j \le T + 1$ and n,

$$e^{\mu(k+1)} P_n(j,k) \le c_2 e^{-(\mu_h - \mu)k + |\lambda_h|T + \mu}.$$

We have $k \ge NT$, and hence, for all sufficiently large N and all n,

$$e^{\mu(k+1)} P_n(j,k) \le c_3 e^{-hk}, \qquad c_3 = \text{const.}$$

Therefore,

$$\begin{aligned}
\Sigma_2 &\le \sum_{k \ge NT} \sum_{n \le k} \sum_{j \le T+1} e^{\mu(k+1)} P_n(j,k)\mathbf{P}(\tau > T - j - 1) \\
&\le c_3(\mathbf{E}\tau + 1) \sum_{k \ge NT} k e^{-hk} \le c_4 NT e^{-hNT}, \qquad c_3, c_4 = \text{const.}
\end{aligned}$$

So, we have established that

$$E(B_4) \le \Sigma_1 + \Sigma_2 \le c_1 e^{-NT} + c_4 NT e^{-hNT}.$$

Assuming for definiteness that $h < 1$, we find that

$$\overline{L}(B_4) \le -hN.$$

This proves (3.7.13).

Combining (3.7.6), (3.7.7), (3.7.12), and (3.7.13), we get, under the conditions of Lemma 3.7.2,

$$\overline{L}(\mathbb{R}) \le \widehat{A}(\mu) + \delta_\varepsilon, \tag{3.7.16}$$

where $\delta_\varepsilon \to 0$ as $\varepsilon \to 0$. This proves Lemma 3.7.2, since ε is arbitrary. □

Lemma 3.7.3 *If* $\mu \in (0, \mu^+)$, $d_- = -\infty$, $|d_+| < \infty$, *then* (3.7.2) *holds.*

Proof Under the conditions of this lemma, we put, for $-N < d_+$, $\varepsilon > 0$,

$$\begin{aligned}
B_1 &= (-\infty, -N), \quad B_2 = [-N, d_+ - \varepsilon], \\
B_3 &= (d_+ - \varepsilon, d_+ + \varepsilon), \quad B_4 = [d_+ + \varepsilon, \infty).
\end{aligned}$$

The values $\overline{L}(B_2)$, $\overline{L}(B_3)$, $\overline{L}(B_4)$ can be estimated from above in the same way as $\overline{L}(B_3)$, $\overline{L}(B_2)$, $\overline{L}(B_1)$ are estimated under the conditions of Lemma 3.7.2. It remains to estimate $\overline{L}(B_1)$. We have

$$E(B_1) = \mathbf{E}\Big(e^{\mu Z(T)}; \frac{Z(T)}{T} < -N\Big) < e^{-\mu NT}.$$

Since $\mu > 0$, for any $b > 0$ we can always satisfy the inequality

$$\overline{L}(B_1) < -G.$$

As in Lemma 3.7.2, this leads to inequality (3.7.16). Lemma 3.7.3 is proved. □

In order to estimate from above $\overline{L}(\mathbb{R})$ in (3.7.2) it remains to consider two possibilities: the case

$$d_- = -\infty, \qquad d_+ = \infty$$

with the sets

$$B_1 = (-\infty, -N), \qquad B_2 = [-N, N], \qquad B_3 = (N, \infty)$$

and the case $|d_-| < \infty$, $d_+ < \infty$ with the sets

$$B_1 = (-\infty, d_- - \varepsilon], \quad B_2 = (d_- - \varepsilon, d_- + \varepsilon), \quad B_3 = [d_- + \varepsilon, d_+ - \varepsilon],$$
$$B_4 = (d_+ - \varepsilon, d_+ + \varepsilon), \quad B_5 = [d_+ + \varepsilon, \infty).$$

The upper estimates for $\overline{L}(B_j)$ for the sets B_j of all these kinds are contained in the proofs of Lemmas 3.7.2, 3.7.3. Hence we get the following

Corollary 3.7.4 *Under the conditions of Theorem* 3.7.1 *and* $\mu \in (0, \mu^+)$,

$$\overline{L}(\mathbb{R}) \le \widehat{A}(\mu).$$

To prove Theorem 3.7.1 in the case $\mu \in (0, \mu^+)$, it remains to estimate $\underline{L}(\mathbb{R})$ from below.

Lemma 3.7.5 *Under the conditions of Theorem* 3.7.1 *and* $\mu \in (0, \mu^+)$,

$$\underline{L}(\mathbb{R}) \ge \widehat{A}(\mu). \tag{3.7.17}$$

Proof We have $\widehat{A}(\mu) = \sup(\alpha\mu - \widehat{D}(\alpha))$, and hence, for any $\varepsilon > 0$, there exists $\alpha_\varepsilon \in \mathbb{R}$ such that

$$\mu\alpha_\varepsilon - \widehat{D}(\alpha_\varepsilon) \ge \widehat{A}(\mu) - \varepsilon.$$

For any $\gamma \in (0, 1)$ by (3.7.5), we have

$$\begin{aligned}\underline{L}(\mathbb{R}) \ge \underline{L}((\alpha_\varepsilon)_\gamma) &\ge \mu\alpha_\varepsilon - \mu\gamma + \underline{\lim}_{T\to\infty} \frac{1}{T} \ln \mathbf{P}\Big(\frac{Z(T)}{T} \in (\alpha_\varepsilon)_\gamma\Big) \\ &\ge \mu\alpha_\varepsilon - \mu\gamma - \widehat{D}(\alpha_\varepsilon) - \delta_\gamma \ge \widehat{A}(\mu) - \mu\gamma - \varepsilon - \delta_\gamma,\end{aligned}$$

where $\delta_\gamma \to 0$ as $\gamma \to 0$. Making γ and ε tend to 0, we get (3.7.17). Lemma 3.7.5 is proved. □

The case $\mu \in (\mu^-, 0)$ is dealt with similarly.

Now let $\mu \notin [\mu^-, \mu^+]$. Then $\widehat{A}(\mu) = A(\mu) = \infty$, and so to prove (3.7.1) it suffices to verify the relation

$$L := \liminf_{t\to\infty} \frac{1}{t} \ln \mathbf{E}e^{\mu Z(t)} = \infty. \tag{3.7.18}$$

We have, for any $\alpha \in \mathbb{R}$ and $\varepsilon > 0$,

$$L \geq \liminf_{t\to\infty} \frac{1}{t} \ln \mathbf{E}\left(e^{\mu Z(t)};\ \frac{Z(t)}{t} \in (\alpha)_\varepsilon\right) \geq \mu\alpha - |\mu|\varepsilon + \liminf_{t\to\infty} \frac{1}{t} \ln \mathbf{P}\left(\frac{Z(t)}{t} \in (\alpha)_\varepsilon\right).$$

Next, applying the lower estimate in the LDP, we get

$$\liminf_{t\to\infty} \frac{1}{t} \ln \mathbf{P}\left(\frac{Z(t)}{t} \in (\alpha)_\varepsilon\right) \geq -\widehat{D}((\alpha)_\varepsilon) \geq -\widehat{D}(\alpha),$$

and hence,

$$L \geq \mu\alpha - \widehat{D}(\alpha) - |\mu|\varepsilon.$$

Since α is arbitrary, by the first of (3.5.62) of Theorem 3.5.11, we get

$$L \geq \sup_\alpha \{\mu\alpha - \widehat{D}(\alpha)\} - |\mu|\varepsilon = \widehat{A}(\mu) - |\mu|\varepsilon = \infty.$$

This proves (3.7.18), and therefore, Theorem 3.7.1. □

Note that a part of the proof of Theorem 3.7.1 can be derived from Theorem 4.3.1 in [57] (see also [118]), where a sequence $Z(T)$ of a more general nature satisfying the large deviation principle is considered. In these papers, it was shown that, under the assumption on the existence of appropriate upper estimates for $\overline{L}(-\infty, -N)$ and $\overline{L}(N, \infty)$, the value of $\lim_{T\to\infty} \frac{1}{T} \ln \mathbf{E}e^{\mu Z(T)}$ is equal to the Legendre transform of the deviation function corresponding to $Z(T)$. Such an approach was employed in [48], where Theorem 3.7.1 was also proved. A similar analysis was used in [93] to examine the case of a multidimensional arithmetic process $Z(t)$.

4

Large Deviation Principles for Trajectories of Compound Renewal Processes

In this chapter, we continue the study of large deviation principles for compound renewal processes. We shall be mostly concerned with probabilities of large deviations of the *trajectories* of CRPs.

The results of §4.1 are preliminary. In §§4.2 and 4.3, we establish "partial" LDPs (local LDPs applicable only in the space of absolutely continuous functions). In §§4.4 and 4.5, "complete" LDPs are obtained under rather restrictive conditions. In §§4.6 and 4.7, it proves possible to obtain LDPs for boundary crossing problems under more broad conditions and with explicit deviation functional. A moderately large deviation principle for CRPs is established in §4.8.

4.1 Conditions for the Fulfillment of the LDP for the Increments of a Process and for Finite-Dimensional Distributionss

4.1.1 LDP for Increments of a CRP

Let the conditions of Theorem 3.4.1 on LDP for a CRP in the state space be satisfied and let $U < T$ unboundedly increase with T so that $T - U \to \infty$. Will in this case the conditions ensuring the LDP for the increments of $Z(T) - Z(U)$ be met? The initial jumps $\xi_{1,U}$ of the projection of the process $Z(t)$ to $[U, T]$ have the following form. We set, as before,

$$\gamma(U) := U - T_{\nu(U)}, \quad \chi(U) := T_{\eta(U)} - U,$$
$$\zeta(U) := \zeta_{\eta(U)}, \quad \xi(U) := (\chi(U), \zeta(U)).$$

In this case, $\xi_{1,U} = \xi(U)$. The distribution of $\xi_{1,U}$ in the general inhomogeneous case is governed by the relations

$$P(u, B) := \mathbf{P}(\chi(U) \geq u,\ \zeta(U) \in B) = \mathbf{P}(\tau_1 \geq U + u,\ \zeta_1 \in B) + \int_0^U \widetilde{H}_1(dt)\mathbf{P}(\tau \geq U - t + u,\ \zeta \in B) \quad (4.1.1)$$

($B \subset \mathbb{R}$), where

$$\widetilde{H}_1(dt) := \sum_{n=1}^{\infty} \mathbf{P}(T_n \in dt) = \widetilde{H}(dt) \quad \text{for} \quad t \neq 0.$$

We have $\int_0^t \widetilde{H}_1(t) \le c_1 t + c_2$ for some c_1, c_2, and hence integration by parts gives

$$\begin{aligned} P(u, B) &\le \mathbf{P}(\tau_1 \ge U + v,\ \zeta_1 \in B) + c_1 \int_0^U \mathbf{P}(\tau \ge U - t + v,\ \zeta \in B)\, dt \\ &\le \mathbf{P}(\tau_1 \ge U + v,\ \zeta_1 \in B) + c_1 \int_0^\infty \mathbf{P}(\tau \ge t + v,\ \zeta \in B)\, dt. \end{aligned}$$

As before (see Lemma 1.1.5), we get

$$\begin{aligned} \psi_{(U)}(\lambda, \mu) &:= \mathbf{E} e^{\lambda\chi(U) + \mu\zeta(U)} \\ &\le \mathbf{E}(e^{\lambda(\tau_1 - U) + \mu\zeta_1};\ \tau_1 \ge U) + c_1 \frac{\psi(\lambda, \mu) - \psi(0, \mu)}{\lambda} \\ &\le \psi_1(\lambda, \mu) + c_1 \mathbf{E}\tau\psi^{(\mathrm{st})}(\lambda, \mu), \end{aligned} \tag{4.1.2}$$

$$\mathcal{A}_1 \cap \mathcal{A}^{(\mathrm{st})} \subset \mathcal{A}_{(U)} := \{(\lambda, \mu) :\ \psi_{(U)}(\lambda, \mu) < \infty\}.$$

The following assertion is a consequence of Remark 3.4.5 to Theorem 3.4.1 (one can represent inequality (4.1.2) in the form (3.4.19)).

Theorem 4.1.1 *The condition*

$$\mathcal{A}^{\le 0} \subset [\mathcal{A}_1 \cap \mathcal{A}^{(\mathrm{st})}] \tag{4.1.3}$$

(*cf.* (3.4.3)), *together with the other conditions of Theorem* 3.4.1, *imply the LDP for* $Z(T) - Z(U)$.

Apparently, if the equality $\widehat{D}(0) = D(0)$ holds, then condition (4.1.3) in Theorem 4.1.1 is also necessary (cf. Remark 3.4.3).

Consider the condition stronger than (4.1.3), which implies the *conditional* LDP for $Z(T) - Z(U)$ with respect to the σ-algebra $\mathfrak{A}_U$ generated by the trajectories of the process $Z(t)$ on $[0, U]$. This condition is closely related to the conditions of asymptotic independence of the increments of the process $Z(t)$. Given $B \subset \mathbb{R}^2$, we have

$$\mathbf{P}(\boldsymbol{\xi}_{1,U} \in B) = \mathbf{P}((\tau_1 - U,\ \zeta_1) \in B;\ \tau_1 \ge U) + \mathbf{P}(\boldsymbol{\xi}(U - \tau_1) \in B;\ \tau_1 < U). \tag{4.1.4}$$

As was already pointed out, the effective domain of the Laplace transform of the first (improper) distribution on the right of (4.1.4) contains $\mathcal{A}_1$. Consider the effective domain of the Laplace transform of the second (also improper) distribution on the right of (4.1.4). The entire dependence of the trajectory of $Z(t)$ on $[U, T]$ on the trajectory on $[0, U]$ is concentrated in the value $\gamma(U)$. For the first jump $\boldsymbol{\xi}_{1,U} =$

$\boldsymbol{\xi}(U) = (\chi(U), \zeta(U))$ of the process $Z(t)$ on $[U,T]$, on the set $\{\gamma(U) \in dt\} \in \mathfrak{A}_U$, we have, for $v \geq 0$, $w \geq 0$,

$$\begin{aligned} \mathbf{P}(\chi(U) \geq v,\ \zeta(U) \geq w |\ \mathfrak{A}_U) &= \mathbf{P}(\tau \geq t+v,\ \zeta \geq w |\ \tau \geq t), \\ \mathbf{P}(\chi(U) \geq v,\ \zeta(U) < -w |\ \mathfrak{A}_U) &= \mathbf{P}(\tau \geq t+v,\ \zeta < -w |\ \tau \geq t). \end{aligned} \tag{4.1.5}$$

On the half-space $\{v \geq 0\} \subset \mathbb{R}^2$ consider the measure $\mathbf{Q}$ defined on the quadrants $\{v \geq 0, w \geq 0\}$ and $\{v \geq 0, w < 0\}$ by

$$\begin{aligned} \mathbf{Q}([v,\infty) \times [w,\infty)) &:= \sup_{t>0} \mathbf{P}(\tau \geq t+v,\ \zeta \geq w |\ \tau \geq t), \\ \mathbf{Q}([v,\infty) \times [-\infty,-w)) &:= \sup_{t>0} \mathbf{P}(\tau \geq t+v,\ \zeta < -w |\ \tau \geq t), \end{aligned} \tag{4.1.6}$$

respectively. If this measure is normalized by the sum

$$Q_0 = \mathbf{Q}([0,\infty) \times [0,\infty)) + \mathbf{Q}([0,\infty) \times (-\infty, 0)),$$

we get the distribution corresponding to (possibly improper) random vector, which we denote by $\boldsymbol{\xi}_\infty = (\tau_\infty, \zeta_\infty)$.

If τ and ζ are independent, then the sum Q_0 is 1, and moreover, $\zeta_\infty \overset{d}{=} \zeta$, τ_∞ and ζ_∞ are also independent. The "tails" of distribution (4.1.6) majorize the corresponding "tails" of the conditional distributions (4.1.5) uniformly with respect to U and all directions.

In what follows, the condition

$$\boldsymbol{\xi}_\infty \in [\mathbf{C}] \tag{4.1.7}$$

will play an important role.

Let

$$\mathcal{A}_\infty := \{(\lambda,\mu) : \psi_\infty(\lambda,\mu) := \mathbf{E}e^{\lambda\tau_\infty + \mu\zeta_\infty} < \infty\}.$$

If (4.1.7) is met, then the domain $\mathcal{A}_\infty$ contains some neighborhood of the origin and lies in the effective domain of the Laplace transform of the initial jumps $\boldsymbol{\xi}_{1,U} = \boldsymbol{\xi}(U)$ for all U and any prehistory from $\mathfrak{A}_U$. Hence, if $\mathcal{A}^{\leq 0} \subset [\mathcal{A}_\infty]$, then the conditions of the form $\mathcal{A}^{\leq 0} \subset [\mathcal{A}_1]$ will be satisfied uniformly for all initial jumps $\boldsymbol{\xi}_{1,U}$ of the projections of the process $Z(t)$ to $[U,T]$.

Taking into account (4.1.4) and the above, we get the following assertion.

Theorem 4.1.2 *The conditions*

$$\boldsymbol{\xi}_\infty \in [\mathbf{C}], \qquad \mathcal{A}^{\leq 0} \subset [\mathcal{A}_1 \cap \mathcal{A}_\infty], \tag{4.1.8}$$

together with the other assumptions of Theorem 3.4.1, *imply the conditional LDP for the trajectories of* $Z(T) - Z(U)$:

$$\lim_{T-U\to\infty} \frac{1}{T-U} \ln \mathbf{P}\Big(\frac{Z(T)-Z(U)}{T-U} \in \Delta[\alpha) |\ \mathfrak{A}_U\Big) = -\widehat{D}(\alpha)$$

as $\Delta = \overline{o}(1)$, $T-U \to \infty$, *which is uniform with respect to all elementary events from* $\mathfrak{A}_U$ *(cf.* (3.4.5)*).*

An assertion analogous to (3.4.6), (3.4.7) also holds for the conditional "integral" LDP.

The condition $\boldsymbol{\xi}_\infty \in [\mathbf{C}]$ is not always met. This condition is not satisfied, for example, if the distribution of τ has large lacunas. The next lemma describes another situation in which the condition $\boldsymbol{\xi}_\infty \in [\mathbf{C}]$ is not satisfied.

Lemma 4.1.3 *Let there exist a constant $c > 0$ and sequences*

$$t_n \uparrow t_\infty \le \infty \quad \textit{and} \quad w_n \begin{cases} \ge ct_n & \textit{if} \quad t_\infty = \infty, \\ \to \infty & \textit{if} \quad t_\infty < \infty, \end{cases}$$

as $n \to \infty$ such that

$$\mathbf{P}(|\zeta| \ge w_n |\, \tau \ge t_n) \ge \begin{cases} e^{o(t_n)} & \textit{if} \quad t_\infty = \infty, \\ p > 0 & \textit{if} \quad t_\infty < \infty. \end{cases} \tag{4.1.9}$$

Then $\zeta_\infty \notin [\mathbf{C}]$.

Condition (4.1.9) indicates that there is a strong dependence between ζ and τ. This condition is always satisfied, for example, for dependencies of the form $\zeta = h\tau + \omega$ with $t_\infty = \infty$. In the case $t_\infty < \infty$, it is satisfied, for example, for the vector (τ, ζ) in which the variable τ is uniformly distributed over $[0, 1]$, $t_n \uparrow 1$, $w_n = -\ln(1-t_n) \to \infty$ as $n \to \infty$, $\zeta = \ln(1 - \tau) + \omega$, ω does not depend on τ, and $\omega \in [\mathbf{C}]$.

Proof of Lemma 4.1.3 For any $\mu > 0$, we have

$$\begin{aligned} \mathbf{E}(e^{\mu|\zeta|} |\, \tau \ge t_n) &\ge \mathbf{E}(e^{\mu|\zeta|},\ |\zeta| \ge w_n \,|\, \tau \ge t_n) \\ &\ge e^{\mu w_n} \mathbf{P}(|\zeta| \ge w_n \,|\, \tau \ge t_n) \ge \begin{cases} e^{\mu c t_n + o(t_n)} & \text{if} \quad t_\infty = \infty, \\ p e^{\mu w_n} & \text{if} \quad t_\infty < \infty. \end{cases} \end{aligned}$$

The right-hand sides of these relations increase unboundedly as $n \to \infty$. This means that $\zeta_\infty \notin [\mathbf{C}]$. The lemma is proved. □

Condition $\boldsymbol{\xi}_\infty \in [\mathbf{C}]$ is always met if τ and ζ are independent and the distribution of τ has the form $\mathbf{P}(\tau \ge v) = e^{-\lambda_+ v} l(v)$, where $l(v)$ is a regularly varying function at infinity. In this case, $\mathbf{P}(\tau \ge t + v \,|\, \tau \ge t) = e^{-\lambda_+ v} \frac{l(t+v)}{l(t)}$, where $\overline{\lim_{t\to\infty}} \frac{l(t+v)}{l(t)} = 1$, so that

$$\zeta_\infty \stackrel{d}{=} \zeta, \qquad \mathbf{P}(\tau_\infty \ge v) \le ce^{-\lambda_+ v}, \qquad c = \text{const}.$$

The following more general assertion holds.

Lemma 4.1.4 *Let $\lambda_+ < \infty$ and let the following conditions be met:*

(1)
$$\begin{aligned} \mathbf{P}(\tau \ge v, \zeta \ge w) &\le \mathbf{P}(\tau \ge v) e^{a(v,w)} \mathbf{P}(\zeta \ge w) e^{b(v,w)}, \\ \mathbf{P}(\tau \ge v, \zeta < -w) &\le \mathbf{P}(\tau \ge v) e^{a(v,w)} \mathbf{P}(\zeta < -w) e^{b(v,w)}, \end{aligned} \tag{4.1.10}$$

where

$$\sup_w \frac{a(v, w)}{v} \to 0 \quad \textit{as} \quad v \to \infty, \qquad \sup_v \frac{b(v, w)}{w} \to 0 \quad \textit{as} \quad w \to \infty.$$

(2)
$$\ln \mathbf{P}(\tau \ge v) = -\lambda_+ v + h(v), \tag{4.1.11}$$
where
$$\sup_t \frac{|h(t+v) - h(t)|}{v} \to 0 \quad as \quad v \to \infty.$$

Then $\mathcal{A}_\infty = \mathcal{A}$ and the domain $\mathcal{A}$ is rectangular.

Proof We have

$$\ln \mathbf{P}(\tau_\infty \ge v, \zeta_\infty \ge w) \le \ln Q_0 + \ln \sup_{t \ge 0} \frac{\mathbf{P}(\tau \ge t+v)}{\mathbf{P}(\tau \ge t)} + o(v)$$
$$+ \ln \mathbf{P}(\zeta \ge w) + o(w) \quad \text{as} \quad v \to \infty, w \to \infty,$$

where

$$\sup_{t \ge 0} \ln \frac{\mathbf{P}(\tau \ge t+v)}{\mathbf{P}(\tau \ge t)} \le -\lambda_+ v + \sup_{t \ge 0} |h(t+v) - h(t)| = -\lambda_+ v + o(v)$$

as $v \to \infty$. It follows that, for $c = \text{const}$,

$$\mathbf{P}(\tau_\infty \ge v, \zeta_\infty \ge w) \le c \exp\{-\lambda_+ v + o(v)\} \mathbf{P}(\zeta \ge w) e^{o(w)}.$$

A similar inequality also holds for $\mathbf{P}(\tau_\infty \ge v, \zeta_\infty < -w)$.

From these relations it follows that $\mathbf{E}e^{\lambda\tau_\infty + \mu\zeta_\infty} < \infty$ for any $\lambda < \lambda_+$ and $\mu \in (\mu_-^{(\zeta)}, \mu_+^{(\zeta)})$. The lemma is proved. □

Let us now formulate the result showing that the condition $\mathbf{P}(|\zeta_\infty| < \infty) > 0$ (and, more so, the condition $\zeta_\infty \in [\mathbf{C}]$) is incompatible with the inequality $\widehat{D}(0) < D(0)$ ($\lambda_+ < D(0)$).

Lemma 4.1.5 *If $\mathbf{P}(|\zeta_\infty| < \infty) > 0$, then $\lambda_+ \ge D(0) = \widehat{D}(0)$.*

Lemma 4.1.5 will be proved in §4.1.2.

Conditions (4.1.10) of "rough" asymptotic independence of the increments of the process $Z(t)$ can be somewhat relaxed. Namely, the following assertion holds.

Theorem 4.1.6 *Let $\xi_1 \in [\mathbf{C}]$, $\xi \in [\mathbf{C}]$, $\mathcal{A}^{\le 0} \subset [\mathcal{A}_1]$ and let at least one of the following conditions be met:*

(1) $\xi_\infty \in [\mathbf{C}]$, $\mathcal{A}^{\le 0} \subset [\mathcal{A}_\infty]$ *(the conditions of Theorem* 4.1.2*);*
(2) *conditions* (4.1.10)*;* (4.1.11) *are met (the conditions of Lemma* 4.1.4*);* (4.1.12)
(3) τ *and* ζ *are independent,* $\tau \in [\mathbf{C}_\infty]$*;*
(4) $\zeta \in [\mathbf{C}_\infty]$.

Then, for any fixed $u \in (0,1)$, the increments $Z(uT)$ and $Z(T) - Z(uT)$ are asymptotically independent in the sense of rough asymptotics and satisfy the "joint" LDP,

$$\lim_{T\to\infty}\frac{1}{T}\ln\mathbf{P}\Big(\frac{Z(uT)}{T}\in\Delta[\alpha),\ \frac{Z(T)-Z(uT)}{T}\in\Delta[\beta)\Big)$$
$$=\lim_{T\to\infty}\frac{1}{T}\ln\Big[\mathbf{P}\Big(\frac{Z(uT)}{T}\in\Delta[\alpha)\Big)\mathbf{P}\Big(\frac{Z(T)-Z(uT)}{T}\in\Delta[\beta)\Big)\Big]$$
$$=-u\mathbf{D}\Big(1,\frac{\alpha}{u}\Big)-(1-u)\mathbf{D}\Big(1,\frac{\beta}{1-u}\Big)$$
$$=-\mathbf{D}(u,\alpha)-\mathbf{D}(1-u,\beta), \tag{4.1.13}$$

where $\Delta=\overline{o}(1)$ *as* $T\to\infty$.

Note that the conditions of Theorem 3.4.1 on LDP for CRPs in the state space are insufficient for the LDP (4.1.13) to hold for finite-dimensional distributions. This is indicated by the following example.

Example 4.1.7 Let us construct the homogeneous CRP ($\xi_1 \underset{d}{=} \xi$, $\mathcal{A}_1=\mathcal{A}$) for which $\widehat{D}(0)=D(0)$, and therefore, the conditions of Theorem 3.4.1 are satisfied, but the LDP (4.1.13) for this process will not hold. Indeed, let $\mathbf{P}(\tau\ge t)=e^{-t}$, $\zeta=\pm\tau$ with probabilities 1/2. Then $\mathbf{E}\zeta=0$, $D(0)=0$, $\lambda_+=1$, $\lambda_+>D(0)=\widehat{D}(0)=0$, and the conditions of Theorem 3.4.1 are satisfied. For $u=1/2$, $\varepsilon\to 0$, $\varepsilon T\to\infty$ as $T\to\infty$, consider the probability

$$P:=\mathbf{P}\Big(\frac{Z(uT)}{T}\in(0)_\varepsilon,\ \frac{Z(T)-Z(uT)}{T}\in(1)_\varepsilon\Big).$$

It is clear that as $T\to\infty$ this probability is not smaller than

$$\frac{1}{3}\mathbf{P}(\tau\ge T(1-\varepsilon))=\frac{1}{3}e^{-T(1-\varepsilon)},\qquad \varliminf_{T\to\infty}\frac{1}{T}\ln P\ge -1.$$

At the same time $\mathbf{D}(1,2)=\infty$, since the rate of growth of $Z(t)$ is not greater than 1 ($\mathbf{P}(Z(T)>T)=0$). Hence equality (4.1.13) does not hold.

Proof of Theorem 4.1.6 From Theorem 3.3.2, Lemma 4.1.5 and any of conditions (4.1.12) it follows that $\lambda_+\ge D(0)=\widehat{D}(0)$. The first assertion of the theorem is a direct consequence of Theorem 4.1.1 and Lemma 4.1.5. The second assertion follows from the first one and Lemma 4.1.4.

Let us prove the third assertion. Consider the homogeneous case. Let B_1, B_2 be two events under the probability sign on the left of (4.1.13), $\tau\in[\mathbf{C}_\infty]$. Then, for $U=uT$, we have

$$\mathbf{P}(B_1B_2)=\mathbf{P}(B_1B_2B_T)+\mathbf{P}(B_1B_2\overline{B}_T), \tag{4.1.14}$$

where $n=[MT]$ and, for any $M>0$,

$$B_T:=\{\eta(U)<n,\ \tau(U)<\delta T\},$$
$$\mathbf{P}(B_1B_2\overline{B}_T)\le\mathbf{P}(\eta(U)\ge n)+n\mathbf{P}(\tau\ge\delta T)\mathbf{P}(T_n<U)+n\mathbf{P}(\tau\ge\delta T). \tag{4.1.15}$$

It is clear that as $T \to \infty$ by choosing an appropriate M the first probability on the right of (4.1.15) can be made smaller than e^{-NT} for any given $N > 0$. The second probability on the right of (4.1.15), for a given $\delta > 0$ and $T \to \infty$, will also be smaller than e^{-NT}, because $\tau \in [\mathbf{C}_\infty]$. It is clear that this estimate will also be preserved if $\delta = \delta_T \to 0$ sufficiently slowly as $T \to \infty$, so that

$$\mathbf{P}(B_1 B_2 \overline{B}_T) \le 2e^{-NT}.$$

Consider now the first term on the right of (4.1.14), which is equal to

$$\mathbf{P}(B_1 B_T)\mathbf{P}(B_2 |\, B_1 B_T).$$

It is clear that, for sufficiently large N,

$$\lim_{T\to\infty} \frac{1}{T} \ln \mathbf{P}(B_1 B_T) = \lim_{T\to\infty} \frac{1}{T} \ln \mathbf{P}(B_1) = -uD\Big(\frac{\alpha}{u}\Big).$$

Next,

$$\mathbf{P}(B_2 |\, B_1 B_T) = \mathbf{P}\Big(\frac{Z^*((1-u)T)}{T} \in \Delta[\beta)\Big),$$

where $Z^*(t)$ is the CRP, whose first jump (τ_1^*, ζ_1^*) is such that

$$\tau_1^* < \delta T, \qquad \xi_1^* \underset{d}{=} \zeta.$$

If we replace this jump by the vector $(\tau_1^{**}, \zeta_1^{**}) = (0, \zeta_1^*)$, we get the process $Z^{**}(t)$ satisfying all the conditions of Theorem 3.4.1 and such that $Z(T) - Z(U) = Z^{**}(T_1)$, where $T_1 = (1-u)T - \tau_1^*$, $\tau_1^* \le \delta T$, so that $T_1 \sim (1-u)T$ as $T \to \infty$. It remains to apply Theorem 3.4.1 and Remark 3.4.4, according to which, for any sequence $T_1 \sim (1-u)T$ as $T \to \infty$,

$$\lim_{T\to\infty} \frac{1}{T} \ln \mathbf{P}\Big(\frac{Z^{**}(T_1)}{T} \in \Delta[\beta)\Big) = -(1-u)D\Big(\frac{\beta}{1-u}\Big).$$

Assertion (4) follows from the LDP for the normalized *trajectories* $z_t(t) = \frac{Z(tT)}{T}$, $t \in [0,1]$, which will be proved below in Theorem 4.5.1.

The changes to be introduced in the proof in the inhomogeneous case are clear. Theorem 4.1.6 is proved. □

From Theorem 4.1.6 in turn we get the LDP for the finite-dimensional distributions of the process $Z(t)$ at the moments $u_j T$, $j = 0, \dots, k$; $0 = u_0 < u_1 < \cdots < u_k = 1$.

Theorem 4.1.8 *Let the conditions of Theorem* 4.1.6 *hold. Then, for any* α_j, $j = 1, \dots, k$,

$$\lim_{T\to\infty} \frac{1}{T} \ln \mathbf{P}\Big(\bigcap_{j=1}^{k}\Big\{\frac{1}{T}\,(Z(u_j T) - Z(u_{j-1}T)) \in \Delta[\alpha_j)\Big\}\Big) = I(f), \quad (4.1.16)$$

where $\Delta = \overline{o}(1)$ *as* $T \to \infty$,

$$I(f) := \int_0^1 D(f'(t))\,dt,$$

and $f(t)$ *is the continuous polygon on* $[0,1]$, $f(0) = 0$, *with nodes* $(u_j, \sum_{i=0}^{j} \alpha_i)$, $j = 0, \dots, k$.

The "integral" LDP for the finite-dimensional distributions of the process $Z(t)$ *also holds: for any measurable sets* $B_j \subset \mathbb{R}$,

$$\overline{\lim_{T\to\infty}} \frac{1}{T} \ln \mathbf{P}\Big(\bigcap_{j=1}^{k}\Big\{\frac{1}{T}(Z(u_jT) - Z(u_{j-1}T)) \in B_j\Big\}\Big) \le -I([B]), \tag{4.1.17}$$

$$\underline{\lim_{T\to\infty}} \frac{1}{T} \ln \mathbf{P}\Big(\bigcap_{j=1}^{k}\Big\{\frac{1}{T}(Z(u_jT) - Z(u_{j-1}T)) \in B_j\Big\}\Big) \ge -I((B)), \tag{4.1.18}$$

where

$$I(B) := \inf_{f\in B} I(f),$$

B *is the class of continuous polygons* $f(t)$ *with nodes* u_j, $j = 1, \dots, k$, *on the interval* $[0,1]$ *such that* $f(0) = 0$, $f(u_j) - f(u_{j-1}) \in B_j$ *for* $j = 1, \dots, k$, *and* $[B]$ *and* (B) *are, respectively, the closure and interior of the set* B *in the uniform metric.*

The integral LDP (4.1.17)–(4.1.18) can be established using the local LDP (4.1.16) as in the proof of Theorem 3.4.1.

It is unknown at present whether the conditions $\mathcal{A}^{\le 0} \subset \mathcal{A}_1$, $\lambda_+ \ge D(0)$ alone are sufficient for the LDP (4.1.16)–(4.1.18) to hold.

4.1.2 Proof of Lemma 4.1.5

The following assertion will be required in the proof of Lemma 4.1.5 and below.

Lemma 4.1.9 *The following equality holds:*

$$l^{(\tau)} := -\overline{\lim_{T\to\infty}} \frac{1}{T} \ln \mathbf{P}(\tau \ge T) = \lambda_+. \tag{4.1.19}$$

Proof By the exponential Chebyshev inequality, for $T \ge \mathbf{E}\tau$, any $\delta > 0$, and with obvious notational conventions, we have

$$\mathbf{P}(\tau \ge T) \le \psi^{(\tau)}(\lambda_+ - \delta) e^{-T(\lambda_+ - \delta)},$$

so that $-l^{(\tau)} \le -\lambda_+ + \delta$. Since δ is arbitrary, we have

$$l^{(\tau)} \ge \lambda_+. \tag{4.1.20}$$

If $\lambda_+ = \infty$, then, clearly, $l^{(\tau)} = \infty$, proving equality (4.1.19).

Let us now prove the inequality converse to (4.1.20) in the case $\lambda_+ < \infty$. Assume, on the contrary, that, for some $\delta > 0$,

$$l^{(\tau)} \geq \lambda_+ + \delta. \tag{4.1.21}$$

By the definition of $l^{(\tau)}$ this inequality means that

$$\mathbf{P}(\tau \geq t) \leq e^{-\lambda_+ t - \delta t + o(t)} \quad \text{as} \quad t \to \infty.$$

This implies that $\mathbf{E}e^{\lambda\tau} < \infty$ for all $\lambda \in [0, \ \lambda_+ + \delta)$, that is, $\lambda_+ \geq \lambda_+ + \delta$. This contradiction proves the lemma. □

Proof of Lemma 4.1.5 Let us show that $\mathbf{P}(|\zeta_\infty| = \infty) = 1$ in the case $\widehat{D}(0) < D(0)$.

Let

$$\lambda_+ = \widehat{D}(0) < D(0).$$

For any fixed $\varepsilon > 0$ and any sequence $t_n \to \infty$, we have

$$\mathbf{P}(\tau \geq t_n, \ |\zeta| \geq \varepsilon t_n) = \mathbf{P}(\tau \geq t_n) - \mathbf{P}(\tau \geq t_n, \ |\zeta| < \varepsilon t_n). \tag{4.1.22}$$

By Lemma 4.1.9, there exists a sequence $t_n \to \infty$ as $n \to \infty$ such that

$$\mathbf{P}(\tau \geq t_n) = e^{-\lambda_+ t_n + o(t_n)}. \tag{4.1.23}$$

Hence, by Remark 3.5.13 to Lemma 3.5.12,

$$\mathbf{P}(\tau \geq t_n, \ |\zeta| < \varepsilon t_n) \leq e^{-(D(0) - \delta(\varepsilon))t_n + o(t_n)}, \tag{4.1.24}$$

where $\delta(\varepsilon) \to 0$ as $\varepsilon \to 0$. We choose $\varepsilon > 0$ to satisfy the inequality $\delta(\varepsilon) < \frac{D(0) - \lambda_+}{2} =: r$. In this case, $D(0) - \delta(\varepsilon) > \lambda_+^{(\tau)} + r$, and the right-hand side in (4.1.24) is majorized by

$$e^{-(\lambda_+ + r)t_n + o(t_n)}.$$

We have $r > 0$, and hence using (4.1.23) we see that the second term on the right of (4.1.22) is negligible as $n \to \infty$ with respect to the first term, and therefore,

$$\begin{gathered} \mathbf{P}(\tau \geq t_n, \ |\zeta| \geq \varepsilon t_n) \sim \mathbf{P}(\tau \geq t_n), \\ \mathbf{P}(|\zeta| \geq \varepsilon t_n | \ \tau \geq t_n) \to 1 \end{gathered} \tag{4.1.25}$$

as $n \to \infty$. Hence, for any $N > 0$,

$$\begin{aligned} \mathbf{P}(|\zeta_\infty| \geq N) &= \mathbf{P}(\zeta_\infty \geq N) + \mathbf{P}(\zeta_\infty \leq -N) \\ &= \sup_{t>0} \mathbf{P}(\zeta \geq N | \ \tau \geq t) + \sup_{t>0} \mathbf{P}(\zeta \leq -N | \ \tau \geq t) \\ &\geq \overline{\lim_{n\to\infty}} \, \mathbf{P}(\zeta \geq N | \ \tau \geq t_n) + \overline{\lim_{n\to\infty}} \, \mathbf{P}(\zeta \leq -N | \ \tau \geq t_n) \\ &\geq \lim_{n\to\infty} \mathbf{P}(|\zeta| \geq N | \ \tau \geq t_n) = 1. \end{aligned}$$

Lemma 4.1.5 is proved. □

To show that the conditions $\lambda_+ < D(0)$ and $\zeta_\infty \in [\mathbf{C}]$ ($\boldsymbol{\xi}_\infty \in [\mathbf{C}]$) are incompatible one could also use Lemma 4.1.3, because in view of (4.1.25) condition (4.1.9) of this lemma is satisfied.

4.2 First Partial Local Large Deviation Principles for the Trajectories of a Compound Renewal Process

For an arbitrary fixed sequence $x = x(T) \sim T$ as $T \to \infty$, consider the family of random processes

$$z_T = z_T(t) = \frac{Z(tT)}{x}, \qquad 0 \le t \le 1.$$

The process z_T takes values in the measurable space $(\mathbb{D}, \mathfrak{B})$, where $\mathbb{D} = \mathbf{D}[0,1]$ is the space of functions $f = f(t)$ on the interval $[0,1]$ without discontinuities of the second kind. It will be convenient to equip the space $\mathbb{D}$ with the uniform metric

$$\rho(f,g) := \sup_{0\le t\le 1} |f(t) - g(t)|,$$

so that $\mathfrak{B}$ is the σ-algebra of Borel sets with respect to this metric. By $(f)_\varepsilon$ we denote the ε-neighborhood of a function $f \in \mathbb{D}$ in the metric ρ. Next, let $\mathbb{C} = \mathbb{C}[0,1]$ be the space of functions $f = f(t)$ continuous on $[0,1]$ and $\mathbb{C}_a \subset \mathbb{C}$ be the class of absolutely continuous functions $f = f(t)$ such that $f(0) = 0$. Our aim in this and the next sections is to prove: (a) two *"partial"* local LDPs (see the definitions below) for the trajectories of a compound renewal process and (b) to establish the "complete" local LDP under more restrictive conditions. In the case (a), we will indicate the conditions (see Theorems 4.2.1, 4.3.1 below) under which the relations

$$\frac{1}{T}\ln \mathbf{P}(\zeta_T \in (f)_\varepsilon) \to \begin{cases} -I(f) & \text{if } \lambda_+ \ge D(0) \ \text{(Theorem 4.2.1)}, \\ -\widehat{I}(f) & \text{if } \lambda_+ < D(0) \ \text{(Theorem 4.3.1)} \end{cases} \tag{4.2.1}$$

hold for any function $f \in \mathbb{C}_a$ and any sequence $\varepsilon = \varepsilon_T = \overline{o}(1)$ (sufficiently slowly converging to 0 as $T \to \infty$), where

$$I(f) := \int_0^1 D(f'(t))\,dt \quad \text{if} \quad f \in \mathbb{C}_a, \tag{4.2.2}$$

and the functional $\widehat{I}(f)$ will be defined below. If $f \in \mathbb{D} \setminus \mathbb{C}_a$ and if only the condition $\boldsymbol{\xi} = (\tau, \zeta) \in [\mathbf{C}]$ is met, then the limits on the right of (4.2.1) may fail to exist. For more details, see the beginning of §4.4.

In the case (b), we will show in Theorem 4.4.1 that, under the additional condition $\zeta \in [\mathbf{C}_\infty]$, for any function $f \in \mathbb{D}$ and any sequence $\varepsilon = \varepsilon_T$ which converges to 0 sufficiently slowly as $T \to \infty$, the *upper relation* in (4.2.1) (corresponding to the case $\lambda_+ \ge D(0)$) holds, in which the functional $I(f)$ is extended to be equal to ∞ on the functions $f \in \mathbb{D} \setminus \mathbb{C}_a$.

So, we first consider the case $\lambda_+ \ge D(0) = \widehat{D}(0)$, $f \in \mathbb{C}_a$.

4.2.1 Main Assertion and Its Proof

Theorem 4.2.1 (First partial local LDP) *Assume that*

$$\xi = (\tau, \zeta) \in\!\!\!\!\in [\mathbf{C}], \qquad \lambda_+ \geq D(0)$$

and that the local LDP (3.4.5) *in the state space holds. Then, for any function* $f \in \mathbb{C}_a$ *and any sequence* $\varepsilon = \varepsilon_T = \overline{o}(1)$ *as* $T \to \infty$,

$$\lim_{T\to\infty} \frac{1}{T} \ln \mathbf{P}(z_T \in (f)_\varepsilon) = -I(f). \tag{4.2.3}$$

We call the assertion of Theorem 4.2.1 the *partial* local LDP because it holds not for all functions $f \in \mathbb{D}$, but only for $f \in \mathbb{C}_a$. The second partial local LDP in the alternative case $\lambda_+ < D(0)$ will be established in §4.3 (Theorem 4.3.1). Complete local LDPs for any $f \in \mathbb{D}$ under the additional condition $\zeta \in\!\!\!\!\in [\mathbf{C}_\infty]$ will be considered in §4.4.

Proof of Theorem 4.2.1 We will assume that the normalizing sequence $x = x(T)$ has the form $x(T) = T$. It is easily seen that in the general case ($x \sim T$ as $T \to \infty$) Theorem 4.2.1 is a corollary of this special case (see Remark 3.4.4 and Lemma 4.5.4 below).

Given an integer $L \geq 1$, by $\mathbf{t}_L = \{0 = t_0 < t_1 < \cdots < t_L = 1\}$ we denote the partition of the half-open interval $(0, 1]$ into L half-open subintervals $(t_{l-1}, t_l]$, $l = 1, \ldots, L$. By $f^{\mathbf{t}_L} = f^{\mathbf{t}_L}(t)$ we denote the continuous polygon constructed with the nodes

$$(t_l, f(t_l)), \qquad l = 0, 1, \ldots, L.$$

The polygon $f^{\mathbf{t}_L}$ linearizes the function $f \in \mathbb{C}_a$, and hence we always have

$$I(f) \geq I(f^{\mathbf{t}_L}) \tag{4.2.4}$$

(see, e.g., [23, §4.2] or [43]). If we consider the so-called "dense" sequences of partitions $\mathbf{t}_L$ for which

$$\max_{1\leq l\leq L} (t_l - t_{l-1}) \to 0 \quad \text{as} \quad L \to \infty,$$

then we have the convergence

$$I(f^{\mathbf{t}_L}) \to I(f) \quad \text{as} \quad L \to \infty \tag{4.2.5}$$

(see, e.g., [23, §4.2]). It is clear that the uniform partitions

$$\mathbf{t}_L^U = \left\{\frac{l}{L}\right\}_{l=0}^{L}$$

form a dense sequence. If the partitions $\mathbf{t}_L$ are embedded (that is, if $\mathbf{t}_L \subset \mathbf{t}_{L+1}$), then, clearly, the convergence in (4.2.5) is monotone:

$$I(f^{\mathbf{t}_L}) \uparrow I(f) \quad \text{as} \quad L \uparrow \infty. \tag{4.2.6}$$

If a fixed function $f \in \mathbb{C}$ contains linear parts (on which $f'(t) = \text{const}$), then the convergence in (4.2.5), (4.2.6) also holds for sequences of “coarser” partitions (depending on f), which are not necessarily “everywhere dense.” It is clear, for example, that if $f(t) = ct$, $t \in [0, 1]$, then $I(f) = I(f^{\mathbf{t}_1})$, and no partitions are required for the convergence in (4.2.5). Hence there is no need to use partitions on the intervals of constancy of the function $f'(t)$ to ensure the convergence in (4.2.5), (4.2.6).

Moreover, the proof that we will use does not allow the presence of partition points on the intervals of constancy of the function f. As a result, we will construct embedded partitions for $f \in \mathbb{C}$ as follows. If f has intervals of constancy, then by (u_r^-, u_r^+), $r = 1, \dots, R$, we denote the longest intervals on which the variation of f is zero ($f'(t) = 0$ a.e.) and enumerate them in order of decreasing lengths. Consider the embedded partitions

$$\mathbf{t}_L = \bigcup_{k=1}^{L} \mathbf{t}_k^U. \tag{4.2.7}$$

So in order that the points of the partition we construct be distributed “uniformly” on the interval [0, 1] with increased number of this partition, we enumerate the elements of $\mathbf{t}_L$ as follows: first, we arrange the elements from $\mathbf{t}_1^U$ in increasing order, then we place the new elements from $\mathbf{t}_2^U$ in increasing order, and so on. For example, the partition $\mathbf{t}_4$ is as follows: $\{0, 1, 1/2, 1/3, 2/3, 1/4, 3/4\}$. Next, consider the intersection $\mathbf{t}_L V$ of the partition $\mathbf{t}_L$ with the set

$$V := [0, 1] \setminus \bigcup_{r=1}^{R} [u_r^-, u_r^+), \tag{4.2.8}$$

which has no intervals of constancy of the function f.

Let L_1 be the smallest number for which the number of points in $\mathbf{t}_{L_1} V$ exceeds L. Given L, we include in the partition $\mathbf{u}_L$ that we construct both the point 0 and the first L points from $\mathbf{t}_{L_1}$ (the point $t = 0$ does not enter the partition $\mathbf{t}_L V$ if one of the numbers u_r^- is 0). It is clear that the nonzero points u_r^- fail to lie in $\mathbf{u}_L$ and that the resulting sequence of embedded partitions is dense in the set V. Hence, for this sequence the convergence in (4.2.6) is preserved:

$$I(f^{\mathbf{u}_L}) \uparrow I(f) \quad \text{as} \quad L \uparrow \infty. \tag{4.2.9}$$

The partition $\mathbf{u}_L$ will be called an *f-partition.*

Consider now the points u_l, $l = 0, \dots, L$, from the partition $\mathbf{u}_L$ arranged in increasing order, and denote by χ_l the overshoot of level u_l by the walk $\{\frac{1}{T} T_k\}$, $k = 0, 1, \dots$:

$$\chi_l := \frac{1}{T} T_{\eta(u_l T)} - u_l, \tag{4.2.10}$$

where the first passage time $\eta(t)$ is defined in §1.1. The time $u_l + \chi_l$ is the time of the first jump of the process $z_T(t)$ after the time $t = u_l$. We set

$$\overline{\chi}_L := \max\{\chi_l : 1 \le l \le L-1\}. \tag{4.2.11}$$

It is clear that, for a given function f, there exists a sequence $\delta_T \downarrow 0$ as $T \uparrow \infty$ such that the event $\{z_T \in (f)_\varepsilon\}$ as $\varepsilon = \varepsilon_T \to 0$ implies the event $\{\overline{\chi}_L < \delta_T\}$. (Recall that the points $\{u_k^-\}$ do not enter the partition $\mathbf{u}_L$. If $\chi_l > c = \text{const} > 0$, then the process $z_T(t)$ remains constant on $[u_l, u_l + c]$, while the function f is not constant, hence the events $\{\chi_l > c\}$ and $z_T \in (f)_\varepsilon$ are incompatible for all sufficiently large T.)

Besides, it is clear that the event $\{z_T \in (f)_\varepsilon\}$ implies the event

$$\{\overline{\zeta}(T) < 2\varepsilon T\}, \quad \text{where} \quad \overline{\zeta}(T) := \max_{0 \le k \le \nu(T)} |\zeta_k|,$$

where $\nu(T)$ is defined in §1.1.

So, we have the inclusion

$$\{z_T \in (f)_\varepsilon\} \subset \Big\{ z_T(u_l) \in (f(u_l))_\varepsilon \text{ for all } \ 1 \le l \le L, \\ \overline{\zeta}(T) < 2\varepsilon T, \ \overline{\chi}_L < \delta_T \Big\}. \tag{4.2.12}$$

The following assertion plays an important role in the proof of Theorem 4.2.1.

Lemma 4.2.2 *Let $f \in \mathbb{C}$, $f(0) = 0$, be an arbitrary function. Then:*

(i) *If the conditions of Theorem* 4.2.1 *are met, then, for the f-partition $\mathbf{u}_L = \{u_l\}_{l=0}^{L}$ and any sequences $\varepsilon = \varepsilon_T \to 0$, $\delta_T \to 0$ as $T \to \infty$,*

$$\overline{\lim_{T\to\infty}} \frac{1}{T} \ln \mathbf{P}\Big(z_T(u_l) \in (f(u_l))_\varepsilon \text{ for all } 1 \le l \le L, \\ \overline{\zeta}(T) < 2\varepsilon T, \overline{\chi}_L < \delta_T \Big) \le -I(f^{\mathbf{u}_L}). \tag{4.2.13}$$

(ii) *If the conditions of Theorem* 3.4.1 *are met, then, for any fixed $\delta > 0$ and any partition $\mathbf{t}_L$,*

$$\underline{\lim_{T\to\infty}} \frac{1}{T} \ln \mathbf{P}\Big(z_T(t_l) \in (f(t_l))_\delta \text{ for all } 1 \le l \le L \Big) \ge -I(f^{\mathbf{t}_L}). \tag{4.2.14}$$

From assertion (ii) of Lemma 4.2.2 one can get the following inequality.

Corollary 4.2.3 *For any fixed $\delta > 0$,*

$$\underline{\lim_{T\to\infty}} \frac{1}{T} \ln \mathbf{P}\Big(z_T(t_l) \in (f(t_l))_\delta \text{ for all } 1 \le l \le L \Big) \ge -I((f^{\mathbf{t}_L})_\delta), \tag{4.2.15}$$

where

$$I(B) := \inf_{f \in B} I(f).$$

Proof of Corollary 4.2.3 For any function $g \in (f^{\mathbf{t}_L})_\delta$ and for a sufficiently small $\gamma > 0$,

$$\Big\{z_T(t_l) \in (f(t_l))_\delta \text{ for all } 1 \le l \le L\Big\} \supset \Big\{z_T(t_l) \in (g(t_l))_\gamma \text{ for all } 1 \le l \le L\Big\}. \tag{4.2.16}$$

Hence

$$\underline{\lim}_{T\to\infty} \frac{1}{T} \ln \mathbf{P}\Big(z_T(t_l) \in (f(t_l))_\delta \text{ for all } 1 \le l \le L\Big) \ge \underline{\lim}_{T\to\infty} \frac{1}{T} \ln \mathbf{P}\Big(z_T(t_l) \in (g(t_l))_\gamma \text{ for all } 1 \le l \le L\Big).$$

Applying (4.2.14) to the right-hand side of the last inequality, we get

$$\underline{\lim}_{T\to\infty} \frac{1}{T} \ln \mathbf{P}\Big(z_T(t_l) \in (f(t_l))_\delta \text{ for all } 1 \le l \le L\Big) \ge -I(g^{\mathbf{t}_L}). \tag{4.2.17}$$

The left-hand side of inequality (4.2.17) does not depend on the choice of the function g within $(f^{\mathbf{t}_L})_\delta$. Hence, replacing the right-hand side of (4.2.17) by the supremum over $g \in (f^{\mathbf{t}_L})_\delta$, we get

$$\underline{\lim}_{T\to\infty} \frac{1}{T} \ln \mathbf{P}\Big(z_T(t_l) \in (f(t_l))_\delta \text{ for all } 1 \le l \le L\Big) \ge -\inf_{g\in(f^{\mathbf{t}_L})_\delta} I(g^{\mathbf{t}_L}).$$

We have $I(g) \ge I(g^{\mathbf{t}_L})$ for any function $g \in \mathbb{D}$, and hence

$$I((f^{\mathbf{t}_L})_\delta) := \inf_{g\in(f^{\mathbf{t}_L})_\delta} I(g) \ge \inf_{g\in(f^{\mathbf{t}_L})_\delta} I(g^{\mathbf{t}_L}),$$

which gives (4.2.15). Corollary 4.2.3 is proved. □

Let us now continue the proof of Theorem 4.2.1 using Lemma 4.2.2 (which will be proved later) and Corollary 4.2.3.

(i) *Upper estimate*. From (4.2.12) and Lemma 4.2.2, we find that

$$\overline{\lim}_{T\to\infty} \frac{1}{T} \ln \mathbf{P}(z_T \in (f)_\varepsilon) \le -I(f^{\mathbf{u}_L}). \tag{4.2.18}$$

The left-hand side in (4.2.18) does not depend on the partition $\mathbf{u}_L$, and therefore, in view of (4.2.9) the right-hand side in (4.2.18) can be replaced by $-I(f)$. This establishes the upper estimate for (4.2.3).

(ii) *Lower estimate*. We claim that

$$\underline{\lim}_{T\to\infty} \frac{1}{T} \ln \mathbf{P}(z_T \in (f)_\delta) \ge -I(f) \tag{4.2.19}$$

for any fixed $\delta > 0$. If $I(f) = \infty$, then this estimate is clear. So in what follows we will assume that $I(f) < \infty$. Assume, to begin with, that, for some $N < \infty$, the random vectors $\boldsymbol{\xi}_1 = (\tau_1, \zeta_1)$, $\boldsymbol{\xi} = (\tau, \zeta)$ satisfy with probability 1 the inequalities

$$\tau_1 \ge \frac{1}{N}, \qquad \tau \ge \frac{1}{N}, \qquad |\zeta_1| \le N, \qquad |\zeta| \le N. \tag{4.2.20}$$

Repeating the arguments used in the derivation of the lower estimate in the local large deviation principle for the trajectories of random walks (see [23, §4.2] or [43]), we find that, for any fixed $\delta > 0$, there exists a number $L^0 = L^0_{\delta,f,N} < \infty$ such that, for all $L \geq L^0$ and all sufficiently large T,

$$\{z_T \in (f)_\delta\} \supset \left\{z_T(l/L) \in (f(l/L))_{\frac{\delta}{2}} \ \text{ for all } \ l = 1, \ldots, L\right\}.$$

Hence, using estimate (4.2.15) from Corollary 4.2.3, we have, for all $L \geq L^0$,

$$Q_\delta(f) := \varliminf_{T\to\infty} \frac{1}{T} \ln \mathbf{P}(z_T \in (f)_\delta) \geq -I((f^{\mathbf{t}_L})_\delta). \tag{4.2.21}$$

Now the required lower estimate (4.2.19) under the additional assumption (4.2.20) follows from (4.2.21) and (4.2.4).

If inequalities (4.2.20) are not satisfied, then we consider the "truncated" random vectors ${}^{(N)}\boldsymbol{\xi}_1$, ${}^{(N)}\boldsymbol{\xi}$, where the distribution of ${}^{(N)}\boldsymbol{\xi}$ is defined by

$$\mathbf{P}({}^{(N)}\boldsymbol{\xi} \in B) = \mathbf{P}\Big(\boldsymbol{\xi} \in B \,|\, \tau \geq \frac{1}{N},\ |\zeta| \leq N\Big) \tag{4.2.22}$$

($B \subset \mathbb{R}^2$). In a similar way, we define the distribution of the truncated vector ${}^{(N)}\boldsymbol{\xi}_1$. The random vectors ${}^{(N)}\boldsymbol{\xi}_i$, $i = 2, 3, \ldots$, are independent copies of ${}^{(N)}\boldsymbol{\xi}$, which are independent of ${}^{(N)}\boldsymbol{\xi}_1$. We denote by ${}^{(N)}z_T = {}^{(N)}z_T(t)$ the process constructed from the truncated vectors. For the truncated vectors ${}^{(N)}\boldsymbol{\xi}_i$ we retain all the notation corresponding to the random vectors $\boldsymbol{\xi}_i$ equipping them with the left superscript (N).

Next, let $M > 0$ be an arbitrary number. We set

$$A_{M,N} := \left\{\tau_i \geq \frac{1}{N},\ |\zeta_i| \leq N \ \text{ for all } \ i = 0, 1, \ldots, [MT]\right\}.$$

Then

$$\begin{aligned} \mathbf{P}(z_T \in (f)_\delta) &\geq \mathbf{P}(z_T \in (f)_\delta,\ \nu(T) \leq TM,\ A_{M,N}) \\ &= \mathbf{P}(A_{M,N})\mathbf{P}({}^{(N)}z_T \in (f)_\delta,\ {}^{(N)}\nu(T) \leq TM). \end{aligned} \tag{4.2.23}$$

Here

$$\frac{1}{T} \ln \mathbf{P}(A_{M,N}) \to M \ln P(N) \tag{4.2.24}$$

as $T \to \infty$, where

$$P(N) := \mathbf{P}\Big(\tau \geq \frac{1}{N},\ |\zeta| \leq N\Big) \to 1 \text{ as } N \to \infty. \tag{4.2.25}$$

The second factor on the right of (4.2.23) is not smaller than

$$\mathbf{P}({}^{(N)}z_T \in (f)_\delta) - \mathbf{P}({}^{(N)}\nu(T) > TM). \tag{4.2.26}$$

Let us first estimate

$${}^{(N)}Q_\delta(f) := \varliminf_{T\to\infty} \frac{1}{T} \ln \mathbf{P}({}^{(N)}z_T \in (f)_\delta).$$

Hence we can apply (4.2.21), which gives

$$^{(N)}Q_\delta(f) \geq -{}^{(N)}I((f^{\mathbf{t}_L})_\delta), \tag{4.2.27}$$

where

$$^{(N)}I((f^{\mathbf{t}_L})_\delta) = \inf_{g \in B_L} {}^{(N)}I(g^{\mathbf{t}_L}), \qquad B_L := (f^{\mathbf{t}_L})_\delta. \tag{4.2.28}$$

Hence, for any $g \in B_L$,

$$^{(N)}Q_\delta(f) \geq -{}^{(N)}I(g^{\mathbf{t}_L}). \tag{4.2.29}$$

Putting $t_l = \frac{l}{L}$ for $l = 0, 1, \ldots, L$, and setting

$$\alpha_l = \left[f\left(\frac{l}{L}\right) - f\left(\frac{l-1}{L}\right)\right]L, \quad \beta_l = \left[g\left(\frac{l}{L}\right) - g\left(\frac{l-1}{L}\right)\right]L$$

for $l = 1, \ldots, L$, we get, for any function $g \in B_L$,

$$^{(N)}Q_\delta(f) \geq -\frac{1}{L}\sum_{l=1}^{L} {}^{(N)}D(\beta_l).$$

The left-hand side of this inequality does not depend on g from the set B_L and since $|\beta_l - \alpha_l| < 2\delta$ for $g \in B_L$, we get

$$\underline{\lim}_{N\to\infty} {}^{(N)}Q_\delta(f) \geq -\frac{1}{L}\sum_{l=1}^{L} \overline{\lim}_{N\to\infty} \inf_{|\beta_l-\alpha_l|<2\delta} {}^{(N)}D(\beta_l), \tag{4.2.30}$$

completing the proof. □

We next will use the following assertion, which will be proved later.

Lemma 4.2.4 *For any $\delta > 0$, $\alpha \in \mathbb{R}$,*

$$\overline{\lim}_{N\to\infty} \inf_{|\beta-\alpha|<\delta} {}^{(N)}D(\beta) \leq D(\alpha). \tag{4.2.31}$$

Inequality (4.2.31) *remains valid for the deviation function* $^{[N]}D(\alpha)$ *of the truncations* $^{[N]}\boldsymbol{\xi}$ *with the distribution*

$$\mathbf{P}(^{[N]}\boldsymbol{\xi} \in B) := \frac{\mathbf{P}(\boldsymbol{\xi} \in B, \tau \leq N, |\zeta| \leq N)}{\mathbf{P}(\tau \leq N, |\zeta| \leq N)}. \tag{4.2.32}$$

Applying (4.2.31) to the right-hand side of (4.2.30), we find that

$$\underline{\lim}_{N\to\infty} {}^{(N)}Q_\delta(f) \geq -\frac{1}{L} \overline{\lim}_{N\to\infty} \sum_{l=1}^{L} {}^{(N)}D(\alpha_l) = -I(f^{\mathbf{t}_L}). \tag{4.2.33}$$

Hence, by the inequality $I(f^{\mathbf{t}_L}) \leq I(f)$, we get the lower estimate

$$\underline{\lim}_{N\to\infty} {}^{(N)}Q_\delta(f) \geq -I(f). \tag{4.2.34}$$

Let us now return to inequality (4.2.23). From this inequality and (4.2.26) it follows that, for any N,

$$Q_\delta(f) \ge r(N) + \varliminf_{T\to\infty} \frac{1}{T} \ln \Big[\mathbf{P}({}^{(N)}z_T \in (f)_\delta) - \mathbf{P}({}^{(N)}\nu(T) > MT)\Big], \qquad (4.2.35)$$

where

$$r(N) := \lim_{T\to\infty} \frac{1}{T} \ln \mathbf{P}(A_{N,M}) \to 0 \qquad (4.2.36)$$

as $N \to \infty$. We further note that ${}^{(N)}\tau \underset{d}{\ge} \tau$, ${}^{(N)}\nu(T) \underset{d}{\le} \nu(T)$. Hence, for all N,

$$\mathbf{P}({}^{(N)}\nu(T) > MT) \le \mathbf{P}(\nu(T) > MT).$$

In (3.4.18) it was shown that, for any $N < \infty$, there exists an M such that

$$\mathbf{P}(\nu(T) > MT) \le e^{-NT}. \qquad (4.2.37)$$

Hence in the case $I(f) < \infty$ one can always choose $M < \infty$ so that, for all N,

$$\mathbf{P}({}^{(N)}\nu(T) \ge MT) \le e^{-3TI(f)}.$$

At the same time, by (4.2.34) the probability $\mathbf{P}({}^{(N)}z_T \in (f)_\delta)$ exceeds $e^{-2TI(f)}$ for all sufficiently large N and T. Hence the second term in square brackets on the right-hand side of (4.2.35) is negligible with respect to the first term. Using (4.2.36) and taking into account (4.2.34), we get from (4.2.35) the required estimate

$$Q_\delta(f) \ge -I(f).$$

This proves (4.2.19), and therefore, Theorem 4.2.1. □

4.2.2 Proofs of Lemmas 4.2.2 and 4.2.4

Proof of Lemma 4.2.2 (i) *The upper estimate.* We argue by induction on L. For $L = 1$, inequality (4.2.13) follows from Theorem 3.4.1. Let us now prove (4.2.13) for $L = 2$. For $L \ge 3$, the transition from $L - 1$ to L in the proof of (4.2.13) can be carried out in exactly the same way as for $L = 2$.

So, consider the partition $u_0 = 0$, $u_1 = u$, $u_2 = 1$, $u \in (0, 1)$, and write

$$\alpha_l := f(u_l) - f(u_{l-1}), \qquad \Delta_l := u_l - u_{l-1}, \quad l = 1, 2. \qquad (4.2.38)$$

Next, let

$$A_1 := \{z_T(u) \in (\alpha_1)_\varepsilon,\ |\zeta_{\eta(uT)}| < 2\varepsilon T,\ \ \chi := \chi_1 < \delta_T\} \qquad (4.2.39)$$

(χ_l is defined in (4.2.10)),

$$A_2 := \{z_T(1) \in (\alpha_1 + \alpha_2)_\varepsilon\},$$

A is the event under the probability sign in (4.2.13) with $L = 2$. Next. let $\mathfrak{A}$ be the σ-algebra generated by the random variables $\eta(uT), \xi_1, \xi_2, \ldots, \xi_{\eta(uT)}$. It is clear that the variables $\zeta_{\eta(uT)}$ and χ are measurable with respect to $\mathfrak{A}$, $A_1 \in \mathfrak{A}$.

The evolution of the process $z_T(t)$ on $(u+\chi,\ 1]$ does not depend on the σ-algebra $\mathfrak{A}$. Therefore, for $L = 2$,

$$\mathbf{P}(A) = \mathbf{P}(A_1 A_2) = \mathbf{E}\mathbf{1}(A_1)\mathbf{E}(\mathbf{1}(A_2)|\,\mathfrak{A}). \tag{4.2.40}$$

We denote by $Z^*(t)$ the homogeneous renewal process on $[0, \infty)$ which does not depend on $Z(t)$, and consider the factor

$$\mathbf{E}(\mathbf{1}(A_2)|\,\mathfrak{A}) = \mathbf{P}(A_2 \mid \mathfrak{A}).$$

This function of elementary outcomes from $\mathfrak{A}$ on the set A_1 can be considered as the probability

$$\begin{aligned} P_T := \mathbf{P}\Big((\alpha_1 + y)T + Z^*((\Delta_2 - v)T) \in (\alpha_1 + \alpha_2)_\varepsilon T\Big) \\ \le \mathbf{P}\Big(Z^*(T^*) \in (\alpha_2 - y)_\varepsilon T^* \times \frac{T}{T^*}\Big) \\ \le \mathbf{P}\Big(Z^*(T^*) \in (\alpha_2)_{3\varepsilon} T^* \times \frac{T}{T^*}\Big), \end{aligned} \tag{4.2.41}$$

where

$$T^* := (\Delta_2 - v)T \sim \Delta_2 T, \qquad \frac{T^*}{T} \sim \Delta_2 \quad \text{as} \quad T \to \infty,$$

for nonrandom v and y such that $|y| < 2\varepsilon_T$, $v \in [0, \delta_T)$. So, we find ourselves under the conditions of Theorem 3.4.1, which gives

$$\overline{\lim_{T^* \to \infty}} \frac{1}{T^*} \ln P_T \le -D\Big(\frac{\alpha_2}{\Delta_2}\Big) \sim -\frac{T}{T^*} \Delta_2 D\Big(\frac{\alpha_2}{\Delta_2}\Big). \tag{4.2.42}$$

In other words, on the entire set A_1 we have

$$\mathbf{P}(A_2|\,\mathfrak{A}) \le \exp\Big\{-T\Delta_2 D\Big(\frac{\alpha_2}{\Delta_2}\Big) + o(T)\Big\}. \tag{4.2.43}$$

Hence by (4.2.40),

$$\mathbf{P}(A) \le \exp\Big\{-T\Delta_2 D\Big(\frac{\alpha_2}{\Delta_2}\Big) + o(T)\Big\} \times \mathbf{P}(Z(uT) \in T(\alpha_1)_\varepsilon), \tag{4.2.44}$$

where another appeal to Theorem 3.4.1 shows that

$$\mathbf{P}(Z(uT) \in T(\alpha_1)_\varepsilon) = \exp\Big\{-T\Delta_1 D\Big(\frac{\alpha_1}{\Delta_1}\Big) + o(T)\Big\}.$$

As a result, we get

$$\mathbf{P}(A) \le \exp\{-TI(f^{\mathbf{u}_2}) + o(T)\}.$$

This proves the required upper estimate (4.2.13) with $L = 2$.

As was already pointed out, the transition from $L-1$ to L for $L \ge 3$ in our induction argument is carried out in the same way – one should consider, as u, the point before the last point u_{L-1} of the partition $\mathbf{u}_L$. This proves the upper estimate (4.2.13).

(ii) The proof of the *lower estimate* in Lemma 4.2.2 resembles that given above. But now, to satisfy the relations

$$\overline{\zeta}(T) = o(T), \quad \overline{\chi}_L = o(T) \quad \text{as} \quad T \to \infty,$$

we will use not a special construction of partitions, but the truncation method. Namely, using induction again we will prove the inequality

$$\varliminf_{T\to\infty} \frac{1}{T} \ln \mathbf{P}\big(z_T(t_l) \in (\alpha_l + \alpha_{l-1})_\delta,\ l = 1, 2\big) \geq -I(f^{\mathbf{t}_2})$$

for any $\delta > 0$, but first in the case when the inequalities

$$\tau_1 \leq N, \qquad \tau \leq N, \qquad |\zeta_1| \leq N, \qquad |\zeta| \leq N \tag{4.2.45}$$

are satisfied for some constant $N < \infty$ with probability 1.

It is clear that in this case

$$\overline{\zeta}(T) \leq N, \qquad \overline{\chi}_L \leq \frac{N}{T}.$$

As in (4.2.38), (4.2.39), for $L = 2$ we put

$$A_1 := \Big\{z_T(u) \in (\alpha_1)_\delta,\ |\zeta_{\eta(uT)}| \leq N,\ \chi := \chi_1 \leq \frac{N}{T}\Big\},$$
$$A_2 := \{z_T(1) \in (\alpha_1 + \alpha_2)_\delta\},$$

A is the event under the probability sign in (4.2.14) with $L = 2$ and the σ-algebra $\mathfrak{A}$ has the same meaning. We have (cf. (4.2.40))

$$\mathbf{P}(A) = \mathbf{P}(A_1A_2) = \mathbf{E}\mathbf{1}(A_1)\mathbf{E}(\mathbf{1}(A_2)|\ \mathfrak{A}).$$

Here in place of (4.2.41), we have, for $T > \frac{N}{\delta}$,

$$P_T \geq \mathbf{P}\left(Z^*(T^*) \in (\alpha_2)_{\delta/2}T^* \times \frac{T}{T^*}\right)$$

with the same notation as in (4.2.41). Next, employing (4.2.42)–(4.2.44), this gives

$$\mathbf{P}(A) \geq \exp\{-TI(f^{\mathbf{t}_2}) + o(T)\}.$$

This establishes the required upper estimate (4.2.14) for $L = 2$. The transition from $L-1$ to L for $L \geq 3$ is carried out in the same way as in Step (i).

If inequalities (4.2.45) are not satisfied, then the above truncation method can again be applied in the same way as in the proof of Theorem 4.2.1, with the only difference that now the truncations are different from (4.2.23). In view of (4.2.45), one should consider the "truncated" vectors ${}^{[N]}\xi_1$, ${}^{[N]}\xi$ with the distribution of ${}^{[N]}\xi$ defined by

$$\mathbf{P}({}^{[N]}\xi \in B) = \mathbf{P}(\xi \in B|\ \tau \leq N,\ |\zeta| \leq N) \tag{4.2.46}$$

($B \subset \mathbb{R}^2$; cf. (4.2.22)). The distribution of the truncation ${}^{[N]}\xi_1$ is defined similarly. Arguing as in the proof of Theorem 4.2.1 and applying the second assertion of Lemma 4.2.4, we get the required lower estimate. Lemma 4.2.2 is proved. □

Let us now proceed with the proof of Lemma 4.2.4 on asymptotic properties of the functions ${}^{(N)}D(\alpha)$ and ${}^{[N]}D(\alpha)$ as $N \to \infty$. To this aim, we need to know the similar properties of the deviation functions ${}^{(N)}\mathbf{\Lambda}(v, \alpha)$, ${}^{[N]}\mathbf{\Lambda}(v, \alpha)$. To keep the calculations as simple as possible and for the sake of greater generality, we change the notation for a while, and, in place of two-dimensional vectors (v, α), we will consider the d-dimensional vectors $\boldsymbol{\alpha} = (\alpha_1, \ldots, \alpha_d)$, $d \geq 1$, and study the asymptotic properties of the deviation functions ${}^{(N)}\mathbf{\Lambda}(\boldsymbol{\alpha})$, ${}^{[N]}\mathbf{\Lambda}(\boldsymbol{\alpha})$, of the truncations ${}^{(N)}\boldsymbol{\xi}$ and ${}^{[N]}\boldsymbol{\xi}$ of the random vectors $\boldsymbol{\xi} = (\xi_1, \ldots, \xi_d)$, respectively, where $\xi_1 \geq 0$ and in the truncation ${}^{(N)}\boldsymbol{\xi}$ we assume that $\xi_1 \leq 1/N$, $|\xi_j| \leq N$ for $j \geq 2$. We denote by $\mathbf{\Lambda}(\boldsymbol{\alpha})$ the deviation function corresponding to $\boldsymbol{\xi}$ and denote by S the convex hull of the support of the distribution of $\boldsymbol{\xi}$, so that $(\mathrm{S}) = (\Lambda^{<\infty})$, where $\Lambda^{<\infty} := \{\boldsymbol{\alpha} : \mathbf{\Lambda}(\boldsymbol{\alpha}) < \infty\}$.

The next assertion describes the required properties of the functions ${}^{(N)}\mathbf{\Lambda}(\boldsymbol{\alpha})$ and ${}^{[N]}\mathbf{\Lambda}(\boldsymbol{\alpha})$.

Lemma 4.2.5 *If $\alpha \in (\mathrm{S})$, then*

$$ {}^{(N)}\mathbf{\Lambda}(\boldsymbol{\alpha}) \to \mathbf{\Lambda}(\boldsymbol{\alpha}) \quad as \quad N \to \infty. \tag{4.2.47} $$

If $\boldsymbol{\alpha} \in \partial \mathrm{S}$, then, for any $\delta > 0$,

$$ {}^{(N)}\mathbf{\Lambda}((\boldsymbol{\alpha})_\delta) \to \mathbf{\Lambda}((\boldsymbol{\alpha})_\delta) \quad as \quad N \to \infty. $$

If $\boldsymbol{\alpha} \notin [\mathrm{S}]$, then

$$ {}^{(N)}\mathbf{\Lambda}(\boldsymbol{\alpha}) = \mathbf{\Lambda}(\boldsymbol{\alpha}) = \infty. $$

The analogous results also hold for ${}^{[N]}\mathbf{\Lambda}(\boldsymbol{\alpha})$.

From Lemma 4.2.5 we get

Corollary 4.2.6 *For any α, there exists $\delta = \delta(\boldsymbol{\alpha}) > 0$ such that*

$$ \begin{aligned} &\lim_{N\to\infty} {}^{(N)}\mathbf{\Lambda}((\boldsymbol{\alpha})_\delta) \to \mathbf{\Lambda}((\boldsymbol{\alpha})_\delta) \leq \mathbf{\Lambda}(\boldsymbol{\alpha}), \\ &\lim_{N\to\infty} {}^{[N]}\mathbf{\Lambda}((\boldsymbol{\alpha})_\delta) \to \mathbf{\Lambda}((\boldsymbol{\alpha})_\delta) \leq \mathbf{\Lambda}(\boldsymbol{\alpha}). \end{aligned} \tag{4.2.48} $$

It is easily checked that in the one-dimensional case $d = 1$ the convergence in (4.2.47) takes place for all α. However, this, generally speaking, is not so for $d > 1$, and in this case we have been able only to derive the "relaxed" version (4.2.48) of this convergence.

Proof of Lemma 4.2.5 To simplify the presentation, we consider in detail only the proof for the truncations ${}^{[N]}\xi$ (the argument for the truncations ${}^{(N)}\xi$ is similar). Assume, for the time being, that $|\boldsymbol{\xi}| = \sum_{i=1}^{d} |\xi_i|$,

$$ \mathbf{P}({}^{[N]}\boldsymbol{\xi} \in B) = \mathbf{P}(\boldsymbol{\xi} \in B \,|\, |\boldsymbol{\xi}| \leq N) $$

(cf. (4.2.46). This change of norms, which does not change anything in the proof, was made only to simplify the presentation. Setting

$$ \psi_N(\boldsymbol{\lambda}) := \mathbf{E}(e^{\langle \boldsymbol{\lambda}, \boldsymbol{\xi} \rangle};\ |\boldsymbol{\xi}| \leq N), \qquad \delta_N := \ln \mathbf{P}(|\boldsymbol{\xi}| \leq N), $$

we have

$$\delta_N \uparrow 0, \qquad \psi_N(\boldsymbol{\lambda}) \uparrow \psi(\boldsymbol{\lambda}), \qquad \mathbf{A}_N(\boldsymbol{\lambda}) := \ln \psi_N(\boldsymbol{\lambda}) \uparrow \mathbf{A}(\boldsymbol{\lambda})$$

as $N \uparrow \infty$. Next

$$^{[N]}\mathbf{A}(\boldsymbol{\lambda}) = \mathbf{A}_N(\boldsymbol{\lambda}) - \delta_N, \ ^{[N]}\boldsymbol{\Lambda}(\boldsymbol{\alpha}) = \sup_{\boldsymbol{\lambda}} \{\langle \boldsymbol{\lambda}, \boldsymbol{\alpha}\rangle - {}^{[N]}A(\boldsymbol{\lambda})\} = \boldsymbol{\Lambda}_N(\boldsymbol{\alpha}) + \delta_N,$$

where

$$\boldsymbol{\Lambda}_N(\boldsymbol{\alpha}) = \sup_{\boldsymbol{\lambda}} \{\langle \boldsymbol{\lambda}, \boldsymbol{\alpha}\rangle - \mathbf{A}_N(\boldsymbol{\lambda})\} \downarrow \quad \text{as} \quad N \uparrow \infty.$$

Since $\langle \boldsymbol{\lambda}, \boldsymbol{\alpha}\rangle - \mathbf{A}_N(\boldsymbol{\lambda})$ and $\langle \boldsymbol{\lambda}, \boldsymbol{\alpha}\rangle - \mathbf{A}(\boldsymbol{\lambda})$ are concave functions, it is easily checked that, for $\alpha \in (\mathrm{S}) = (\Lambda_{<\infty})$, we have

$$\boldsymbol{\Lambda}_N(\boldsymbol{\alpha}) \downarrow \boldsymbol{\Lambda}(\boldsymbol{\alpha}), \qquad ^{[N]}\Lambda(\alpha) \to \boldsymbol{\Lambda}(\alpha)$$

as $N \uparrow \infty$. This fact can also be derived from convergence theorems in convex analysis.

Now let $\alpha \in \partial \mathrm{S}$. For $\beta \in (\alpha)_\delta \cap (\mathrm{S})$, from the first assertion of the lemma we have

$$\varlimsup_{N\to\infty} {}^{[N]}\boldsymbol{\Lambda}((\boldsymbol{\alpha})_\delta) \le \varlimsup_{N\to\infty} {}^{[N]}\boldsymbol{\Lambda}(\boldsymbol{\beta}) = \boldsymbol{\Lambda}(\boldsymbol{\beta}).$$

Hence the left-hand side of this inequality (which does not depend on $\boldsymbol{\beta}$) is majorized by $\inf\{\boldsymbol{\Lambda}(\boldsymbol{\beta}) : \boldsymbol{\beta} \in (\boldsymbol{\alpha})_\delta \cap (\mathrm{S})\}$. The function $\boldsymbol{\Lambda}(\boldsymbol{\beta})$ is continuous on $\partial\mathrm{S}$ from inside of the set S (along the rays), and so we can take the infimum over the set $\boldsymbol{\beta} \in (\boldsymbol{\alpha})_\delta \cap [\mathrm{S}]$. This proves the upper estimate for the second assertion of the lemma, inasmuch as $\boldsymbol{\Lambda}(\boldsymbol{\beta}) = \infty$ as $\boldsymbol{\beta} \notin [\mathrm{S}]$. The lower estimate is clear, because $^{[N]}\boldsymbol{\Lambda}(\boldsymbol{\alpha}) = \boldsymbol{\Lambda}_N(\boldsymbol{\alpha}) + \delta_N \ge \boldsymbol{\Lambda}(\boldsymbol{\alpha}) + \delta_N$.

The last assertion of the lemma follows the embedding $^{[N]}\Lambda^{<\infty} \subset \Lambda^{<\infty}$, and Lemma 4.2.5 is proved. □

Let us go back to our former notation and give the proof of Lemma 4.2.4, which is somewhat more cumbersome than that of Lemma 4.2.5.

Proof of Lemma 4.2.4 We denote by

$$^{(\infty)}D_\delta(\alpha) := \varlimsup_{N\to\infty} {}^{(N)}D((\alpha)_\delta)$$

the left-hand side of (4.2.31). We need to show that, for any fixed $\delta > 0$, $\alpha \in \mathbb{R}$,

$$^{(\infty)}D_\delta(\alpha) \le D(\alpha) = \mathbf{D}(1, \alpha) \tag{4.2.49}$$

(see (4.2.31)). We set

$$^{(\infty)}\mathbf{D}_{(\omega)}(1, \alpha) := \varlimsup_{N\to\infty} {}^{(N)}\mathbf{D}((1, \alpha)_\omega).$$

For a fixed $\alpha \in \mathbf{R}$, inequality (4.2.49) will be obtained as a corollary of the following two assertions:

(A) *for any* $\delta > 0$, $\gamma \in (0, \frac{1}{2})$, *there exists* $\omega = \omega(\delta, \gamma) > 0$ *such that*

$$^{(\infty)}D_\delta(\alpha) \le \frac{1}{1-\gamma}{}^{(\infty)}\mathbf{D}_{(\omega)}(1, \alpha); \tag{4.2.50}$$

(B) *for any* $\omega > 0$,

$$^{(\infty)}\mathbf{D}_{(\omega)}(1, \alpha) \le \mathbf{D}(1, \alpha). \tag{4.2.51}$$

Let us prove (A). Given real numbers $\gamma \in (0, \frac{1}{2})$, $h \in (-\gamma, \gamma)$, we have

$$^{(N)}\mathbf{D}(1, \beta) = \frac{1}{1-h}{}^{(N)}\mathbf{D}(1-h, (1-h)\beta) \le \frac{1}{1-\gamma}{}^{(N)}\mathbf{D}(1-h, (1-h)\beta). \tag{4.2.52}$$

The left-hand side of (4.2.52) does not depend on h, and hence the inequality will be preserved if the right-hand side in (4.2.52) is replaced by the infimum over $|h| < \gamma$:

$$^{(N)}\mathbf{D}(1, \beta) \le \frac{1}{1-\gamma} \inf_{|h|<\gamma} {}^{(N)}\mathbf{D}(1-h, (1-h)\beta). \tag{4.2.53}$$

From (4.2.53) it follows that

$$\begin{aligned} ^{(N)}\mathbf{D}(1, (\alpha)_\delta) &= \inf_{|\beta-\alpha|<\delta} {}^{(N)}\mathbf{D}(1, \beta) \\ &\le \frac{1}{1-\gamma} \inf_{\substack{|h|<\gamma \\ |\beta-\alpha|<\delta}} {}^{(N)}\mathbf{D}(1-h, (1-h)\beta) \\ &= \frac{1}{1-\gamma} \inf_{(v,b)\in B(\gamma,\delta)} {}^{(N)}\mathbf{D}(v, b), \end{aligned} \tag{4.2.54}$$

where $v := 1 - h$, $b := (1-h)\beta$,

$$B(\gamma, \delta) := \{(v, b) = (1-h, (1-h)\beta) : |h| < \gamma, \ |\beta - \alpha| < \delta\}.$$

Hence inequality (4.2.54) becomes

$$^{(N)}\mathbf{D}(1, (\alpha)_\delta) \le \frac{1}{1-\gamma} \inf_{(v,b)\in B(\gamma,\delta)} {}^{(N)}\mathbf{D}(v, b). \tag{4.2.55}$$

For any fixed $\delta > 0$ and $\gamma \in (0, \frac{1}{2})$, there exists $\omega = \omega(\gamma, \delta) > 0$ such that $(1, \alpha)_\omega \subset B(\gamma, \delta)$, and hence, from (4.2.55), we get the inequality

$$^{(N)}\mathbf{D}(1, (\alpha)_\delta) \le \frac{1}{1-\gamma}{}^{(N)}\mathbf{D}((1, \alpha)_\omega),$$

which yields (4.2.50).

Let us now establish assertion (B). Recall that the functions $\mathbf{D}(v, b)$ and $\mathbf{D}_\Lambda(v, b)$ are related via (3.2.18), from which we get, for any $\delta > 0$,

$$\mathbf{D}_\Lambda((v, b)_\delta) \le \mathbf{D}(v, b) \le \mathbf{D}_\Lambda(v, b). \tag{4.2.56}$$

Hence, for any $(v, b) \in (1, \alpha)_\omega$, $\theta > 0$,

$$^{(N)}\mathbf{D}((1, \alpha)_\omega) \le {}^{(N)}\mathbf{D}(v, b) \le {}^{(N)}\mathbf{D}_\Lambda(v, b) \le {}^{(N)}\mathbf{\Lambda}\Big(\frac{v}{\theta}, \frac{b}{\theta}\Big)\theta.$$

The left-hand side of the last chain of inequalities does not depend on the choice of $(\frac{v}{\theta}, \frac{b}{\theta})$ within the set $(\frac{1}{\theta}, \frac{\alpha}{\theta})_{\frac{\omega}{\theta}}$, and hence the right-hand side can be replaced by the supremum over this set. As a result, we get

$$^{(N)}\mathbf{D}((1,\alpha)_\omega) \leq {}^{(N)}\mathbf{\Lambda}\Big(\Big(\frac{1}{\theta}, \frac{\alpha}{\theta}\Big)_{\frac{\omega}{\theta}}\Big)\theta,$$

and hence, making $N \to \infty$, we have, for an arbitrary $\theta > 0$,

$$^{(\infty)}\mathbf{D}_{(\omega)}(1,\alpha) \leq \overline{\lim_{N\to\infty}}\, {}^{(N)}\mathbf{\Lambda}\Big(\Big(\frac{1}{\theta}, \frac{\alpha}{\theta}\Big)_{\frac{\omega}{\theta}}\Big)\theta. \tag{4.2.57}$$

Let us now use Corollary 4.2.6, from which it follows that the right-hand side in (4.2.57) is

$$\mathbf{\Lambda}\Big(\Big(\frac{1}{\theta}, \frac{\alpha}{\theta}\Big)_{\frac{\omega}{\theta}}\Big)\theta.$$

Now using definition (3.2.12) of the function $\mathbf{D}_\Lambda$ and taking into account the first inequality in (4.2.56), we check that

$$\inf_{\theta>0} \mathbf{\Lambda}\Big(\Big(\frac{1}{\theta}, \frac{\alpha}{\theta}\Big)_{\frac{\omega}{\theta}}\Big)\theta \leq \mathbf{D}_\Lambda((1,\alpha)_\omega),$$

$$^{(\infty)}D_{(\omega)}(1,\alpha) \leq \mathbf{D}_\Lambda((1,\alpha)_\omega) \leq \mathbf{D}(1,\alpha).$$

This proves assertion (B) and the first assertion of Lemma 4.2.4. The second assertion is verified in the same way. Lemma 4.2.4 is proved. □

4.3 The Second Partial Local Large Deviation Principle

Recall that $(\mathbb{D}, \rho)$ is the metric space of functions $f = f(t)$ without discontinuities of the second kind and equipped with the uniform metric ρ. In this space, we study the family of processes

$$z_T = z_T(t) := \frac{Z(tT)}{x}, \qquad t \in [0,1],$$

where $x \sim T$ as $T \to \infty$, and $Z(t)$ is the compound renewal process defined in §1.1.

The first partial local large deviation principle was established in Theorem 4.2.1 under the assumption $\lambda_+ \geq D(0)$.

4.3.1 Main Results

We now consider the alternative case

$$\lambda_+ < D(0). \tag{4.3.1}$$

In this case, the most probable trajectories of the event $\{z_T(1) \in (\alpha)_\varepsilon\}$ involve large jumps τ (comparable with T). In this section, the assertions are of less "collective" character by the following reasons.

(1) The condition $\lambda_+ < D(0)$ itself, as was already noted, is quite special – it implies the strong asymptotic dependence of τ and ζ in the large deviation zone. More precisely, in Lemma 3.5.12 it was shown that under (4.3.1), for sufficiently small $\varepsilon > 0$,

$$\lim_{t\to\infty} \mathbf{P}(|\zeta| \geq \varepsilon t |\ \tau \geq t) = 1$$

(see (3.5.65)).

(2) In the case $\lambda_+ < D(0)$ the additional condition

$$\ln \mathbf{P}(\tau \geq T) \sim -\lambda_+ T \quad \text{as} \quad T \to \infty \tag{4.3.2}$$

on the "rough" asymptotics of the distribution of τ is imposed (see condition $[\overline{\boldsymbol{\lambda}}_+]$).

(3) Additional conditions on the distribution of $\boldsymbol{\xi}_1 = (\tau_1, \zeta_1)$ will be imposed depending on the form of the function f.

To formulate the main assertion, we will introduce some new notation. We set

$$\begin{aligned} t^{1]} &= t^{1]}(f) := \min\{u : \operatorname{Var}_{[u,1]} f(t) = 0\}, \\ t^{[0} &= t^{[0}(f) := \max\{u : \operatorname{Var}_{[0,u]} f(t) = 0\}. \end{aligned} \tag{4.3.3}$$

In terms of the f-partitions $\{(u_r^-, u_r^+)_{r=1}^R\}$, which were introduced in the proof of Theorem 4.2.1, in the case $t^{1]} < 1$ there exists $r \leq R$ such that $u_r^+ = 1$, $t^{1]} = u_r^-$, where $[u_r^-, u_r^-]$ are intervals of constancy of the function f. Similarly, in the case $t^{[0} > 0$, there exists an $r \leq R$ such that $u_r^- = 0$, $t^{[0} = u_r^+$.

For $f \in \mathbb{C}_a$, we set

$$\widehat{I}(f) := \int_0^{t^{1]}} D(f'(t))\, dt + (1 - t^{1]})\widehat{D}(0),$$

so that

$$\widehat{I}(f) = I(f) = \int_0^1 D(f'(t))\, dt, \ \text{if } \lambda_+ \geq D(0)\ (\widehat{D}(0) = D(0)). \tag{4.3.4}$$

Theorem 4.3.1 (Second partial local LDP) *Assume that* $\boldsymbol{\xi}_1 = (\tau_1, \zeta_1) \in [\mathbf{C}]$, $\boldsymbol{\xi} = (\tau, \zeta) \in [\mathbf{C}]$ *and* $\mathcal{A}^{\leq 0} \subset [\mathcal{A}_1]$. *If* $t^{1]} < 1$, $\lambda_+ < D(0)$, *then we also assume that the "smoothness" condition* (4.3.2) *is met; if, in addition,* $f(t) \equiv 0$, *then we also assume that* $\lambda_+^{(\tau_1)} := \sup\{\lambda : \mathbf{E}e^{\lambda\tau_1} < \infty\} \geq \lambda_+$. *Then, for* $f \in \mathbb{C}_a$ *and any sequence* $\varepsilon = \varepsilon_T$ *which converges to* 0 *sufficiently slowly as* $T \to \infty$,

$$\lim_{T\to\infty} \frac{1}{T} \ln \mathbf{P}(z_T \in (f)_\varepsilon) = -\widehat{I}(f). \tag{4.3.5}$$

So, in view of (4.3.4) equality (4.3.5) is an extension of Theorem 4.2.1. Theorem 4.2.1 is already proved, and hence Theorem 4.3.1 will be verified only in the case $\lambda_+ < D(0)$.

Corollary 4.3.2 *If $f \in \mathbb{C}_a$ has no interval of constancy at the end of* $[0, 1]$, *that is,* $t^{1]}(f) = 1$, *then, for any relation between* λ_+ *and* $D(0)$, *the left-hand side in* (4.3.5) *is* $-I(f)$.

As corollaries from Theorems 4.2.1 and 4.3.1 we also get partial local LDPs for the trajectories of the CRP $z_T^{(q)}$ with linear drift,

$$z_T^{(q)} = z_T^{(q)}(t) := \frac{1}{x} Z^{(q)}(tT), \quad 0 \le t \le 1,$$

where

$$Z^{(q)}(t) := Z(t) + q\mathrm{e}(t), \qquad \mathrm{e}(t) := t; \quad 0 \le t < \infty.$$

Corollary 4.3.3 *Let the conditions of Theorems* 4.2.1 *or* 4.3.1 *be met. Then, for any* $q \in \mathbb{R}$, *the family* $\{z_T^{(q)}\}$ *satisfies the partial local LDP* (4.3.5) *with the deviation functional* $\widehat{I}^{(q)}(f) := \widehat{I}(f - q\mathrm{e})$.

To prove Corollary 4.3.3 it suffices to use the identity

$$\mathbf{P}(z_T^{(q)} \in (f)_\varepsilon) = \mathbf{P}(z_T \in (f - q\mathrm{e})_\varepsilon),$$

and then apply Theorem 4.2.1 or 4.3.1 to the right-hand side of this identity.

In parallel with the process $Z^{(q)}$, we now consider the very close process $Z^{[q]}$ governed by the sequence

$$\xi_n^{[q]} = (\tau_n^{[q]}, \zeta_n^{[q]}) := (\tau_n, \zeta_n + q\tau_n), \quad n \ge 0.$$

Here the processes $Z^{(q)}$ and $Z^{[q]}$ are said to be close, since they have equal values at the moments of jumps $T_{1,k}$. The difference is that the process $Z^{(q)}$ (but not $Z^{[q]}$) gains continuously with rate q the part $q\tau_k$ of the increment $\zeta_k^{[q]} = \zeta_k + q\tau_k$ on the half-open interval $(T_{1,k-1}, T_{1,k}]$. The natural question here is whether these processes have the same LDP. To answer this question, we evaluate, first of all, the second deviation function $D^{[q]}$ corresponding to the vector $\boldsymbol{\xi}^{[q]}$. Under the obvious notational conventions, we have

$$\begin{aligned}\mathbf{A}^{[q]}(\lambda, \mu) &:= \ln \mathbf{E}e^{\lambda\tau + \mu(\zeta + q\tau)} = \mathbf{A}(\lambda + q\mu, \mu),\\ \boldsymbol{\Lambda}^{[q]}(v, \alpha) &:= \sup_{(\lambda,\mu)} \{\lambda v + \mu\alpha - \mathbf{A}^{[q]}(\lambda, \mu)\} = \boldsymbol{\Lambda}(v, \alpha - qv).\end{aligned}$$

Hence the second deviation function for $\xi^{[q]}$ has the form

$$\mathbf{D}_\Lambda^{[q]}(v, \alpha) = \inf_{\theta > 0} \theta \boldsymbol{\Lambda}^{[q]}\Big(\frac{v}{\theta}, \frac{\alpha}{\theta}\Big) = \inf_{\theta > 0} \theta \boldsymbol{\Lambda}\Big(\frac{v}{\theta}, \frac{\alpha - qv}{\theta}\Big) = \mathbf{D}_\Lambda(v, \alpha - qv),$$

so that

$$\mathbf{D}_\Lambda^{[q]}(1, \alpha) = \mathbf{D}_\Lambda(1, \alpha - q), \qquad D^{[q]}(\alpha) = D(\alpha - q).$$

As we have already seen, the relation between λ_+ and $D(0)$, that is, between λ_+ and $D(-q)$ in our case, plays an important role in determination of the deviation function corresponding to the process Z. The function $D(-q)$ is a nonnegative convex function, which assumes its minimum 0 at the point $q = -a = -\frac{a_\zeta}{a_\tau}$. Hence the equation

$$D(-q) = \lambda_+$$

has two solutions $q_- < q_+$ and the condition $\lambda_+ \geq D^{[q]}(0)$ has the form

$$q \in [q_-, q_+]. \tag{4.3.6}$$

So, under the conditions of Theorem 3.4.1, the deviation functions for the processes $Z^{(q)}$ and $Z^{[q]}$ have the same form $D(\alpha - q)$ if (4.3.6) is met. The same applies to the conclusion of Corollary 4.3.3, but with one additional condition: if the function f is not linear at the end of the interval [0, 1], that is, if $t^{1]}(f') = 1$ (f' has no interval of constancy at the end of [0, 1]), then we always have

$$\lim_{T\to\infty} \frac{1}{T} \ln \mathbf{P}(z_T^{[q]} \in (f)_\varepsilon) = \lim_{T\to\infty} \frac{1}{T} \ln \mathbf{P}(z_T^{(q)} \in (f)_\varepsilon) = -I(f - q\mathrm{e}) \tag{4.3.7}$$

(cf. Corollary 4.3.3). If f has a linear part at the end of [0, 1] and if for $q = f'(1)$ condition (4.3.6) is not satisfied, then relation (4.3.7) will not hold.

4.3.2 On the Most Probable Trajectories

We note two facts showing that the usual properties of the trajectories of renewal processes (similar to those of random walks), which take place in the case $\lambda_+ \geq D(0)$ ($\widehat{D}(0) = D(0)$), are violated in the case $\lambda_+ < D(0)$ in which large jumps of τ may have an effect on the formation of the most probable trajectory in the event $\{z_T(1) \in (0)_\varepsilon\}$.

1. Assume, for definiteness, that $\alpha > 0$, $\widehat{D}(\alpha) < D(\alpha)$ and $f(t) = \alpha t$. Hence by (4.3.5),

$$\lim_{T\to\infty} \frac{1}{T} \ln \mathbf{P}(z_T \in (f)_\varepsilon) = -D(\alpha).$$

However, by Theorem 3.4.1,

$$\lim_{T\to\infty} \frac{1}{T} \ln \mathbf{P}(z_T(1) \in (\alpha)_\varepsilon) = -\widehat{D}(\alpha) > -D(\alpha). \tag{4.3.8}$$

This means that the rectilinear trajectory $f(t) = \alpha t$ is not the most probable in the event $\{z_T(1) \in (\alpha)_\varepsilon\}$. If $t_\alpha < 1$ is such that

$$\widehat{D}(\alpha) = \mathbf{D}(t_\alpha, \alpha) + (1 - t_\alpha)\lambda_+,$$

then the continuous polygon connecting the points $(0, 0)$, (t_α, α), $(1, \alpha)$ will be the most probable trajectory determining the probability (4.3.8). This is impossible in the case, $\widehat{D}(\alpha) = D(\alpha)$, or in the case of random walks generated by sums of random variables (see, e.g., [23]). This is the essential difference between the properties of the trajectories of $z_T(\cdot)$ with $\lambda_+ = \widehat{D}(0) < D(0)$ from those of the trajectories of the above processes, for which the most probable trajectories are always linear.

2. Another characteristic property of the processes $Z(t)$ in the case

$$\lambda_+ \geq D(0) \qquad (\widehat{D}(0) = D(0))$$

is that "narrow pencils" of trajectories of such processes on neighboring intervals are "asymptotically independent" in the sense of rough asymptotics. Indeed, let $\lambda_+ \geq \widehat{D}(0) = D(0)$, let the process $Z(t)$ satisfy the conditions $\mathcal{A}^{\leq 0} \subset [\mathcal{A}_1 \cap \mathcal{A}^{(\mathrm{st})}]$ or $\mathcal{A}^{\leq 0} \subset [\mathcal{A}_1 \cap \mathcal{A}_\infty]$ (see §§3.4, 4.1), and let f_1, f_2 be two arbitrary functions from $\mathbb{C}_a$, $f_1(0) = f_2(0) = 0$. For any fixed $u \in (0,1)$ we set

$$f(t) = \begin{cases} f_1(t) & \text{for} \quad t \in [0,u], \\ f_1(u) + f_2(t-u) & \text{for} \quad t \in [u,1], \end{cases}$$

and denote $B_\varepsilon = \{z_T \in (f)_\varepsilon\}$, where $\varepsilon = \varepsilon_T = \overline{o}(1)$ as $T \to \infty$,

$$B_{1,\varepsilon} = \Big\{z_T(t) \in \big(f_1(t)\big)_\varepsilon \quad \text{for} \quad t \in [0,u]\Big\},$$
$$B_{2,\varepsilon} = \Big\{z_T(u+t) - z_T(u) \in \big(f_2(t)\big)_\varepsilon \quad \text{for} \quad t \in [0,1-u]\Big\}.$$

We clearly have

$$\mathbf{P}(B_{1,\varepsilon}B_{2,\varepsilon}) \leq \mathbf{P}(B_{2\varepsilon}) \leq \mathbf{P}(B_{1,4\varepsilon}B_{2,4\varepsilon}).$$

Therefore,

$$\frac{\mathbf{P}(B_{\varepsilon/2})}{\mathbf{P}(B_{1,\varepsilon})} \leq \mathbf{P}(B_{2,\varepsilon} |\, B_{1,\varepsilon}) \leq \frac{\mathbf{P}(B_{2\varepsilon})}{\mathbf{P}(B_{1,\varepsilon})},$$

and by Theorem 4.2.1,

$$\overline{\lim_{T\to\infty}} \frac{1}{T} \ln \mathbf{P}(B_{2,\varepsilon} |\, B_{1,\varepsilon})$$
$$\leq -\int_0^1 D(f'(t))\,dt + \int_0^u D(f'(t))\,dt = -\int_u^1 D(f'(t))\,dt. \qquad (4.3.9)$$

The lower estimate is obtained similarly and has the same right-hand side. As a result,

$$\lim_{T\to\infty} \frac{1}{T} \ln \mathbf{P}(B_{2,\varepsilon} |\, B_{1,\varepsilon})$$
$$= -\int_u^1 D(f'(t))\,dt = -\int_0^{1-u} D(f_2'(t))\,dt = \lim_{T\to\infty} \frac{1}{T} \ln \mathbf{P}(B_{2,\varepsilon}).$$

Here the last equality is satisfied in view of the conditions $\mathcal{A}^{\leq 0} \subset [\mathcal{A}_1 \cap \mathcal{A}^{(\mathrm{st})}]$ or $\mathcal{A}^{\leq 0} \subset [\mathcal{A}_1 \cap \mathcal{A}_\infty]$ (see Theorems 4.1.1, 4.1.2). Thus, the events $B_{1,\varepsilon}$ and $B_{2,\varepsilon}$ are asymptotically independent in the sense of rough asymptotics.

However, there is no such asymptotics if $\lambda_+ = \widehat{D}(0) < D(0)$. Indeed, we choose f_1 and f_2 such that $f'(1) \neq 0$ and the function f has an interval of constancy (u^-, u^+) containing the point u, $u^- > 0$, $u_+ < 1$. In this case, (4.3.9) does not hold by Theorem 4.3.1, since the representations for $\mathbf{P}(B_\varepsilon)$ and $\mathbf{P}(B_{2,\varepsilon})$ are preserved, but

$$\lim_{T\to\infty} \frac{1}{T} \ln \mathbf{P}(B_{1,\varepsilon}) = -\int_0^{u^-} D(f_1'(t))\,dt - (u-u^-)\lambda_+ > -\int_0^{u} D(f_1'(t))\,dt.$$

Note that the above asymptotic independence of "narrow pencils" of trajectories along functions from $\mathbb{C}_a$ does not yet imply, in general, the asymptotic independence of the increments of the process (the increments of $Z(U)$ and $Z(T) - Z(U)$, in our case), because the effect of large vertical jumps of ζ_j was excluded in the consideration of the above "narrow pencils." Theorem 4.1.6 gives conditions for asymptotic weak dependence of the coordinates of τ and ζ sufficient for the asymptotic independence of the increments of $Z(U)$ and $Z(T) - Z(U)$. Recall that one of these conditions follows from Theorem 4.5.1 that will be established below (the principal part of this condition involves the assumptions $\zeta_1 \in [\mathbf{C}_\infty]$ and $\zeta \in [\mathbf{C}_\infty]$).

For more details on the most probable trajectories, see §4.7.1.

4.3.3 Auxiliary Assertions

Let us return to Theorems 3.4.1 and 4.2.1 and obtain an assertion that supplements these theorems and will be required in the proof of Theorem 4.3.1.

Lemma 4.3.4 *Assume that the local LDP* (3.4.5) *holds and that the condition* $\lambda_+ < D(0)$ *is met. Assume, in addition, that at least one of the two conditions is satisfied*:

$$\text{either } \mathcal{A}^{\le 0} \subset [\mathcal{A}_1] \text{ or } \lambda_+^{(\tau_1)} < D(0),\ \ D(0) \le D^{(1)}(0), \tag{4.3.10}$$

where the function $D^{(1)}$ *is defined in the same way as* D*, but for the vector* $\boldsymbol{\xi}_1$. *Then*

$$\overline{\lim_{T\to\infty}} \frac{1}{T} \ln \mathbf{P}\Big(\frac{1}{T} Z(T) \in (\alpha)_\varepsilon, |\zeta_{\eta(T)}| < T\delta\Big) \le -D(\alpha) \tag{4.3.11}$$

for any $\alpha \in \mathbb{R}$ *and any sequences* $\varepsilon = \varepsilon_T \to 0$, *and* $\delta = \delta_T \to 0$ *as* $T \to \infty$.

Relation (4.3.11) and Theorem 3.4.1 show that in the case $\lambda_+ < D(0)$ the principal contribution into the probability of the event $\{\frac{1}{T} Z(T) \in (\alpha)_\varepsilon\}$ as $T \to \infty$ and $D(\alpha) > \widehat{D}(\alpha)$ comes from the trajectories for which $|\zeta_{\eta(T)}| \ge cT$ with some $c > 0$.

The proof of Lemma 4.3.4 is similar to that of the upper estimate in assertion (i) of Theorem 3.4.1. Let $\alpha \ne 0$. We start with the following analog of (3.2.4):

$$\mathbf{P}\Big(\frac{1}{T} Z(T) \in \Delta[\alpha], |\zeta_{\eta(T)}| < T\delta\Big) = \int_0^T H_1(dt, T\Delta[\alpha])\mathbf{P}(\tau > T - t,\ |\zeta| < T\delta). \tag{4.3.12}$$

Using Theorem 3.2.1 on the asymptotics of the renewal measure H_1, we get, as before,

$$-\frac{1}{T} \ln \mathbf{P}\Big(\frac{1}{T} Z(T) \in \Delta[\alpha]\Big) \sim \inf_t \{D(t,\alpha) + l^*(T,t)\},$$

which has the form of (3.2.2), and in which now the function $l(T,t)$ in (3.2.2) is replaced by the function

$$l^*(T,t) := -\frac{1}{T}\ln \mathbf{P}\Big(\tau > (T(1-t)),\ |\zeta| < T\delta\Big).$$

Earlier using the condition $[\overline{\lambda}_+]$ we arrived the relation $l(T,t) \sim (1-t)\lambda_+$ as $T \to \infty$, which implied that the right-hand side in (3.2.2) is $\widehat{D}(\alpha)$. Now by Lemma 3.5.8 we have

$$l^*(T,t) \geq (1-t)D(0).$$

This gives us the upper estimate

$$-\inf_{t\in[0,1]}\{\mathbf{D}(t,\alpha) + (1-t)\mathbf{D}(1,0)\}$$

for the left-hand side in (4.3.11). Since $\mathbf{D}(t,\alpha)$ is semiadditive (see §3.3.1), we have

$$\mathbf{D}(t,\alpha) + (1-t)\mathbf{D}(1,0) = \mathbf{D}(t,\alpha) + \mathbf{D}(1-t,0) \geq \mathbf{D}(1,\alpha) = D(\alpha),$$

which proves (4.3.11).

Now let $\alpha = 0$. In this case, the proof is similar, but now in place of (4.3.12) we need to use the analog of representation (3.2.4). In this case, the additional term

$$\mathbf{P}(\tau_1 > T,\ |\zeta_1| < T\varepsilon)$$

will appear on the right of (4.3.12). The following analog of Lemma 3.5.8 will be used to estimate this term.

Lemma 4.3.5 *If conditions* (4.3.10) *hold, then*

$$\overline{\lim_{T\to\infty}}\,\frac{1}{T}\ln \mathbf{P}(\tau_1 > T,\ |\zeta_1| < T\varepsilon) \leq -D(0).$$

Proof Assume that the first of conditions (4.3.10) is satisfied. Then, for any $(\lambda,\mu) \in (\mathcal{A}^{\leq 0}) \subset \mathcal{A}_1$, $(v,\alpha) \in \mathbb{R}^2$,

$$\frac{1}{T}\mathbf{\Lambda}_1(Tv,T\alpha) \geq \lambda v + \mu\alpha - \frac{1}{T}\mathbf{\Lambda}_1(\lambda,\mu).$$

Hence for $v = v_T \to v$, $\alpha = \alpha_T \to 0$ as $T \to \infty$, we have

$$\underline{\lim}_{T\to\infty}\frac{1}{T}\mathbf{\Lambda}_1(Tv_T,T\alpha_T) \geq \lambda v. \tag{4.3.13}$$

The left-hand side of (4.3.13) does not depend on (λ,μ) within the set $(\mathcal{A}^{\leq 0})$, and so the right-hand side of (4.3.13) can be replaced by the supremum over $(\lambda,\mu) \in (\mathcal{A}^{\leq 0})$, which gives

$$\underline{\lim}_{T\to\infty}\frac{1}{T}\mathbf{\Lambda}_1(Tv_T,T\alpha_T) \geq \sup_{(\lambda,\mu)\in(\mathcal{A}^{\leq 0})}\{\lambda v\} = \mathbf{D}(v,0).$$

Therefore, for any $v > 0$ and $\varepsilon = \varepsilon_T \to 0$, we get, as $T \to \infty$,

$$\varliminf_{T\to\infty} \frac{1}{T} \inf_{\substack{|u-Tv|<T\varepsilon \\ |\beta|<T\varepsilon}} \mathbf{\Lambda}_1(u,\beta) \geq \mathbf{D}(v,0). \tag{4.3.14}$$

Using the exponential Chebyshev-type inequality for convex open sets (see [23], §4.4) and employing (4.3.14), we get the following upper estimate: for any $v > 0$, as $\varepsilon = \varepsilon_T \to 0$,

$$\varlimsup_{T\to\infty} \frac{1}{T} \ln \mathbf{P}(|\tau_1 - Tv| < T\varepsilon,\ |\zeta_1| < T\varepsilon) \leq -\mathbf{D}(v,0). \tag{4.3.15}$$

Next, for any $N < \infty$, there exists an $M < \infty$ such that

$$\varlimsup_{T\to\infty} \frac{1}{T} \ln \mathbf{P}(\tau_1 > TM) \leq -N. \tag{4.3.16}$$

From (4.3.15), (4.3.16) we get the required upper estimate

$$\varlimsup_{T\to\infty} \frac{1}{T} \ln \mathbf{P}(\tau_1 > T,\ |\zeta_1| < T\varepsilon) \leq -\inf_{v\geq 1} \mathbf{D}(v,0) = -\mathbf{D}(1,0).$$

If the second condition in (4.3.10) is satisfied, then the conclusion of Lemma 4.3.5 easily follows from Lemma 3.5.8. This proves Lemmas 4.3.4 and 4.3.5. □

4.3.4 Proof of Theorem 4.3.1

The scheme of the proof of Theorem 4.3.1 has many common points with that of Theorem 4.2.1, but there are also significant differences. Now $\widehat{D}(0) < D(0)$ and as the set V (cf. (4.2.8)), which does not contain intervals of constancy, we will consider the set

$$V = [0,1] \setminus \bigcup_{r=1}^{R} (u_r^-, u_r^+),$$

which contains all points $u_r^\pm$. We define the partition $\mathbf{t}_L$ as the union of the right-hand side of (4.2.7) with the set consisting of the first $\min\{R, L/2\}$ points $u_r^\pm$, $r = 1, 2, \ldots$ We let L_1 denote the smallest integer number for which the number of points in $\mathbf{t}_{L_1} V$ exceeds L. The further construction of f-partitions is the same as that of the f-partitions under the conditions of Theorem 4.2.1, so we have the convergence (4.2.9). For the new f-partition, the event $\{z_T \in (f)_\varepsilon\}$ does not in general imply the event $\{\overline{\chi}_L \leq \delta_T\}$ (see (4.2.11)). However, this inclusion will take place if by $\overline{\chi}_L$ we now understand the value

$$\overline{\chi}_L := \max\{\chi_l \ \text{ over all } \ u_l \notin \{u_r^-\},\ l \leq L-1\}.$$

Hence in place of (4.2.12) a different inclusion will be used. Given points $u_l \in \{u_r^-\}$, we set

$$\gamma_l := u_l - \frac{1}{T} \tau_{\nu(u_l T)},$$

$$\overline{\gamma}_L := \max\{\gamma_l \ \text{ over all } \ u_l \in \{u_r^-\},\ u_l \neq 0,\ l \leq L\}.$$

We clearly have

$$\{z_T \in (f)_\varepsilon\} \subset \{\overline{\gamma}_L < \delta_T\}.$$

It is plain that the inclusion

$$\{z_T \in (f)_\varepsilon\} \subset \{\overline{\zeta}(T) \le 2\varepsilon T\}$$

remains valid. So, in place of (4.2.12) we now have the inclusion

$$\{z_T \in (f)_\varepsilon\}$$
$$\subset \{z_T(u_l) \in (f(u_l))_\varepsilon \text{ for all } 1 \le l \le L,\ \overline{\zeta}(T) \le 2\varepsilon T,\ \overline{\chi}_L < \delta_T,\ \overline{\gamma}_L < \delta_T\}.$$

The next result now plays the role of Lemma 4.2.2.

Lemma 4.3.6 *Let the conditions of Theorem* 4.3.1 *hold and let* $f \in \mathbb{C}_a$, $f(0) = 0$, *be an arbitrary function. Then, for any* f*-partition* $\mathbf{u}_L = \{u_l\}_{l=0}^L$ *and any sequences* $\varepsilon = \varepsilon_T \to 0$, $\delta_T \to 0$ *as* $T \to \infty$,

$$\varlimsup_{T\to\infty} \frac{1}{T} \ln \mathbf{P}\Big(z_T(u_l) \in (f(u_l))_\varepsilon \ \text{for all}\ 1 \le l \le L;$$
$$\overline{\zeta}(T) \le 2\varepsilon T,\ \overline{\chi}_L < \delta_T,\ \overline{\gamma}_L < \delta_T\Big) \le -\widehat{I}(f^{\mathbf{u}_L}). \tag{4.3.17}$$

The second assertion of Lemma 4.2.2, which gives the required lower estimate, is completely preserved and is not reproduced here.

Proof of Lemma 4.3.6 This is similar to that of the upper estimate in Lemma 4.2.2. We again use induction on L. For $L = 1$, estimate (4.3.17) follows from Theorem 3.4.1. Let us go from $L = 1$ to $L = 2$. Now we will distinguish between the cases $u_1 = u \notin \{u_r^-\}$ and $u \in \{u_r^-\}$.

If $\alpha_2 \ne 0$, then we necessarily have $u \notin \{u_r^-\}$, and the proof of Lemma 4.2.2 will be completely the same if in addition to Theorem 3.4.1 one should also use Lemma 4.3.4.

If $\alpha_2 = 0$, then $u \in \{u_r^-\}$, $\alpha_1 \ne 0$ (see (4.2.38)) and the event A_1 in (4.2.39) should be replaced by the event

$$A_1 := \{z_T(u) \in (\alpha_1)_\varepsilon,\ |\zeta_{\eta(uT)}| < 2\varepsilon T,\ \gamma < \delta_T\},$$

where $\gamma := u - \frac{1}{T} Z_{\nu(uT)}$. Let $\mathfrak{A}$ be the same as in (4.2.40). It is clear that $A_1 \in \mathfrak{A}$ and that equality (4.2.40) is preserved. Consider the factor

$$\mathbf{E}(\mathbf{1}(A_2)|\,\mathfrak{A}) = \mathbf{P}(A_2|\,\mathfrak{A}).$$

This function on the set A_1 can be considered as the probability (for $T^* = (1-u)T$)

$$\mathbf{P}\Big(Z^*(T^*) \in (\alpha_2 - y)_\varepsilon T^* \times \frac{T}{T^*} |\mathfrak{A}\Big)$$
$$\le \mathbf{P}\Big(Z^*(T^*) \in (0)_{3\varepsilon} T^* \times \frac{T}{T^*} |\gamma = v\Big) =: P_v(T^*), \tag{4.3.18}$$

where $v < \delta_T$, $Z^*(t)$ is the renewal process with the initial jump $(\tau_1, \zeta_1) = (\chi, 0)$,

$$\mathbf{P}(\chi > t) := \frac{\mathbf{P}(\tau > v + t)}{\mathbf{P}(\tau > v)}.$$

Since τ satisfies the smoothness condition $[\overline{\lambda}_+]$, we have, for $v < \delta_T$,

$$\mathbf{P}(\tau > v) \geq \mathbf{P}(\tau > \delta_T) = e^{o(T)},$$

and so, if t is comparable with T, then

$$\mathbf{P}(\tau_1 > t) = \mathbf{P}(\chi > t) \leq \mathbf{P}(\tau > t) e^{o(T)}$$

and

$$\mathbf{P}(\tau_1 > t) \leq \mathbf{P}(\tau > t) e^{o(t)} \quad \text{as} \quad t \to \infty. \tag{4.3.19}$$

It is easily seen that all the assertions on rough asymptotics of the renewal measure $H_0(T\Delta[t), T\Delta[y))$ for the renewal process with the initial jump $(\tau_1, 0)$ satisfying (4.3.19) will have the same form as for the renewal function H corresponding to the homogeneous renewal process and will satisfy the inequality $\lambda_+^{(\tau_1)} \geq \lambda_+$. Hence using (3.4.14) and the arguments after (3.4.14), we, as in the proof of Theorem 3.4.1, arrive at (3.4.5) or, equivalently, we prove that in our case, for the right-hand side of (4.3.18),

$$\lim_{T^* \to \infty} \frac{1}{T^*} \ln P_v(T^*) = -\widehat{D}(0)$$

on the set A_1. This, in turn, means that on the set A_1

$$\lim_{T^* \to \infty} \frac{1}{T^*} \ln \mathbf{P}(A_2 | \mathfrak{A}) = -\widehat{D}(0) = -\lambda_+.$$

As a result, similarly to (4.2.44), we get

$$\mathbf{P}(A) \leq e^{-T\widehat{I}(f^{\mathbf{u}_2}) + o(T)}.$$

This proves the required upper estimate (4.3.17) for $L = 2$. The induction step is the same as in the proof of Lemma 4.2.2. This proves estimate (4.3.17), and therefore, Lemma 4.3.6. □

Based on Lemma 4.3.6, let us complete the proof of Theorem 4.3.1.

(i) *Upper estimate.* From (4.2.13) and using Lemma 4.3.6, we get

$$\overline{\lim_{T \to \infty}} \frac{1}{T} \ln \mathbf{P}(z_T \in (f)_\varepsilon) \leq -\widehat{I}(f^{\mathbf{u}_L}). \tag{4.3.20}$$

The left-hand side of (4.3.20) does not depend on the partition $\mathbf{u}_L$, and hence by (4.2.9) the right-hand side in (4.3.20) can be replaced by $-\widehat{I}(f)$. This establishes the upper estimate for (4.3.5).

(ii) *Lower estimate.* From assertion (ii) of Lemma 4.2.2 it follows that

$$\underline{\lim_{T \to \infty}} \frac{1}{T} \ln \mathbf{P}(z_T \in (f)_\varepsilon) \geq -I(f). \tag{4.3.21}$$

This establishes the required lower estimate for (4.3.5) for the functions $f \in \mathbb{C}_a$ satisfying $\widehat{I}(f) = I(f)$.

Now let $\widehat{I}(f) < I(f)$, that is, the function f has an interval of constancy $[u, 1]$, $u := t^{1]}(f)$. Then, for any sequence $\varepsilon = \varepsilon_T \to 0$ as $T \to \infty$, there exists a sequence $\delta_T \to 0$ such that

$$\Big\{ \sup_{0\le t\le u} |z_T(t) - f(t)| < \varepsilon \Big\} = \Big\{ \sup_{0\le t\le u} |z_T(t) - f(t)| < \varepsilon, \gamma(uT) \le T\delta_T \Big\},$$

where, as before,

$$\gamma(uT) := Tu - T_{\nu(Tu)}$$

is the defect up to the level Tu of the walk T_k. Hence

$$\begin{aligned}\mathbf{P}(z_T \in (f)_\varepsilon) &\ge \mathbf{P}\Big(\sup_{0\le t\le u} |z_T(t) - f(t)| < \varepsilon,\ \gamma(uT) \le T\delta_T,\ \chi(Tu) > T(1-u)\Big) \\ &\ge \mathbf{P}\Big(\sup_{0\le t\le u} |z_T(t) - f(t)| < \varepsilon\Big)\mathbf{P}(\tau > T(1-u) + T\delta_T). \qquad (4.3.22)\end{aligned}$$

To estimate from below the first factor on the right of the last equality, we will employ (4.3.21), which has already been established, and to estimate from below the second factor we will use the rough smoothness condition $[\overline{\lambda}_+]$. As a result we get that the left-hand side of (4.3.22) is not smaller than

$$\exp\Big\{-T\Big[\int_0^u D(1, f'(t))\, dt + \lambda_+(1-u)\Big] + o(T)\Big\} = \exp\{-T\widehat{I}(f) + o(T)\}.$$

This proves the required lower estimate for (4.3.5), and therefore, Theorem 4.3.1. □

4.4 Complete Local Large Deviation Principle

As has already been pointed out, the local LDPs derived in §§4.2, 4.3 are "partial" principles, because they hold not for all functions $f \in \mathbb{D}$, but only for $f \in \mathbb{C}_a$. Using such local LDPs it seems impossible to obtain an "integral" LDP for the asymptotics of $\mathbf{P}(z_T \in B)$, $B \subset \mathbb{D}$, because the asymptotics of $\mathbf{P}(z_T \in (f)_\varepsilon)$ for $f \notin \mathbb{C}_a$ remains unknown. If $f \notin \mathbb{C}_a$ and if only condition [**C**] holds (but not [$\mathbf{C}_\infty$]), then the limit in (4.2.3) (finite or infinite) may fail to exist. Assume, for example, that τ_1 and τ are arithmetic, ζ does not depend on τ and is exponentially distributed on the right half of the axis:

$$\mathbf{P}(\zeta > v) = ce^{-bv} \quad \text{for} \quad v > 0, \quad b > 0, \quad c > 0, \quad \mathbf{E}\zeta = 0,$$

$$f(t) = \begin{cases} 0 & \text{for} \quad t < \frac{1}{2}, \\ 1 & \text{for} \quad t \ge \frac{1}{2}. \end{cases}$$

Then, for *even* integer $T \to \infty$, the trajectory $z_T(t)$ has a jump at the point $t = \frac{1}{2}$ with probability close to $\frac{1}{\mathbf{E}\tau}$, and the limit

$$\lim_{T\to\infty} \frac{1}{T} \ln \mathbf{P}(z_T \in (f)_\varepsilon) = -D(0) - b$$

exists for $\varepsilon = \varepsilon_T = \overline{o}(1)$. If $T \to \infty$ along *odd* Ts, then the probability of having a jump at the point $t = \frac{1}{2}$ is zero, and the corresponding limit is $-\infty$. This appears because the uniform metric is "nonharmonized" with the nature of the space $\mathbb{D}$: for the functions f with discontinuity at the point t_0 the neighborhood $(f)_\varepsilon$ in the uniform metric turns out to be too "thin" and, for sufficiently small ε, any function $g \in (f)_\varepsilon$ must have a discontinuity at the same point t_0. This "inconvenience" disappears to a great extent if ζ is required to satisfy condition $[\mathbf{C}_\infty]$. Then the presence of a jump for f in (4.2.3) always makes the limit in (4.2.3) to be $-\infty$, and in this case, there holds the "more complete" local LDP for all $f \in \mathbb{D}$.

On the other hand, to the uniform metric ρ there correspond the thinnest boundaries of the sets B and the broadest class of functionals continuous in this metric. Here we compare the uniform metric with the metrics in $\mathbb{D}$ considered in Chapter 4 of [23] (and, in particular, with the metric $\rho_{\mathbb{D}}$ of §1.7, which is a some relaxation of the Skorokhod metric), which are free from the above "inconvenience." On the other hand, if one deals with the "integral" LDP in $(\mathbb{D}, \rho)$ with the uniform metric ρ, then the continuity condition (required for the coincidence of the upper and lower limits) will be satisfied for the broadest class of sets (see Corollary 4.5.3).

In this section, we will obtain a "complete" local LDP capable of deriving a "complete integral" LDP (see §4.5); however, we shall only be able to do so under the condition $\zeta_1 \in [\mathbf{C}_\infty]$, $\zeta \in [\mathbf{C}_\infty]$. In this case, $\lambda_+ \geq D(0)$ ($\widehat{D}(\alpha) = D(\alpha)$) by Theorem 3.3.2 (see (3.3.14)).

Theorem 4.4.1 (Complete local LDP) *Let the conditions of Theorem* 4.2.1 *hold and let, in addition,* $\zeta_1 \in [\mathbf{C}_\infty]$, $\zeta \in [\mathbf{C}_\infty]$. *Then, for any function* $f \in \mathbb{D}$ *and any sequence* $\varepsilon = \varepsilon_T$ *which converges to* 0 *sufficiently slowly as* $T \to \infty$,

$$\lim_{T\to\infty} \frac{1}{T} \ln \mathbf{P}(z_T \in (f)_\varepsilon) = -I(f), \tag{4.4.1}$$

where

$$I(f) := \begin{cases} \int_0^1 D(f'(t))\,dt & \text{if } f \in \mathbb{C}_a, \\ \infty & \text{if } f \in \mathbb{D} \setminus \mathbb{C}_a. \end{cases}$$

As a direct corollary to Theorem 4.4.1, we get the complete local LDP for the trajectories of the compound renewal processes $z_T^{(q)}$ with linear drift

$$z_T^{(q)} = z_T^{(q)}(t) := \frac{1}{x} Z^{(q)}(tT), \quad 0 \leq t \leq 1,$$

where

$$Z^{(q)}(t) := Z(t) + q\mathrm{e}(t), \qquad \mathrm{e}(t) := t;\ 0 \leq t < \infty.$$

Corollary 4.4.2 *Let the conditions of Theorem* 4.4.1 *hold. Then, for any* $q \in \mathbb{R}$, *the family* $\{z_T^{(q)}\}$ *satisfies the corresponding LDP* (4.4.1) *with the deviation functional* $I^{(q)}(f) := I(f - q\mathrm{e})$.

Proof of Theorem 4.4.1 The proof of Theorem 4.4.1 is similar to that of Theorem 4.2.1, but additional considerations are required for functions f from $\mathbb{D} \setminus \mathbb{C}_a$. Namely, we need the following two assertions.

(1) *If f has a discontinuity, then*

$$\lim_{T\to\infty} \frac{1}{T} \ln \mathbf{P}(z_T \in (f)_\varepsilon) = -\infty. \tag{4.4.2}$$

(2) *If $f \in \mathbb{C} \setminus \mathbb{C}_a$, then* (4.4.2) *also holds.*

We start with the first assertion. If f has a discontinuity of size $2h$ at a point $t \in (0, 1)$, then, for sufficiently small ε, any function $g \in (f)_\varepsilon$ has a discontinuity of size at least h at the point t. If $\mathbf{P}(T_k = t) = 0$ for all k, then $\mathbf{P}(z_T \in (f)_\varepsilon) = 0$, and in this case (4.4.2) clearly holds. If

$$\mathbf{P}\Big(\bigcup_{k=0}^{\infty}\{T_k = t\}\Big) > 0,$$

then for integer MT and $\overline{\zeta}_n := \max_{k\le n} |\zeta_k|$, we have

$$\mathbf{P}(z_T \in (f)_\varepsilon) \le \mathbf{P}(\eta(Tt) > MT) + \mathbf{P}(\overline{\zeta}_{MT} \ge hT). \tag{4.4.3}$$

As we have already seen, for any given $N < \infty$, there exists an $M < \infty$ such that (see (4.2.37))

$$\mathbf{P}(\eta(T) \ge MT) \le e^{-NT}.$$

Besides, by the assumption $\zeta \in [\mathbf{C}_\infty]$ for all sufficiently large T,

$$\mathbf{P}(\overline{\zeta}_{MT} \ge hT) \le MT\mathbf{P}(|\zeta| \ge hT) \le MTe^{-NT}.$$

Together with (4.4.3) this means that the left-hand side in (4.4.2) is majorized by $-N$ for any N. This proves assertion (1).

Let us now verify assertion (2) for $f \in \mathbb{C} \setminus \mathbb{C}_a$. Using (4.2.12) and employing Lemma 4.2.2, we have, in this case,

$$\overline{\lim_{T\to\infty}} \frac{1}{T} \ln \mathbf{P}(z_T \in (f)_\varepsilon) \le -I(f^{\mathbf{u}_L}).$$

According to §4.2 of [23], we have

$$I(f^{\mathbf{u}_L}) \to J(f) := \begin{cases} \infty & \text{if } \mathrm{Var} f = \infty, \\ I(f_a) + \mu_+^{(\zeta)} \mathrm{Var}\, f_s^+ - \mu_-^{(\zeta)} \mathrm{Var}\, f_s^- & \text{if } \mathrm{Var} f < \infty \end{cases} \tag{4.4.4}$$

as $L \to \infty$, where f_a and f_s are, respectively, the absolutely continuous and singular components of the function $f = f_a + f_s$, $f_s^\pm$ are, respectively, the increasing and decreasing components of the function f_s, and $\mu_\pm^{(\zeta)}$ are the end-points of the largest interval $(\mu_-^{(\zeta)}, \mu_+^{(\zeta)})$ on which the function $\mathbf{E}e^{\mu\zeta}$ is finite. We have $\zeta \in [\mathbf{C}_\infty]$, and hence $\mu_{\pm}^{(\zeta)} = \pm\infty$. As a result, $J(f) = \infty$ if $\operatorname{Var} f_s > 0$. This proves assertion (2), and therefore, Theorem 4.4.1. □

4.5 Integral Large Deviation Principle for Trajectories of a Compound Renewal Process

4.5.1 Main Result and Its Proof

In the present section, we shall show that, under the conditions of Theorem 4.4.1, the process $z_T = z_T(t)$, $0 \le t \le 1$, will satisfy the "integral" LDP.

The next theorem is the main assertion of this section.

Theorem 4.5.1 (Compete integral LDP) *Let the conditions of Theorem* 4.4.1 *hold. Then the (integral) LDP holds for the family of processes* $\{z_T\}$*: for any measurable set* $B \subset \mathbb{D}$,

$$\overline{\lim_{T\to\infty}} \frac{1}{T} \ln \mathbf{P}(z_T \in B) \le -I([B]), \tag{4.5.1}$$

$$\underline{\lim_{T\to\infty}} \frac{1}{T} \ln \mathbf{P}(z_T \in B) \ge -I((B)), \tag{4.5.2}$$

where $[B]$ *and* (B) *are, respectively, the closure and interior of a set* B *in the uniform metric,*

$$I(B) := \inf_{f\in B} I(f).$$

As before, as a simple corollary to the LDP for z_T we get an analogous result for the family

$$z_T^{(q)} := \frac{1}{x} Z^{(q)}(tT), \quad 0 \le t \le 1$$

corresponding to the compound renewal process $Z^{(q)}(t) = Z(t) + qt$ with linear drift qt.

Corollary 4.5.2 *Let the conditions of Theorem* 4.5.1 *hold. Then, for any* $q \in \mathbb{R}$, *the LDP with deviation functional* $I^{(q)}(f) := I(f - q\mathrm{e})$, *where* $\mathrm{e} = \mathrm{e}(t) = t$, $0 \le t \le 1$, *holds for the family* $\{z_T^{(q)}\}$.

As was pointed out before Theorem 4.4.1, to the uniform metric ρ there corresponds the broadest class of sets satisfying the continuity condition $I((B)) = I([B])$. In particular, to the uniform metric there corresponds the broadest class of functionals G on $\mathbb{D}$ continuous with respect to this metric. Let

$$B_y := \{f \in \mathbb{D} : G(f) < y\}.$$

Then, for ρ-continuous functionals G, the set $B_y = (B_y)$ is open and the set $\{f \in \mathbb{D} : G(f) \le y\} = [B_y]$ is closed.

Corollary 4.5.3 *Let the conditions of Theorem* 4.4.1 *hold and let G be a functional continuous in the uniform metric. Then, for almost all y, the limit*

$$\lim_{T\to\infty}\frac{1}{T}\ln\mathbf{P}(G(z_T)<y)=-I(B_y) \tag{4.5.3}$$

exists.

Proof Since the function $I(B_y)$ is not increasing in y, it is almost everywhere continuous, and hence, for almost all y,

$$I(B_y)=I([B_y]).$$

Now (4.5.3) follows from (4.5.1) and (4.5.2), proving the corollary. □

We give some definitions, which will be used in the proof of Theorem 4.5.1.

A functional $I=I(f)$ will be said to be *compact* if, for any $N\ge 0$, the set $\{f\in\mathbb{D}: I(f)\le N\}$ is compact in the space $(\mathbb{D},\rho)$ equipped with the uniform metric ρ.

Processes $z_T(t)$ and $\widetilde{z}_T(t)$, $t\in[0,1]$, will be said to be *asymptotically equivalent* (with respect to the LDP) if, for any $\delta>0$,

$$\lim_{T\to\infty}\frac{1}{T}\ln\mathbf{P}(\rho(z_T,\,\widetilde{z}_T)\ge\delta)=-\infty. \tag{4.5.4}$$

The next lemma clarifies the meaning of this definition.

Lemma 4.5.4 *If processes z_T, $\widetilde{z}_T$ are asymptotically equivalent and if the process z_T satisfies the integral* (*local*) *LDP with lower-continuous compact deviation functional $I=I(f)$, then the process $\widetilde{z}_T$ also satisfies the integral* (*local*) *LDP with the same deviation functional.*

Proof of Lemma 4.5.4 Let the process z_T satisfy the LDP.

(i) *Upper estimate.* We use the following clear inequality

$$\mathbf{P}(\widetilde{z}_T\in B)\le\mathbf{P}(z_T\in(B)_\delta)+\mathbf{P}(\rho(z_T,\,\widetilde{z}_T)\ge\delta),$$

in which $(B)_\delta$ is the δ-neighborhood of a set B. Applying the upper estimate for the LDP for z_T and employing (4.5.4), we get, for any $\delta>0$,

$$\overline{\lim_{T\to\infty}}\frac{1}{T}\ln\mathbf{P}(\widetilde{z}_T\in B)\le -I\big([(B)_\delta]\big)\le -I((B)_{2\delta}).$$

Making $\delta\to 0$, this gives

$$\overline{\lim_{T\to\infty}}\frac{1}{T}\ln\mathbf{P}(\widetilde{z}_T\in B)\le -I(B+),$$

where $I(B+):=\lim_{\delta\to 0}I((B)_\delta)$.

Next, assume, for definiteness, that ζ takes values of both signs. From assertion (iv) of Theorem 3.3.2 it follows that the convex lower-continuous function $\mathbf{D}(\alpha)$ for $\zeta\in[\mathbf{C}_\infty]$ increases as $\alpha\to\pm\infty$ faster than any linear function. Hence the functional $I(f)$ is a *convex lower-continuous compact functional* (see, e.g., [23], §4.2).

It remains to note that

$$I(B+) = I([B])$$

for a lower-continuous compact functional I (see [23], §4.2).

This proves the upper estimate

$$\overline{\lim_{T\to\infty}} \frac{1}{T} \ln \mathbf{P}(\widetilde{z}_T \in B) \le -I([B])$$

in the LDP for $\widetilde{z}_T$.

(ii) *Lower estimate.* For any fixed $\delta > 0$, we have

$$\mathbf{P}(\widetilde{z}_T \in (f)_\delta) \ge \mathbf{P}(z_T \in (f)_{\delta/2}) - \mathbf{P}(\rho(z_T, \widetilde{z}_T) \ge \delta/2).$$

Applying the lower estimate for the LDP for z_T and using (4.5.4), we get the lower estimate in the local LDP for $\widetilde{z}_T$,

$$\underline{\lim_{T\to\infty}} \frac{1}{T} \ln \mathbf{P}(\widetilde{z}_T \in (f)_\delta) \ge -I((f)_{\frac{\delta}{2}}) \ge -I(f).$$

As a result, we get the lower estimate for the LDP for $\widetilde{z}_T$ (see [23], § 4.2),

$$\underline{\lim_{T\to\infty}} \frac{1}{T} \ln \mathbf{P}(\widetilde{z}_T \in B) \ge -I((B)).$$

The part of the lemma for the local LDP can be verified similarly. Lemma 4.5.4 is proved. □

Proof of Theorem 4.5.1 We can assume without loss of generality that the normalizing sequence $x = x(T)$ (from which the process $z_T(t)$ is constructed) has the form $x = T$ (see also the proof of Theorem 4.2.1). If this is not so, then instead of the process $z_T(t)$ one can consider the asymptotically equivalent process $\frac{x}{T} z_T(t)$, for which the normalizing sequence has the required form.

It will be convenient to prove the LDP not directly for the original process z_T, but for the asymptotically equivalent process $\widetilde{z}_T$, which we define in two stages.

We first construct the "aggregated" renewal process $Z^*(t)$ from the jumps (τ_1^*, ζ_1^*), (τ_i^*, ζ_i^*), $i \ge 1$, where the above random variables are defined as follows:

$$\tau_1^* := T_\eta, \quad \zeta_1^* := Z_\eta, \qquad \eta := \eta(1) = \min\{k \ge 0 : T_k > 1\}. \tag{4.5.5}$$

The random vectors (τ_i^*, ζ_i^*) for $i \ge 2$ are independent copies of the random vector (τ^*, ζ^*), which is also defined by relations (4.5.5), but for the homogeneous sequence $\{T_k, Z_k\}$. It is clear that the processes Z and Z^* can be constructed on the same probability space, so that $Z(t)$ and $Z^*(t)$ have equal values at τ_1^*, $\tau_1^* + \tau_2^*, \ldots$.

At the second stage, we "smooth" the process $Z^*(t)$ by constructing from this process the continuous polygon $\widetilde{Z}(t)$ with nodes $(0, 0)$, $(\tau_1^*, Z^*(\tau_1^*){=}\zeta_1^*)$, $(\tau_1^*{+}\tau_2^*, Z^*(\tau_1^*+\tau_2^*) = \zeta_1^* + \zeta_2^*), \ldots$ and so on. The process

$$\widetilde{z}_T(t) := \frac{1}{T}\widetilde{Z}(Tt), \quad t \in [0, 1]$$

will be considered in what follows as an asymptotically equivalent process for the process z_T.

The following steps are required for the proof of Theorem 4.5.1 with the process thus constructed.

I. Show that the processes z_T and $\widetilde{z}_T$ are asymptotically equivalent.

II. Prove the LDP for the process $\widetilde{z}_T$.

We start with the first step. We need to know some properties of the jumps ζ_i^*, $i \geq 1$.

Lemma 4.5.5 *If* $\zeta_1 \in [\mathbf{C}_\infty]$, $\zeta \in [\mathbf{C}_\infty]$, *then* $\zeta_i^* \in [\mathbf{C}_\infty]$, $i \geq 1$.

Proof Consider the nonnegative martingale

$$X_0 = 1, \qquad X_k = \frac{e^{\lambda Z_k}}{\psi_1(\lambda)\psi^{k-1}(\lambda)},$$

where $\psi_1(\lambda) := \mathbf{E}e^{\lambda\zeta_1}$, $\psi(\lambda) := \mathbf{E}e^{\lambda\zeta}$. The variable η is a Markov moment with respect to the sequence $\{(\tau_i, \zeta_i);\ i \geq 1\}$. Hence

$$\mathbf{E}X_\eta \leq 1 \tag{4.5.6}$$

(see, e.g., Corollary 15.2.2 in [21]). Next, by the Cauchy inequality and (4.5.6),

$$\begin{aligned}
\mathbf{E}e^{\lambda Z_\eta} &= \mathbf{E}\frac{e^{\lambda Z_\eta}}{\sqrt{\psi_1(2\lambda)\psi^{\eta-1}(2\lambda)}}\sqrt{\psi_1(2\lambda)\psi^{\eta-1}(2\lambda)} \\
&\leq \Big[\mathbf{E}\frac{e^{2\lambda Z_\eta}}{\psi_1(2\lambda)\psi^{\eta-1}(2\lambda)}\Big]^{1/2}\big[\mathbf{E}\psi_1(2\lambda)\psi^{\eta-1}(2\lambda)\big]^{1/2} \\
&\leq \Big[\frac{\psi_1(2\lambda)}{\psi(2\lambda)}\Big]^{1/2}\big[\mathbf{E}\psi^{\eta}(2\lambda)\big]^{1/2}.
\end{aligned} \tag{4.5.7}$$

Further, we note that $\eta \in [\mathbf{C}_\infty]$. Indeed, by the exponential Chebyshev inequality,

$$\mathbf{P}(\eta > n+1) \leq \mathbf{P}(T_n \leq 1) \leq e^{-n\Lambda^{(\tau)}(\frac{1}{n})},$$

where $\Lambda^{(\tau)}(v) \to \infty$ as $v \to 0$. Hence $\mathbf{E}e^{\lambda Z_\eta} < \infty$ for any λ. Now from (4.5.7) we get

$$\zeta_1^* \underset{d}{=} Z_\eta \in [\mathbf{C}_\infty].$$

Setting $\psi_1(\lambda) = \psi(\lambda)$, we also get in the homogeneous case

$$\zeta^* \underset{d}{=} Z_\eta \in [\mathbf{C}_\infty].$$

The lemma is proved. □

It is also easily seen that

$$\mathbf{P}(\tau_1^* \geq t) \leq \mathbf{P}(\tau_1 \geq t) + \int_0^1 \widetilde{H}(dv)\mathbf{P}(\tau \geq t - v) \leq \mathbf{P}(\tau_1 \geq t) + \widetilde{H}(1)\mathbf{P}(\tau \geq t-1),$$

and, therefore, τ_1^* satisfies the same conditions as τ_1 and τ.

Corollary 4.5.6 *Under the conditions of Lemma* 4.5.5,

$$\overline{Z}_\eta := \max_{k \le \eta} |Z_k| \in [\mathbf{C}_\infty], \tag{4.5.8}$$

$$\max_{t \le \tau_1^*} |\widetilde{Z}(t) - Z(t)| \in [\mathbf{C}_\infty], \tag{4.5.9}$$

$$\max_{t \le \tau_1^*} |Z^*(t) - Z(t)| \in [\mathbf{C}_\infty]. \tag{4.5.10}$$

Proof Inclusion (4.5.8) follows from the inequality

$$\overline{Z}_\eta \le \sum_{k=1}^{\eta} |Z_k - Z_{k-1}|$$

and Lemma 4.5.5 for the terms $|Z_1| = |\zeta_1|$, $|Z_k - Z_{k-1}| = |\zeta_k|$ with $k \ge 2$.

Inclusion (4.5.9) follows from (4.5.8) and the inequality

$$\max_{t \le \tau_1^*} |\widetilde{Z}(t) - Z(t)| \le \overline{Z}_\eta + |Z_\eta|.$$

The proof of (4.5.10) is similar. Corollary 4.5.6 is proved. □

We can now establish the principal assertion in part I of the proof of Theorem 4.5.1.

Lemma 4.5.7 *Under the conditions of Lemma* 4.5.5, *the processes* z_T, $\widetilde{z}_T$, *and* $z_T^*(t) := \frac{1}{T} Z^*(Tt)$ *are asymptotically equivalent.*

Proof Since $\tau_i^* \ge 1$, we have

$$P_T := \mathbf{P}\Big(\max_{t \le T} |\widetilde{Z}(t) - Z(t)| \ge \delta T \Big) \le \mathbf{P}(W_1 > \delta T) + T\mathbf{P}(W \ge \delta T),$$

where

$$W_1 := \max_{t \le \tau_1^*} |\widetilde{Z}(t) - Z(t)|, \qquad W := \max_{t \le \tau^*} |\widetilde{Z}(t) - Z(t)|,$$

and τ_1^*, τ^* are defined in (4.5.5). By the exponential Chebyshev-type inequality,

$$P_T \le e^{-\Lambda^{(W_1)}(\delta T)} + Te^{-\Lambda^{(W)}(\delta T)},$$

where $\Lambda^{(W_1)}$, $\Lambda^{(W)}$ are the deviation functions for W_1 and W, respectively. By Lemma 4.5.7 we have $\Lambda^{(W_1)}(v) \gg v$, $\Lambda^{(W)}(v) \gg v$ as $v \to \infty$, and hence $\frac{1}{T} \ln P_T \to -\infty$ as $T \to \infty$. This means that the processes z_T and $\widetilde{z}_T$ are asymptotically equivalent. That z_T and z_T^* are equivalent is proved similarly. The lemma is proved. □

Let us now proceed with Part II of the proof and establish the LDP for the process $\widetilde{z}_T$. In accordance with general theorems on LDPs in metric spaces (see, e.g., [23], Chapter 4), to prove the LDP for the trajectories of $\widetilde{z}_T(t)$, it suffices to prove the following two facts.

1. The local LDP for $\widetilde{z}_T(t)$.
2. The "exponential density" of the distribution $\widetilde{z}_T$, that is, we need to show that,

for any $N < \infty$, there exists a compact set K in the space $(\mathbb{C}, \rho)$ such that, for all sufficiently large T,

$$\mathbf{P}(\widetilde{z}_T \notin K) \le e^{-TN}. \tag{4.5.11}$$

That the local LDP holds for $\widetilde{z}_T(t)$ follows from Theorem 4.4.1 and Lemmas 4.5.5 and 4.5.7. Inequality (4.5.11) is a consequence of the following assertion.

Lemma 4.5.8 *Inequality* (4.5.11) *holds under the conditions of Theorem* 4.5.1.

So, to prove Theorem 4.5.1 it remains to verify Lemma 4.5.8. Let us prove this lemma. From the above it follows that the presence of the inhomogeneity $\boldsymbol{\xi}_1 \neq_d \boldsymbol{\xi}$, with $\boldsymbol{\xi}_1 \in [\mathbf{C}_\infty]$, $\boldsymbol{\xi} \in [\mathbf{C}_\infty]$ does not alter the required properties of the trajectories of $Z(t)$, $\widetilde{Z}(t)$ in comparison with the homogeneous case. So, in what follows, to simplify the presentation, we will always assume that $\tau_1 \underset{d}{=} \tau$. Besides, for clarity, we also assume that T are integer. In parallel with the aggregated process z_T^* and its continuous version $\widetilde{z}_T$, we consider the trajectory of the random walk $s_n(t)$ for $n = T$ constructed as the continuous polygon with nodes $(\frac{k}{n}, \frac{Z_k^*}{x})$, where

$$Z_0^* = 0, \qquad Z_k^* = \sum_{j=1}^{k} \zeta_j^*, \quad k = 0, 1, \ldots, n.$$

We further note that if $\zeta^* \in [\mathbf{C}_\infty]$, then

$$K_v := \{f \in \mathbb{C} : \ f(0) = 0, \ \ I^*(f) \le v\}$$

is a compact subset in $(\mathbb{C}, \rho)$, where

$$I^*(f) = \begin{cases} \int_0^1 \Lambda^*(f'(t))\,dt & \text{if} \quad f \in \mathbb{C}_a; \\ \infty & \text{if} \quad f \in \mathbb{D} \setminus \mathbb{C}_a, \end{cases}$$

and Λ^* is the deviation function for the random variable ζ^* (see, e.g., [23], §4.2).

Assume first that $\mathbf{E}\zeta = 0$, so that $\mathbf{E}\zeta^* = \mathbf{E}\zeta\mathbf{E}\eta = 0$. To simplify the notation, we will drop the superscript "$*$" on τ_i^*, ζ_i^*, Λ^*, I^*, $\eta^*(t)$, and assume that $\tau_i \ge 1$ and $\zeta \in [\mathbf{C}_\infty]$.

We need to estimate

$$\mathbf{P}(\widetilde{z}_T \notin K_v) \le \mathbf{P}(I(\widetilde{z}_T) > v) \le \mathbf{P}\Big(\sum_{k=1}^{\eta(T)} \tau_k \Lambda\Big(\frac{\zeta_k}{\tau_k}\Big) > vT\Big).$$

Since $\eta(T) \le T$ and since by the assumption $T = n$ is an integer number, we have

$$\mathbf{P}(\widetilde{z}_T \notin K_v) \le \mathbf{P}\Big(\sum_{k=1}^{n} \tau_k \Lambda\Big(\frac{\zeta_k}{\tau_k}\Big) > vn\Big).$$

Further, since Λ is convex and since $\Lambda(0) = 0$, $\tau_k \ge 1$, we have $\frac{1}{\tau_k} \le 1$, and moreover,

$$\Lambda\Big(\frac{\zeta_k}{\tau_k}\Big) \le \Big(1 - \frac{1}{\tau_k}\Big)\Lambda(0) + \frac{1}{\tau_k}\Lambda(\zeta_k) = \frac{1}{\tau_k}\Lambda(\zeta_k),$$

so that

$$\mathbf{P}(\widetilde{z}_T \notin K_v) \le \mathbf{P}\Big(\sum_{k=1}^{n} \Lambda(\zeta_k) > vn\Big). \tag{4.5.12}$$

From Lemma 4.4.4 in [23] it follows that the right-hand side in (4.5.12) is majorized by

$$e^{-n(v-\mathbf{E}\gamma)},$$

where $\gamma := \Lambda(\zeta)$, $\mathbf{E}\gamma < \infty$. This proves (4.5.11) with $v = N + \mathbf{E}\gamma$.

Now let $a_\zeta := \mathbf{E}\zeta \ne 0$. Then

$$\widetilde{z}_T(t) = \widetilde{z}^0_T(t) + a_\zeta \pi_T(t),$$

where the process $\widetilde{z}^0_T$ is constructed in the same way as $\widetilde{z}_T$, but from the jumps $\zeta^0 = \zeta - a_\zeta$, $\mathbf{E}\zeta^0 = 0$; the process $\pi_T(t)$ is the continuous polygon constructed from the simple renewal process with the same jump points T_k (at which now jumps of size 1 take place) as for the process z^*_T. If $\widetilde{z}^0_T \in K_{(1)}$, $a_\zeta \pi_T(t) \in K_{(2)}$, where $K_{(1)}$, $K_{(2)}$ are compact sets, then it is clear that

$$\widetilde{z}^0_T + a_\zeta \pi_T(t) \in K_{(3)},$$

where $K_{(3)}{:=}K_{(2)} \cup K_{(2)}$ is also a compact set. Assume, for definiteness, that $a_\zeta > 0$. Then, since $\tau_k \ge 1$, we have $0 \le a_\zeta \pi'_T(t) \le a_\zeta$ for all $t \in [0, 1]$, and therefore, $a_\zeta \pi_T(t) \in K_{(2)}$, where

$$K_{(2)} := \{f : f(0) = 0,\ 0 \le f'(t) \le a_\zeta\}$$

is a compact set. Next,

$$\mathbf{P}(\widetilde{z}^0_T \in K_{(1)},\ a_\zeta \pi_T(t) \in K_{(2)}) \le \mathbf{P}(\widetilde{z}_T \in K_{(3)}),$$
$$\mathbf{P}(\widetilde{z}_T \notin K_{(3)}) \le \mathbf{P}(\widetilde{z}^0_T \notin K_{(1)}).$$

It now remains to use the first part of the proof. Lemma 4.5.8, and therefore, Theorem 4.5.1, are proved. □

Note that Lemma 4.5.8 can be verified using different approaches. Let $T = n$ be an integer number and let $s_n(t)$ be the continuous polygon constructed from the nodes $(\frac{k}{n}, \frac{Z^*_k}{x})$, $k = 0, 1, \ldots, n$. It is easily checked that

$$\omega(\Delta, \widetilde{z}_T) \le \omega(\Delta, s_n), \tag{4.5.13}$$

where $\omega(\Delta, f)$ is the modulus of continuity of the function f:

$$\omega(\Delta, f) := \sup\Big\{|f(u+v) - f(u)| :\ 0 \le u < u+v \le 1,\ v \le \Delta\Big\}.$$

Next, using the condition $\zeta \in [\mathbf{C}_\infty]$, one can explicitly construct (from the deviation function of the random variable ζ) the function $\delta(v) \downarrow 0$ as $v \downarrow 0$ and the corresponding compact set

$$K^{(\delta)} := \{f \in \mathbb{C} :\ f(0) = 0,\ \omega(\Delta, f) \le \delta(\Delta)\ \text{ for } 0 \le \Delta \le 1\},$$

for which relation (4.5.11) holds for the process s_n, and therefore, by (4.5.13), also for the process $\widetilde{z}_T$.

The "highly probable" compact set for the process s_n can be also constructed as the set $\{I^*(f) \le v\}$.

4.5.2 On the Relaxation of the Conditions of Theorem 4.5.1

We have seen in previous sections that the proof of the LDP in the space $\mathbb{D}$ with the uniform metric ρ involves the very restrictive condition $\zeta \in [\mathbf{C}_\infty]$. The relaxation of this condition to the "minimal" condition $\boldsymbol{\xi} \in [\mathbf{C}]$, which was used in the LDP for the processes $Z(t)$ in the state space (Theorem 3.4.1), can be carried out in the following alternative directions:

(a) replacement of the uniform metric by a weaker metric;

(b) reduction of the class of CRPs under consideration;

(c) reduction of the class of the sets B (or the functionals f) for which the LDP holds (this direction is illustrated in §§4.6 and 4.7).

Let us briefly consider these approaches.

(a) The replacement of the uniform metric by a weaker metric involves the restriction of the class of functionals continuous in this new metric – and the class of sets B for which the limit

$$\lim_{T\to\infty} \frac{1}{T} \ln \mathbf{P}(z_T \in B)$$

exists; that is, this approach is related to direction (c). A similar situation occurs in the study of LDPs for random walks, for example, for the walks $\{Z_k\}$ (or, equivalently, for $Z(k)$ for $\tau_k \equiv 1$). For such walks, first the LDP for the trajectories was established under the condition $\zeta \in [\mathbf{C}_\infty]$ (see [15], and also [23], Chapter 4 and the references given therein). Later, in order to relax this condition, in [43] (see also [23], Chapter 4 and the references given there) substantial changes were incorporated both in the statement of the problem and in the proof by introduction of the new metric $\rho_{\mathbb{D}}$, which is weaker than ρ, which implies the change of the deviation function.

The new metric $\rho_{\mathbb{D}}$ was described in §1.6. The deviation functional $J(f)$ corresponding to this metric has the form (4.4.4). If the condition $\zeta \in [\mathbf{C}_\infty]$ is not satisfied, then this functional is generally different from $I(f)$, so that the condition $\zeta \in [\mathbf{C}_\infty]$ is in a sense necessary for the LDP to hold with the functional $I(f)$ (see also (4.5.1) and (4.5.2) in Theorem 4.5.1).

The LDP in the space $(\mathbb{D}, \rho_{\mathbb{D}})$ with the deviation functional (4.4.4) is called in [23] and [43] the *extended* LDP. In [88], for the compound Poisson process (that is, for a CRP with exponentially distributed τ), an extended LDP was established in which in fact all three alternative approaches (a)–(c) were used simultaneously. The following LDP is established in the space $(\mathbb{V}, \rho_{\mathbb{D}})$, where $\mathbb{V}$ is the space of functions of bounded variation.

Theorem 4.5.9 *Let $Z(t)$ be a compound Poisson process, $\zeta \in$ [**C**]. Then the normalized processes $z_T(t) = \frac{Z(tT)}{T}$, $t \in [0, 1]$, satisfy the extended LDP in the space $(\mathbb{V}, \rho_{\mathbb{D}})$ with the deviation functional $J(f)$, that is,*

$$\overline{\lim_{T\to\infty}} \frac{1}{T} \ln \mathbf{P}(z_T \in B) \le -J(B+),$$

$$\underline{\lim_{T\to\infty}} \frac{1}{T} \ln \mathbf{P}(z_T \in B) \ge -J((B)),$$

where (B) is the interior of a set B,

$$J(B+) := \lim_{\varepsilon\to 0} J((B)_\varepsilon),$$

$(B)_\varepsilon$ is the ε-neighborhood of a set B, $J(B) = \inf_{f\in B} J(f)$,

$$J(f) = \begin{cases} \infty & \text{if } \operatorname{Var} f = \infty, \\ \int_0^1 \Lambda(f'_a(t))\,dt + \mu_+^{(\zeta)} \operatorname{Var} f_s^+ - \mu_-^{(\zeta)} \operatorname{Var} f_s^- & \text{if } \operatorname{Var} f < \infty, \end{cases}$$

where f_a and f_s are, respectively, the absolutely continuous and singular components of the function $f = f_a + f_s$, $f_s^\pm$ are, respectively, the increasing and decreasing components of the function f_s, $\mu_\pm^{(\zeta)}$ are the end-points of the largest interval $(\mu_-^{(\zeta)}, \mu_+^{(\zeta)})$ on which the function $\mathbf{E}e^{\mu Z(1)}$ is finite, $\Lambda(\alpha)$ is the deviation function corresponding to the random variable $Z(1)$.

In other words, the processes $z_T(t)$ satisfy the same extended LDP as the normalized random walks $\{\frac{Z(k)}{T}\}_{k=1}^\infty$.

In [90], this result was extended to general processes with independent increments in the space $(\mathbb{D}, \rho_{\mathbb{D}})$ which satisfy condition [**C**].

We omit the proof of Theorem 4.5.9, which involves considerable difficulties beyond the scope of the present book.

In applied problems concerned with the determination of the asymptotics of $\mathbf{P}(z_T \in B)$ the sets B are quite simple and one can hardly expect that the infimum in $\inf_{f\in B} J(f)$ is attained on a function $f \notin \mathbb{C}_a$. If $f \in \mathbb{C}_a$, then $J(f) = I(f)$, and there arises the possibility for realization of the third alternative approach (c), under which the LDPs (4.5.1) and (4.5.2) remain valid only under condition [**C**]. As a class of events B, for which this approach is reasonable, we mention, for example, the class of events related to crossing or noncrossing a remote boundary by the trajectory of $z_T(t)$. The next two sections will be devoted to evaluating the limit $\lim_{T\to\infty} \frac{1}{T} \ln \mathbf{P}(z_T \in B)$ in the above boundary crossing problems under condition [**C**] only.

4.6 Large Deviation Principles for the First Boundary Crossing Problem

As we have already seen, the evaluation of the asymptotics

$$\frac{1}{T} \ln \mathbf{P}(z_T \in B), \quad z_T = z_T(t) = \frac{Z(tT)}{T}, \quad t \in [0, 1] \text{ as } T \to \infty \tag{4.6.1}$$

for arbitrary measurable sets B, say, in the space $\mathbb{D}$ of functions without discontinuities of the second kind involves considerable difficulties and restrictive assumptions (see Theorems 4.4.1 and 4.5.1).

In this and the subsequent sections *we will consider sets B corresponding to boundary crossing problems of two types*. Recall that

$$\frac{Z(t)}{t} \underset{a.s.}{\to} a := \frac{a_\tau}{a_\zeta} \quad \text{as} \quad t \to \infty, \quad \text{where} \quad a_\tau := \mathbf{E}\tau, \quad a_\zeta := \mathbf{E}\zeta.$$

I. In the so-called *first boundary crossing problem* in the large deviation zone one should find the asymptotics of $\mathbf{P}(z_T \in B_g)$, where

$$B_g = \left\{ f \in \mathbb{D} : \sup_{t \in [0,1]} (f(t) - g(t)) \geq 0 \right\} \tag{4.6.2}$$

for a given function $g(t) > at$, $t \in [0, 1]$, from the space $\mathbb{D}$.

II. In the *second boundary crossing problem* one searches the asymptotics of $\mathbf{P}(z_T \in B_{g_- g_+})$, where the set $B_{g_- g_+}$ has the form

$$B_{g_- g_+} = \{ f \in \mathbb{D} : g_-(t) \geq f(t) \geq g_+(t) \text{ for all } t \in [0, 1] \}, \tag{4.6.3}$$

and the given functions $g_-(t) < g_+(t)$, $g_-(0) < 0 < g_+(0)$, from $\mathbb{D}$ have finitely many jumps and are such that the line $z = at$ crosses at least one of the curves $g_\pm$.

Before proceeding with the study of the asymptotics (4.6.1) with the above sets B, we require some auxiliary results.

Recall that if condition $[\lambda_+] \cup [\overline{\lambda}_+]$ is met and the admissible inhomogeneity conditions

$$\mathcal{A}^{\leq 0} \subset [\mathcal{A}_1], \qquad \lambda_+^{(\tau_1)} \geq \min[\lambda_+, D(0)] \tag{4.6.4}$$

are satisfied, then the LDPs, as obtained above for the homogeneous and inhomogeneous CRPs, are equal (the second condition in (4.6.4) is required only in rare special cases when the zero trajectory appears in the analysis).

4.6.1 Level Lines

In order to describe the asymptotics of

$$\frac{1}{T} \ln \mathbf{P}(z_T \in B_g) \tag{4.6.5}$$

in the first boundary crossing problem (see (4.6.2)), we need to introduce the concept of *level lines*. In the context of random walks, such lines were introduced and studied in [14] and [23], §3.7. Using these lines, in [14] and [23], rough and exact asymptotics of $\mathbf{P}(z_T \in B_g)$ were obtained, when z_T is the trajectory of a random walk ($\tau \equiv 1$). For CRPs, the *lines $\mathfrak{l}_\alpha(t)$ of level α* are defined in the same fashion – these are the lines (functions) for which $\mathfrak{l}_\alpha(1) = \alpha$, while the limit as $T \to \infty$ of the sequence

$$\frac{1}{tT} \ln \mathbf{P}\big(z_T(t) \in \Delta[\mathfrak{l}_\alpha(t))\big)$$

for $\Delta = \Delta_T = \overline{o}(1)$ and all $t \in (0, 1]$ keeps the same value (equal to $D(\alpha)$). From Theorem 3.4.1 it follows that under conditions $[\boldsymbol{\lambda}_+]$ and (4.6.4) (the second condition in (4.6.4) becomes the inequality $\lambda_+^{(\tau_1)} \geq D(0)$) such a function has to satisfy the equation

$$tD\left(\frac{\mathrm{l}_\alpha(t)}{t}\right) = D(\mathrm{l}_\alpha(1)) = D(\alpha). \tag{4.6.6}$$

If $D(d_+) = \infty$ (that is, the case $\mathbf{P}(\zeta/\tau = d_+) > 0$ is excluded), then from the properties of $D(\alpha)$ it follows that in the domain (a, d_+) this function is continuous and monotone increasing from 0 to ∞. Therefore, the branch of the function D in the domain (a, d_+) always has the inverse function $D^{(-1)}$.

Definition 4.6.1 The function

$$\mathrm{l}_\alpha(t) := tD^{(-1)}\left(\frac{D(\alpha)}{t}\right), \qquad t \in (0, 1], \quad a < \alpha < d_+ \tag{4.6.7}$$

is called the *line of level* α.

At $t = 0$, the function $\mathrm{l}_\alpha(t)$ is defined by right continuity: $\mathrm{l}_\alpha(0) = \mathrm{l}_\alpha(+0)$.

Some properties of level lines are given in the next lemma.

Lemma 4.6.2 *Let* $\alpha \in (a, d_+)$, $D(d_+) = \infty$ $(\mathbf{P}(\zeta/\tau = d_+) = 0)$. *Then*

(i) $$\mathrm{l}_\alpha(0) = \frac{D(\alpha)}{\mu^+} \quad \left(\mu^+ = \lim_{\alpha\to\infty} \frac{D(\alpha)}{\alpha}\right), \quad \frac{\mathrm{l}_\alpha(t)}{t} \to \infty \text{ as } t \to 0. \tag{4.6.8}$$

(ii) *The functions* $\mathrm{l}_\alpha(t) - at$ *are increasing and concave on* $[0, 1]$ $(\mathrm{l}''_\alpha(t) \leq 0$ *a.e.*). *For each t,* $\mathrm{l}_\alpha(t)$ *is increasing with respect to* α.

(iii) *The function* $\mathrm{l}_\alpha(t)$ *is glued (keeping the derivative continuous) from the following three pieces in general*:

(1) *the linear function*

$$\mathrm{l}_\alpha(t) = \frac{D(\alpha)}{\mu^+} + t\left(\alpha^+ - \frac{D(\alpha^+)}{\mu^+}\right) \quad \text{for } t \in [0, t_+], \quad t_+ = t_+(\alpha) := \frac{D(\alpha)}{D(\alpha^+)} \tag{4.6.9}$$

(*this piece is absent if* $\mu^+ = \infty$ *or* $D(\alpha^+) = \infty$);

(2) *the "intermediate" piece on* $[t_+, t^+]$, *where* $t^+ := \frac{D(\alpha)}{D(\alpha_+)}$, *which may include intervals of analyticity of the function* $\mathrm{l}_\alpha(t)$ (*this piece is absent if* $\mu^+ = \mu_+$ *or, equivalently, if* $\mathbf{A}(\lambda, \mu) > 0$ *for* $(\lambda, \mu) \in \partial\mathcal{A}$, $\mu \in (0, \mu^+)$);

(3) *a strictly concave analytic function in* $(t^+, 1]$.

(iv) *We set for brevity* $\mathrm{l}_\alpha(t) = \mathrm{l}$. *Then, for* $\alpha \in (a, \alpha_+)$,

$$\mathrm{l}'_\alpha(t) = \frac{A(\mu(\mathrm{l}/t))}{\mu(\mathrm{l}/t)} = -\frac{\lambda(\mathrm{l}/t)}{\mu(\mathrm{l}/t)}, \tag{4.6.10}$$

$$\mathrm{l}'_\alpha(1) = -\frac{\lambda(\alpha)}{\mu(\alpha)} \in (0, \alpha). \tag{4.6.11}$$

From the lemma it follows that in the case $D(\alpha_+) = \infty$ the level lines $\mathrm{l}_\alpha(t)$ are strictly concave and analytic on the whole interval $(0, 1)$.

Proof If, in order to simplify the argument, we put $a = 0$, then the proof of assertions (i), (ii) of Lemma 4.6.2 repeats that of the corresponding assertions in Theorem 3.7.2 in [23], which was concerned with level lines of random walks (in [23] we have to replace $\Lambda(\alpha)$ and λ_+ by $D(\alpha)$ and μ^+, respectively). The case $a \neq 0$ requires no essential changes in the proof.

Some difference from Theorem 3.7.2 in [23] appears in the proof of assertions (iii), (iv) of the lemma. By Theorem 3.5.2, the function $D(\alpha)$ for $\alpha > a$ is glued in general from the three pieces described in assertions (1)–(3) of Theorem 3.5.2 (recall that the right-hand branch of the function $\Lambda(\alpha)$ is glued from two pieces). We have $D'(\alpha) > 0$ on (a, ∞), and hence the inverse function $D^{(-1)}(v)$ of the right-hand branch of $D(\alpha)$ on $(0, \infty)$ is also glued from at most three pieces at the points $v_+ = D(\alpha_+)$ and $v^+ = D(\alpha^+)$ (see Theorem 3.5.2). The function $D^{(-1)}(v)$ is analytic on $(0, v_+)$ and on (v^+, ∞) it is linear,

$$D^{(-1)}(v) = \alpha^+ + \frac{v - D(\alpha^+)}{\mu^+}.$$

Hence by definition (4.6.7), the function $\mathfrak{l}_\alpha(t)$ is analytic for $\frac{D(\alpha)}{t} < v_+$ or, equivalently, for

$$t > \frac{D(\alpha)}{D(\alpha_+)} = t^+.$$

For $t \in (t^+, t_+)$, $t_+ := \frac{D(\alpha)}{D(\alpha^+)}$, the function $\mathfrak{l}_\alpha(t)$ is of "intermediate" character, while for $\frac{D(\alpha)}{t} > v^+$ (or, equivalently, for $t < t_+$), by (4.6.7) we have

$$\mathfrak{l}_\alpha(t) = t\Big[\alpha^+ + \frac{D(\alpha)/t - D(\alpha^+)}{\mu^+}\Big],$$

which is equivalent to (4.6.9).

(iv) Differentiating (4.6.6) with respect to t, we get

$$D(\mathfrak{l}/t) + t\mu(\mathfrak{l}/t)\Big(\frac{\mathfrak{l}'_\alpha(t)}{t} - \frac{\mathfrak{l}}{t^2}\Big) = 0,$$

$$\mathfrak{l}'_\alpha(t)\mu(\mathfrak{l}/t) = \frac{\mathfrak{l}}{t}\mu(\mathfrak{l}/t) - D(\mathfrak{l}/t) = A(\mu(\mathfrak{l}/t)) > 0,$$

which implies (4.6.10). Since $\mathfrak{l}_\alpha(1) = \alpha$, at the point $t = 1$ we have

$$\mathfrak{l}'_\alpha(1) = \frac{A(\mu(\alpha))}{\mu(\alpha)} = -\frac{\lambda(\alpha)}{\mu(\alpha)}.$$

Next, $A(\mu)/\mu < A'(\mu)$ for $a > 0$, $\mu > 0$. Therefore,

$$\frac{A(\mu(\alpha))}{\mu(\alpha)} < A'(\mu(\alpha)) = \alpha.$$

This proves (4.6.11), and therefore, Lemma 4.6.2. □

As in §3.7.2 in [23], from Lemma 4.6.2 one can easily derive that, for small $\alpha - a$, v and for each $t \in (0, 1]$,

$$D(\alpha) \approx \frac{(\alpha - a)^2}{2\sigma^2}, \qquad D^{(-1)}(v) \approx a + \sigma\sqrt{2v}, \qquad \mathfrak{l}_\alpha(t) \approx at + (\alpha - a)\sqrt{t}.$$

4.6.2 Inequalities for the Distribution of the Maximum Value of a CRP

To prove the main theorem (Theorem 4.6.4), we need to estimate the distribution of

$$\overline{Z}(T) := \sup_{t \le T} Z(t).$$

These estimates are of independent interest. The following assertion is an analog of Theorem 1.1.1 in [23] on estimates for the distribution of the maximum of sums of random variables.

Theorem 4.6.3 *Assume that conditions* $[\boldsymbol{\lambda}_+] \cup [\overline{\boldsymbol{\lambda}}_+]$, (4.6.4) *are met. Then*
(i) *As* $T \to \infty$, *for all* $x > 0$, $\mu \ge 0$,

$$\mathbf{P}(\overline{Z}(T) \ge x) \le \exp\Big\{ -\mu x + T \max\big(0, \widehat{A}(\mu)\big) + o(T)\Big\}. \tag{4.6.12}$$

(ii) *Let* $a < 0$,

$$\mu_0^{(\zeta)} := \sup\{\mu : \mathbf{A}(0, \mu) = A_\zeta(\mu) \le 0\},$$
$$\mu_0 := \sup\{\mu : A(\mu) \le 0\}.$$

Assume that $A(\mu^+) \le 0$. *Then*

$$\mathbf{P}(\overline{Z}(T) \ge x) \le \min\big(e^{-\mu_0^{(\zeta)} x}, e^{-\mu^+ x}\big) \tag{4.6.13}$$

(for independent τ *and* ζ *we always have* $\mu_0^{(\zeta)} = \mu_0$*).*
If $A(\mu^+) > 0$, *then*

$$\begin{aligned} &A(\mu_0) = 0, \quad \alpha_0 := A'(\mu_0) \in (0, \infty), \\ &D(\alpha) \ge \mu_0 \alpha, \quad D(\alpha_0) = \mu_0 \alpha_0 \end{aligned} \tag{4.6.14}$$

and moreover, as $T \to \infty$, $\alpha = x/T$,

$$\mathbf{P}(\overline{Z}(T) \ge x) \le \begin{cases} e^{-\mu_0 x} & \text{for all} \quad \alpha, \\ e^{-TD(\alpha) + o(T)} & \text{for} \quad \alpha \ge \alpha_0. \end{cases} \tag{4.6.15}$$

(iii) *If* $a \ge 0$, *then, for* $\alpha = x/T \ge a$, *as* $T \to \infty$,

$$\mathbf{P}(\overline{Z}(T) \ge x) \le e^{-TD(\alpha) + o(T)}. \tag{4.6.16}$$

(iv) (a) *If condition* $[\boldsymbol{\lambda}_+]$ *is met,* $a < 0$, $A(\mu^+) \le 0$, *then*

$$\mathbf{P}(\overline{Z}(T) \ge x) = e^{-\mu_0 x + o(x)} \tag{4.6.17}$$

as $x \to \infty$, $T \to \infty$.

(b) *If*

$$\alpha = \frac{x}{T} > a \geq 0$$

and

$$a < 0, \qquad A(\mu^+) > 0, \qquad \alpha > \alpha_0,$$

then

$$\mathbf{P}(\overline{Z}(T) \geq x) = e^{-TD(\alpha)+o(T)}. \tag{4.6.18}$$

A comparison of assertion (iv) with inequalities (4.6.13), (4.6.16), and the second inequality in (4.6.15) shows that these inequalities are exponentially unimprovable. This also means that the LDP for $\mathbf{P}(\overline{Z}(T) \geq x)$ holds.

Theorem 4.6.3 will be refined in §5.6 of the next chapter.

Proof (i) We set

$$\eta_x = \min\{t : Z(t) \geq x\}.$$

Since the CRP $Z(t)$ can exceed the level $x > 0$ only by a jump, it follows that η_x is a Markov moment. Hence, for $t < T$, the event $\{\eta_x \in dt\}$ and the random variable $Z(T) - Z(t)$ are independent, the process $Z(t+u) - Z(t)$ is a homogeneous CRP on the set $\{\eta_x \in dt\}$, and

$$\mathbf{E}e^{\mu Z(T)} \geq \int_0^T \mathbf{E}(e^{\mu Z(T)}; \eta_x \in dt) \geq \int_0^T \mathbf{P}(\eta_x \in dt)\mathbf{E}(e^{\mu(x+Z(T)-Z(t))} \mid \eta_x = t). \tag{4.6.19}$$

Using Theorem 3.7.1 (see (3.7.1)) and the conditions of the theorem, we have

$$\mathbf{E}(e^{\mu(Z(T)-Z(t))} \mid \eta_x = t) = e^{(T-t)\widehat{A}(\mu)+o(T-t)} \quad \text{as} \quad T-t \to \infty,$$

and hence from (4.6.19) we get

$$e^{T\widehat{A}(\mu)+o(T)} \geq \begin{cases} e^{\mu x+o(T)}\,\mathbf{P}(\eta_x \leq T) & \text{if} \quad \widehat{A}(\mu) \geq 0, \\ e^{\mu x+\widehat{A}(\mu)T+o(T)}\,\mathbf{P}(\eta_x \leq T) & \text{if} \quad \widehat{A}(\mu) < 0. \end{cases}$$

Now (4.6.12) follows, since $\mathbf{P}(\eta_x \leq T) = \mathbf{P}(\overline{Z}(T) \geq x)$.

(ii) We have

$$\overline{Z}(T) \leq \overline{Z} := \sup_{t\geq 0} Z(t) = \sup_{k\geq 0} Z_k,$$

and hence in the case $a < 0$, an appeal to Theorem 1.1.1 in [23] shows that

$$\mathbf{P}(\overline{Z}(T) \geq x) \leq \mathbf{P}(\overline{Z}(\infty) \geq x) = \mathbf{P}(\overline{Z} \geq x) \leq e^{-\mu_0^{(\zeta)}x}. \tag{4.6.20}$$

Besides, for $A(\mu^+) \leq 0$ from (4.6.12) with $\mu = \mu^+$, one can easily show that

$$\mathbf{P}(\overline{Z}(\infty) \geq x) \leq e^{-\mu^+ x}. \tag{4.6.21}$$

As a result, we get (4.6.13).

If $A(\mu^+) > 0$, then

$$A(\mu_0) = 0, \qquad \alpha_0 = A'(\mu_0) \in (0, \infty).$$

The function $D(\alpha)$ is the Legendre transform of $A(\mu)$, and hence

$$D(\alpha) \geq \alpha\mu_0 - A(\mu_0) = \alpha\mu_0. \tag{4.6.22}$$

Besides, $\mu_0 = \mu(\alpha_0)$, and

$$D(\alpha_0) = \alpha_0\mu_0.$$

This proves (4.6.14).

If $\alpha = x/n \geq \alpha_0$, then $\mu(\alpha) \geq \mu_0$, $A(\mu(\alpha)) \geq A(\mu_0) = 0$. For such α, applying (4.6.12), we find that

$$\mathbf{P}(\overline{Z}(T) \geq x) \leq e^{-T(\alpha\mu - A(\mu)) + o(T)},$$

which for $\mu = \mu(\alpha)$ gives the second inequality in (4.6.15). The first inequality in (4.6.15) follows from (4.6.12) with $\mu = \mu_0$ and since $\widehat{A}(\mu_0) = A(\mu_0) = 0$.

(iii) For $a \geq 0$, $\mu \geq 0$, we get $A(\mu) \geq 0$. Besides, $\mu(a) = 0$, and therefore, $\mu(\alpha) \geq 0$ for $\alpha \geq a$. Hence, substituting $\mu = \mu(\alpha)$ into (4.6.12), we get (4.6.16).

(iv) (a) In the case $[\lambda_+]$, $a < 0$, $A(\mu^+) \leq 0$, for any $t \leq 1$, by Theorem 3.4.1 we have

$$\mathbf{P}(\overline{Z}(T) \geq x) \geq \mathbf{P}(Z(tT) \geq x) = e^{-tTD(\frac{x}{tT}) + o(tT)}, \tag{4.6.23}$$

where for $\gamma = \frac{x}{tT}$,

$$tTD\Big(\frac{x}{tT}\Big) = \frac{x}{\gamma} D(\gamma).$$

For $\gamma \geq \alpha^+ := A'(\mu^+ - 0)$, by assertion (iii), (3) of Theorem 3.5.2, we get $\mu(\gamma) = \mu^+$, and

$$D(\gamma) = D(\alpha^+) + \mu^+(\gamma - \alpha^+) = -A(\mu^+) + \gamma\mu^+.$$

Hence, for any $\varepsilon > 0$, one can take a sufficiently small t so that $\gamma \geq \alpha^+$ and

$$\frac{D(\gamma)}{\gamma} \leq \mu^+ + \varepsilon.$$

For such γ from (4.6.23) we get

$$\mathbf{P}(\overline{Z}(T) \geq x) \geq e^{-x(\mu^+ + \varepsilon) + o(tT)}. \tag{4.6.24}$$

We have $tT = x/\gamma$, $o(tT) = o(x)$, $\varepsilon > 0$ is arbitrary, the left-hand side of (4.6.24) does not depend on ε, and hence

$$\mathbf{P}(\overline{Z}(T) \geq x) \geq e^{-x\mu^+ + o(x)}.$$

This together with (4.6.13) proves (4.6.17).

(b) Under the conditions of assertion (b) the proof is quite clear. If $\alpha > a \geq 0$, then by Theorem 3.4.1,

$$\mathbf{P}(\overline{Z}(T) \geq x) \geq \mathbf{P}(Z(T) \geq x) = e^{-TD(\alpha)+o(T)}.$$

Now a comparison with (4.6.16) gives (4.6.18).

The argument is similar if $a < 0$, $\alpha > \alpha_0$.

Theorem 4.6.3 is proved. □

4.6.3 Large Deviation Principles for the First Boundary Crossing Problem

Let us now formulate the main result of this section on the asymptotics of sequence (4.6.5). For definiteness, the function $g(t) > at$, $t \in [0, 1]$, in (4.6.2), (4.6.5) will be assumed to be lower-continuous, that is, for each $t \in (0, 1)$,

$$g(t) = \lim_{\varepsilon \to 0} \inf_{u \in (t)_\varepsilon} g(u), \tag{4.6.25}$$

where $(t)_\varepsilon = (t - \varepsilon, t + \varepsilon)$. By α^g we denote α for which the level line $\mathfrak{l}_\alpha(t)$ is tangent to the curve $g(t)$ for the first time as α grows from zero. Let t^g be the least value of t such that

$$\mathfrak{l}_{\alpha^g}(t^g) = g(t^g)$$

(by (4.6.25) such t always exists).

Theorem 4.6.4 *Let conditions* $[\lambda_+]\cup[\overline{\lambda}_+]$ *be met and let the admissible inhomogeneity conditions* (4.6.4) *be satisfied. Next, assume that* $D(d_+) = \infty$ ($\mathbf{P}(\zeta/\tau = d_+) = 0$), $t^g > 0$. *Then*

$$\lim_{T \to \infty} \frac{1}{T} \ln \mathbf{P}(z_T \in B_g) = -D(\alpha^g). \tag{4.6.26}$$

Remark 4.6.5 (a) If $\min_{t \in [0,1]} g(t) > 0$, then the second condition in (4.6.4) seems to be superfluous, because the function $f_0(t) \equiv 0$ does not lie in the set B_g.

(b) Condition $D(d_+) = \infty$ is superfluous if $\alpha^g < d_+$. For a description of the level lines in the case $\alpha^g = d_+$, $D(d_+) < \infty$, see Theorem 3.7.2 in [23] in which the function Λ is replaced by D. We omit the case $D(d_+) < \infty$, since it is rather particular and complicates the proof.

Proof of Theorem 4.6.4 Let conditions $[\lambda_+] \cup [\overline{\lambda}_+]$ and (4.6.4) be met.

Upper estimate. We set

$$\eta_g = \min\{t : z_T(t) \geq g(t)\}.$$

Then

$$\mathbf{P}(z_T \in B_g) = \mathbf{P}(\eta_g \leq 1).$$

We derive upper estimates for this probability in three stages. Let, for brevity, $\mathfrak{l}_{\alpha^g}(t) = \mathfrak{l}(t)$.

I. Given some $t_* \in (0, t^g)$ (which will be chosen later), we estimate $\mathbf{P}(\eta_g \le t_*)$. Since $g(t) > \mathrm{l}(t)$ for all $t \in [0, t^g)$ and since $\mathrm{l}(t)$ is a smooth function, there exist $t_* \in (0, t^g)$ and $g_0 > 0$ such that

$$\begin{gathered} \min_{t\in[0,t_*]} g(t) > g_0 > \mathrm{l}(0), \\ \mathbf{P}(\eta_g \le t_*) \le \mathbf{P}\big(\max_{t\le t_*} z_T(t) \ge g_0\big) = \mathbf{P}\big(\overline{Z}(t_*T) \ge Tg_0\big). \end{gathered} \tag{4.6.27}$$

Assume first that $a \ge 0$. Then, for t_* such that $g_0/t_* \ge a$, using assertion (iii) of Theorem 4.6.3, we get

$$\mathbf{P}\big(\overline{Z}(t_*T) \ge Tg_0\big) \le e^{-Tt_*D\left(\frac{g_0}{t_*}\right)+o(T)}, \tag{4.6.28}$$

where $\gamma = g_0/t_*$ can be made large by choosing an appropriate t_*.

We have $\frac{D(\gamma)}{\gamma} \to \mu^+$ as $\gamma \to \infty$, and so in the case $\mu^+ < \infty$ the value $t_*D(\frac{g_0}{t_*})$ can be made arbitrarily close to

$$g_0\mu^+ > \mathrm{l}(0)\mu^+ = D(\alpha^g) \tag{4.6.29}$$

by choosing a suitable t_* (see (4.6.8)). In view of (4.6.28), this means that

$$\mathbf{P}(\eta_g \le t_*) = o\big(e^{-TD(\alpha^g)}\big). \tag{4.6.30}$$

In the case $\mu^+ = \infty$, $t_*D(\frac{g_0}{t_*})$ can be made arbitrarily large by choosing a suitable t_*, and hence in view of (4.6.28) we again get (4.6.30).

Now let

$$a < 0, \qquad A(\mu^+) > 0.$$

In this case, we need to employ the second inequality in (4.6.15) and repeat the above arguments (with a replaced by $\alpha_0 = A'(\mu_0)$), which again give (4.6.30).

Now consider the case

$$a < 0, \qquad A(\mu^+) \le 0.$$

Here by (4.6.13), (4.6.27), and (4.6.29), we have

$$\mathbf{P}(\eta_g \le t_*) \le \mathbf{P}\big(\overline{Z}(t_*T) \le Tg_0\big) \le e^{-T\mu^+ g_0} < e^{-T\mu^+ \mathrm{l}(0)} = e^{-TD(\alpha^g)}. \tag{4.6.31}$$

II. Let us now estimate the probability

$$\mathbf{P}(\eta_g \in (t_*, 1]) \le \mathbf{P}(\eta_\mathrm{l} \in (t_*, 1]) \tag{4.6.32}$$

in the case

$$\mathrm{l}'(1) \ge 0.$$

In this case, the trajectory $z_T(t)$ can intersect the boundary $\mathrm{l}(t)$ only by vertical jumps; therefore η_l is a Markov moment.

Without loss of generality we assume that

$$p := \mathbf{P}(\tau \ge 1) > 0, \tag{4.6.33}$$

and denote by κ the index of the jump of the walk $\{T_k\}$ on which the value of Z_k exceeds that of $T\mathrm{l}(\frac{T_k}{T})$ for the first time. We have

$$\mathbf{P}(\eta_{\mathrm{l}} \in (t_*, 1]) = \frac{1}{p}\mathbf{P}(\eta_{\mathrm{l}} \in (t_*, 1], \tau_{\kappa+1} \geq 1).$$

But the event $\{\eta_{\mathrm{l}} \in (t_*, 1], \tau_{\kappa+1} \geq 1\}$ implies the union of the events

$$\bigcup_{t_*T \leq k \leq T} \left\{ z_T\left(\frac{k}{T}\right) \geq \mathrm{l}\left(\frac{k}{T}\right) - \frac{b_*}{T} \right\}, \tag{4.6.34}$$

where $b_* = \max_{t\in(t_*,1)} \mathrm{l}'(t) = \mathrm{l}'(t_*) > 0$. Therefore,

$$\mathbf{P}(\eta_{\mathrm{l}} \in (t_*, 1]) \leq \frac{1}{p} \sum_{t_*T \leq k \leq T} \mathbf{P}\left(z_T\left(\frac{k}{T}\right) \geq \mathrm{l}\left(\frac{k}{T}\right) - \frac{b_*}{T} \right). \tag{4.6.35}$$

But, for any $k \geq t_*T$ and $t = \frac{k}{T}$, the definition of level lines yields (see Theorem 3.4.1)

$$\mathbf{P}(z_T(t) > \mathrm{l}(t) - b_*/T) = e^{-tTD(\frac{\mathrm{l}(t)}{t})+o(T)} = e^{-TD(\alpha^g)+o(T)}.$$

Hence from (4.6.35) it follows that

$$\varlimsup_{T\to\infty} \frac{1}{T} \ln \mathbf{P}(\eta_g \in (t_*, 1]) \leq -D(\alpha^g). \tag{4.6.36}$$

III. Next, let us estimate probability (4.6.32) under the condition

$$\mathrm{l}'(1) < 0.$$

Since $\mathrm{l}'(t)$ is continuous and monotone decreasing, for $\mathrm{l}'(t_*) > 0$ there exists a point $t^* < 1$ such that $\mathrm{l}'(t^*) = 0$. If $\mathrm{l}'(t_*) \leq 0$, then we set $t^* = t_*$. If the interval (t_*, t^*) is nonempty, then $\mathbf{P}(\eta_{\mathrm{l}} \in (t_*, t^*])$ is estimated as in Step II. It remains to estimate

$$\mathbf{P}(\eta_{\mathrm{l}} \in (t^*, 1]).$$

On the interval $(t^*, 1]$, the function $\mathrm{l}(t)$ is decreasing and the trajectory $z_T(t)$ can cross $\mathrm{l}(t)$ not only with a vertical jump (an event A) but also with a horizontal jump (an event B). In this case, the probability $\mathbf{P}(\eta_{\mathrm{l}} \in (t^*, 1];\ A)$ is again estimated as in Step II (with b_* replaced by 0).

Consider the probability $\mathbf{P}(\eta_{\mathrm{l}} \in (t^*, 1];\ B)$. Let $\mathrm{l}^{(-1)}(u)$ be the inverse of the monotone continuous function $\mathrm{l}(t)$ on $(t^*, 1)$ and let τ_κ be a jump crossing the decreasing boundary $T\mathrm{l}(\frac{t}{T})$ for $t \in (t^*, 1]$. If $\mathrm{l}(1) < 0$, then the value $\kappa = 1$ is possible. In this case,

$$\mathbf{P}(\eta_{\mathrm{l}} \in (t^*, 1];\ B) = \mathbf{P}(\tau_1 > T\mathrm{l}^{(-1)}(0)) + \mathbf{P}(\eta_{\mathrm{l}} \in (t^*, 1];\ B, \kappa > 1),$$

where

$$\ln \mathbf{P}(\tau_1 > T\mathrm{l}^{-1}(0)) \leq -T\lambda_+^{(\tau_1)}\mathrm{l}^{(-1)}(0) + o(T),$$
$$\lambda_+^{(\tau_1)} = \sup\{\lambda : \mathbf{E}^{\lambda\tau_1} < \infty\},$$

and by the second condition in (4.6.4), for $t = \mathrm{I}^{(-1)}(0)$ we get

$$\lambda_+^{(\tau_1)} \geq D(0), \qquad \lambda_+^{(\tau_1)}\mathrm{I}^{(-1)}(0) \geq D(0)\mathrm{I}^{-1}(0) = tD\Big(\frac{\mathrm{I}(t)}{t}\Big) = D(\alpha^g).$$

Therefore,

$$\ln \mathbf{P}(\tau_1 > T\mathrm{I}^{(-1)}(0)) \leq -TD(\alpha^g) + o(T).$$

It remains to consider

$$\mathbf{P}(\eta_\mathrm{I} \in (t^*, 1];\ B, \kappa > 1).$$

If $(u, x) = (T_{\kappa-1}, Z(T_{\kappa-1}))$, $x < T\mathrm{I}(\frac{t}{T})$, and if $t \leq T$ is the point from which a horizontal jump of length τ_κ starts and crosses the decreasing boundary $T\mathrm{I}(t/T)$, $t \leq T$, then $\tau_\kappa \geq T\mathrm{I}^{(-1)}(x/T) - u$ and the distribution of the jump τ_κ is the conditional distribution

$$\mathbf{P}(\tau > v \,|\, \tau > T\mathrm{I}^{(-1)}(x/T) - u)$$

(the entire prehistory of the process $Z(t)$ up to $T_{\kappa-1}$ obviously does not affect the distribution of τ_κ). But, for any s,

$$\mathbf{P}(\tau \geq v \,|\, \tau \geq s) = \left\{\begin{matrix} 1 & \text{for} \quad v \leq s \\ \dfrac{\mathbf{P}(\tau \geq v)}{\mathbf{P}(\tau \geq s)} & \text{for} \quad v < s \end{matrix}\right\} \geq \mathbf{P}(\tau \geq v),$$

so that $\tau_\kappa \underset{p}{\geq} \tau$. Therefore, assuming again without loss of generality that condition (4.6.33) holds, we obtain

$$\begin{aligned}\mathbf{P}(\eta_\mathrm{I} \in (t^*, 1], B, \kappa > 1) &= \frac{\mathbf{P}(\eta_l \in (t^*, 1], B, \tau_\kappa \geq 1, \kappa > 1)}{\mathbf{P}(\tau_\kappa \geq 1 \mid \eta_l \in (t^*, 1], B, \kappa > 1)} \\ &\leq \frac{1}{p}\mathbf{P}(\eta_\mathrm{I} \in (t^*, 1], B, \tau_\kappa \geq 1, \kappa > 1). \qquad (4.6.37)\end{aligned}$$

The subsequent argument is quite similar to that given at Step II of the proof. The event under the probability sign on the right-hand side of (4.6.37) implies the union of the events

$$\bigcup_{Tt^* \leq k \leq T} \Big\{z_T\Big(\frac{k}{T}\Big) \geq \mathrm{I}\Big(\frac{k}{T}\Big) - \frac{b^*}{T}, \kappa > 1\Big\},$$

where $b^* = -\mathrm{I}'(1) > 0$. Therefore, by the definition of level lines and using Theorem 3.4.1, we get

$$\begin{aligned}\mathbf{P}(\eta_\mathrm{I} \in (t^*, 1], B, \kappa > 1) &\leq \frac{1}{p} \sum_{Tt^* \leq k \leq T} \mathbf{P}\Big(z_T\Big(\frac{k}{T}\Big) \geq \mathrm{I}\Big(\frac{k}{T}\Big) - \frac{b_*}{T}\Big) \\ &\leq \frac{(1-t^*)T}{p} e^{-TD(\alpha^g)+o(T)} = e^{-TD(\alpha^g)+o(T)}. \qquad (4.6.38)\end{aligned}$$

Combining the above results, we get the required upper estimate

$$\mathbf{P}(\eta_g \leq 1) \leq \mathbf{P}(\eta_\mathrm{I} \leq 1) \leq e^{-TD(\alpha^g)+o(T)}. \qquad (4.6.39)$$

The *lower estimate* is almost obvious. We have

$$\{z_T(t_g) \geq g(t_g)\} \subset \{\eta_g \leq 1\},$$

and hence, by Theorem 3.4.1,

$$\mathbf{P}(\eta_g \leq 1) \geq \mathbf{P}(z_T(t_g) \geq g(t_g)) = e^{-Tt_g D\left(\frac{g(t_g)}{t_g}\right)+o(T)}$$
$$= e^{-Tt_g D\left(\frac{l(t_g)}{t_g}\right)+o(T)} = e^{-TD(\alpha^g)+o(T)}.$$

This proves (4.6.26).

Theorem 4.6.4 is proved. □

4.7 Large Deviation Principles for the Second Boundary Crossing Problem

4.7.1 Most Probable (Shortest) Trajectories

To begin with, in addition to the results given in 4.3.2, we consider in more detail the problem of most probable trajectories.

By the *most probable (shortest) trajectories* of a normalized CRP $z_T(t)$ connecting the point $(0,0)$ with small neighborhood $(1,(\alpha)_\varepsilon)$ of the point $(1,\alpha)$, we mean trajectories (functions) $f(t) \in \mathbb{C}_a$, $f(0) = 0$, $f(1) = \alpha$, such that

$$-I_P(f) := \lim_{T\to\infty} \frac{1}{T} \ln \mathbf{P}(z_T(\cdot) \in (f)_\varepsilon) = -\widehat{D}(\alpha)$$
$$\left(= \lim_{T\to\infty} \frac{1}{T} \ln \mathbf{P}(z_T(1) \in (\alpha)_\varepsilon)\right), \tag{4.7.1}$$

where $\varepsilon = \varepsilon_T$ converges to 0 sufficiently slowly as $T \to \infty$.

In this section, we will prove the following assertion. By $f_\alpha(t)$ we denote the function

$$f_\alpha(t) := \alpha t, \qquad t \in [0,1],$$

and by $\mathcal{R}$, the set of α for which $\widehat{D}(\alpha) = D(\alpha)$ (so that $\mathcal{R} = \mathbb{R} \setminus (\beta_-, \beta_+)$; see Lemma 3.5.8). Next, by $\mathcal{L}$ we denote some interval (if it exists) on which the function $D(\alpha)$ is linear.

Theorem 4.7.1 *Let conditions* $[\lambda_+] \bigcup [\overline{\lambda}_+]$ *be met and let the admissible inhomogeneity conditions* (4.6.4) *be satisfied. Then*

(i) *If* $\alpha \in (\mathcal{R})$ *and* $\alpha \notin \mathcal{L}$ *for any interval* $\mathcal{L}$ *of linearity then the rectilinear trajectory* $f_\alpha(t)$ *is a unique shortest trajectory between* $(0,0)$ *and* $(1,\alpha)$.

(ii) *If* $\alpha \in (\mathcal{R}) \bigcap \mathcal{L}$ *for some* $\mathcal{L}$, *then, in addition, to* f_α, *there exists a continuum of shortest trajectories.*

(iii) *If* $\alpha \notin \mathcal{R}$, $t_\alpha \in (0,1)$, *then* $\alpha/v \in \mathcal{R}$ *for any* $v \leq t_\alpha$. *If, besides,* $\alpha/t_\alpha \notin \mathcal{L}$, *then there exists a unique shortest trajectory* g_{t_α} *(which will be constructed below; see* (4.7.4)*) which is different from* f_α.

(iv) *If* $\alpha \notin \mathcal{R}$, $t_\alpha \in (0,1)$, $\frac{\alpha}{t_\alpha} \in \mathcal{L}$, *then, in addition to* g_{t_α}, *there exists a continuum of shortest trajectories.*

(v) *If* $\alpha = 0 \notin \mathcal{L}$, *then the trajectory* $f_0(t)$ *is a unique shortest trajectory.*

Remark 4.7.2 If $\lambda_+ < D(0)$, $\alpha > 0$, $\mathbf{P}(\zeta > 0) > 0$, then $t_\alpha \in (0,1)$. This is so because in this case $\beta_+ \in (0,\infty)$, $t_\alpha = \frac{\alpha}{\beta_+}$ (see (3.5.54)). A similar result holds in the case $\alpha < 0$, $\mathbf{P}(\zeta < 0) < 0$. So, for ζ that take values of both signs and for $\alpha \neq 0$, we always have $t_\alpha \in (0,1)$.

From assertions (iii), (iv) of the theorem it follows that (in contrast to random walks) a shortest trajectory for a CRP cannot be rectilinear.

Proof of Theorem 4.7.1 Let conditions (4.6.4) be met. Then from Theorems 4.2.1, 4.3.1 it follows that

$$I_P(f) = I(f) \quad \text{under } [\boldsymbol{\lambda}_+], \tag{4.7.2}$$

$$I_P(f) = \widehat{I}(f) \quad \text{under } [\overline{\boldsymbol{\lambda}}_+] \tag{4.7.3}$$

(see (4.7.1), (4.2.3), (4.3.5)). We first consider the second case (4.7.3).

(i) If $\alpha \in (\mathcal{R})$, $\alpha \notin \mathcal{L}$, then the set of values of $g'(t)$ for $g \in \mathbb{C}_a$, $g \neq f_\alpha$, $g(0) = 0$, $g(1) = \alpha$, will define the domain of strict convexity of the function D, and for $t^{1]} = 1$ (see (4.3.3)) we have

$$\widehat{I}(g) = \int_0^1 D(g'(t))\,dt > D(\alpha) = \widehat{I}(f_\alpha).$$

If $t^{1]} \in (0,1)$, then $\alpha \in (\mathcal{R})$ implies that $t_\alpha = 1$ (see the proof of assertion (i) of Theorem 3.5.9), and

$$\widehat{I}(g) = \int_0^{t^{1]}} D(g'(t))\,dt + (1-t^{1]})\lambda_+ \geq t^{1]} D\Big(\frac{\alpha}{t^{1]}}\Big) + (1-t^{1]})\lambda_+ > \widehat{D}(\alpha) = D(\alpha).$$

We have $\widehat{I}(f_\alpha) = D(\alpha)$ ($\alpha \neq 0$, because $\alpha \in (\mathcal{R})$), and therefore, the first assertion of the theorem follows from the above inequalities and (4.7.1).

(ii) If $\alpha \in (\mathcal{R}) \bigcap \mathcal{L}$, then there exist points $\alpha_1 < \alpha < \alpha_2$, $\alpha_1 \neq 0$, from $(\mathcal{R}) \bigcap \mathcal{L}$ such that, for $p := \frac{\alpha - \alpha_1}{\alpha_2 - \alpha_1}$, for the function

$$g(t) = \begin{cases} \alpha_2 t & \text{for } t \leq p, \\ \alpha - \alpha_1(1-t) & \text{for } t \in [p,1], \end{cases}$$

we have

$$\widehat{I}(g) = \int_0^1 D(g'(t))dt = pD(\alpha_2) + (1-p)D(\alpha_1) = D(\alpha) = \widehat{I}(f_\alpha).$$

This implies assertion (ii).

(iii) Let us prove the first part of assertion (iii).

If $\lambda_+ < D(0)$, $\alpha \notin \mathcal{R}$, $t_\alpha \in (0,1)$, then

$$\widehat{D}(\alpha) = \mathbf{D}(t_\alpha, \alpha) - (1-t_\alpha)\lambda_+ < D(\alpha).$$

The point t_α is characterized as follows:

$$\mathbf{D}'_{(1)}(t_\alpha, \alpha) = \lambda_+, \qquad \mathbf{D}'_{(1)}(t, \alpha) \lessgtr \lambda_+ \quad \text{for} \quad t \lessgtr t_\alpha.$$

Hence, by (3.5.54) (see Theorem 3.5.9),

$$\frac{\alpha}{t_\alpha} = \beta_\pm \lessgtr 0 \quad \text{for} \quad \alpha \lessgtr 0.$$

Therefore, for $v \le t_\alpha$, we have

$$\frac{\alpha}{v} \in \mathcal{R} = \mathbb{R} \setminus (\beta_-, \beta_+).$$

This proves the first part of assertion (iii).

Next, consider the trajectory

$$g(t) = g_v(t) = \begin{cases} \dfrac{\alpha t}{v} & \text{for} \quad t \le v, \\ \alpha & \text{for} \quad t \in [v, 1]. \end{cases} \tag{4.7.4}$$

For this trajectory, we have $t^{1]} = v$,

$$\widehat{I}(g_v) = vD\left(\frac{\alpha}{v}\right) + (1 - v)\lambda_+ = \mathbf{D}(v, \alpha) + (1 - v)\lambda_+.$$

Hence, for $v = t_\alpha$,

$$\widehat{I}(g_{t_\alpha}) = \mathbf{D}(t_\alpha, \alpha) + (1 - t_\alpha)\lambda_+ = \widehat{D}(\alpha) < D(\alpha) = \widehat{I}(f_\alpha).$$

For $v \neq t_\alpha$, we have

$$\widehat{I}(g_v) > \widehat{D}(\alpha).$$

Thus, the shortest trajectory has on $[t_\alpha, 1]$ a rectilinear horizontal part at level α.

If $\frac{\alpha}{t_\alpha} \notin \mathcal{L}$, then as in Step (i) of the proof we verify that the shortest trajectory connecting $(0, 0)$ and (t_α, α) is rectilinear and unique (recall that $\frac{\alpha}{t_\alpha} \in \mathcal{R}$).

(iv) If $\frac{\alpha}{t_\alpha} \in \mathcal{L}$, then, as in Step (ii) of the proof, we verify that the shortest trajectory $g_{t_\alpha}(t)$ is not unique.

(v) If $\alpha = 0$, then in the case $\lambda_+ < D(0)$ we have $t^{1]} = 0$ and $I_P(f_0) = \lambda_+ = \widehat{D}(0)$. In the case $\lambda_+ \ge D(0)$,

$$I_P(f_0) = D(0) = \widehat{D}(0).$$

Thus, f_0 is a shortest trajectory. We have $0 \notin \mathcal{L}$, and hence, for any function $f \neq f_0$, we see that $I_P(f) > I_P(f_0)$ and the shortest trajectory f_0 is unique.

In case (4.7.2), the proof of the theorem proceeds as above, but simpler. For example, $\mathcal{R} = \mathbb{R}$ for $\lambda_+ \ge D(0)$ and the possibility $\alpha \notin \mathcal{R}$ is excluded ($\widehat{D}(\alpha) = D(\alpha)$ for all α).

The theorem is proved. □

4.7.2 The Second Boundary Crossing Problem

In this section, we will find the asymptotics of

$$\frac{1}{T}\ln \mathbf{P}(z_T \in B_{g_-g_+})$$

as $T \to \infty$, where the set $B_{g_-g_+}$ is defined in (4.6.3). We assume that the upper boundary $g_+(t)$ is lower-continuous and that the lower boundary $g_-(t)$ is upper-continuous. The conditions imposed on the distribution of the CRP $Z(t)$ in this section are more restrictive than in §4.6. In particular, they imply the fulfillment of condition $[\boldsymbol{\lambda}_+]$.

Consider the set $B_{g_-g_+}$ and find in it the most probable trajectory – this is a function $f_{g_-g_+}(t)$ such that $\inf_{f\in B_{g_-g_+}} I(f)$ (see (4.7.1), (4.7.2)) is attained at $f = f_{g_-g_+}$. We can construct this function in two stages using the following properties of the convex function $D(\alpha)$.

(a) For every function $f \in \mathbb{C}_a$ passing through the points (t_1,α_1), (t_2,α_2), $t_2 > t_1$, we have

$$\int_{t_1}^{t_2} D(f'(t))\,dt \ge (t_2-t_1)\,D\Big(\frac{\alpha_2-\alpha_1}{t_2-t_1}\Big), \tag{4.7.5}$$

that is, under the conditions of Theorem 4.1.2, the most probable ("shortest") trajectory among all the trajectories connecting the points (t_1,α_1) and (t_2,α_2), is rectilinear.

(b) Among all the trajectories connecting the points (t_1,α_1), (t_2,α), where α varies, the most probable one is the line connecting the points (t_1,α_1) and $(t_2,\alpha_1+a(t_2-t_1))$.

The first stage of the construction of the trajectory $f_{g_-g_+}$ is as follows. We fix some point $\alpha \in [g_-(1), g_+(1)]$ and consider the "tense string" $g_\alpha(t)$ connecting the points $(0,0)$ and $(1,\alpha)$ and not crossing the boundaries $g_\pm(t)$.

At the second stage, we minimize with respect to α the value $I(g_\alpha)$ (which describes the asymptotics of $-\ln \mathbf{P}(z_T \in (g_\alpha)_\varepsilon)$ as $T \to \infty$, $\varepsilon = \varepsilon_T = \overline{o}(1)$), that is, we search for α_1 for which $I(g_{\alpha_1}) = \min_\alpha I(g_\alpha)$, so that $f_{g_-g_+} = g_{\alpha_1}$. It is clear that if $\alpha_1 \in (g_-(1),\ g_+(1))$, then $g'_{\alpha_1}(1-0) = a$,

$$I(g_{\alpha_1}) = I(B_{g_-g_+}) = \int_0^{t_1} D(g'_{\alpha_1}(t))\,dt,$$

where $t_1 = \max\{t : g_{\alpha_1}(t) = g_-(t) \text{ or } g_{\alpha_1}(t) = g_+(t)\} < 1$.

It is also clear that every deviation from the function $f_{g_-g_+}$ in the domain $B_{g_-g_+}$ (its warping in this domain) increases the value of the functional $I(f)$ in comparison with $I(f_{g_-g_+})$.

Theorem 4.7.3 *Let $B_{g_-g_+}$ be the set defined in* (4.6.3), *let functions $g_\pm \in \mathbb{D}$ have at most finitely many discontinuity points, and let g_- be upper-continuous and g_+ be lower-continuous.*

Assume further that the conditions $\mathcal{A}^{\le 0} \subset [\mathcal{A}_1]$ and $\lambda_+^{(\tau_1)} \ge D(0)$ are met and at least one of the conditions holds:

(1) $\xi_\infty \in [\mathbf{C}]$, $\mathcal{A}^{\le 0} \subset [\mathcal{A}_\infty]$ (*the conditions of Theorem* 4.1.2);

(2) *conditions* (4.1.10) *and* (4.1.11) *are met* (*the conditions of Lemma* 4.1.4);
(3) τ *and* ζ *are independent,* $\tau \in [\mathbf{C}_\infty]$;
(4) $\zeta \in [\mathbf{C}_\infty]$.
Then there exists

$$\lim_{T\to\infty} \frac{1}{T} \ln \mathbf{P}(z_T \in B_{g_-g_+}) = -I(B_{g_-g_+}) = -I(f_{g_-g_+}). \tag{4.7.6}$$

If the function $f(t) \equiv 0$ does not belong to $B_{g_-g_+}$, then the condition $\lambda_+^{(\tau_1)} \geq D(0)$ is apparently superfluous.

As noted above, each of conditions (1)–(4) of Theorem 4.7.3 implies condition $[\boldsymbol{\lambda}_+]$.

Proof The proof is divided into several steps.

(i) Assume first that the functions $g_\pm$ are continuous.

Upper estimate. Given an integer K, consider the uniform partition of the interval $[0,1]$ into K intervals (t_{j-1}, t_j), $t_j = j/K$, $j = 1, \ldots, K$. Let

$$B_K := \left\{ f \in \mathbb{D} : \; f(t_j) \in \left[g_-(t_j), g_+(t_j)\right] \quad \text{for all} \;\; j = 1, \ldots, K \right\}.$$

By Theorem 4.1.2,

$$\lim_{T\to\infty} \frac{1}{T} \ln \mathbf{P}(z_T \in B_K) = -I(B_K L_K), \tag{4.7.7}$$

where L_K the set of continuous polygons with the nodes at t_j, $j = 0, \ldots, K$.

Since $B_{g_-g_+} \subset B_K$, we have

$$\mathbf{P}(z_T \in B_{g_-g_+}) \leq \mathbf{P}(z_T \in B_K) \tag{4.7.8}$$

for any K, and now by (4.7.7),

$$\overline{\lim_{T\to\infty}} \frac{1}{T} \ln \mathbf{P}(z_T \in B_{g_-g_+}) \leq -I(B_K L_K). \tag{4.7.9}$$

The lower boundary $g_-^{(K)}$ of the set $B_K L_K$ is a continuous polygon connecting the points $(t_j, g_-(t_j))$, $j = 0, \ldots, K$. It is clear that $g_-^{(K)} \to g_-$ in the uniform metric as $K \to \infty$. A similar convergence holds for the upper boundary: $g_+^{(K)} \to g_+$ as $K \to \infty$. Therefore, if we denote by f^K the most probable trajectory in the set $B_K L_K$, then $f^K \to f_{g_-g_+}$ in the uniform metric as $K \to \infty$. Besides, the tense strings f^K, $f_{g_-g_+}$ are functions from $\mathbb{C}_a$ and their derivatives converge. Therefore, by the properties of the functional I,

$$I(B_K L_K) = I(f^K) \to I(f_{g_-g_+}) = I(B_{g_-g_+}) \tag{4.7.10}$$

as $K \to \infty$ (see also [23, §4.2]). From (4.7.10), (4.7.8) we get

$$\overline{\lim_{T\to\infty}} \frac{1}{T} \ln \mathbf{P}(z_T \in B_{g_-g_+}) \leq -I(B_{g_-g_+}). \tag{4.7.11}$$

Lower estimate. Let $\delta > 0$ be a small number. Consider the functions

$$g_-^\delta(t) := g_-(t) + \delta, \qquad g_+^\delta(t) := g_+(t) - \delta$$

and the closed set

$$B^{\delta}_{g_-g_+} \subset B_{g_-g_+}$$

of functions from $\mathbb{D}$ lying between the boundaries $g^{\delta}_{\pm}$ (the boundaries can be touched). Let f^{δ}_K be the most probable trajectory in $B^{\delta}_{g_-g_+} L_K$. Then $(f^{\delta}_K)_\delta \subset B_{g_-g_+}$. Therefore,

$$\begin{aligned} \{z_T \in (f^{\delta}_K)_\delta\} &\subset \{z_T \in B_{g_-g_+}\}, \\ \mathbf{P}(z_T \in B_{g_-g_+}) \geq \mathbf{P}(z_T &\in (f^{\delta}_K)_\delta) \geq \mathbf{P}(z_T \in (f^{\delta}_K)_\varepsilon), \end{aligned} \tag{4.7.12}$$

where $\varepsilon = \varepsilon_T \to 0$ as $T \to \infty$. If ε_T tends to 0 sufficiently slowly, then by Theorem 4.1.2 and using (4.7.12) we get

$$\varliminf_{T\to\infty} \frac{1}{T} \ln \mathbf{P}(z_T \in B_{g_-g_+}) \geq \lim_{T\to\infty} \frac{1}{T} \mathbf{P}(z_T \in (f^{\delta}_K)_\varepsilon) = -I(f^{\delta}_K). \tag{4.7.13}$$

As before, we establish that $f^{\delta}_K \to f_{g_-g_+}$ in the uniform metric as $\delta \to 0$, $K \to \infty$, and hence

$$I(f^{\delta}_K) \to I(f_{g_-g_+}) = I(B_{g_-g_+}).$$

Since δ and K are arbitrary and since the left-hand side of (4.7.13) is independent of δ and K, we get

$$\varliminf_{T\to\infty} \frac{1}{T} \ln \mathbf{P}(z_T \in B_{g_-g_+}) \geq -I(B_{g_-g_+}).$$

This together with (4.7.11) justifies (4.7.6).

(ii) Assume now that g_+ has a unique finite discontinuity at the point $u \in (0, 1)$. For definiteness, we assume that

$$g_+(u) = g_+(u-0) < g_+(u+0) < \infty.$$

Consider the new upper boundaries

$$g^{-\delta}_+(t) := \begin{cases} g_+(t) & \text{for } t \notin [u-\delta, u], \\ g_+(u-\delta) + (t-u+\delta)\dfrac{(g_+(u+0) - g_+(u-\delta))}{\delta} & \text{for } t \in [u-\delta, u]; \end{cases}$$

and

$$g^{+\delta}_+(t) := \begin{cases} g_+(t) & \text{for } t \notin [u, u+\delta], \\ g_+(u) + (t-u)\dfrac{(g_+(u+\delta) - g_+(u))}{\delta} & \text{for } t \in [u, u+\delta]. \end{cases}$$

It is clear that $g^{-\delta}_+(t) \geq g_+(t) \geq g^{+\delta}_+(t)$ for all t and that the new boundaries are continuous. Hence by Step (i),

$$\begin{aligned} \varlimsup_{T\to\infty} \frac{1}{T} \ln \mathbf{P}(z_T \in B_{g_-g_+}) &\leq \lim_{T\to\infty} \frac{1}{T} \ln \mathbf{P}(z_T \in B_{g_-g}) = -I(f_{g_-g}) \text{ for } g = g^{-\delta}_+; \\ \varliminf_{T\to\infty} \frac{1}{T} \ln \mathbf{P}(z_T \in B_{g_-g_+}) &\geq \lim_{T\to\infty} \frac{1}{T} \ln \mathbf{P}(z_T \in B_{g_-g}) = -I(f_{g_-g}) \text{ for } g = g^{+\delta}_+. \end{aligned}$$

It is easily seen that

$$f_{g-g} \to f_{g-g_+} \quad \text{for } g = g_+^{\pm\delta} \text{ and } \delta \to 0$$

in the uniform metric, so that

$$I(f_{g-g}) \to I(f_{g-g_+}),$$

which implies (4.7.6).

(iii) Consider now the case of infinite discontinuity at the point u: $g_+(u+0) = \infty$. For any (large) N, we put

$$\Delta_N = \begin{cases} \inf\{\Delta > 0 : g_+(u+\Delta) < N\} & \text{if } g_+(1) < \infty, \\ 1-u \quad \text{if} \quad g_+(v) = \infty & \text{for all } v \in (u,1]; \end{cases}$$

$$\overline{z}(N) = \max_{0 \le v \le \Delta_N} z_T(u+v).$$

Then

$$\mathbf{P}(z_T \in B_{g-g_+}) = \mathbf{P}(z_T \in B_{g-g_+}, \overline{z}(N) \le N) + \mathbf{P}(z_T \in B_{g-g_+}, \overline{z}(N) > N). \quad (4.7.14)$$

Let us estimate the last term on the right of (4.7.14). It is clear that this term is majorized by

$$\mathbf{P}(\overline{z}(N) \ge N) \le \mathbf{P}(\overline{Z}_0(\Delta_N T) \ge (N - g_+(u))T),$$

where Z_0 is a homogeneous CRP. Without loss of generality we assume that $a = 0$. By Theorem 4.6.3, for sufficiently large N and as $T \to \infty$,

$$\ln \mathbf{P}(\overline{z}(N) \ge N) \le -\Delta_N T D\Big(\frac{N}{2\Delta_N}\Big) + o(T), \quad (4.7.15)$$

where by Theorem 3.5.2 (see (3.5.21)), for $\mu^+ < \infty$ and sufficiently large N,

$$\Delta_N D\Big(\frac{N}{2\Delta_N}\Big) \ge \frac{\mu^+ N}{3}.$$

Hence by choosing an appropriate N we can make the left-hand side in (4.7.15) smaller than $-TI(g_{g-g_+})$. This is certainly so for $\mu^+ = \infty$.

The first term on the right-hand side of (4.7.14) is again the probability of the form $\mathbf{P}(z_T \in B_{g-g_+^N})$, but for the new upper boundary

$$g_+^N(t) = \min(g_+(t), N),$$

which has a finite discontinuity at the point u. It is clear that, for a sufficiently large N, the functions $f_{g-g_+^N}$ and f_{g-g_+} are equal. Hence from Step (ii) and from (4.7.14) (in which, according to the above the second term on the right is $o(\exp\{-TD(f_{g-g_+})\})$ we get (4.7.6).

(iv) A similar analysis applies to the cases when the upper boundary g_+ has several jumps or when the lower boundary g_- has jumps.

Theorem 4.7.3 is proved.

If conditions $\zeta \in [\mathbf{C}_\infty]$, $\zeta_1 \in [\mathbf{C}_\infty]$ are met, then (4.7.6) can also be derived from Theorem 4.5.1 (see (4.5.1) and (4.5.2)).

4.8 Moderately Large Deviation Principles for Trajectories of Compound Renewal Processes

4.8.1 Main Results

Consider the "asymptotically centered" processes

$$Y(t) - at = Z_{\eta(t)} - aT_{\eta(t)} + a\chi(t),$$
$$Z(t) - at = Z_{\nu(t)} - aT_{\nu(t)} - a\gamma(t).$$

In terms of the random variables

$$\xi = \zeta - a\tau, \qquad \xi_i = \zeta_i - a\tau_i, \qquad S_n = \sum_{i=1}^{n} \xi_i = Z_n - aT_n,$$

these processes can be written as

$$Y(t) - at = S_{\eta(t)} + a\chi(t), \qquad Z(t) - at = S_{\nu(t)} - a\gamma(t). \tag{4.8.1}$$

Let, as before,

$$\sigma_\xi^2 = \mathbf{E}\xi^2, \qquad \sigma^2 = \frac{\sigma_\xi^2}{a_\tau}.$$

In this section, as in §1.6, we are concerned with the limit laws for the normalized processes

$$y_T(u) := \frac{Y(uT) - auT}{\sigma\sqrt{T}}, \quad z_T(u) := \frac{Z(uT) - auT}{\sigma\sqrt{T}}, \tag{4.8.2}$$
$$u \in [0, 1] \quad \text{as } T \to \infty,$$

but now in the zone of moderate large deviations. We will establish the moderately large deviation principle (MLDP) under the assumption that either of the following two moment conditions is satisfied:

condition $[\mathbf{C}]$ or
broader moment condition $[\mathbf{C}^V]$ describing distributions with "heavier tails."

In order to state condition $[\mathbf{C}^V]$, we introduce the class $\mathcal{L}_\beta$, $\beta \in (0, 1)$, of functions $V(t)$ with the following properties:

1. *$V(t) = t^\beta l(t)$ is a regularly varying function ($l(t)$ is a slowly varying function (SVF) as $t \to \infty$);*

2. *if $t \to \infty$, $v = o(t)$, then*

$$V(t + v) - V(t) = \beta v \frac{V(t)}{t}(1 + o(1)) + o(1).$$

Now the aforementioned condition that is broader than [**C**] is as follows:

$$[\mathbf{C}^V] \qquad \mathbf{P}(\tau \ge t) \le e^{-V(t)}, \qquad \mathbf{P}(|\xi| \ge t) \le e^{-V(t)} \quad \text{for} \quad t > 0,$$

where $V(t) \in \mathcal{L}_\beta$, $\beta \in (0,1)$.

The class of distributions $\mathcal{S}e$ such that $\mathbf{P}(\xi \ge t) = e^{-V(t)}$, $V \in \mathcal{L}_\beta$ is called the *class of semiexponential distributions* (see, e.g., [38], Chapter 5). For this class, the large deviation probabilities were studied in detail for the sums S_n (see, e.g., Chapter 5 in [38], Chapter 5 in [23] and [42]). There exist several other distribution classes that are close to $\mathcal{S}e$ and for which their large deviation probabilities for S_n have also been studied (see, e.g., [76], [116], [3], [108], [58], [94], and [105]).

Under condition [**C**] we consider the deviations $x = x(T)$ for processes (4.8.2) specified by

$$x \to \infty, \qquad x = o(\sqrt{T}) \quad \text{as} \quad T \to \infty. \tag{4.8.3}$$

If condition [**C**] is not satisfied, but if condition [$\mathbf{C}^V$] is met, then we consider the more narrow class of moderately large deviations $x = x(T)$ with the property

$$x \to \infty, \qquad x = o(\widehat{x}(T)) \quad \text{as} \quad T \to \infty, \tag{4.8.4}$$

where the regularly varying sequence $\widehat{x}(T)$ grows slower than $\sqrt{T}$ as $T \to \infty$. This sequence is defined as follows. Given a function $V \in \mathcal{L}_\beta$ from condition [$\mathbf{C}^V$], we consider the function

$$v(t) := V(t)t^{-2} = t^{\beta-2} l(t) \tag{4.8.5}$$

and define

$$\widehat{x}(T) = v^{(-1)}(1/T)T^{-1/2}, \tag{4.8.6}$$

where $v^{(-1)}(u)$ is the generalized inverse function of $v(t)$:

$$v^{(-1)}(u) := \sup\{t \ge 0 : \ v(t) \ge u\}. \tag{4.8.7}$$

It is known (see, e.g., [38], Theorem 1.1.3) that $v^{(-1)}(1/T)$ has the form

$$v^{(-1)}\left(\frac{1}{T}\right) = T^{1/(2-\beta)} l_1(T) = o(T),$$

so that

$$\widehat{x}(T) = T^{\beta/2(2-\beta)} l_1(T),$$

where $l_1(T)$ is an SVF as $T \to \infty$. In the special case when the SVF $l(t)$ satisfies the additional condition

$$l(tl^{1/(2-\beta)}(t)) \sim l(t) \quad \text{as} \quad t \to \infty,$$

the SVF $l_1(t)$ has the form

$$l_1(t) \sim l^{1/(2-\beta)}(t^{1/(2-\beta)}) \quad \text{as} \quad t \to \infty,$$

so in this case, one has

$$\widehat{x}(T) = T^{\frac{\beta}{2(2-\beta)}}\, l^{\frac{1}{2-\beta}}(T^{\frac{1}{2-\beta}}) \quad \text{as} \quad T \to \infty. \tag{4.8.8}$$

Thus, we assume that the following condition is met:

either condition [**C**] or condition [$\mathbf{C}^V$] is satisfied, and the deviations of $x = x(T)$ are such that, as $T \to \infty$,

$$x \to \infty, \;\; x(T) = \begin{cases} o(\sqrt{T}) & \text{if condition } [\mathbf{C}] \text{ is met;} \\ o(\widehat{x}(T)) & \text{if condition } [\mathbf{C}^V] \text{ is met.} \end{cases} \tag{4.8.9}$$

Consider the measurable space $(\mathbb{D}, \mathfrak{B}(\mathbb{D}, \rho))$ of functions without discontinuities of the second kind endowed with the uniform metric ρ and the Borel σ-algebra.

We set

$$I_0(f) := \begin{cases} \frac{1}{2}\int_0^1 (f'(t))^2\, dt & \text{if} \quad f(0) = 0, \quad f \in \mathbb{C}_a; \\ \infty, & \text{otherwise,} \end{cases}$$

where $\mathbb{C}_a \subset \mathbb{C}$ is the space of absolutely continuous functions. The functional $I_0(f)$ has the following properties: $I_0(f)$ *is convex, lower semicontinuous*, and *for any* $v \geq 0$, *the sets* $\{f : I(f) \leq v\}$ *are compact in* $(\mathbb{D}, \rho)$ (for more details, see, e.g., §4.2 in [23]). Let $(f)_\varepsilon$ be the ε-neighborhood of $f \in \mathbb{D}$ in the uniform metric.

Theorem 4.8.1 (Local MLDP for the trajectories of y_T, z_T) *Let τ and ξ satisfy condition* [**C**] *or condition* [$\mathbf{C}^V$]. *Then, for any function $f \in \mathbb{D}$, $f(0) = 0$, there exists a sequence $\varepsilon = \varepsilon_T$ in the deviation zone* (4.8.9) *that tends sufficiently slowly to* 0 *as $T \to \infty$ and such that*

$$\lim_{T\to\infty} x^{-2} \ln \mathbf{P}(y_T \in x(f)_\varepsilon) = -I_0(f). \tag{4.8.10}$$

The same assertion also holds for the processes z_T.

Now we will turn to the "integral" MLDP for y_T, z_T. Given a set $B \in \mathfrak{B}(\mathbb{C}, \rho)$, we denote by $[B]$ and (B), respectively, its closure and interior in the uniform metric. We put

$$I_0(B) := \inf_{f \in B} I_0(f).$$

Theorem 4.8.2 (MLDP for the trajectories of y_T, z_T) *Let τ and ξ satisfy condition* [**C**] *or condition* [$\mathbf{C}^V$]. *Then, in the deviation zone* (4.8.9), *for any $B \in \mathfrak{B}(\mathbb{D}, \rho)$,*

$$\underline{\lim}_{T\to\infty} x^{-2} \ln \mathbf{P}(y_T \in xB) \geq -I_0((B)). \tag{4.8.11}$$

$$\overline{\lim}_{T\to\infty} x^{-2} \ln \mathbf{P}(y_T \in xB) \leq -I_0([B]). \tag{4.8.12}$$

The same assertions also hold for the processes z_T.

It is easily seen that Theorem 4.8.2 implies the following MLDP in the state space of the processes y_T, z_T.

Corollary 4.8.3 *Let τ and ξ satisfy condition* **[C]** *or* **[C^V]**. *Then, for any $r \in \mathbb{R}$, there exists an $\varepsilon = \overline{o}(1)$ as $T \to \infty$ such that, in the deviation zone* (4.8.9),

$$\lim_{T\to\infty} x^{-2} \ln \mathbf{P}(y_T(1) \in x(r)_\varepsilon) = -\frac{r^2}{2}. \tag{4.8.13}$$

The same assertion also holds for the processes z_T.

Proof For $r > 0$, for the set

$$B_r := \{f \in \mathbb{D} : f(1) \geq r, f(0) = 0\}$$

and for the function

$$f_r(t) = rt \quad \text{for} \quad t \in [0, 1],$$

we have

$$I_0((B_r)) = I_0([B_r]) = I_0(f_r) = \frac{r^2}{2},$$

and so, by Theorem 4.8.2,

$$\lim_{T\to\infty} x^{-2} \ln \mathbf{P}(y(1) \geq zx) = -\frac{r^2}{2}.$$

Hence, for any $\varepsilon \in (0, 1)$,

$$\mathbf{P}(y_T(1) \in x(r)_\varepsilon) = \mathbf{P}(y_T(1) > x(r-\varepsilon)) - \mathbf{P}(y_T(1) \geq x(r+\varepsilon))$$
$$= \exp\left\{-\frac{x^2(r-\varepsilon)^2}{2} + o(x^2)\right\}$$

as $x \to \infty$. The same equalities also hold true if $\varepsilon = \varepsilon_T \to 0$ sufficiently slowly as $T \to \infty$. This means that (4.8.13) is satisfied for such ε.

A similar argument shows that equality (4.8.13) also holds true for $r < 0$, $r = 0$. Corollary 4.8.3 is proved. □

From this corollary one can also derive the "ordinary" (integral) MLDP.

4.8.2 Proofs

Let us prove Theorems 4.8.1 and 4.8.2 for the processes y_T. At the end of this section we will discuss possible extensions of these results to the processes z_T.

Setting

$$S_t = S_{[t]}, \qquad s_T(u) = \frac{S_{uT}}{\sigma_\xi \sqrt{T}},$$

$$\theta_T(u) = \frac{\eta(uT)}{T}, \qquad \vec{s}_T(u) = s_T(\theta_T(u)), \quad u \geq 0,$$

we will assume without loss of generality that

$$a_\tau = 1, \quad \text{so that} \quad \sigma = \sigma_\xi.$$

Hence (see (4.8.1))

$$y_T(u) = \frac{S_{\eta(uT)}}{\sigma\sqrt{T}} + \frac{a\chi(uT)}{\sigma\sqrt{T}} = \vec{s}_T(u) + \frac{a\chi(uT)}{\sigma\sqrt{T}}. \tag{4.8.14}$$

For $s_T(u)$, the MLDP is known (see below). Hence, the MLDP for y_T will be established if we show that the processes $\vec{s}_T(u)$ and $s_T(u)$ are sufficiently close to each other and estimate the last term on the right of (4.8.14).

Our arguments here are based on the following three assertions concerning the MLDP for the random walks $s_T = s_T(u)$. These assertions follow from Theorems 5.2.1, 5.2.2 and Corollary 5.2.1 in the book [23].

Theorem 4.8.4 (Local MLDP for the trajectories of s_T) *Let ξ satisfy condition* [**C**] *or condition* [$\mathbf{C}^V$]. *Then, for any function $f \in \mathbb{D}$, $f(0) = 0$, there exists an $\varepsilon = \varepsilon_T = \overline{o}(1)$ as $T \to \infty$, such that, in the deviation zone* (4.8.9),

$$\lim_{T\to\infty} x^{-2} \ln \mathbf{P}(s_T \in x(f)_\varepsilon) = -I_0(f). \tag{4.8.15}$$

(It is worth noting that in [23] the probabilities of the form $\mathbf{P}(\frac{S_{uT}}{\sigma_\xi x} \in B)$ as $x \gg \sqrt{T}$ for the processes $\frac{S_{uT}}{\sigma_\xi x}$, which differ from the processes $s_T(u)$ in the normalization, were studied. This is why the statements in [23] are of a somewhat different form.)

Theorem 4.8.5 (MLDP for the trajectories of s_T) *Let ξ satisfy condition* [**C**] *or condition* [$\mathbf{C}^V$]. *Then, in the deviation zone* (4.8.9), *for any $B \in \mathfrak{B}(\mathbb{D}, \rho)$,*

$$\overline{\lim_{T\to\infty}}\, x^{-2} \ln \mathbf{P}(s_T \in xB) \le -I_0([B]), \tag{4.8.16}$$

$$\underline{\lim_{n\to\infty}}\, x^{-2} \ln \mathbf{P}(s_T \in xB) \ge -I_0((B)). \tag{4.8.17}$$

Theorem 4.8.6 *Let $s_T^*(u)$ be random processes in $(\mathbb{D}, \rho)$ satisfying the condition*

$$x^{-2} \ln \mathbf{P}(\rho(s_T, s_T^*) > xh) \to -\infty \quad as \quad T \to \infty \tag{4.8.18}$$

for any $h > 0$ in the deviation zone (4.8.9). *Then assertions* (4.8.15)–(4.8.17) *of Theorems* 4.8.4 *and* 4.8.5 *remain valid for the processes s_T^* as well.*

In view of Theorems 4.8.4–4.8.6, to prove Theorems 4.8.1 and 4.8.2 it suffices to check that (for $s_T^* = y_T$) one has

$$x^{-2} \ln \mathbf{P}(\rho(y_T, s_T) > xh) \to -\infty \quad \text{as} \quad T \to \infty. \tag{4.8.19}$$

We have

$$\rho(y_T, s_T) \le \rho(y_T, \vec{s}_T) + \rho(\vec{s}_T, s_T). \tag{4.8.20}$$

To estimate the terms on the right of (4.8.20), we require some auxiliary results. Given $b > 0$, $\delta > 0$, consider the event

$$A_{T,b,\delta} := \Big\{ \sup_{t \le (1+b)T} |\eta(t) - t| < \delta T \Big\} = \Big\{ \sup_{u \le 1+b} |\theta_T(u) - u| < \delta \Big\}.$$

Lemma 4.8.7 *Let τ satisfy condition* [**C**] *or condition* [$\mathbf{C}^V$]. *Then, for all sufficiently large T,*

$$\ln \mathbf{P}(\overline{A}_{T,b,\delta}) \le \begin{cases} -c\delta^2 T & \text{if [C] is satisfied,} \\ -\delta^{\beta'} V(T) & \text{if } [\mathbf{C}^V] \text{ is satisfied and any } \beta' > \beta, \end{cases} \tag{4.8.21}$$

where $c \ge 1/3\sigma_\tau^2$ for sufficiently small b and δ.

Proof We have

$$\mathbf{P}(\overline{A}_{T,b,\delta}) \le P_1 + P_2,$$

where

$$P_1 := \mathbf{P}\Big(\max_{t \le (1+b)T} (\eta(t) - t) > \delta T\Big), \quad P_2 := \mathbf{P}\Big(\min_{t \le (1+b)T} (\eta(t) - t) < -\delta T\Big).$$

It is easily seen that

$$P_1 \le \mathbf{P}\big(\min_{k \le T(1+b+\delta)} (T_k - k) < -\delta T\big) = \mathbf{P}\big(\max_{k \le T(1+b+\delta)} (k - T_k) > \delta T\big). \tag{4.8.22}$$

Since the random variable $1 - \tau$ is bounded from above, the deviation rate function $\Lambda^{(1-\tau)}(v)$ is well defined for $v > 0$, and

$$\Lambda^{(1-\tau)}(v) \sim \frac{v^2}{2\sigma_\tau^2} \quad \text{as} \quad v \to 0.$$

Therefore,

$$P_1 \le \exp\Big\{-T(1+b+\delta)\Lambda^{(1-\tau)}\Big(\frac{\delta}{1+b+\delta}\Big)\Big\} \le \exp\{-c_1\delta^2 T\},$$

where for small δ the constant c_1 is close to $1/2(1+b+\delta)\sigma_\tau^2$, and hence, for sufficiently small b and δ, it is greater than $1/3\sigma_\tau^2$.

Under condition [**C**], the estimate

$$P_2 \le \mathbf{P}\big(\max_{k \le T(1+b+\delta)} (T_k - k) > \delta T\big)$$

is obtained similarly.

Now let τ satisfy condition [$\mathbf{C}^V$]. Then by Corollary 5.2.1 (i) in [38] and from the properties of slowly varying functions (see, e.g., §1.1 in [38] or [11]), one has

$$P_2 \le c_2 T(1+b+\delta)\exp\{-V(\delta T)\},$$
$$\ln P_2 \le -(\delta T)^\beta l(\delta T) + \ln T + O(1) \le -\delta^{\beta'} V(T)$$

for any $\beta' > \beta$ and all sufficiently large T. Lemma 4.8.7 is proved. □

Lemma 4.8.8 *Let τ satisfy condition* [**C**] *or condition* [$\mathbf{C}^V$]. *Then, in the deviation zone* (4.8.9), *for all sufficiently large T,*

$$\ln \mathbf{P}\big(\sup_{t \le T} \chi(t) > hx\sqrt{T}; A_{T,0,\delta}\big) \le \begin{cases} -chx\sqrt{T} & \text{if [C] is met, } c = \text{const}, \\ -h^{\beta'} V(x\sqrt{T}) & \text{if } [\mathbf{C}^V] \text{ is met, for any } \beta' > \beta. \end{cases}$$

Proof Setting

$$G_h := \{ \sup_{t \le T} \chi(t) > hx\sqrt{T} \},$$

we have

$$\mathbf{P}(G_h A_{T,0,\delta}) \le \mathbf{P}\Big(\max_{k \le T(1+\delta)} \tau_k > hx\sqrt{T}\,\Big)$$

$$\le T(1+\delta)\mathbf{P}(\tau > hx\sqrt{T}) \le \begin{cases} T(1+\delta)\exp\{ -\Lambda^{(\tau)}(hx\sqrt{T})\} & \text{if } [\mathbf{C}] \text{ is met;} \\ T(1+\delta)\exp\{ -V(hx\sqrt{T})\} & \text{if } [\mathbf{C}^V] \text{ is met.} \end{cases}$$

Under $[\mathbf{C}]$, we have

$$\Lambda^{(\tau)}(v) \ge c_3 v - c_4 \quad \text{for some } c_3 > 0,\ c_4 > 0 \text{ and all } v.$$

Hence, for sufficiently large T,

$$\ln \mathbf{P}(G_h A_{T,0,\delta}) \le -c_5 hx\sqrt{T}, \qquad c_5 = \text{const} > 0.$$

If $[\mathbf{C}^V]$ is met, then for $h < 1$, any $\beta' > \beta$, and all sufficiently large T,

$$\ln \mathbf{P}(G_h A_{T,0,\delta}) \le -h^{\beta'} V(x\sqrt{T}).$$

Lemma 4.8.8 is proved. □

Let us return to the proof of Theorems 4.8.1 and 4.8.2. As was already pointed out, by Theorem 4.8.6, in which we put $s_T^* = y_T$, in order to prove Theorems 4.8.1 and 4.8.2 it suffices to verify (4.8.19) for any $h > 0$. In view of (4.8.20), this result follows if

$$x^{-2} \ln \mathbf{P}(\rho(y_T, \bar{s}_T) > xh) \to -\infty, \tag{4.8.23}$$

$$x^{-2} \ln \mathbf{P}(\rho(\bar{s}_T, s_T) > xh) \to -\infty \quad \text{as } T \to \infty. \tag{4.8.24}$$

(1) Let us prove the first of these relations. We have

$$\rho(y_T, \bar{s}_T) = \sup_{t \le T} \frac{|a|\chi(t)}{\sigma\sqrt{T}},$$

and so for $a = 0$ (4.8.23) clearly holds. If $a \ne 0$, then

$$\mathbf{P}(\rho(y_T, \bar{s}_T) > xh) \le \mathbf{P}(\overline{A}_{T,0,\delta}) + \mathbf{P}\big(\sup_{t \le T} \chi(t) > h_1 x\sqrt{T}, A_{T,0,\delta}\big), \tag{4.8.25}$$

where $h_1 = \frac{h\sigma}{|a|}$.

If condition $[\mathbf{C}]$ is met, then by Lemma 4.8.7 we have in (4.8.25) for any $\delta > 0$,

$$x^{-2} \ln \mathbf{P}(\overline{A}_{T,0,\delta}) \le -c\delta^2 T x^{-2} \to -\infty \quad \text{as } T \to \infty \tag{4.8.26}$$

since $x = o(\sqrt{T})$. By virtue of Lemma 4.8.8, the product of x^{-2} and the logarithm of the second term in (4.8.25) does not exceed

$$-ch\frac{\sqrt{T}}{x} \to -\infty \quad \text{as } T \to \infty.$$

This proves (4.8.23) for each fixed $\delta > 0$ under condition **[C]**. As a result, relation (4.8.23) is also preserved for $\delta = \delta_T$ that tend 0 sufficiently slowly as $T \to \infty$ (for example, for $\delta_T = (\frac{x^2}{T})^{1/3}$; see (4.8.26)).

Assume now that condition $[\mathbf{C}^V]$ is satisfied. Using Lemma 4.8.7, we have

$$x^{-2} \ln \mathbf{P}(\overline{A}_{T,0,\delta}) \le -\delta^{\beta'} x^{-2} V(T), \tag{4.8.27}$$

where for $x < \widehat{x}(T)$ by (4.8.8) the power part of the factor $x^{-2}V(T)$ on the right of (4.8.27) is greater than

$$T^{\beta - \frac{2}{2-\beta} + 1} = T^{\frac{\beta(1-\beta)}{2-\beta}} \to \infty \quad \text{as } T \to \infty.$$

It follows that the left-hand side (4.8.27) tends to $-\infty$ as $T \to \infty$ for any fixed $\delta > 0$ and if $\delta = \delta_T$ tends to 0 sufficiently slowly as $T \to \infty$.

For the second term in (4.8.25) we have by Lemma 4.8.8,

$$\begin{aligned} x^{-2} \ln \mathbf{P}(G_h A_{T,0,\delta}) &\le -x^{-2} h^{\beta'} V(x\sqrt{T}) = -T h^{\beta'} v(x\sqrt{T}) \\ &\ll -T h^{\beta'} v(\widehat{x}(T)\sqrt{T}) = -T h^{\beta'} v(v^{(-1)}(1/T)) = -h^{\beta'} \end{aligned} \tag{4.8.28}$$

as $T \to \infty$ (see (4.8.6)). This means that the left-hand side in (4.8.28) converges to $-\infty$ as $T \to \infty$. Thus (4.8.23) is proved.

(2) Let us now prove (4.8.24). On the set $A_{T,b,\delta}$, the trajectories of $s_T(u)$ and $\bar{s_T}(u) = s_T(\theta_T(u))$ differ only by a displacement of the argument by a quantity not exceeding δ. Therefore,

$$\rho(s_T, \bar{s_T}) \le \omega_\delta(s_T), \tag{4.8.29}$$

where $\omega_\delta(f)$ is the modulus of continuity of the function f,

$$\begin{aligned} \omega_\delta(f) &= \sup \left\{ |f(t) - f(t')| : \ |t - t'| \le \delta; \ t, t' \in [0,1] \right\}; \\ \mathbf{P}(\rho(\bar{s_T}, s_T) > xh) &\le \mathbf{P}(\overline{A}_{T,b,\delta}) + \mathbf{P}(\omega_\delta(s_T) > hx). \end{aligned} \tag{4.8.30}$$

Here, as before, the term $x^{-2} \ln \mathbf{P}(\overline{A}_{T,b,\delta})$ can be estimated using (4.8.26) and (4.8.27). Let us now consider the second term in (4.8.30). It is easily seen that if $1/\delta$ is integer, then

$$\{\omega_\delta(f) > r\} \subset \left\{ \max_{k \le 1/\delta} \sup_{u \le \delta} |f(k\delta + u) - f(k\delta)| > \frac{r}{2} \right\}.$$

Hence (we assume for simplicity that T and δT are both integers), we find

$$\mathbf{P}(\omega_\delta(s_T) > hx) \le \frac{1}{\delta} \mathbf{P}(\max_{k \le \delta T} |S_k| > h' x \sqrt{T}) \tag{4.8.31}$$

for $h' = \frac{h\sigma}{2}$.

If condition [**C**] is satisfied, then, for $x = o(\sqrt{T})$,

$$\mathbf{P}(\max_{k\le\delta T} S_k > h'x\sqrt{T}) \le \exp\Big\{-\delta T \Lambda^{(\xi)}\Big(\frac{h'x\sqrt{T}}{\delta T}\Big)\Big\}$$
$$\le \exp\Big\{-\delta T \frac{h'x^2}{2\sigma_\xi^2\delta^2 T}(1+o(1))\Big\} \le \exp\Big\{-\frac{h''x^2}{\delta}\Big\}, \quad (4.8.32)$$

where $h'' = \frac{h'}{3\sigma_\xi^2}$. The probability $\mathbf{P}(\min_{k\le\delta T} S_k < -h'x\sqrt{T})$ is estimated similarly. Hence, by (4.8.29)–(4.8.32),

$$x^{-2}\ln \mathbf{P}(\rho(\bar{s}_T, s_T) > xh) \le \ln 2 - \ln\delta - \frac{h''}{\delta}. \quad (4.8.33)$$

Now we choose $\delta = \delta_T \to 0$ so that $\delta_T T \to \infty$ and (4.8.26) holds. Then the right-hand side in (4.8.33) converges to $-\infty$ as $T\to\infty$ and (4.8.26) and (4.8.24) are satisfied.

Assume now that condition [$\mathbf{C}^V$] is met. Using inequality (4.8.31) and employing Corollary 5.2.1 in [38], we have

$$\mathbf{P}(\max_{k\le\delta T} S_k > h'x\sqrt{T}) \le c\delta T \exp\{-V(h'x\sqrt{T})\};$$
$$x^{-2}\ln\mathbf{P}(\max_{k\le\delta T} S_k > h'x\sqrt{T}) \le x^{-2}\ln c\delta T - c_1 T(x\sqrt{T})^{-2}V(x\sqrt{T}), \quad (4.8.34)$$

where

$$(x\sqrt{T})^{-2}V(x\sqrt{T}) = v(x\sqrt{T}) \quad (4.8.35)$$

(see (4.8.5)). The first term on the right-hand side of (4.8.34) is $o(\ln\delta T)$. The absolute value of the second term is majorized as $x\to\infty$, $x\le\ln T$, by

$$c_1 T v(\sqrt{T}\ln T) = c_1\frac{V(\sqrt{T}\ln T)}{(\ln T)^2} \gg \ln\delta T$$

as $T\to\infty$. Hence, for $x\le\ln T$ and $\delta = \overline{o}(1)$, $\delta T\to\infty$, the right-hand side of (4.8.34) converges to $-\infty$ as $T\to\infty$.

Now let $x > \ln T$, $x = o(\widehat{x}(T))$. In this case, the first term on the right-hand side of (4.8.34) converges to 0 as $T\to\infty$, while for the second term we have (see (4.8.6) and (4.8.35))

$$v(x\sqrt{T}) \gg v(\widehat{x}(T)) = v\Big(v^{(-1)}\Big(\frac{1}{T}\Big)\Big) = \frac{1}{T}$$

(here we can assume without loss of generality that the function $v(z)$ is monotone and continuous). Hence, using (4.8.34), we have

$$x^{-2}\ln\mathbf{P}(\max_{k\le\delta T} S_k > h'x\sqrt{T}) \to -\infty \quad \text{as} \quad T\to\infty.$$

The probability $\mathbf{P}(\min_{k\le\delta T} S_k < -h'x\sqrt{T})$ is estimated similarly. This proves (4.8.24), and therefore, Theorems 4.8.1 and 4.8.2 for the processes y_T.

In order to prove Theorems 4.8.1 and 4.8.2 for the processes z_T, one should use representation (4.8.2) and repeat the above arguments with $\eta(t)$ and $\chi(t)$ replaced by $\nu(t)$ and $\gamma(t)$, respectively. The above assertions remain valid. Theorems 4.8.1 and 4.8.2 are proved. □

Theorems 4.8.1–4.8.5 show that the MLDPs for random walks and for compound renewal processes are essentially of the same form, which, however, cannot be said about the LDPs in the zone of "proper" large deviations.

4.8.3 Rough (Logarithmic) Invariance Principle for CRPs in the Moderately Large Deviation Zone

Let $w(u)$ be the standard Wiener process on $[0, 1]$, and let $\mathcal{W}$ be the measure on $\mathfrak{B}(\mathbb{D}, \rho_{\mathbb{C}})$ corresponding to this process ($\mathcal{W}$ is concentrated on the space $\mathbb{C}$).

Theorem 4.8.9 *Let a set $B \subset \mathfrak{B}_{\mathbb{D}}(\mathbb{D}, \rho)$ be such that*

$$I_0((B)) = I_0([B]) > 0,$$
$$\mathcal{W}((vB)) = \mathcal{W}([vB]) > 0 \quad \textit{for any fixed } v.$$

Then, as $T \to \infty$, for all

$$x \geq x_0 > 0, \qquad x = \begin{cases} o(\sqrt{T}) & \textit{if } [\mathbf{C}] \textit{ holds,} \\ o(\widehat{x}(T)) & \textit{if } [\mathbf{C}^V] \textit{ holds} \end{cases}$$

the asymptotic equivalence

$$\ln \mathbf{P}(y_T(\cdot) \in xB) = \ln \mathcal{W}(xB)(1 + o(1))$$

holds.

In the moderately large deviation zone (as $x \to \infty$), this assertion is a corollary to Theorem 5.4.1 of [23] (on the asymptotic behavior of $\mathcal{W}(xB)$) and Theorems 4.8.1 and 4.8.2. In the normal deviation zone this result is a consequence of the invariance principle for the processes $y_T(u)$, which was established in §1.5.

Theorems 4.8.1 and 4.8.2 can be extended to the inhomogeneous case using the same line of arguments, which was employed in §1.5.3. It is easily seen that the conditions $\xi_1 \in [\mathbf{C}]$ and $\xi_1 \in [\mathbf{C}^V]$, respectively, for the deviations of $x = o(\sqrt{T})$ and $x = o(\widehat{x}(1))$, are sufficient that the results of Theorems 4.8.1 and 4.8.2 would remain valid in the inhomogeneous case.

5

Integro-Local Limit Theorems under the Cramér Moment Condition

5.1 Introduction

In this chapter, we shall continue the study of integro-local probabilities

$$\mathbf{P}(Z(t) \in \Delta[x)), \qquad \mathbf{P}(Y(t) \in \Delta[x)), \tag{5.1.1}$$

which was initiated in Chapter 2 in the normal deviation zone. Now, assuming that the vector (τ, ζ) satisfies the Cramér moment condition [**C**], we will study the integro-local probability in a wider zone, which in analogy with random walks, can be called the Cramér deviation zone. This zone includes the zones of normal, moderately large, and "usual" large deviations. We shall use the notation adopted in Chapter 3.

So, in this chapter, unless otherwise stated, we shall assume that the Cramér condition is satisfied in the following form:

$$[\mathbf{C}]\ \mathbf{E}e^{v|\xi_1|} < \infty, \quad \mathbf{E}e^{v|\xi|} < \infty \text{ with some } v > 0.$$

Besides, we will assume that the random vector $\boldsymbol{\xi} = (\tau, \zeta)$ is *nonlattice. To avoid repetition, these two conditions will not be specifically mentioned in the statements of the main results.*

The local theorems in the arithmetic case differ only in that their statements and proofs are substantially simpler. They can be found in [93].

In this chapter, we shall be mostly concerned with the exact asymptotics of the integro-local probabilities (5.1.1), where $\Delta[x)$ is the half-open interval $[x,\ x + \Delta)$, $\Delta = \Delta_T = \overline{o}(1)$ (i.e., Δ converges to 0 sufficiently slowly as $T \to \infty$), and $\alpha = \frac{x}{T}$ lies in some compact set (a closed interval) K, which will be defined below.

It is easily seen (see Corollary 5.3.6 and comments to it) that from knowledge of the asymptotics of (5.1.1) one can easily derive integral limit theorems for CRPs.

The study of the asymptotics of the probabilities in (5.1.1) is a rather difficult problem requiring knowledge of integro-local theorems for the renewal measure corresponding to the sequence $\{\tau_j, \zeta_j\}$. But this knowledge makes it possible to obtain fairly complete results. The approach based on the asymptotics of the renewal measure seems to be the only possible one. For the time being, no other approaches to obtaining limit theorems (and, in particular, integral theorems) in the large deviation zone are in sight.

The main results of this chapter were established in [47]. In [92], these results were carried over to the multidimensional case.

5.2 Main Results

5.2.1 Integro-Local Theorem for the Process $Z(t)$

In addition to the aforementioned moment and structural conditions, we shall assume that the normalized deviation $\alpha = \frac{x}{T}$ belongs (depending on the problem under consideration) to a compact set of one of two types:

$$\begin{aligned} &\text{a closed interval} \quad K^{<} \subset (\alpha_-, \alpha_+) \qquad \text{or} \\ &\text{a closed interval} \quad K^{>} \subset (\alpha_-, \alpha_+) \setminus [\beta_-, \beta_+] \end{aligned} \tag{5.2.1}$$

($\alpha_\pm$ are defined in Theorem 3.5.2, and $\beta_\pm$, in Lemma 3.5.8), which are closed intervals containing the neighborhood of the point $\alpha = a$. By $\mathcal{A}_{K^{\lessgtr}}$ we shall denote the closed part of the curve $\partial\mathcal{A}^{\leq 0}$ in $\mathbb{R}^2$ defined as $\mathcal{A}_{K^{\lessgtr}} := \{(\lambda(\alpha'), \mu(\alpha')) \,:\, \alpha' \in K^{\lessgtr}\}$, which clearly contains the intersection of the neighborhood of the point $(0, 0)$ with $\partial\mathcal{A}^{\leq 0}$. Next, we set

$$\mathbf{w} = [w_1, w_2), \qquad B(^{\lessgtr}u, v, \mathbf{w}) := \{\gamma(T) \lessgtr u,\ \chi(T) \geq v,\ \zeta(T) \in \mathbf{w}\}$$

($\chi(t), \gamma(t), \zeta(t)$ were introduced in §1.1).

In the results that follow we single out two most important particular cases: either the processes Z, Y are homogeneous or they are processes with stationary increments. The latter processes will be denoted by $Z^{(st)}$, $Y^{(st)}$. As was already pointed out (see §1.1.2), if (τ_1, ζ_1) has the Laplace transform

$$\begin{aligned} \psi_1(\lambda, \mu) = \psi^{(st)}(\lambda, \mu) &:= \frac{\psi(\lambda, \mu) - \psi(0, \mu)}{a_\tau \lambda}, \\ \psi(\lambda, \mu) &= \mathbf{E} e^{\lambda\tau + \mu\zeta}, \end{aligned} \tag{5.2.2}$$

then the process $Z(t)$ is a CRP *with stationary increments*. For if the distribution of $(\chi(t), \zeta(t))$ does not depend on t (as well as the distribution of the increments $Z(t + u) - Z(t)$),

$$\begin{aligned} \mathcal{A}^{(st)} &:= \{(\lambda, \mu) : \psi^{(st)}(\lambda, \mu) < \infty\} = \mathcal{A} \cap \mathcal{A}^{(\zeta)} \subset \mathcal{A}, \\ \mathcal{A}^{(\zeta)} &:= \{\mu : \psi^{(\zeta)}(\mu) < \infty\}. \end{aligned}$$

We set

$$\begin{aligned} I_Z(\alpha, {}^{<}u, v, \mathbf{w}) &= \int_0^u e^{\lambda(\alpha)y} \mathbf{P}(\tau > y + v,\ \zeta \in \mathbf{w})\, dy, \\ I_Z(\alpha, {}^{>}u, v, \mathbf{w}) &= \int_u^\infty e^{\lambda(\alpha)y} \mathbf{P}(\tau > y + v,\ \zeta \in \mathbf{w})\, dy. \end{aligned}$$

Theorem 5.2.1 *Let $\alpha = \frac{x}{T} \in K^{\lessgtr}$, $\Delta = \overline{o}(1)$ as $T \to \infty$ and let the "admissible inhomogeneity" condition*

$$\mathcal{A}_{K^{\lessgtr}} \subset (\mathcal{A}_1) \tag{5.2.3}$$

be satisfied. Then:

(i) *For some $\varepsilon > 0$, as $T \to \infty$,*

$$\begin{aligned} P^{\lessgtr}_{T,\alpha} &:= \mathbf{P}(Z(T) \in \Delta[x);\ B(^{\lessgtr}u, v, \boldsymbol{w})) \\ &= \mathbf{I}_{\{x \in (-\Delta, 0]\}} \mathbf{P}(\tau_1 \geq T + v, \zeta_1 \in \boldsymbol{w}) \\ &\quad + \frac{\Delta}{\sqrt{T}} \psi_1 C(\alpha) e^{-TD(\alpha)} I_Z(\alpha, {}^{\lessgtr}u, v, \boldsymbol{w})(1 + o(1)) + O(e^{-T(D(\alpha)+\varepsilon)}), \end{aligned} \tag{5.2.4}$$

where $\psi_1 = \psi_1(\lambda(\alpha), \mu(\alpha))$, $C(\alpha) := C(1, \alpha)$, $C(\theta, \alpha)$ is the function defined in (5.3.9), which is positive and continuous in the cone $\mathcal{D} = \{(\theta, \alpha) : \alpha/\theta \in [\alpha_-, \alpha_+)\}$.

For the probabilities $\mathbf{P}(Z(T) \in \Delta[x), B(^{<}u, v, \boldsymbol{w}))$, the remainder terms $o(1)$ and $O(\cdot)$ are uniform in $\alpha \in K^{<}$, $u \leq u_0$ for any fixed $u_0 > 0$ and all $v \geq 0$, $\boldsymbol{w}$.

For the probabilities, $\mathbf{P}(Z(T) \in \Delta[x), B(^{>}u, v, \boldsymbol{w}))$ the above uniform convergence holds for $\alpha \in K^{>}$ and under the same conditions on the remaining parameters.

(ii) *If $Z(t)$ is a homogeneous CRP, then the first term on the right of (5.2.4) can be omitted (it is included in the remaining terms).*

(iii) *Let $Z(t) = Z^{(st)}(t)$ be a process with stationary increments and let the condition*

$$\mathcal{M}_{K^{\lessgtr}} := \{\mu(\alpha') : \alpha' \in K^{\lessgtr}\} \subset (\mu_-^{(\zeta)}, \mu_+^{(\zeta)}) \tag{5.2.5}$$

be met. Then the admissible inhomogeneity condition (5.2.3) for $\mathcal{A}_1 = \mathcal{A}^{(st)}$ is satisfied and the first term on the right of (5.2.4) can be removed.

Condition (5.2.5) is always satisfied if the domain $(\mathcal{A})$ is rectangular,

$$(\mathcal{A}) = (\lambda < \lambda_+) \times (\mu_-^{(\zeta)} < \mu < \mu_+^{(\zeta)}) \tag{5.2.6}$$

(as in the case of independent τ and ζ).

By the integro-local Theorem 2.3.1 in the normal deviation zone (see (2.3.3)),

$$\mathbf{P}(Z(T) - aT \in \Delta[x)) = \frac{\Delta}{\sigma\sqrt{2\pi T}} e^{-\frac{x^2}{2T\sigma^2}} + O(T^{-\beta})$$

for $\beta > 1/2$ uniformly in x. Moreover, for α near the point a (see (3.5.22) in Theorem 3.5.2),

$$D(\alpha) = \frac{(\alpha - a)^2 D''(a)}{2} + O(|\alpha - a|^3) = \frac{(\alpha - a)^2}{2\sigma^2} + O(|\alpha - a|^3).$$

Comparing this with (5.2.4), we see that, for α from a small neighborhood of the point a,

$$\psi_1 C(\alpha) I_Z(\alpha, {}^{>}0, 0, \mathbb{R}) = \frac{1 + o(1)}{\sigma\sqrt{2\pi}},$$

where $\psi_1 = \psi_1(\lambda(\alpha), \mu(\alpha)) \to \psi_1(0,0) = 1$ and $I_Z(\alpha, {}^{>}0, 0, \mathbb{R}) \to a_\tau$, so that $C(\alpha) \to \frac{1}{a_\tau \sigma \sqrt{2\pi}}$ as $\alpha \to a$.

Remark 5.2.2 The integrals

$$I_Z(\alpha, {}^{<}u, v, \mathbf{w}) \quad \text{and} \quad I_Z(\alpha, {}^{>}u, v, \mathbf{w})$$

are considered separately, because the integral $I_Z(\alpha, {}^{>}u, v, \mathbf{w})$ for $\alpha \in [\beta_-, \beta_+]$ (in the case $\lambda_+ < D(0)$) is divergent, since for such α we have $\lambda_+ < \lambda(\alpha)$. At the same time, the integral $I_Z(\alpha, {}^{<}u, v, \mathbf{w})$ is always finite. Hence, for $\alpha \in (\beta_-, \beta_+)$, we have $P^{>}_{T,\alpha} \gg P^{<}_{T,\alpha}$ as $T \to \infty$. This corresponds to the fact that for $\alpha \in (\beta_-, \beta_+)$ the deviation function in the local LDP for $Z(T)$ is $\widehat{D}(\alpha) < D(\alpha)$ and the principal contribution to the integro-local probabilities comes from the trajectories for which $\gamma(T)$ is comparable with T; see §3.4.

It is also worth pointing out that the compact set $K^{>}$ can be defined in terms of the fundamental function $\widehat{A}(\mu)$. We let $(\widehat{\mu}_-, \widehat{\mu}_+)$ denote the maximal domain of analyticity of the function $\widehat{A}(\mu)$ that contains the point $\mu = 0$. We also set

$$\widehat{\alpha}_- = A'(\widehat{\mu}_- + 0), \qquad \widehat{\alpha}_+ = A'(\widehat{\mu}_+ - 0).$$

Then the interval $(\widehat{\alpha}_-, \widehat{\alpha}_+)$ is the domain of analyticity of the function $\widehat{D}(\alpha)$ that contains the point $\alpha = a$, and $K^{>}$ can be defined to be any closed interval in $(\widehat{\alpha}_-, \widehat{\alpha}_+)$ containing the point $\alpha = a$.

If $\lambda_+ > D(0)$, then $\widehat{\mu}_\pm = \mu_\pm$, $\widehat{\alpha}_\pm = \alpha_\pm$ and the compact sets $K^{<}$ and $K^{>}$ are indistinguishable by their definitions (the forbidden set $[\beta_-, \beta_+]$ is empty).

The condition $\lambda_+ > D(0)$ is clearly satisfied if $\lambda_+ = \infty$ or $a = 0$ (in this case, $D(0) = 0$). According to Theorem 3.3.1, this condition is also satisfied if the boundary $\partial\mathcal{A}^{\leq 0}$ in $\mathbb{R}^2$ lies in the open half-plane $\{\lambda < \lambda^+\}$ (this is always so if $\mathbf{A}(\lambda_+, \mu) > 0$ for all μ); see assertion (iv) of Theorem 3.3.2.

Remark 5.2.3 It is clear that if, for any v, $\mathbf{w}$,

$$\mathbf{P}(\tau_1 \geq T + v,\ \zeta_1 \in \mathbf{w}) = o\Big(\frac{1}{\sqrt{T}} e^{-TD(0)} I_Z(0, 0, v, \mathbf{w})\Big) \tag{5.2.7}$$

as $T \to \infty$, then first term on the right of (5.2.4) can be removed. If

$$\mathbf{P}(\tau_1 \geq T) = o\Big(\frac{1}{\sqrt{T}} e^{-TD(0)}\Big) \quad \text{as} \quad T \to \infty, \tag{5.2.8}$$

then condition (5.2.7) is satisfied for $v = 0$ and $\mathbf{w} = \mathbb{R}$. A sufficient condition for (5.2.8) to hold is that

$$\lambda_+^{(\tau_1)} > D(0). \tag{5.2.9}$$

In case (5.2.9), the first term on the right-hand side of (5.2.4) can be included in the remainder term $O(e^{-T(D(\alpha)+\varepsilon)})$.

Remark 5.2.4 Integrating by parts $I_Z(\alpha, {}^{>}0, 0, \mathbb{R})$, we get

$$I_Z(\alpha, {}^{>}0, 0, \mathbb{R}) = \frac{\psi^{(\tau)}(\lambda(\alpha)) - 1}{\lambda(\alpha)}, \quad \text{where} \quad \psi^{(\tau)}(\lambda) := \mathbf{E}e^{\lambda\tau}. \tag{5.2.10}$$

From (5.2.10) and (5.2.2) it follows that

$$I_Z(\alpha, {}^{>}0, 0, \mathbb{R}) = a_\tau \psi^{(st)}(\lambda(\alpha), 0). \tag{5.2.11}$$

If $Z(t)$ is a CRP with stationary increments ($\psi_1(\lambda, \mu) = \psi^{(st)}(\lambda, \mu)$) and if the domain $(\mathcal{A})$ is rectangular (see (5.2.6)), then for the set $\mathcal{A}^{(st)}$ we have

$$(\mathcal{A}^{(st)}) = \{\lambda < \lambda_+\} \times \{\mu_-^{(\zeta)} < \mu < \mu_+^{(\zeta)}\} = (\mathcal{A}).$$

Since $\mathcal{A}_{K^{\lessgtr}} \subset (\mathcal{A}) = (\mathcal{A}^{(st)})$, the admissible inhomogeneity condition (5.2.3) is always satisfied in this case.

Remark 5.2.5 The significance of the presence of the forbidden set $[\beta_-, \beta_+]$ in the conditions of Theorem 5.2.1 was explained above. The significance of the admissible inhomogeneity condition (5.2.3) is clarified by the presence of the factor $\psi_1(\lambda(\alpha), \mu(\alpha))$ on the right of (5.2.4).

The example of integro-local theorems for random walks (see, e.g., [21], Chapter 9) shows that outside the Cramér domain (α_-, α_+) the asymptotics of $\mathbf{P}(Z(T) \in \Delta[x))$ is different. So, the above means that the conditions of Theorem 5.2.1 are close to the minimal ones.

Remark 5.2.6 In what follows, we will have in mind that the compact sets $K^{\lessgtr} \subset \mathbb{R}$ are chosen to be as large as possible but in such a way that the admissible inhomogeneity condition (5.2.3) is satisfied (if $(\mathcal{A}) \subset (\mathcal{A}_1)$, then condition (5.2.3) ceases to be restrictive).

Remark 5.2.7 Assume that in the case $\lambda_+ > D(0)$, in addition to equality (3.7.1) of Theorem 3.7.1, near the point $\mu = \mu(\alpha)$, $\alpha \in K^{<}$, the more precise representation

$$\mathbf{E}e^{\mu Z(T)} = c(\mu)e^{TA(\mu)}(1 + o(1)) \tag{5.2.12}$$

holds as $T \to \infty$ for some smooth function $c(\mu)$. In this case, if one employs the inversion formula for finding the asymptotics of $\mathbf{P}(Z(T) \in \Delta[x))$, then using the Cramér transform at the point $\mu(\alpha)$ and employing standard techniques (see the proof of Theorem 9.3.1 in [21]), we get the value

$$\frac{c(\mu(\alpha))\Delta e^{-TD(\alpha)}}{\sqrt{2\pi T A''(\mu(\alpha))}}(1 + o(1)),$$

which coincides with the asymptotics on the right of (5.2.4) (for $u = v = 0$, $\mathbf{w} = \mathbb{R}$) with an appropriately chosen function $c(\mu)$. This supports the assumption (5.2.12), which will be proved in §5.6.

Remark 5.2.8 When going through the proof of Theorem 5.2.1, one can easily check that the uniform convergence in u in the conclusion of the theorem will also remain valid in the domain $[0, \delta T]$ (which is larger than $[0, u_0]$) for sufficiently small $\delta > 0$.

Corollary 5.2.9 *Let the conditions of Theorem 5.2.1 be satisfied. Assume that $\Delta > 0$ is fixed. Then, for some $\varepsilon > 0$,*

$$\mathbf{P}(Z(T) \in \Delta[x);\ B(^{\lessgtr}u, v, \mathbf{w})) \\ = \frac{1 - e^{-\mu(\alpha)\Delta}}{\mu(\alpha)\sqrt{T}} \psi_1 C(\alpha) e^{-TD(\alpha)} I_Z(\alpha, {}^{\lessgtr}u, v, \mathbf{w})(1 + o(1)) + O(e^{-T(D(\alpha)+\varepsilon)}), \tag{5.2.13}$$

where the remainder terms are uniform in $\alpha \in K^{\lessgtr}$, $u \le u_0$, v, $\mathbf{w}$.

If, moreover, $\alpha \to a$ and $\Delta \ge \Delta_0 > 0$ is such that $|\alpha - a|\Delta \to 0$ as $T \to \infty$, then the factor $\frac{1-e^{-\mu(\alpha)\Delta}}{\mu(\alpha)}$ can be replaced by Δ.

Proof of Corollary 5.2.9 Let $\Delta > 0$ be fixed, let $N \to \infty$ (sufficiently slowly as $T \to \infty$), and let

$$\Delta_N := \frac{\Delta}{N}, \quad x_k = x + k\Delta_N, \quad k = 0, 1, \dots, N-1;$$
$$\Delta_N[x_k) = [x_k, x_k + \Delta_N).$$

Then, for $\alpha_k := \alpha + \frac{k\Delta_N}{T}$, as $T \to \infty$,

$$TD(\alpha_k) = TD(\alpha) + k\Delta_N \mu(\alpha) + O\Big(\frac{(k\Delta_N)^2}{T}\Big).$$

Since $\Delta[x) = \bigcup_{k=0}^{N-1} \Delta_N[x_k)$, $\Delta_N \to 0$ and since

$$\sum_{k=0}^{N-1} \Delta_N e^{-TD(\alpha_k)} = e^{-TD(\alpha)} \int_0^\Delta e^{-\mu(\alpha)y} dy (1 + o(1)) \\ = e^{-TD(\alpha)} \frac{1 - e^{-\mu(\alpha)\Delta}}{\mu(\alpha)} (1 + o(1)),$$

it follows from Theorem 5.2.1 that (5.2.13) is satisfied for the left-hand side of (5.2.13), which is equal to

$$\sum_{k=0}^{N-1} \mathbf{P}(Z(T) \in \Delta_N[x_k),\ B(^{\lessgtr}u, v, \mathbf{w})).$$

The second assertion of the corollary is clear, inasmuch as $\mu(\alpha)\Delta$ and $(\alpha - a)\Delta$ have the same order of smallness as $\alpha \to a$. The corollary is proved. □

In a similar manner, from Theorem 5.2.1 and using the large deviation principle for $Z(T)$, one can derive integro-local limit theorems for any Δ (and, in particular, integral theorems). For example, for $\alpha \ge \alpha_0 > a$ we have $\mu(\alpha) \ge \mu_0 > 0$ and (5.2.13) remains valid for any Δ (in particular, one can put $\Delta = \infty$).

5.2.2 Integro-Local Theorem for the Process $Y(t)$

The analog of Theorem 5.2.1 for the process $Y(t)$ is slightly different. We set

$$B({}^{\lessgtr}u, v) := \{\gamma(T) \lessgtr u, \quad \chi(T) \geq v\} = B({}^{\lessgtr}u, v, \mathbb{R}),$$

$$I_Y(\alpha, {}^{<}u, v) := \int_0^u e^{\lambda(a)y}\mathbf{E}(e^{\mu(\alpha)\zeta};\ \tau \geq y + v)\,dy,$$

$$I_Y(\alpha, {}^{>}u, v) := \int_u^\infty e^{\lambda(a)y}\mathbf{E}(e^{\mu(\alpha)\zeta};\ \tau \geq y + v)\,dy,$$

$$\mu_-^{(\zeta)} := \inf\{\mu :\ \mathbf{E}e^{\mu\zeta} < \infty\}, \qquad \mu_+^{(\zeta)} := \sup\{\mu :\ \mathbf{E}e^{\mu\zeta} < \infty\}.$$

Below, in Theorem 5.2.10, as in Theorem 5.2.1, we will consider the normalized deviations $\alpha = \frac{x}{T} \in K^{\lessgtr}$, where the compact sets $K^{\lessgtr}$ are defined in (5.2.1).

Theorem 5.2.10 *Let $\alpha \in K^{>}$ and let condition* (5.2.5) *hold. Then:*

(i) *If $\lambda(\alpha) \neq 0$, then*

$$I_Y(\alpha, {}^{>}0, 0) = \frac{1 - \psi^{(\zeta)}(\mu(\alpha))}{\lambda(\alpha)} < \infty, \tag{5.2.14}$$

where $\psi^{(\zeta)}(\mu) := \psi(0, \mu)$ (cf. (5.2.10)).

If $\lambda(\alpha) = 0$, then

$$I_Y(\alpha, {}^{>}0, 0) = \psi'_{(1)}(0, \mu(\alpha)) < \infty \quad (= \mathbf{E}\tau, \ \text{if}\ \alpha = a). \tag{5.2.15}$$

(ii) *Let the admissible inhomogeneity condition* (5.2.3) *hold. Then, for any fixed $\Delta > 0$,*

$$\mathbf{P}(Y(T) \in \Delta[x);\ B({}^{\lessgtr}u, v)) = P_1(v, \Delta[x)) + \frac{e^{\mu(\alpha)\Delta} - 1}{\mu(\alpha)\sqrt{T}}\psi_1 C(\alpha)e^{-TD(\alpha)}\left[I_Y(\alpha, {}^{\lessgtr}u, v) + o(1)\right] \tag{5.2.16}$$

as $T \to \infty$, where

$$P_1(v, \Delta[x)) := \mathbf{P}(\tau_1 > T + v,\ \zeta_1 \in \Delta[x)),$$

the remainder term $o(1)$ being uniform in $\alpha \in K^{\lessgtr}$, $v \geq 0$, $u \leq u_0$.

(iii) *If the process $Y(t)$ is homogeneous, then* (5.2.16) *holds, in which the term $P_1(v, \Delta[x))$ on the right of* (5.2.16) *can be omitted.*

(iv) *If $Y(t) = Y^{(st)}(t)$ is the process with stationary increments, then the admissible inhomogeneity condition $(\mathcal{A}_{K^{\lessgtr}}) \subset (\mathcal{A}^{(st)})$ and relations* (5.2.16) *hold, in which the term $P_1(v, \Delta[x))$ can be omitted.*

It is clear that if $\Delta = \Delta_T \to 0$ sufficiently slowly as $T \to \infty$, then the factor $\frac{e^{\mu(\alpha)\Delta}-1}{\mu(\alpha)}$ in (5.2.16) can be replaced by Δ. Remarks 5.2.5 and 5.2.6 to Theorem 5.2.1 are fully applicable here.

Remark 5.2.11 It is clear from (5.2.16) that in order to neglect the term $P_1(v, \Delta[x))$ in (5.2.16) in the general case, one should require that

$$\mathbf{P}(\tau_1 > T,\ \zeta_1 \in \Delta[x)) = o\left(\frac{e^{-TD(\alpha)}}{\sqrt{T}}\right) \quad \text{as} \quad T \to \infty. \tag{5.2.17}$$

Thus, for the initial jump (τ_1, ζ_1) to have no effect on the asymptotics in integro-local theorems for the CRPs $Z(t)$ and $Y(t)$, one should, in addition to the "principal" admissible inhomogeneity condition (5.2.3), require that conditions (5.2.7) and (5.2.17) be satisfied in Theorems 5.2.1 and 5.2.10, respectively.

For condition (5.2.17) to hold, it suffices that

$$\mathbf{\Lambda}_1(T, \alpha T) \geq \mathbf{\Lambda}(T, \alpha T) + o(T) \quad \text{as} \quad T \to \infty, \tag{5.2.18}$$

where $\mathbf{\Lambda}_1(T, \alpha T)$ is the deviation function corresponding to (τ_1, ζ_1) (in the homogeneous case, this condition is always satisfied).

Indeed, suppose for simplicity that the rectangle $[T, \infty) \times \Delta[x)$ is tangent to the level surface of the deviation function $\mathbf{\Lambda}_1$ at a point (T, x). Then

$$\mathbf{P}(\tau_1 \geq T,\ \zeta_1 \in \Delta[x)) \leq e^{-\mathbf{\Lambda}_1(T, \alpha T)}. \tag{5.2.19}$$

From Lemma 5.3.4, which we will prove below, it follows that, for some $\varepsilon > 0$ and all sufficiently large T,

$$\frac{1}{T}\mathbf{\Lambda}(T, \alpha T) \geq \mathbf{D}(1, \alpha) + \varepsilon = D(\alpha) + \varepsilon$$

and therefore, by (5.2.18),

$$\mathbf{\Lambda}_1(T, \alpha T) \geq \mathbf{\Lambda}(T, \alpha T) + o(T) \geq TD(\alpha) + \varepsilon T + o(T) \geq TD(\alpha) + \frac{\varepsilon}{2}T$$

as $T \to \infty$. Now (5.2.17) follows from (5.2.19) and the above inequality.

Another condition

$$(0, \mu(\alpha)) \in (\mathcal{A}_1) \quad \text{for all} \quad \alpha \in K^{\lessgtr}, \tag{5.2.20}$$

under which (5.2.17) holds, will be given below in Lemma 5.5.1.

According to Theorem 5.2.1, condition (5.2.5) is always satisfied if the domain $\mathcal{A}$ is rectangular.

Remark 5.2.12 Note that as in (5.2.11)

$$I_Y(\alpha, {}^{>}0, 0) = \frac{\psi(\lambda(\alpha), \mu(\alpha)) - \psi(0, \mu(\alpha))}{\lambda(\alpha)} = a_\tau \psi^{(st)}(\lambda(\alpha), \mu(\alpha)).$$

For CRPs with stationary increments and the rectangular domain $(\mathcal{A})$, the admissible inhomogeneity condition (5.2.3) is always satisfied.

5.2.3 Integro-Local Theorem for Finite-Dimensional Distributions of the Process $Z(t)$

If the conditions

$$\lambda_+ > D(0), \qquad \mu_\pm = \mu^\pm, \qquad A'(\mu_\pm \mp 0) = \pm\infty \tag{5.2.21}$$

are met, then the forbidden set $[\beta_1, \beta_+]$ is empty, $\alpha_\pm = \pm\infty$, and hence as compact sets $K^\lessgtr$ one can take the compact set $K = [-N, N]$ for any $N > 0$; we assume that the number N is sufficiently large so that $a \in K$, $(0,0) \in \mathcal{A}_K$. As was already pointed out, $\lambda(\alpha) \le \lambda(0) = D(0)$ for $\alpha \in K$. Hence by (5.2.21) there exists a $\delta > 0$ such that

$$\lambda(\alpha) < \lambda_+ - \delta \quad \text{for all} \quad \alpha.$$

We shall require the following condition.

$[\widehat{\mathbf{P}}_K]$ *There exist a distribution $\widehat{P}$ in $\mathbb{R}^2$ and a constant $c < \infty$ such that, for all $v > 0$, w and all sufficiently large t,*

$$e^{-\delta t}\mathbf{P}(\tau > t + v,\ \zeta \in w \mid \tau > t) \le c\widehat{P}((v,\infty),\ w); \tag{5.2.22}$$

besides,

$$\widehat{\psi}(\lambda, \mu) < \infty \quad \textit{for} \quad (\lambda, \mu) \in \mathcal{A}_K, \tag{5.2.23}$$

where $\widehat{\psi}$ is the Laplace transform of $\widehat{P}$.

If τ and ζ are independent, then it is easily checked that condition $[\widehat{\mathbf{P}}_K]$ is always met whenever τ satisfies the large deviation principle

$$\ln \mathbf{P}(\tau > t) \ge -\Lambda^{(\tau)}(t) + o(t), \tag{5.2.24}$$

which for $\lambda_+ < \infty$ is the same,

$$\ln \mathbf{P}(\tau > t) \sim -\lambda_+ t \quad \text{as} \quad t \to \infty.$$

Indeed, if (5.2.24) holds, then

$$\frac{e^{-\delta t}\mathbf{P}(\tau > t + v)}{\mathbf{P}(\tau > t)} \le \exp\big\{-\delta t - \Lambda^{(\tau)}(t+v) + \Lambda^{(\tau)}(t) + o(t)\big\}. \tag{5.2.25}$$

Since

$$\Lambda^{(\tau)}(t+v) - \Lambda^{(\tau)}(t) \ge \lambda^{(\tau)}(t)v, \quad \lambda^{(\tau)}(t) := (\Lambda^{(\tau)})'(t) \to \lambda_+ \text{ as } t \to \infty,$$

we have, for $\lambda_+ < \infty$, any $\varepsilon > 0$, and all sufficiently large t, the left-hand side in (5.2.25) is majorized by

$$\exp\big\{-(\lambda_+ - \varepsilon)v\big\}.$$

Hence if we put $\widehat{P}((v,\infty),\ \mathbf{w}) = e^{-(\lambda_+ - \varepsilon)v}\mathbf{P}(\zeta \in \mathbf{w})$, then, for sufficiently small $\varepsilon > 0$, the compact set $\mathcal{A}_K$ will be contained in the effective domain of $\widehat{\psi}(\lambda, \mu)$, and so (5.2.23) is satisfied. For $\lambda_+ = \infty$, the arguments are simpler.

In addition to Theorem 5.2.1, the following integro-local theorem for the finite-dimensional distributions of the process $Z(t)$ holds. Given

$$0 = u_0 < u_1 < \cdots < u_M = 1, \qquad \alpha_1, \ldots, \alpha_M, \quad \alpha_0 = 0,$$

we set

$$\gamma_j := \frac{\alpha_j - \alpha_{j-1}}{u_j - u_{j-1}}, \quad j = 1, \ldots, M,$$

$$J_Z(\gamma) := I_Z(\gamma, 0, 0, \mathbb{R}) = \int_0^\infty e^{\lambda(\gamma)y}\mathbf{P}(\tau > y)\, dy,$$

where the function $I_Z(\cdot, \cdot, \cdot, \cdot)$ is from Theorem 5.2.1. Next, consider the measure

$$P_\gamma((v, \infty), \mathbf{w}) := \int_0^\infty e^{\lambda(\gamma)y}\mathbf{P}(\tau > y + v, \zeta \in \mathbf{w})\, dy$$

and let $\psi^{(\gamma)}(\lambda, \mu)$ be the Laplace transform of P_γ:

$$\psi^{(\gamma)}(\lambda, \mu) := \int_0^\infty \int_{-\infty}^\infty e^{\lambda v + \mu z} P_\gamma(dv, dz).$$

The value γ_0 is not defined, and it is convenient for us to put

$$\psi^{(\gamma_0)}(\lambda, \mu) := \int_0^\infty \int_{-\infty}^\infty e^{\lambda v + \mu z}\mathbf{P}(\tau_1 \in dv, \zeta_1 \in dz).$$

With this notation, the following theorem holds.

Theorem 5.2.13 *Let conditions* (5.2.21), $[\widehat{\mathbf{P}}_K]$, *the admissible inhomogeneity condition* $\mathcal{A}_K \subset (\mathcal{A}_1)$, *and condition* (5.2.7) *be met and let* $\alpha_1 = 0$. *Then, for* $x_k = \alpha_k T$,

$$\mathbf{P}\Big(\bigcap_{k=1}^{M}\{Z(u_k T) \in \Delta[x_k)\}\Big)$$
$$= \frac{\Delta^M}{T^{M/2}} e^{-TI(f)} \prod_{j=1}^{M} \frac{\psi^{(\gamma_{j-1})}(\lambda(\gamma_j), \mu(\gamma_j))}{\sqrt{u_j - u_{j-1}}} C(\gamma_j) J_Z(\gamma_j)(1 + o(1)), \quad (5.2.26)$$

where $f = f(t)$, $0 \le t \le 1$ *is the continuous polygon with nodes at the points* (u_j, α_j), $0 \le j \le M$,

$$I(f) := \int_0^1 D(f'(t))\, dt,$$

and $\Delta = \Delta_T \to 0$ *sufficiently slowly as* $T \to \infty$. *The remainder term* $o(1)$ *in* (5.2.26) *is uniform in* $(\gamma_1, \ldots, \gamma_M) \in K^M = \prod_{j=1}^{M} \{|\gamma_j| \le N\}$ *for any* $N > 0$.

In addition to condition $[\widehat{\mathbf{P}}_K]$, we can indicate another condition (possibly more simple and transparent), which also secures the integro-local theorem for the finite-dimensional distributions of the process $Z(t)$. This condition is as follows.

[h] *There exists a function* $h(v) = o(v)$ *as* $v \to \infty$ *such that, for any* $\varepsilon > 0$ *and* $\lambda < \lambda_+ - \varepsilon$,

$$\int_v^\infty e^{\lambda t} \mathbf{P}(\tau > t,\ \zeta \in \boldsymbol{w})\, dt \le e^{\lambda v + h(v)} \mathbf{P}(\tau > v, \zeta \in \boldsymbol{w}).$$

This condition is satisfied if there exist t_0 and $q < 1$ (depending on λ) such that the measure

$$p(dt, \mathbf{w}) := e^{(\lambda_+ - \varepsilon)t} \mathbf{P}(\tau > t, \zeta \in \mathbf{w})\, dt$$

has the property

$$p(d(t + t_0), \mathbf{w}) \le q p(dt, \mathbf{w}),$$

for all $\mathbf{w}$ and all sufficiently large t. In this case, it is clear that

$$\int_v^\infty e^{\lambda t} \mathbf{P}(\tau > t, \zeta \in \mathbf{w})\, dt \le \frac{t_0}{1-q} e^{\lambda v} \mathbf{P}(\tau > v, \zeta \in \mathbf{w}).$$

If τ and ζ are independent, then condition **[h]** is satisfied if it holds for $\mathbf{w} = \mathbb{R}$.

Theorem 5.2.14 *Let the conditions of Theorem* 5.2.13 *be satisfied with condition* $[\widehat{\mathbf{P}}_K]$ *replaced by condition* **[h]**. *Then* (5.2.26) *remains valid.*

From (5.2.26) we see that even under the above assumptions (5.2.21), $[\widehat{\mathbf{P}}_K]$, and **[h]**, the increments of the process $Z(t)$ on large intervals are not asymptotically independent in the large deviation zone in the sense of integro-local theorems (for random walks this independence holds). This is indeed so because of the presence of the factors $\psi^{(\gamma_{j-1})}(\lambda(\gamma_j), \mu(\gamma_j))$ depending on two neighboring normalized increments γ_{j-1} and γ_j. In the normal and moderately large deviation zones this dependence ceases to hold.

5.2.4 Normal and Moderately Large Deviations

If

$$\alpha = \frac{x}{T} \to a \quad \text{as} \quad T \to \infty,$$

then $(\lambda(\alpha), \mu(\alpha)) \to (0, 0)$ as $T \to \infty$,

$$\psi_1(\lambda(\alpha), \mu(\alpha)) \to 1, \tag{5.2.27}$$

$$I_Z(\alpha, {}^{>}u, v, \mathbf{w}) \to \int_u^\infty \mathbf{P}(\tau \ge y + v,\ \zeta \in \mathbf{w})\, dy,$$

$$I_Y(\alpha, {}^{>}u, v) \to \int_u^\infty \mathbf{P}(\tau \ge y + v)\, dy = \mathbf{E}\tau \mathbf{P}(\chi \ge u + v), \tag{5.2.28}$$

where χ is the overshoot of the random walk $\{T_n\}$ over the infinite barrier.

Next, since the function $D(\alpha)$ is analytic on the interval (α_-, α_+) containing the point $\alpha = a$, we have

$$e^{-TD(\alpha)} = e^{-\frac{(x-Ta)^2}{2\sigma^2 T}(1+o(1))} \quad \text{as} \quad T \to \infty,$$

where

$$\sigma^2 = (D''(a))^{-1} = a_\tau^{-1}\mathbf{E}(\zeta - a\tau)^2$$

(see (3.5.22)). In this case, the admissible inhomogeneity conditions in Theorems 5.2.1 and 5.2.10 are satisfied. All that allows us to state the integro-local theorem in the normal and moderately large deviation zones in the following form.

Corollary 5.2.15 *Let $\alpha = \frac{x}{T} \to a$ as $T \to \infty$. Then*

$$C(\alpha) \sim C(a) = \frac{1}{\mathbf{E}\tau\sigma\sqrt{2\pi}} \quad \textit{as} \quad T \to \infty \tag{5.2.29}$$

and, for any fixed $\Delta > 0$ and some $\varepsilon > 0$, the following assertions hold:

I. $\mathbf{P}(Z(T) \in \Delta[x);\ B(^{>}u, v, \mathbb{R}))$

$$= \frac{\Delta}{\sqrt{2\pi T}\sigma} e^{-TD(\alpha)}\mathbf{P}(\chi \geq u + v)(1 + o(1)) + O(e^{-T\varepsilon}), \tag{5.2.30}$$

where the remainder terms $o(1)$, $O(e^{-T\varepsilon})$ are uniform in $u \leq u_0$, $v \geq 0$.

If $y := x - aT = o(T^{\frac{2}{3}})$ as $T \to \infty$, then the factor $e^{-TD(\alpha)}$ in the right-hand side of (5.2.30) can be replaced by $e^{-\frac{y^2}{2T\sigma^2}}$.

II. $\mathbf{P}(Y(T) \in \Delta[x);\ B(^{>}u, v, \mathbb{R}))$

$$= \frac{\Delta}{\sqrt{2\pi T}\sigma} e^{-TD(\alpha)}(\mathbf{P}(\chi \geq u + v) + o(1)), \tag{5.2.31}$$

where the remainder term $o(1)$ is uniform in $u \leq u_0$ and $v \geq 0$.

Proof of Corollary 5.2.15 I. From Theorem 5.2.1 (in view of Corollary 5.2.9 and (5.2.27), (5.2.28)) it follows that, for any fixed $\Delta > 0$,

$$\mathbf{P}(Z(T) \in \Delta[x);\ B(^{>}u, v, \mathbb{R})) = \mathbf{I}_{\{x \in (-\Delta, 0]\}}\mathbf{P}(\tau_1 > T + v)$$
$$+ \frac{\Delta}{\sqrt{T}} e^{-TD(\alpha)}C(a)\mathbf{E}\tau\mathbf{P}(\chi \geq u + v)(1 + o(1)) + O(e^{-T\varepsilon}), \tag{5.2.32}$$

where the remainder terms are uniform in $u \leq u_0$, $v \geq 0$. We have $\mathbf{P}(\tau_1 > T+v) = o(e^{-T\varepsilon})$ as $T \to \infty$ and some $\varepsilon > 0$, and hence we can get rid of the first term on the right of (5.2.32). So, to establish (5.2.30) it suffices to verify (5.2.29).

From (5.2.32) it follows that, for Δ, which converges to 0 sufficiently slowly as $T \to \infty$, the "predensities"

$$\frac{\mathbf{P}(Z(T) \in \Delta[x))}{\Delta}$$

are approaching (in the sense of the relative convergence) the function

$$\frac{C(\alpha)\mathbf{E}\tau}{\sqrt{T}} e^{-\frac{y^2}{2T\sigma^2}}. \tag{5.2.33}$$

On the other hand, the distribution of $\frac{Z(T)-aT}{\sigma\sqrt{T}}$ converges as $T \to \infty$ to the standard normal distribution (see Theorem 1.3.3). Comparing the value of $\mathbf{P}(Z(T) - aT \in [-b\sqrt{T},\ b\sqrt{T}])$ for $b > 0$, obtained via (5.2.33) and one obtained from the normal approximation, we get

$$C(a)\mathbf{E}\tau = \frac{1}{\sqrt{2\pi}\sigma},$$

thereby proving (5.2.29).

II. Assertion (5.2.31) can be verified in the same way with the help of Theorem 5.2.10. Corollary 5.2.15 is proved. □

5.3 Integro-Local Theorems for the Renewal Measure

In order to prove the assertions formulated in §5.2 we shall need integro-local theorems for the renewal measure corresponding to the vector (τ, ζ).

We set, as before,

$$\psi(\lambda,\mu) := \mathbf{E}e^{\lambda\tau+\mu\zeta}, \qquad \psi_1(\lambda,\mu) := \mathbf{E}e^{\lambda\tau_1+\mu\zeta_1},$$
$$\mathbf{A}(\lambda,\mu) := \ln\psi(\lambda,\mu), \quad \mathbf{A}_1(\lambda,\mu) := \ln\psi_1(\lambda,\mu), \quad (\lambda,\mu) \in \mathbb{R}^2;$$
$$\mathcal{A} := \{(\lambda,\mu) : \mathbf{A}(\lambda,\mu) < \infty\}, \qquad \mathcal{A}_1 := \{(\lambda,\mu) : \mathbf{A}_1(\lambda,\mu) < \infty\}.$$

It is clear that by condition [**C**] the interiors $(\mathcal{A})$, $(\mathcal{A}_1)$ of the sets $\mathcal{A}$, $\mathcal{A}_1$ contain the point $(\lambda,\mu) = (0,0)$ and are the domains of analyticity of the functions $\mathbf{A}(\lambda,\mu)$, $\mathbf{A}_1(\lambda,\mu)$, respectively.

Given a function $F(u,v)$ of two variables u and v, the subscripts (1) and (2) will be used to denote the derivatives with respect to the first and second argument, respectively. For example,

$$F'_{(1)}(u_1,v_1) = \frac{\partial}{\partial u}F(u,v_1)\Big|_{u=u_1}, \quad F''_{(1)}(u_1,v_1) = \frac{\partial^2}{\partial u^2}F(u,v_1)\Big|_{u=u_1},$$
$$F''_{(2)}(u_1,v_1) = \frac{\partial^2}{\partial v^2}F(u_1,v)\Big|_{v=v_1}, \quad F''_{(2,1)}(u_1,v_1) = \frac{\partial}{\partial u}\frac{\partial}{\partial v}F(u,v)\Big|_{(u,v)=(u_1,v_1)}.$$

By $\mathbf{F}' = \mathbf{F}'(u,v)$ and $\mathbf{F}'' = \mathbf{F}''(u,v)$ we will denote, respectively, the vector

$$\mathbf{F}' = \mathbf{F}'(u,v) = (F'_{(1)}(u,v), F'_{(2)}(u,v))$$

and the matrix

$$\mathbf{F}'' = \mathbf{F}''(u,v) = \|F''_{(i,j)}(u,v)\|_{i,j=1,2}.$$

The determinant of the matrix $\mathbf{F}''$ will be denoted by $|\mathbf{F}''|$.

Recall that, as before, that single-symbol vectors will be highlighted in bold.

We next will use the well-known integro-local theorems for the sums

$$\mathbf{S}_n := \sum_{j=1}^{n} \boldsymbol{\xi}_j = (T_n,\ Z_n)$$

(see, e.g., [23], §2.9 or Theorem 5.3.3 below). In our analysis, an important role will be played by the *deviation function*

$$\mathbf{\Lambda}(\theta, \alpha) := \sup_{(\lambda,\mu)} \{\lambda\theta + \mu\alpha - \mathbf{A}(\lambda, \mu)\}, \tag{5.3.1}$$

corresponding to the random vector $\boldsymbol{\xi} = (\tau, \zeta)$. This function is obtained as the Legendre transform of the convex lower-continuous function $\mathbf{A}(\lambda, \mu)$. Hence the function $\mathbf{\Lambda}(\theta, \alpha)$ is also convex and lower continuous.

In parallel with the sets $\mathcal{A}_1$ and $\mathcal{A}$, we shall also consider the domain $\mathcal{L}$ of analyticity of the function $\mathbf{\Lambda}(\theta, \alpha)$. This domain contains the points (θ, α) for which the system of equations

$$\begin{cases} \mathbf{A}'_{(1)}(\lambda, \mu) = \theta, \\ \mathbf{A}'_{(2)}(\lambda, \mu) = \alpha, \end{cases} \tag{5.3.2}$$

for the point (λ, μ) at which the supremum in (5.3.1) is attained has a solution $(\boldsymbol{\lambda}(\theta, \alpha), \boldsymbol{\mu}(\theta, \alpha))$ lying in $(\mathcal{A})$, so that

$$\mathcal{L} = \{\mathbf{A}'(\lambda, \mu) :\ (\lambda, \mu) \in (\mathcal{A})\}.$$

Since the function $\mathbf{A}(\lambda, \mu)$ is strictly convex in $(\mathcal{A})$, this solution is always unique (see, e.g., [73], Chapter 1). This means that the conditions

$$(\theta, \alpha) \in \mathcal{L} \quad \text{and} \quad (\boldsymbol{\lambda}(\theta, \alpha), \boldsymbol{\mu}(\theta, \alpha)) \in (\mathcal{A})$$

are equivalent. It is clear that the point $(\theta, \alpha) = (a_\tau, a_\zeta)$, where

$$a_\tau := \mathbf{E}\tau, \qquad a_\zeta := \mathbf{E}\zeta,$$

always lies in $\mathcal{L}$; for this point, $(\boldsymbol{\lambda}(a_\tau, a_\zeta), \boldsymbol{\mu}(a_\tau, a_\zeta)) = (0, 0) \in (\mathcal{A})$.

Lemma 5.3.1 *Let $(\theta, \alpha) \in \mathcal{L}$. Then*

$$\mathbf{\Lambda}'_{(1)}(\theta, \alpha) = \boldsymbol{\lambda}(\theta, \alpha), \qquad \mathbf{\Lambda}'_{(2)}(\theta, \alpha) = \boldsymbol{\mu}(\theta, \alpha). \tag{5.3.3}$$

Proof We have

$$\begin{aligned} \mathbf{\Lambda}(\theta, \alpha) &= \boldsymbol{\lambda}(\theta, \alpha)\theta + \boldsymbol{\mu}(\theta, \alpha)\alpha - \mathbf{A}(\boldsymbol{\lambda}(\theta, \alpha), \boldsymbol{\mu}(\theta, \alpha)), \qquad (5.3.4)\\ \mathbf{\Lambda}'_{(1)}(\theta, \alpha) &= \boldsymbol{\lambda}(\theta, \alpha) + \boldsymbol{\lambda}'_{(1)}(\theta, \alpha)\theta + \boldsymbol{\mu}'_{(1)}(\theta, \alpha)\alpha, \\ &\quad - \mathbf{A}'_{(1)}(\boldsymbol{\lambda}(\theta, \alpha), \boldsymbol{\mu}(\theta, \alpha))\boldsymbol{\lambda}'_{(1)}(\theta, \alpha) - \mathbf{A}'_{(2)}(\boldsymbol{\lambda}(\theta, \alpha), \boldsymbol{\mu}(\theta, \alpha))\boldsymbol{\mu}'_{(1)}(\theta, \alpha). \end{aligned}$$

In view of (5.3.2) this implies the first equality in (5.3.3). A similar argument establishes the second equality, so that

$$\mathbf{\Lambda}'(\theta, \alpha) := (\mathbf{\Lambda}'_{(1)}(\theta, \alpha), \mathbf{\Lambda}'_{(2)}(\theta, \alpha)) = (\boldsymbol{\lambda}(\theta, \alpha), \boldsymbol{\mu}(\theta, \alpha)). \tag{5.3.5}$$

The lemma is proved. □

If τ and ζ are independent, then

$$\mathbf{A}(\lambda, \mu) = A^{(\tau)}(\lambda) + A^{(\zeta)}(\mu),$$

where

$$A^{(\tau)}(\lambda) := \ln \mathbf{E} e^{\lambda\tau}, \qquad A^{(\zeta)}(\mu) := \ln \mathbf{E} e^{\mu\zeta}.$$

Hence $\mathbf{A}'_{(1)}(\lambda, \mu)$ does not depend on μ, $\mathbf{A}'_{(2)}(\lambda, \mu)$ does not depend on λ, and the domains $(\mathcal{A})$ and $\mathcal{L}$ are rectangular.

Let, as before, $\mathcal{D}$ be the open cone of analyticity of the function $\mathbf{D}(\theta, \alpha)$ (see (3.5.37)). For $(\theta, \alpha) \in \mathcal{D}$, we let, for brevity (see (3.5.39) and (3.5.40)),

$$\begin{gathered}(\widehat{\lambda}, \widehat{\mu}) := \Big(\lambda\Big(\frac{\alpha}{\theta}\Big), \mu\Big(\frac{\alpha}{\theta}\Big)\Big) = \mathbf{D}'(\theta, \alpha) = \Big(-A\Big(\mu\Big(\frac{\alpha}{\theta}\Big)\Big), \mu\Big(\frac{\alpha}{\theta}\Big)\Big), \\ (\widehat{\theta}, \widehat{\alpha}) := \mathbf{A}'(\widehat{\lambda}, \widehat{\mu})\end{gathered} \tag{5.3.6}$$

(the functions $\mu(\alpha)$, $\lambda(\alpha)$ are defined in (3.5.14), (3.5.38), respectively), so that the vectors $(\widehat{\lambda}, \widehat{\mu})$, $(\widehat{\theta}, \widehat{\alpha})$ are functions of one argument $\frac{\alpha}{\theta}$. Note that, for $(\theta, \alpha) \in \mathcal{D}$, we have (see (3.3.2) and (3.5.38))

$$\begin{gathered}\mathbf{D}(\theta, \alpha) = \theta D\Big(\frac{\alpha}{\theta}\Big) = \theta\Big(\lambda\Big(\frac{\alpha}{\theta}\Big) + \frac{\alpha}{\theta}\mu\Big(\frac{\alpha}{\theta}\Big)\Big) = \theta\widehat{\lambda} + \alpha\widehat{\mu}, \\ \mathbf{A}(\widehat{\lambda}, \widehat{\mu}) = 0, \ (\widehat{\lambda}, \widehat{\mu}) \in (\mathcal{A}).\end{gathered} \tag{5.3.7}$$

Below (see Lemma 5.3.4) it will be shown that for $(\theta, \alpha) \in \mathcal{D}$ the minimum over $r \in (0, \infty)$ of the function

$$L(r) = L_{\theta,\alpha}(r) := r\mathbf{\Lambda}\Big(\frac{\theta}{r}, \frac{\alpha}{r}\Big) \tag{5.3.8}$$

is attained at the unique point

$$r_{\theta,\alpha} = \frac{\theta}{\widehat{\theta}} = \frac{\alpha}{\widehat{\alpha}}.$$

We set

$$Q(\theta, \alpha) := (\theta, \alpha)\mathbf{\Lambda}''(\widehat{\theta}, \widehat{\alpha})(\theta, \alpha)^*, \ \ C(\theta, \alpha) := \sqrt{\frac{r_{\theta,\alpha}}{2\pi}\frac{|\mathbf{\Lambda}''(\widehat{\theta}, \widehat{\alpha})|}{Q(\theta, \alpha)}}, \tag{5.3.9}$$

where the superscript "*" means transposition, so that

$$(\theta, \alpha)\mathbf{\Lambda}''(\widehat{\theta}, \widehat{\alpha})(\theta, \alpha)^* := \theta^2\mathbf{\Lambda}''_{(1)}(\widehat{\theta}, \widehat{\alpha}) + 2\theta\alpha\mathbf{\Lambda}''_{(1,2)}(\widehat{\theta}, \widehat{\alpha}) + \alpha^2\mathbf{\Lambda}''_{(2)}(\widehat{\theta}, \widehat{\alpha}).$$

As was already pointed out, in order to prove the main results of this chapter we shall need integro-local theorems for the renewal measure

$$H(B) := \sum_{n=0}^{\infty} \mathbf{P}(\mathbf{S}_n \in B),$$

corresponding to the random walk $\{\mathbf{S}_n\}$. To be able to distinguish between the homogeneous and the inhomogeneous ($\xi_1 \underset{d}{\neq} \xi$) cases, we denote, as before, by $\widetilde{H}(B)$ the renewal measure in the inhomogeneous case. Let

$$\delta[t) := [t, t+\delta), \qquad \Delta[x) := [x, x+\Delta)$$

be the half-open intervals of length δ and Δ, respectively. We also set

$$B = B_{t,x} := \delta[t) \times \Delta[x). \tag{5.3.10}$$

Consider the parameter T, which characterizes by the equality $(t, x) = (T\theta, T\alpha)$, together with the point (θ, α), the distance from the origin to the parallelepiped $B_{t,x}$. We shall assume that the point $(\theta, \alpha) := \frac{1}{T}(t, x)$ lies in some compact set $\mathbf{K}_{\mathcal{D}} \subset \mathbb{R}^2$ separated from $(0, 0)$ and embedded into the open cone $\mathcal{D}$.

Theorem 5.3.2 *Let $(\theta, \alpha) \in \mathbf{K}_{\mathcal{D}}$ and let the "admissible inhomogeneity" conditions*

$$\mathcal{A}_{\mathbf{K}} \subset \mathcal{A}_1, \text{ where } \mathcal{A}_{\mathbf{K}} := \left\{\left(\lambda\Big(\frac{\alpha}{\theta}\Big), \mu\Big(\frac{\alpha}{\theta}\Big)\right) : (\alpha, \theta) \in \mathbf{K}_{\mathcal{D}}\right\} \tag{5.3.11}$$

be met. Assume that $\delta = \delta_T \to 0$ and $\Delta = \Delta_T \to 0$ sufficiently slowly as $T \to \infty$. Then

$$\widetilde{H}(B_{t,x}) = \frac{\delta\Delta}{\sqrt{T}} \psi_1(\widehat{\lambda}, \widehat{\mu}) C(\theta, \alpha) e^{-T\mathbf{D}(\theta,\alpha)}(1 + o(1)), \tag{5.3.12}$$

where the remainder term $o(1)$ is uniform in $(\theta, \alpha) \in \mathbf{K}_{\mathcal{D}}$.

Most of Theorem 5.3.2 was proved in [39] (see Theorem 5). We will give here a slightly different version of the proof of this theorem for the following reasons.

(1) In [39], the Cramér condition for the characteristic function for the random vector (τ, ζ) was used. To prove (5.3.12) it suffices to invoke Theorem 5.3.3 (see below), in which this condition is not involved.

(2) Theorem 5 in [39] establishes asymptotic expansions, which are not required here. This simplifies the proof.

(3) Theorem 5 in [39] deals only with the homogeneous case.

The proof of Theorem 5.3.2 is based on the integro-local theorem for the sums $\mathbf{S}_n$ as $n \to \infty$. Recall that in parallel with the functions $\lambda(\alpha)$, $\mu(\alpha)$ of one argument α (see (3.5.14) and (3.5.38)), which are the partial derivatives of the second deviation function $\mathbf{D}(\theta, \alpha)$ at the point $\theta = 1$ (see (3.5.39) and (3.5.40)), we in (5.3.3) consider the functions $\boldsymbol{\lambda}(\theta, \alpha)$, $\boldsymbol{\mu}(\theta, \alpha)$ of two variables θ, α. These functions appear as the partial derivatives of the deviation function $\mathbf{\Lambda}(\theta, \alpha)$. The latter functions will be required in the following integro-local theorem, which is a consequence of Theorem 2.8.2 in [23].

Theorem 5.3.3 *Let $\mathbf{K}_{\Lambda}$ be a fixed compact set from the domain of analyticity $\mathcal{L}$ of the function $\mathbf{\Lambda}(\gamma, \beta)$. Next, let points t and x be such that $(\gamma, \beta) := (\frac{t}{n}, \frac{x}{n}) \in \mathbf{K}_{\Lambda}$, and besides, $(\boldsymbol{\lambda}(\gamma, \beta), \boldsymbol{\mu}(\gamma, \beta)) \in (\mathcal{A}_1)$. Then*

$$\mathbf{P}(T_n \in \delta[t),\ Z_n \in \Delta[x)) = \frac{\delta\Delta}{n}\frac{\psi_1}{\psi} C_1(\gamma, \beta) e^{-n\mathbf{\Lambda}(\gamma,\beta)}(1 + o(1)), \tag{5.3.13}$$

if $\delta = \delta_n \to 0$, $\Delta = \Delta_n \to 0$ *sufficiently slowly as* $n \to \infty$, *where* ψ_1, ψ *are, respectively, the values of the functions* $\psi_1(\cdot,\cdot)$, $\psi(\cdot,\cdot)$ *at the point* $(\boldsymbol{\lambda}(\gamma,\beta), \boldsymbol{\mu}(\gamma,\beta))$,

$$C_1(\gamma,\beta) := \frac{\sqrt{|\boldsymbol{\Lambda}''(\gamma,\beta)|}}{2\pi},$$

and the remainder term $o(1)$ *is uniform in* $(\gamma,\beta) \in \mathbf{K}_{\boldsymbol{\Lambda}}$.

Below, representation (5.3.13) will be used in deriving the asymptotics of the renewal measure $\widetilde{H}(B_{t,x})$ for the remote sets $B_{t,x}$ (see Theorem 5.3.2), that is, the asymptotics of the sum of probabilities (5.3.13). The following assertion will be needed in the proof of Theorem 5.3.2.

Lemma 5.3.4 *Let* $(\theta,\alpha) \in \mathcal{D}$ *and let the vectors* $(\widehat{\lambda},\widehat{\mu})$, $(\widehat{\theta},\widehat{\alpha})$ *be defined in* (5.3.6). *Then*:

I. *The minimum of the function* $L(r) = r\boldsymbol{\Lambda}(\frac{\theta}{r},\frac{\alpha}{r})$ *(see* (5.3.8)*) is attained at the unique point*

$$r_{\theta,\alpha} = \frac{\theta}{\widehat{\theta}} = \frac{\alpha}{\widehat{\alpha}}.$$

The functions $\widehat{\lambda}$, $\widehat{\mu}$ *admit the representations*

$$\widehat{\lambda} = \boldsymbol{\lambda}\Big(\frac{\theta}{r_{\theta,\alpha}}, \frac{\alpha}{r_{\theta,\alpha}}\Big), \qquad \widehat{\mu} = \boldsymbol{\mu}\Big(\frac{\theta}{r_{\theta,\alpha}}, \frac{\alpha}{r_{\theta,\alpha}}\Big), \tag{5.3.14}$$

where the functions $\boldsymbol{\lambda}(\theta,\alpha)$, $\boldsymbol{\mu}(\theta,\alpha)$ *are the solutions of system* (5.3.2). *Besides,*

$$L(r_{\theta,\alpha}) = \mathbf{D}(\theta,\alpha), \qquad L'(r)\Big|_{r=r_{\theta,\alpha}} = 0, \tag{5.3.15}$$

$$L''(r)\Big|_{r=r_{\theta,\alpha}} = \frac{1}{r^3_{\theta,\alpha}}\,Q(\theta,\alpha) = \frac{1}{r_{\theta,\alpha}}\,Q(\widehat{\theta},\widehat{\alpha}) > 0. \tag{5.3.16}$$

II. *For a given* $\varepsilon > 0$, *there exists an* $\varepsilon_1 > 0$ *such that*

$$\min_{|r-r_{\theta,\alpha}|\ge\varepsilon} L(r) \ge L(r_{\theta,\alpha}) + \varepsilon_1.$$

Proof of Lemma 5.3.4 1. The deviation function $\boldsymbol{\Lambda}(\theta,\alpha)$ is convex (see, e.g., [21], Chapter 9):

$$p\boldsymbol{\Lambda}(\theta_1,\alpha_1) + q\boldsymbol{\Lambda}(\theta_2,\alpha_2) \ge \boldsymbol{\Lambda}(p\theta_1 + q\theta_2, p\alpha_1 + q\alpha_2)$$

for any $p > 0$, $q > 0$, $p + q = 1$, (θ_1,α_1), $(\theta_2,\alpha_2) \in \mathbb{R}^2$. Hence the function $L(r)$ is convex in $(0,\infty)$:

$$pL(r_1) + qL(r_2) \ge L(pr_1 + qr_2)$$

for any $r_1 > 0$, $r_2 > 0$, $p > 0$, $q > 0$, $p + q = 1$.

2. Putting $(\theta_r,\alpha_r) := \frac{1}{r}(\theta,\alpha)$ and differentiating the function $L(r)$, we have

$$L'(r) = \boldsymbol{\Lambda}(\theta_r,\alpha_r) - \boldsymbol{\Lambda}'_{(1)}(\theta_r,\alpha_r)\theta_r - \boldsymbol{\Lambda}'_{(2)}(\theta_r,\alpha_r)\alpha_r.$$

Now an appeal to (5.3.5) and (5.3.4) shows that

$$L'(r) = \mathbf{\Lambda}(\theta_r,\alpha_r) - \boldsymbol{\lambda}(\theta_r,\alpha_r)\theta_r - \boldsymbol{\mu}(\theta_r,\alpha_r)\alpha_r - \mathbf{A}(\boldsymbol{\lambda}(\theta_r,\alpha_r),\boldsymbol{\mu}(\theta_r,\alpha_r)). \quad (5.3.17)$$

3. Differentiating the function $L'(r)$ in (5.3.17), we have

$$\begin{aligned} L''(r) = &-\mathbf{A}'_{(1)}(\boldsymbol{\lambda}(\theta_r,\alpha_r),\boldsymbol{\mu}(\theta_r,\alpha_r))\Big[\boldsymbol{\lambda}'_{(1)}(\theta_r,\alpha_r)\Big(-\frac{\theta}{r^2}\Big) + \boldsymbol{\lambda}'_{(2)}(\theta_r,\alpha_r)\Big(-\frac{\alpha}{r^2}\Big)\Big] \\ &- \mathbf{A}'_{(2)}(\boldsymbol{\lambda}(\theta_r,\alpha_r),\boldsymbol{\mu}(\theta_r,\alpha_r))\Big[\boldsymbol{\mu}'_{(1)}(\theta_r,\alpha_r)\Big(-\frac{\theta}{r^2}\Big) + \boldsymbol{\mu}'_{(2)}(\theta_r,\alpha_r)\Big(-\frac{\alpha}{r^2}\Big)\Big]. \end{aligned}$$

Since by (5.3.5),

$$\begin{aligned} \boldsymbol{\lambda}'_{(1)}(\theta_r,\alpha_r) &= \mathbf{\Lambda}''_{(1)}(\theta_r,\alpha_r), & \boldsymbol{\lambda}'_{(2)}(\theta_r,\alpha_r) &= \mathbf{\Lambda}''_{(1,2)}(\theta_r,\alpha_r), \\ \boldsymbol{\mu}'_{(1)}(\theta_r,\alpha_r) &= \mathbf{\Lambda}''_{(2,1)}(\theta_r,\alpha_r), & \boldsymbol{\mu}'_{(2)}(\theta_r,\alpha_r) &= \mathbf{\Lambda}''_{(2)}(\theta_r,\alpha_r), \end{aligned}$$

and by (5.3.2),

$$\mathbf{A}'_{(1)}(\boldsymbol{\lambda}(\theta_r,\alpha_r),\boldsymbol{\mu}(\theta_r,\alpha_r)) = \theta_r, \qquad \mathbf{A}'_{(2)}(\boldsymbol{\lambda}(\theta_r,\alpha_r),\boldsymbol{\mu}(\theta_r,\alpha_r)) = \alpha_r,$$

we have

$$L''(r) = \frac{1}{r}(\theta_r,\alpha_r)\mathbf{\Lambda}''(\theta_r,\alpha_r)(\theta_r,\alpha_r)^* > 0, \quad (5.3.18)$$

where the matrix $\mathbf{\Lambda}''(\theta_r,\alpha_r)$ is positive definite (see, e.g., [23], §1.2).

4. From assertions 1–3 we conclude that *if a number $\rho > 0$ is such that the vector $(\theta_\rho,\alpha_\rho) := \frac{1}{\rho}(\theta,\alpha)$ satisfies the two conditions*

$$\mathbf{A}(\boldsymbol{\lambda}(\theta_\rho,\alpha_\rho),\boldsymbol{\mu}(\theta_\rho,\alpha_\rho)) = 0, \qquad (\theta_\rho,\alpha_\rho) \in \mathcal{L}, \quad (5.3.19)$$

then the point $r = \rho$ is a unique point from the half-axis $(0,\infty)$ *at which the function $L(r)$ attains its minimum, that is,*

$$L(\rho) = \inf_{r>0} L(r), \qquad L(r) > L(\rho) \quad \text{for} \quad r > 0, \quad r \neq \rho.$$

5. Recalling that $(\widehat{\lambda},\widehat{\mu}) = \big(\lambda(\frac{\alpha}{\theta}),\mu(\frac{\alpha}{\theta})\big)$ (see (5.3.6)), let us verify that the number

$$\rho := \frac{\theta}{\mathbf{A}'_{(1)}(\widehat{\lambda},\widehat{\mu})} \quad (5.3.20)$$

satisfies both conditions in (5.3.19). To this end, we differentiate with respect to $t \in (\alpha_-,\alpha_+)$ each of the two equalities (see (3.5.38))

$$\mathbf{A}(\lambda(t),\mu(t)) = 0, \qquad -A(\mu(t)) = \lambda(t).$$

This gives us

$$\mathbf{A}'_{(1)}(\lambda(t),\mu(t))\lambda'(t) + \mathbf{A}'_{(2)}(\lambda(t),\mu(t))\mu'(t) = 0, \;\; -A'(\mu(t))\mu'(t) = \lambda'(t).$$

Using the second equality to exclude from the first equality the function $\lambda'(t)$, we get the identity

$$\mathbf{A}'_{(1)}(\lambda(t),\mu(t))A'(\mu(t)) = \mathbf{A}'_{(2)}(\lambda(t),\mu(t)).$$

Substituting here $t = \frac{\alpha}{\theta} \in (\alpha_-, \alpha_+)$ and using the notation $(\widehat{\lambda}, \widehat{\mu}) = \left(\lambda(\frac{\alpha}{\theta}), \mu(\frac{\alpha}{\theta})\right)$, we find that

$$\mathbf{A}'_{(1)}(\widehat{\lambda}, \widehat{\mu}) A'(\widehat{\mu}) = \mathbf{A}'_{(2)}(\widehat{\lambda}, \widehat{\mu}).$$

Next, using the equality $A'(\widehat{\mu}) = \alpha/\theta$ (recall that $A'(\mu(\alpha)) \equiv \alpha$), we get in view of (5.3.20) that

$$(\mathbf{A}'_{(1)}(\widehat{\lambda}, \widehat{\mu}), \mathbf{A}'_{(2)}(\widehat{\lambda}, \widehat{\mu})) = (\theta_\rho, \alpha_\rho). \tag{5.3.21}$$

We have $(\widehat{\lambda}, \widehat{\mu}) \in (\mathcal{A})$ (see (5.3.7)), and hence from (5.3.21) we obtain

$$(\theta_\rho, \alpha_\rho) \in \mathcal{L}. \tag{5.3.22}$$

Thus, we have verified that the second condition in (5.3.19) is satisfied for the number ρ defined by (5.3.20).

Turning back to system (5.3.2), we find that, for $(\theta, \alpha) = (\theta_\rho, \alpha_\rho)$,

$$(\mathbf{A}'_{(1)}(\boldsymbol{\lambda}(\theta_\rho, \alpha_\rho), \boldsymbol{\mu}(\theta_\rho, \alpha_\rho)), \mathbf{A}'_{(2)}(\boldsymbol{\lambda}(\theta_\rho, \alpha_\rho), \boldsymbol{\mu}(\theta_\rho, \alpha_\rho))) = (\theta_\rho, \alpha_\rho) \in \mathcal{L}. \tag{5.3.23}$$

As was already pointed out, system (5.3.2) for $(\theta, \alpha) \in \mathcal{L}$ has a unique solution $(\boldsymbol{\lambda}(\theta, \alpha), \boldsymbol{\mu}(\theta, \alpha))$ (see, e.g., [73], Chapter 1). Hence, comparing the identities (5.3.21) and (5.3.23), we infer that

$$(\widehat{\lambda}, \widehat{\mu}) = (\boldsymbol{\lambda}(\theta_\rho, \alpha_\rho), \boldsymbol{\mu}(\theta_\rho, \alpha_\rho)). \tag{5.3.24}$$

Next

$$\mathbf{A}(\widehat{\lambda}, \widehat{\mu}) = 0$$

(see (5.3.7)), and hence the number ρ, as defined by (5.3.20), obeys the first condition in (5.3.19). Now equality (5.3.14) follows from (5.3.24)). Next, substituting $r = r_{\theta,\alpha}$ into equalities (5.3.17) and (5.3.18), using (5.3.14), and employing the assertion obtained in Step 4 in the proof of the lemma, we arrive at (5.3.15) and (5.3.16). This establishes assertion I of the lemma.

6. Assertion II is a consequence of assertion I. The proof of Lemma 5.3.4 is complete. □

Proof of Theorem 5.3.2 In order to slightly simplify the presentation, we shall assume in the proof of Theorem 5.3.2, in fact, without loss of generality that instead of the condition $(\theta, \alpha) \in \mathbf{K}_{\mathcal{D}}$ (that is, the condition of Theorem 5.3.2), we have the convergence

$$(\theta, \alpha) := \left(\frac{t}{T}, \frac{x}{T}\right) \to (\theta^0, \alpha^0) \quad \text{as} \quad T \to \infty, \tag{5.3.25}$$

where (θ^0, α^0) is any fixed point from $\mathbf{K}_{\mathcal{D}}$ or from $\mathcal{D} = (\mathcal{D})$. Note that the admissible inhomogeneity condition (5.3.11) can be written in the form

$$\left(\lambda\left(\frac{\alpha^0}{\theta^0}\right), \mu\left(\frac{\alpha^0}{\theta^0}\right)\right) \in (\mathcal{A}_1). \tag{5.3.26}$$

If we justify (5.3.12) for $(\theta, \alpha) \to (\theta^0, \alpha^0) \in \mathcal{D}$, then in view of (5.3.12) the standard argument (to the effect that $\mathbf{K}_{\mathcal{D}}$ has convergent subsequences) can be employed to show that the remainder term $o(1)$ in (5.3.12) is uniform in $(\theta, \alpha) \in \mathbf{K}_{\mathcal{D}}$.

The proof is in two steps. In Step I we will prove Theorem 5.3.2 in the homogeneous case. The inhomogeneous case will be treated in Step II.

I. Recall that the renewal measure in the homogeneous case (when $\boldsymbol{\xi}_1 = (\tau_1, \zeta_1) =_d (\tau, \zeta) = \boldsymbol{\xi}$) is denoted by $H(B)$, so that

$$H(B_{t,x}) = \sum_{n=1}^{\infty} \mathbf{P}(T_n \in \delta[t),\ Z_n \in \Delta[x)) \tag{5.3.27}$$

for $\boldsymbol{\xi}_1 \underset{d}{=} \boldsymbol{\xi}$. Next, the set $B_{t,x}$, as defined in (5.3.10), is separated from $(0, 0)$. We split the series on the right of (5.3.27) into three sums

$$\sum_{n\in\mathcal{N}_1}, \quad \sum_{n\in\mathcal{N}_2}, \quad \sum_{n\in\mathcal{N}_3}$$

over the intervals

$$\mathcal{N}_1 := \{1 \le n < c_1 T\}, \quad \mathcal{N}_2 := \{c_1 T \le n \le c_2 T\}, \quad \mathcal{N}_3 := \{n > c_2 T\},$$

respectively, where $c_1 := r_{\theta^0,\alpha^0} - b$ and $c_2 := r_{\theta^0,\alpha^0} + b$, $b > 0$.

Let us verify that if the difference $c_2 - c_1 = 2b$ is sufficiently small, then there exists a compact set $\mathbf{K}_\Lambda \subset \mathcal{L}$ such that

$$(\gamma, \beta) := \left(\frac{t}{n}, \frac{x}{n}\right) = \left(\frac{\theta}{r}, \frac{\alpha}{r}\right) \in \mathbf{K}_\Lambda \tag{5.3.28}$$

for all sufficiently large T for $n \in \mathcal{N}_2$ and $r := \frac{n}{T}$.

Indeed, since $(\theta^0, \alpha^0) \in \mathcal{D}$, we have

$$(\gamma_0, \beta_0) := \left(\frac{\theta^0}{r_{\theta^0,\alpha^0}}, \frac{\alpha^0}{r_{\theta^0,\alpha^0}}\right) \in \mathcal{L}$$

(see (5.3.22) in the proof of Lemma 5.3.4). Therefore, there exists an $\varepsilon_1 > 0$ such that the compact set

$$\mathbf{K}_\Lambda := \{(u, v) : |(u, v) - (\gamma_0, \beta_0)| \le \varepsilon_1\}$$

is wholly contained in the domain $\mathcal{L}$,

$$\mathbf{K}_\Lambda \subset \mathcal{L}.$$

Since $(\theta, \alpha) \to (\theta^0, \alpha^0)$ as $T \to \infty$ (see (5.3.25)) and since the function $r_{\theta,\alpha}$ is analytic is some neighborhood of the point (θ^0, α^0), there exist $T_0 < \infty$ and $b > 0$ such that, for all $T \ge T_0$ and $n \in \mathcal{N}_2$,

$$\left|(\gamma, \beta) - (\gamma_0, \beta_0)\right| \le \varepsilon_1$$

(recall that the constant $b > 0$ appears in the definition of $\mathcal{N}_2$). Therefore, (5.3.28) holds.

Relation (5.3.28) allows us to employ Theorem 5.3.3. For $r := \frac{n}{T} \in [c_1, c_2]$ and $\psi_1 = \psi$, we get

$$\sum_{n\in\mathcal{N}_2} = \delta\Delta \sum_{n\in\mathcal{N}_2} \frac{C_1(\gamma,\beta)}{Tr} e^{-n\boldsymbol{\Lambda}(\gamma,\beta)}(1+o(1))$$
$$= \delta\Delta \sum_{n:\, r\in[c_1,c_2]} \frac{C_1(\frac{\theta}{r},\frac{\alpha}{r})}{Tr} e^{-Tr\boldsymbol{\Lambda}(\frac{\theta}{r},\frac{\alpha}{r})}(1+o(1)).$$

Since the variation of the function $r\boldsymbol{\Lambda}(\frac{\theta}{r},\frac{\alpha}{r}) = L(r)$ near the point $r = r_{\theta,\alpha} \in [c_1, c_2]$ on the interval of length $\frac{1}{T}$ is $o(\frac{1}{T})$, we have

$$\sum_{n\in\mathcal{N}_2} = \delta\Delta \int_{c_1}^{c_2} \frac{1}{r} C_1\Big(\frac{\theta}{r}\cdot\frac{\alpha}{r}\Big) e^{-TL(r)} dr (1+o(1)). \tag{5.3.29}$$

Let us use the equality $L(r_{\theta,\alpha}) = \mathbf{D}(\theta,\alpha)$ (see the first formula in (5.3.15)) and use the well-known Laplace method for evaluation of integrals of the form (5.3.29). As a result, for $(\widehat{\theta},\widehat{\alpha}) = (\frac{\theta}{r_{\theta,\alpha}}, \frac{\alpha}{r_{\theta,\alpha}})$, we have

$$\sum_{n\in\mathcal{N}_2} = \delta\Delta \frac{C_1(\widehat{\theta},\widehat{\alpha})\sqrt{2\pi}}{\sqrt{TL''(r_{\theta,\alpha})}r_{\theta,\alpha}} e^{-T\mathbf{D}(\theta,\alpha)}(1+o(1)).$$

By Lemma 5.3.4 (see (5.3.16)) the right-hand side of this equality coincides with the right-hand side of (5.3.12) with $\psi_1 = \psi$.

Let us now estimate $\sum_{n\in\mathcal{N}_1}$ and $\sum_{n\in\mathcal{N}_3}$. By the multidimensional exponential Chebyshev-type inequality (see Theorem 1.3.2 in [23]) we get

$$\mathbf{P}(T_n \in \delta[t), Z_n \in \Delta[x))$$
$$\le \exp\Big\{-T \inf\Big\{r\boldsymbol{\Lambda}\Big(\frac{\theta'}{r},\frac{\alpha'}{r}\Big) : r = \frac{n}{T},\ \theta \le \theta' \le \theta + \frac{\delta}{T},\ \alpha \le \alpha' \le \alpha + \frac{\Delta}{T}\Big\}\Big\}.$$

By assertion II of Lemma 5.3.4, there exist $\varepsilon_2 > 0$ and $T_0 < \infty$ such that, for all $T \ge T_0$, $n \in \mathcal{N}_1 \cup \mathcal{N}_3$,

$$\inf\Big\{r\boldsymbol{\Lambda}\Big(\frac{\theta'}{r},\frac{\alpha'}{r}\Big) : r = \frac{n}{T},\ \theta \le \theta' \le \theta + \frac{\delta}{T},\ \alpha \le \alpha' \le \alpha + \frac{\Delta}{T}\Big\} \ge \mathbf{D}(\theta,\alpha) + \varepsilon_2.$$

Therefore, for $T \ge T_0$,

$$\mathbf{P}(T_n \in \delta[t), Z_n \in \Delta[x)) \le e^{-T(\mathbf{D}(\theta,\alpha)+\varepsilon_2)}. \tag{5.3.30}$$

Hence by (5.3.30),

$$\sum_{n\in\mathcal{N}_1} \le c_1 T e^{-T(D(\theta,\alpha)+\varepsilon_2)} = o\Big(\frac{\delta\Delta}{\sqrt{T}} e^{-T\mathbf{D}(\theta,\alpha)}\Big).$$

To estimate the sum $\sum_{n\in\mathcal{N}_3}$, we split it into two parts: the sum $\sum'$ over the interval $c_2T \le n \le T^2$ and the sum $\sum''$ over the interval $n > T^2$. The sum $\sum'$ is estimated in the same way as the sum $\sum_{n\in\mathcal{N}_1}$. For the sum $\sum''$, we have

$$\sum{}'' \le \sum_{n>T^2} \mathbf{P}(T_n \le 2t) \le \sum_{n>T^2} e^{-n\Lambda^{(\tau)}(\frac{2t}{n})}, \tag{5.3.31}$$

where, for $n > T^2$ and $\frac{2\theta}{T} \le \mathbf{E}\tau$,

$$\Lambda^{(\tau)}\Big(\frac{2t}{n}\Big) = \Lambda^{(\tau)}\Big(\frac{2\theta T}{n}\Big) \ge \Lambda^{(\tau)}\Big(\frac{2\theta}{T}\Big)$$

(since the function $\Lambda^{(\tau)}(t)$ is decreasing for $t < \mathbf{E}\tau$). We have $\Lambda^{(\tau)}(0) = \infty$, $\Lambda^{(\tau)}(\frac{2\theta}{T}) \to \infty$ as $T \to \infty$, and hence

$$\sum{}'' \le \sum_{n\ge T^2} e^{-n\Lambda^{(\tau)}(\frac{2\theta}{T})} \le 2e^{-T^2\Lambda^{(\tau)}(\frac{2\theta}{T})} = o\Big(\frac{\delta\Delta}{\sqrt{T}} e^{-TD(\theta,\alpha)}\Big). \tag{5.3.32}$$

This proves the assertion of Theorem 5.3.2 in the homogeneous case.

II. Let us consider the inhomogeneous case. We need the following assertion.

Lemma 5.3.5 *Let conditions* (5.3.25) *and* (5.3.26) *hold. Then, for some* $c > 0$, $C < \infty$, $T_0 < \infty$, *and for all* $T \ge T_0$,

$$\Big|\widetilde{H}(\delta[t), \Delta[x)) - \mathbf{E}\Big(H(\delta[t-\tau_1), \Delta[x-\zeta_1)); \ |\boldsymbol{\xi}_1| \le \ln^2 T\Big)\Big| \le Ce^{-T\mathbf{D}(\theta,\alpha) - c\ln^2 T}. \tag{5.3.33}$$

The proof of Lemma 5.3.5 will be given later.

In view of this lemma, to prove (5.3.12) in Theorem 5.3.2 in the general case it suffices to find the asymptotics of

$$\mathbf{E}(H(\delta[t-\tau_1), \Delta[x-\zeta_1)); \ |\boldsymbol{\xi}_1| \le \ln^2 T).$$

Using (5.3.33) and the results of Step I of the proof, we get

$$\begin{aligned}
&\widetilde{H}(\delta[t), \Delta[x)) \\
&\quad = \delta\Delta\frac{1}{\sqrt{T}}\mathbf{E}\Big(C_1\Big(\theta - \frac{\tau_1}{T}, \alpha - \frac{\zeta_1}{T}\Big)e^{-T\mathbf{D}(\theta - \frac{\tau_1}{T},\ \alpha - \frac{\zeta_1}{T})}(1 + o(1)); \ |\boldsymbol{\xi}_1| \le \ln^2 T\Big) \\
&\qquad + o\Big(\frac{\delta\Delta}{\sqrt{T}}\, e^{-T\mathbf{D}(\theta,\alpha)}\Big).
\end{aligned} \tag{5.3.34}$$

Recall that $\mathbf{D}'(\theta,\alpha) = (\widehat{\lambda}, \widehat{\mu})$ (see (5.3.6)). The function $\mathbf{D}(\theta,\alpha)$ is convex, and hence, for any (τ_1, ζ_1), we have

$$-T\mathbf{D}\Big(\theta - \frac{\tau_1}{T}, \alpha - \frac{\zeta_1}{T}\Big) \le -T\mathbf{D}(\theta,\alpha) + \widehat{\lambda}\tau_1 + \widehat{\mu}\zeta_1.$$

Now, since in the domain $|\boldsymbol{\xi}_1| \le \ln^2 T$,

$$-T\mathbf{D}\Big(\theta - \frac{\tau_1}{T}, \alpha - \frac{\zeta_1}{T}\Big) = -T\mathbf{D}(\theta,\alpha) + \widehat{\lambda}\tau_1 + \widehat{\mu}\zeta_1 + o(1)$$

as $T \to \infty$, we see that the right-hand side of (5.3.34) coincides with the right-hand side of (5.3.12).

So, it remains only to prove Lemma 5.3.5.

Proof of Lemma 5.3.5 To this end, we note that since $(0,0) \notin B_{t,x}$, we have

$$\mathbf{P}\big((T_0, Z_0) \in B_{t,x}\big) = 0,$$

and so for $B_{t,x} = \delta[t) \times \Delta[x)$ we get

$$\widetilde{H}(B_{t,x}) = \sum_{n=1}^{\infty} \mathbf{P}((T_n, Z_n) \in B_{t,x}) = \sum_{n=0}^{\infty} \mathbf{P}((\tau_1 + T_n', \zeta_1 + Z_n') \in B_{t,x}),$$

where (T_n', Z_n'), $n = 0, 1, \ldots$ is a sequence of sums of homogeneous independent summands which do not depend on (τ_1, ζ_1). Hence the renewal measure of the set $B_{t,x}$ in the inhomogeneous case can be written as

$$\widetilde{H}(B_{t,x}) = \widetilde{H}(\delta[t), \Delta[x)) = \mathbf{E}H(\delta[t - \tau_1),\ \Delta[x - \zeta_1)). \tag{5.3.35}$$

By (5.3.35) we have

$$\begin{aligned}\widetilde{H}(\delta[t), \Delta[x)) = {}& \mathbf{E}\Big(H(\delta[t - \tau_1),\ \Delta[x - \zeta_1));\ |\boldsymbol{\xi}_1| \le \ln^2 T\Big)\\ &+ \mathbf{E}\Big(H(\delta[t - \tau_1),\ \Delta[x - \zeta_1));\ |\boldsymbol{\xi}_1| > \ln^2 T\Big).\end{aligned}$$

Hence

$$\Big|\widetilde{H}(\delta[t), \Delta[x)) - \mathbf{E}\Big(H(\delta[t - \tau_1), \Delta[x - \zeta_1));\ |\boldsymbol{\xi}_1| \le \ln^2 T\Big)\Big| = \sum_{n=1}^{\infty} P_n,$$

where

$$P_n := \mathbf{P}(T_n \in \delta[t),\ Z_n \in \Delta[x),\ |\boldsymbol{\xi}_1| > \ln^2 T),$$

and so to prove the lemma it suffices to show that

$$\sum_{n=1}^{\infty} P_n \le C e^{-T\mathbf{D}(\theta,\alpha) - c\ln^2 T}. \tag{5.3.36}$$

For all sufficiently large T, we have $(\widehat{\lambda}, \widehat{\mu}) \in (\mathcal{A})$. Hence for such T, for any $n \ge 1$ by (5.3.7), we have

$$\begin{aligned}P_n &= \mathbf{E}(e^{-\widehat{\lambda}T_n - \widehat{\mu}Z_n + \widehat{\lambda}T_n + \widehat{\mu}Z_n};\ T_n \in \delta[t),\ Z_n \in \Delta[x),\ |\boldsymbol{\xi}_1| > \ln^2 T)\\ &\le \exp\{-\widehat{\lambda}t - \widehat{\mu}x + |\widehat{\lambda}|\delta + |\widehat{\mu}|\Delta\}\mathbf{E}(e^{\widehat{\lambda}T_n + \widehat{\mu}Z_n}; |\boldsymbol{\xi}_1| > \ln^2 T)\\ &= \exp\{-T\mathbf{D}(\theta,\alpha) + |\widehat{\lambda}|\delta + |\widehat{\mu}|\Delta\}\mathbf{E}(e^{\widehat{\lambda}\tau_1 + \widehat{\mu}\zeta_1};\ |\boldsymbol{\xi}_1| > \ln^2 T)\prod_{k=2}^{n}\mathbf{E}e^{\widehat{\lambda}\tau_k + \widehat{\mu}\zeta_k}\\ &\le \exp\{-T\mathbf{D}(\theta,\alpha) + |\widehat{\lambda}|\delta + |\widehat{\mu}|\Delta\}\mathbf{E}(e^{\widehat{\lambda}\tau_1 + \widehat{\mu}\zeta_1};\ |\boldsymbol{\xi}_1| > \ln^2 T).\end{aligned}$$

In the last equality we used the fact that by (5.3.7) the vector $(\widehat{\lambda}, \widehat{\mu})$ lies on the boundary of the set $\mathcal{A}^{\le 0}$, and

$$\mathbf{E}e^{\widehat{\lambda}\tau_k+\widehat{\mu}\zeta_k} = 1$$

for $k \ge 2$. Applying further Hölder's inequality, we get, for $p > 0$, $q > 0$, $p + q = 1$,

$$\mathbf{E}(e^{\widehat{\lambda}\tau_1+\widehat{\mu}\zeta_1};\ |\xi_1| > \ln^2 T) \le \left(\mathbf{E}e^{\frac{1}{p}\widehat{\lambda}\tau_1+\frac{1}{p}\widehat{\mu}\zeta_1}\right)^p \mathbf{P}^q(|\xi_1| > \ln^2 T).$$

Next, in view of (5.3.26), for all sufficiently large T, we have $(\widehat{\lambda}, \widehat{\mu}) \in (\mathcal{A}_1)$, and hence choosing $\frac{1}{p} > 1$ to be sufficiently close to 1, we find that

$$\mathbf{E}e^{\frac{1}{p}\widehat{\lambda}\tau_1+\frac{1}{p}\widehat{\mu}\zeta_1} \le C_0 < \infty. \tag{5.3.37}$$

For $q = 1 - p$ and some $C_1 < \infty$ and $c_1 > 0$ from condition [**C**] we have

$$\mathbf{P}^q(|\xi_1| > \ln^2 T) \le C_1 e^{-c_1 \ln^2 T}. \tag{5.3.38}$$

Hence, for any $n \ge 1$,

$$P_n \le C_0^p C_1 e^{-T\mathbf{D}(\theta,\alpha)-c_1\ln^2 T+|\widehat{\lambda}|\delta+|\widehat{\mu}|\Delta}. \tag{5.3.39}$$

Next,

$$\sum_{n=1}^{\infty} P_n \le T^2 \sup_{n\ge 1} P_n + \sum_{n\ge T^2} P_n, \tag{5.3.40}$$

where the first term on the right is estimated using (5.3.39), and the second one, via inequalities (5.3.30) and (5.3.31), from which it follows that

$$\limsup_{T\to\infty} \frac{1}{T} \ln \sum_{n\ge T^2} P_n \le \limsup_{T\to\infty} \frac{1}{T} \ln \sum_{n\ge T^2} \mathbf{P}(T_n < t + \delta) = -\infty. \tag{5.3.41}$$

From inequalities (5.3.39)–(5.3.41), for some $c > 0$ and $C < \infty$ we have (5.3.36), and hence the required assertion (5.3.33). This proves Lemma 5.3.5, and therefore, Theorem 5.3.2. □

Consider now the case when the distribution $P_{1,T}$ of the initial vector $\xi_1 = \xi_{1,T}$ *depends on* T. This situation occurs when studying the increments $Z(U+t) - Z(U)$, $t > 0$, where U depends on T; see the proofs of Theorems 5.2.13 and 5.2.14 below. It will be interesting to find conditions under which the assertion of Theorem 5.3.2 remains valid in this case.

Let $\psi_1(\lambda, \mu) = \psi_{1,T}(\lambda, \mu)$ be the Laplace transform of the distribution $P_{1,T}$. Assume that there exists a set $\widehat{\mathcal{A}}_1$, containing the neighborhood of the point (0, 0) and not depending on T such that, for all sufficiently large T,

$$\sup_{(\lambda,\mu)\in\widehat{\mathcal{A}}_1} \psi_{1,T}(\lambda, \mu) < c_1 < \infty \tag{5.3.42}$$

(it is the uniform version of condition [**C**] for $\xi_1 = \xi_{1,T}$).

Corollary 5.3.6 *Let the conditions of Theorem* 5.3.2 *be satisfied, in which the admissible inhomogeneity condition has the same form as in* (5.3.26)

$$\Big(\lambda\Big(\frac{\alpha^0}{\theta^0}\Big), \mu\Big(\frac{\alpha^0}{\theta^0}\Big)\Big) \in (\widehat{\mathcal{A}}_1), \tag{5.3.43}$$

but now it pertains to the set $\widehat{\mathcal{A}}_1$*, which is defined in* (5.3.42)*. Then the assertion of Theorem* 5.3.2 *remains valid for the initial jumps* $\xi_{1,T}$ *depending on T and satisfying* (5.3.42).

The proof of Theorem 5.3.2 in the inhomogeneous case is based on Lemma 5.3.5. Looking through its proof, we see that under the conditions of Corollary 5.3.6 the main estimates (5.3.37) and (5.3.38) hold uniformly in T, that is, the constants c and C in inequalities (5.3.33) and (5.3.36) are independent of T. This proves Corollary 5.3.6.

5.4 Proof of Theorem 5.2.1 and Its Generalization

5.4.1 Proof of Theorem 5.2.1

Assertion (i) Let us prove the integro-local assertion (5.2.4) for the process $Z(t)$. Similarly to the above, we can without loss of generality reduce the condition $\alpha \in K^{\lessgtr}$ to the convergence $\alpha \to \alpha^0 \in K^{\lessgtr}$, and reduce the admissible inhomogeneity condition to the condition $(\lambda(\alpha^0), \mu(\alpha^0)) \in (\mathcal{A}_1)$.

To simplify the presentation, we will consider in detail only the more difficult problem on the asymptotic behavior of $P^{>}_{T,\alpha}$ in (5.2.4) corresponding to the event $B({}^{>}u, v, \mathbf{w})$. The modifications required in the proof of the simpler problem on the asymptotics of $P^{<}_{T,\alpha}$ will be indicated at the end of the proof.

To simplify the notation in the study of the asymptotic properties of $P^{>}_{T,\alpha}$ we drop, for the moment, the superscripts $\lessgtr$ and will write K instead of $K^{>}$, $B(u, v, \mathbf{w})$ instead of $B({}^{>}u, v, \mathbf{w})$, $I_Z(\alpha, u, v, \mathbf{w})$ instead of $I_Z(\alpha, {}^{>}u, v, \mathbf{w})$, and so on.

For $u < T$, we have

$$\mathbf{P}(Z(T) \in \Delta[x),\ B(u, v, \mathbf{w})) = P_{Z,1} + P_{Z,2}, \tag{5.4.1}$$

where

$$\begin{aligned} P_{Z,1} &:= \mathbf{P}(Z(T) \in \Delta[x),\ B(u, v, \mathbf{w}),\ \tau_1 > T) \\ &= \mathbf{I}_{\{0\in\Delta[x)\}}\mathbf{P}(\tau_1 \ge T+v,\ \zeta_1 \in \mathbf{w}),\ \ \mathbf{I}_{\{0\in\Delta[x)\}} = \mathbf{I}_{\{x\in(-\Delta,0]\}}, \qquad (5.4.2)\\ P_{Z,2} &:= \mathbf{P}(Z(T) \in \Delta[x),\ B(u, v, \mathbf{w}),\ \tau_1 \le T) \\ &= \int_0^{T-u} \widetilde{H}(dy, \Delta[x))\mathbf{P}(\tau > T - y + v,\ \zeta \in \mathbf{w}). \qquad (5.4.3)\end{aligned}$$

We first estimate the part

$$J_1 := \int_{T(1-p)}^{T-u} \widetilde{H}(dy, \Delta[x))\mathbf{P}(\tau > T - y + v,\ \zeta \in \mathbf{w})$$

of the integral in the right-hand side of (5.4.3), which is taken over the domain $(T(1-p), T-u)$ for a fixed $p \in (0,1)$. For future reference, we formulate the upper estimate for the integral J_1 as a lemma.

Lemma 5.4.1 *For any $\varepsilon > 0$, $u \le u_0$ and $T \to \infty$,*

$$J_1 \le \frac{\Delta}{\sqrt{T}} \psi_1 C(1,\alpha) e^{-TD(\alpha)} (1+o(1)) \times \int_u^\infty e^{\lambda(\alpha)y} \mathbf{P}(\tau > y+v-\varepsilon,\ \zeta \in \mathbf{w}), \quad (5.4.4)$$

where the remainder term $o(1)$ is uniform in $u \le u_0$, does not depend on v and $\mathbf{w}$, and $C(\theta,\alpha)$ is defined in (5.3.9).

Proof of Lemma 5.4.1 To simplify the presentation, we first prove (5.4.4) with $\mathbf{w} = \mathbb{R}$. Let $p > 0$ be fixed and such that the integro-local Theorem 5.3.2 for the renewal measure is applicable at the points (t', x), $t' \in (T(1-p), T)$. Setting $y_k = k\delta$, $k = 0, 1, \ldots$, we get the upper estimate

$$J_1 \le \sum_{u \le y_k \le pT} \widetilde{H}(\delta[T-y_k), \Delta[x))\mathbf{P}(\tau > y_k + v - \delta). \quad (5.4.5)$$

By Theorem 5.3.2,

$$J_1 \le \sum_{u \le y_k \le pT} \frac{\delta\Delta}{\sqrt{T}} \psi_1(\lambda(\alpha), \mu(\alpha)) C(1,\alpha)$$
$$\times \exp\Big\{-T\mathbf{D}\Big(1 - \frac{y_k}{T}, \alpha\Big)\Big\}(1+o(1))\mathbf{P}(\tau > y_k + v - \delta), \quad (5.4.6)$$

where $o(1)$ is uniform in $u \le u_0$ and does not depend on v. The function $\mathbf{D}(v,\alpha)$ is convex, and hence

$$\mathbf{D}\Big(1 - \frac{y_k}{T}, \alpha\Big) \ge \mathbf{D}(1,\alpha) - \frac{y_k}{T} \mathbf{D}'_{(1)}(1,\alpha). \quad (5.4.7)$$

By Lemma 3.5.7 for $\alpha \in (\alpha_-, \alpha_+) \setminus [\beta_-, \beta_+]$ we have

$$\mathbf{D}'_{(1)}(1,\alpha) = \lambda(\alpha) < \lambda_+. \quad (5.4.8)$$

Since the point α^0 is separated from the forbidden set $[\beta_-, \beta_+]$ (see Lemma 3.5.8), the value $\lambda(\alpha)$ is separated from λ_+, that is,

$$\lambda(\alpha) < \lambda_+ - \varepsilon \quad \text{with some} \quad \varepsilon > 0. \quad (5.4.9)$$

Since the subsequent calculations are simpler in the case $\lambda_+ = \infty$, in (5.4.9) and below we shall consider only the more difficult case $\lambda_+ < \infty$ (the arguments in the case $\lambda_+ = \infty$ are similar, but simpler).

From (5.4.6), (5.4.7), and (5.4.9) it follows that

$$J_1 \le \frac{\Delta}{\sqrt{T}} \psi_1 C(1,\alpha) e^{-TD(\alpha)} \sum_{u \le y_k \le pT} \delta e^{\lambda(\alpha) y_k} \mathbf{P}(\tau > y_k + v - \delta)(1+o(1)), \quad (5.4.10)$$

where the remainder term $o(1)$ is uniform in $u \le u_0$ and does not depend on v. Next, let, for definiteness, $\lambda(\alpha) \ge 0$. Then

$$\delta e^{y_k \lambda(\alpha)} \mathbf{P}(\tau > y_k + v - \delta) \le \int_{y_k}^{y_{k+1}} e^{\lambda(\alpha) y} \mathbf{P}(\tau > y + v - 2\delta)\, dy$$

and hence from (5.4.10) we get

$$J_1 \le \frac{\Delta}{\sqrt{T}} \psi_1 C(1,\alpha) e^{-TD(\alpha)} \times \int_u^{pT} e^{\lambda(\alpha) y} \mathbf{P}(\tau > y + v - 2\delta)\, dy (1 + o(1)), \quad (5.4.11)$$

where the remainder term $o(1)$ is uniform in $u \le u_0$ and does not depend on v, and the integral on the right of (5.4.11) is convergent by (5.4.8).

From (5.4.11) it follows that, for all sufficiently large T, the integral J_1 is not greater than the right-hand side in (5.4.4) for $\mathbf{w} = \mathbb{R}$. If the event $\{\zeta \in \mathbf{w}\}$ is added under the probability sign in inequalities (5.4.5), (5.4.6), (5.4.10), and (5.4.11) , then all the assertions in these results on the independence of v will obviously turn to the results on the independence of v and $\mathbf{w}$. Hence, for all sufficiently large T, the integral J_1 is majorized by the right-hand side of (5.4.4) uniformly in $\alpha \in K$, $u \le u_0$, v, $\mathbf{w}$. This proves Lemma 5.4.1. □

The lower estimate of the same form for J_1 can be derived in the same way: one should replace $\mathbf{P}(\tau > y_k + v - \delta)$ by $\mathbf{P}(\tau > y_k + v + \delta)$ on the right of (5.4.5) and, instead of (5.4.7), employ the asymptotic representation

$$\mathbf{D}\left(1 - \frac{y_k}{T}, \alpha\right) = \mathbf{D}(1,\alpha) - \frac{y_k}{T} \mathbf{D}'_{(1)}(1,\alpha) + O\left(\left(\frac{y_k}{T}\right)^2\right)$$

for small $\frac{y_k}{T}$ (for small p).

As a result, from these estimates we get

$$J_1 = \frac{\Delta}{\sqrt{T}} \psi_1 C(1,\alpha) e^{-TD(\alpha)} (1 + o(1)) \int_u^{\infty} e^{\lambda(\alpha) y} \mathbf{P}(\tau > y + v - \varepsilon,\ \zeta \in \mathbf{w}). \quad (5.4.12)$$

Let us now estimate from above the second part J_2 of the integral in (5.4.3). We have

$$\begin{aligned} J_2 &:= \int_0^{T(1-p)} \widetilde{H}(dy, \Delta[x)) \mathbf{P}(\tau > T - y + v,\ \zeta \in \mathbf{w}) \\ &\le \int_0^{T(1-p)} \widetilde{H}(dy, \Delta[x)) \mathbf{P}(\tau > T - y). \end{aligned} \quad (5.4.13)$$

For $y_k := \delta k$, $k = 0, 1, \ldots$, $v \ge 0$, we get

$$\begin{aligned} J_2 &\le \sum_{k=0}^{[T(1-p)/\delta]} \widetilde{H}(\delta[y_k), \Delta[x)) \mathbf{P}(\tau > T - y_k) \\ &\le \sum_{k=0}^{[T(1-p)/\delta]} \widetilde{H}(\delta[y_k), \Delta[x)) \mathbf{P}(\tau > T - y_k). \end{aligned} \quad (5.4.14)$$

We now need the following assertion (cf. Theorem 5.3.2).

Lemma 5.4.2 *For some $C < \infty$, $T_0 < \infty$, the inequality*

$$\widetilde{H}(\delta[y), \Delta[x)) \le CT^2 e^{-\lambda(\alpha)y-\mu(\alpha)x} \tag{5.4.15}$$

holds for all

$$T \ge T_0, \quad 0 \le y \le T(1-p), \quad 0 < \delta \le 1, \quad 0 < \Delta \le 1.$$

Lemma 5.4.2 will be proved below in this section, but now using this assertion we will continue to estimate J_2 from above.

From the exponential Chebyshev-type inequality it follows that

$$\limsup_{T\to\infty} \frac{1}{T} \ln \mathbf{P}(\tau \ge T) \le -\lambda_+. \tag{5.4.16}$$

From (5.4.15) and (5.4.16) we get the upper estimate for logarithm of J_2 in the form

$$\limsup_{T\to\infty} \frac{1}{T} \ln J_2 \le - \min_{0\le w\le 1-p} \{\lambda(\alpha^0)w + \mu(\alpha^0)\alpha^0 + (1-w)\lambda_+\}. \tag{5.4.17}$$

The condition $\alpha^0 \notin [\beta_-, \beta_+]$ implies that $\lambda_+ - \lambda(\alpha^0) > 0$, and hence the infimum on the right of (5.4.17) is attained at $w = 1-p$, so that the right-hand side in (5.4.17) equals

$$-\lambda(\alpha^0) - \mu(\alpha^0)\alpha^0 - p(\lambda_+ - \lambda(\alpha^0)) = -\mathbf{D}(1, \alpha^0) - p(\lambda_+ - \lambda(\alpha^0)).$$

Hence

$$\limsup_{T\to\infty} \frac{1}{T} \ln J_2 \le -\mathbf{D}(1,\alpha^0) - p(\lambda_+ - \lambda(\alpha^0)) \le -D(\alpha) - \frac{p}{2}(\lambda_+ - \lambda(\alpha^0)). \tag{5.4.18}$$

From (5.4.18) it follows that, for some $\varepsilon > 0$ and as $T \to \infty$,

$$J_2 = O(e^{-T(D(\alpha)+\varepsilon)}) = o\Big(\frac{\Delta}{\sqrt{T}} e^{-TD(\alpha)}\Big). \tag{5.4.19}$$

So, the sum $P_{Z,2} = J_1 + J_2$ is equal to the right-hand side of (5.4.12). In view of (5.4.1)–(5.4.3) this proves the required assertion (5.2.4). Therefore, to verify (5.2.4) it suffices to complete the proof of Lemma 5.4.2.

Proof of Lemma 5.4.2 For any $n \ge 1$, $\Delta \le 1$, $\delta \le 1$ we have

$$\begin{aligned}
&\mathbf{P}(T_n \in \delta[y),\ Z_n \in \Delta[x)) \\
&\quad = \mathbf{E}\Big(\exp\{-\lambda(\alpha)T_n - \mu(\alpha)Z_n + \lambda(\alpha)T_n + \mu(\alpha)Z_n\};\ T_n \in \delta[y),\ Z_n \in \Delta[x)\Big) \\
&\quad \le \exp\{-\lambda(\alpha)y - \mu(\alpha)x\} e^{|\lambda(\alpha)|\delta + |\mu(\alpha)|\Delta} \mathbf{E} e^{\lambda(\alpha)T_n + \mu(\alpha)Z_n} \\
&\quad \le \exp\{-\lambda(\alpha)y - \mu(\alpha)x\} e^{|\lambda(\alpha)| + |\mu(\alpha)|} \psi_1(\lambda(\alpha), \mu(\alpha)) \psi^{n-1}(\lambda(\alpha), \mu(\alpha)).
\end{aligned}$$

We have $\alpha \to \alpha^0$ as $T \to \infty$, and hence, for all sufficiently large T,

$$(\lambda(\alpha), \mu(\alpha)) \in (\mathcal{A}), \qquad \psi(\lambda(\alpha), \mu(\alpha)) = 1,$$

Next, by the admissible inhomogeneity condition $(\lambda(\alpha^0), \mu(\alpha^0)) \in (\mathcal{A}_1)$, we obtain

$$\limsup_{T\to\infty} \psi_1(\lambda(\alpha), \mu(\alpha)) < \infty.$$

Hence, for some $T_0 < \infty$, $C_1 < \infty$ and for all $T \geq T_0, n \geq 1$,

$$\mathbf{P}(T_n \in \delta[y],\ Z_n \in \Delta[x]) \leq C_1 e^{-\lambda(\alpha)y-\mu(\alpha)x}.$$

Therefore, for all $T \geq T_0$,

$$\widetilde{H}(\delta[y], \Delta[x]) \leq C_1 T^2 e^{-\lambda(\alpha)y-\mu(\alpha)x} + \sum_{n>T^2} \mathbf{P}(T_n \in \delta[y],\ Z_n \in \Delta[x]). \tag{5.4.20}$$

Next, for $y + \delta \leq T(1-p) + \delta \leq T$ we have

$$\mathbf{P}(T_n \in \delta[y],\ Z_n \in \Delta[x]) \leq \mathbf{P}(T_n \leq T) \leq \mathbf{P}\Big(\sum_{k=2}^{n} \tau_k \leq T\Big),$$

and now an appeal to the exponential Chebyshev inequality shows that, for $n \geq T^2+1$,

$$\mathbf{P}\Big(\sum_{k=2}^{n} \tau_k \leq T\Big) \leq e^{-(n-1)\Lambda^{(\tau)}(\frac{T}{n-1})} \leq e^{-(n-1)\Lambda^{(\tau)}(\frac{1}{T})},$$

where $\Lambda^{(\tau)}(\frac{1}{T}) \to \Lambda^{(\tau)}(0) = \infty$ as $T \to \infty$. Hence (cf. (5.3.31) and (5.3.32) in which the homogeneous case is considered) we get

$$\limsup_{T\to\infty} \frac{1}{T} \ln \sum_{n>T^2} \mathbf{P}(T_n \in \delta[y],\ Z_n \in \Delta[x]) = -\infty. \tag{5.4.21}$$

Now (5.4.15) follows from (5.4.20) and (5.4.21). Lemma 5.4.2, and therefore, relation (5.2.4) are proved. □

Assertion (ii) Let us verify assertion (ii) of Theorem 5.2.1 for the homogeneous processes $Z(t)$. Consider the first summand on the right of (5.2.4). In the case $0 \notin K$ this summand is absent (for sufficiently small Δ). In the alternative case $0 \in K$, we have $0 \notin [\beta_-, \beta_+]$ (see Lemma 3.5.8) and, therefore, the set $[\beta_-, \beta_+]$ is empty, $\lambda_+^{(\tau_1)} = \lambda_+ > D(0)$. In this case, the sufficient condition (5.2.9) of Remark 5.2.3 is satisfied, in view of which this summand behaves as $O(e^{-T(D(0)+\varepsilon)})$.

Assertion (iii) Let $Z(t) = Z^{(st)}(t)$. If $\alpha \in K$, then for α' near the point α, we have $\psi(\lambda(\alpha'), \mu(\alpha')) = 1$, and hence by (5.2.10),

$$\psi(0, \mu(\alpha')) < \infty, \qquad \psi^{(st)}(\lambda(\alpha'), \mu(\alpha')) < \infty.$$

This means that $(\mathcal{A}_K) \subset (\mathcal{A}^{(st)})$ and that the admissible homogeneity condition is satisfied for $\mathcal{A}_1 = \mathcal{A}^{(st)}$. The rest of the argument, showing that the first summand on the right-hand side of (5.2.4) can be neglected, repeats that given in the proof of assertion (ii), since again, as in (ii), we have $\lambda_+^{(\tau_1)} = \lambda_+$.

If $\mathcal{A}$ is rectangular, then

$$\mathcal{A}_K = \{(\lambda(\alpha), \mu(\alpha)) : \alpha \in K\} \subset (\mathcal{A}) = (-\infty, \lambda_+) \times (\mu_-^{(\zeta)}, \mu_+^{(\zeta)}),$$

which implies (5.2.5).

The proof of (5.2.4) in Theorem 5.2.1 for the probability $P^{<}_{T,\alpha}$ is much simpler than the proof given above. The integral from 0 to $T - u$ in (5.4.3) has to be replaced by the integral from $T - u$ to T. To this integral there corresponds in (5.2.4) the integral

$$I_Z(\alpha, {}^{<}u, v, \mathbf{w}) = \int_0^u e^{\lambda(\alpha)y} \mathbf{P}(\tau > y + v, \zeta \in \mathbf{w})\, dy,$$

which is always finite, and hence there is no need to invoke the condition $\alpha \notin [\beta_-, \beta_+]$ to ensure the inequality $\lambda(\alpha) < \lambda_+ - \varepsilon$ and check that the integral $I_Z(\alpha, {}^{>}u, v, \mathbf{w})$ is finite, as was done for the probability $P^{>}_{T,\alpha}$. The study of the asymptotic behavior of the integral in the range from $T - u$ to T in (5.4.3), which is obtained with the above change, is much simpler.

Theorem 5.2.1 is proved. □

In a similar manner, but with many simplifications, we can consider the case when the random variable τ is arithmetic, ζ is nonlattice, and the parameter $T = n$ is integer. As a result, we obtain, in particular, the following assertion.

Corollary 5.4.3 *Let the random variable ζ be nonlattice. In the homogeneous case,*

$$\mathbf{P}(Z(n) \in \Delta[x), \gamma(n) = l) = \frac{\Delta C(\alpha)}{\sqrt{n}} e^{-nD(\alpha)} \mathbf{P}(\tau > l)(1 + o(1))$$

for $\alpha \in K^{<}$, any fixed l, and $\Delta = \overline{o}(1)$, as $n \to \infty$.

5.4.2 Extension of Results to the Case when the Distribution of (τ_1, ζ_1) Depends on a Parameter

Consider now the case when the distribution $P_1 = P_{1,U}$ of the initial vector $\boldsymbol{\xi}_1 = \boldsymbol{\xi}_{1,U}$ depends on some parameter U so that

$$P_{1,U} \Rightarrow P_1 \quad \text{as} \quad U \to \infty. \tag{5.4.22}$$

Let U depend on T so that $U \to \infty$ as $T \to \infty$. The question is, under what conditions on $P_{1,U}$ as $T \to \infty$, will the assertion of Theorem 5.2.1 remain valid?

We need the following condition.

[$\widehat{\mathbf{P}}_1$] *There exist a distribution $\widehat{P}_1$ in $\mathbb{R}^2$ and a constant $c < \infty$ such that*

1. $$P_{1,U}((t, \infty) \times \Delta[x)) \le c\widehat{P}_1((t, \infty) \times \Delta[x)), \tag{5.4.23}$$

for all $t \ge 0$, x, any fixed $\Delta > 0$, and all sufficiently large U (or T).

2. $$\widehat{\psi}_1(\lambda, \mu) < \infty \quad \textit{for} \quad (\lambda, \mu) \in \mathcal{A}_{K^{\lessgtr}}$$

where $\widehat{\psi}_1(\lambda, \mu)$ is the Laplace transform of the distribution $\widehat{P}_1$ and, as before,

$$\mathcal{A}_K = \{(\lambda(\alpha), \mu(\alpha)) : \alpha \in K\}.$$

The following assertion holds.

Theorem 5.4.4 *Let ψ_1 and $\psi_{1,T}$ be the Laplace transforms of the distributions P_1 and $P_{1,T}$, respectively. Next, let the conditions of Theorem* 5.2.1 *be met, in which the admissible inhomogeneity condition* (5.2.3) *is replaced by conditions* (5.4.22) *and* $[\widehat{\mathbf{P}}_1]$. *Then $\psi_{1,U}(\lambda,\mu) \to \psi_1(\lambda,\mu)$ for $(\lambda,\mu) \in \mathcal{A}_{K^{\lessgtr}}$, $T \to \infty$, and besides,*

$$\mathbf{P}(Z(T) \in \Delta[x), B(^{\lessgtr}u, v, \mathbf{w})) = \frac{\Delta}{\sqrt{T}}\psi_1 C(\alpha)e^{-TD(\alpha)}I_Z(\alpha, {}^{\lessgtr}u, v, \mathbf{w})(1 + o(1)) \tag{5.4.24}$$

for $\Delta = \overline{o}(1)$ as $T \to \infty$ (cf. Theorem 5.2.1).

Proof Let K be either of the sets $K^{\lessgtr}$. The condition $[\widehat{\mathbf{P}}_1]$ implies that

$$\psi_{1,U}(\lambda,\mu) \le c\widehat{\psi}_1(\lambda,\mu) \tag{5.4.25}$$

for $(\lambda,\mu) \in \mathcal{A}_K$ and all sufficiently large U. Indeed, integrating by parts two times, we get

$$\begin{aligned}\psi_{1,U}(\lambda,\mu) &= \int_{-\infty}^{\infty}\int_0^{\infty} e^{\lambda t+\mu x}P_{1,U}(dt,dx)\\ &= \int_{-\infty}^{\infty} e^{\mu x}\Big[e^{\lambda t}P_{1,U}((0,\infty)\times dx) + \int_0^{\infty}\lambda e^{\lambda t}P_{1,U}((t,\infty)\times dx)\,dt\Big]\\ \le c&\int_{-\infty}^{\infty} e^{\mu x}\Big[e^{\lambda t}\lambda t\widehat{P}_1((0,\infty)\times dx) + \int_0^{\infty}\lambda e^{\lambda t}\widehat{P}_1((t,\infty)\times dx)\,dt\Big] = c\widehat{\psi}_1(\lambda,\mu).\end{aligned}$$

From (5.4.25) we have

$$\sup_{(\lambda,\mu)\in\mathcal{A}_K} \psi_{1,U}(\lambda,\mu) \le c_1 < \infty \tag{5.4.26}$$

for all sufficiently large U. The set $\mathcal{A}_K$ contains a neighborhood of the point $(0,0)$ (see Remark 5.2.3), and hence by (5.4.26) condition (5.3.42) is satisfied with $\widehat{\mathcal{A}}_1 = \mathcal{A}_K$. Under the conditions of Theorem 5.3.2, we have

$$\left(\lambda\Big(\frac{\alpha^0}{\theta^0}\Big), \mu\Big(\frac{\alpha^0}{\theta^0}\Big)\right) \in (\mathcal{A}_K) = (\widehat{\mathcal{A}}_1),$$

and hence condition (5.3.43) of Corollary 5.3.6 with $\widehat{\mathcal{A}}_1 = \mathcal{A}_K$ is also satisfied. Further, from (5.4.22) and (5.4.25) it follows that

$$\psi_{1,U}(\lambda,\mu) \to \psi_1(\lambda,\mu) \tag{5.4.27}$$

as $T \to \infty$ and $(\lambda,\mu) \in (\mathcal{A}_K)$. From Corollary 5.3.6 and (5.4.27) we see that under (5.4.22) and $[\widehat{\mathbf{P}}_1]$ the assertion of Theorem 5.3.2 remains valid, and therefore, all the arguments in the proof of Theorem 5.2.1 will apply. Theorem 5.4.4 is proved. □

Let us now invoke Theorem 5.4.4 in order to obtain an integro-local theorem for the increments

$$Z_U(t) = Z(U+t) - Z(U),$$

where $Z(t)$ is a homogeneous CRP. In this case, the initial jump $\xi_1 = \xi_{1,U}$ depends on U and is equal to

$$\xi_{1,U} = (\chi(U), \zeta(U)).$$

Assume, for definiteness, that $\lambda_+ \geq D(0)$ and $K = K^<$. Let us show that the distribution $P_{1,U}$ of the vector $\xi_{1,U}$ satisfies the condition $[\widehat{\mathbf{P}}_1]$ in which as the distribution $\widehat{P}_1$ one should take the distribution $P^{(st)}$ of the vector $\xi^{(st)}$ with the Laplace transform

$$\psi^{(st)}(\lambda, \mu) = \frac{1}{\lambda a_\tau}\,(\psi(\lambda, \mu) - \psi(0, \mu)).$$

Indeed, from the analysis of §1.1 it follows that $P_{1,U} \Rightarrow P^{(st)}$ as $U \to \infty$, so that $P_1 = P^{(st)}$. Inequality (5.4.23) with $\widehat{P}_1 = P_1$ can be easily derived from the proof of Lemma 1.1.1. We thus have the following result.

Corollary 5.4.5 *Let $Z(t)$ be a homogeneous CRP, let $\lambda_+ \geq D(0)$, let $\alpha = x/T \in K$, and let $\Delta = \overline{o}(1)$ as $T \to \infty$. Then*

$$\mathbf{P}(Z(U+T) - Z(U) \in \Delta[x)) = \frac{\Delta}{\sqrt{T}}\,\psi^{(st)} C(\alpha) e^{-TD(\alpha)} I_Z(\alpha, 0, 0, \mathbb{R})(1 + o(1))$$

as $U \to \infty$, $T \to \infty$.

5.5 Proofs of Theorems 5.2.10–5.2.14

5.5.1 Proof of Theorem 5.2.10

Let us again confine ourselves to the consideration of the more difficult case concerning the asymptotics of $\mathbf{P}(Y(T) \in \Delta[x), B(^>u, v))$.

(i) For $\lambda = \lambda(\alpha)$ and $\mu = \mu(\alpha)$ we have

$$I_Y(\alpha, {}^>0, 0) = \int_0^\infty e^{\lambda y} \mathbf{E}(e^{\mu\zeta};\ \tau > y)\,dy = \int_{-\infty}^\infty e^{\mu z} M(\lambda, dz),$$

where

$$M(\lambda, dz) := \int_0^\infty e^{\lambda y} \mathbf{P}(\zeta \in dz,\ \tau > y)\,dy.$$

For $\lambda \neq 0$, setting $U(y) := \mathbf{P}(\zeta \in dz,\ \tau > y)$, and $V(y) := \frac{e^{\lambda y}}{\lambda}$, we have

$$M(\lambda, dz) = \int_0^\infty U(y) dV(y) = U(y)V(y)\,\Big|_0^\infty + \frac{1}{\lambda}\int_0^\infty e^{\lambda y} \mathbf{P}(\zeta \in dz,\ \tau \in dy),$$

where

$$U(y)V(y)\,\Big|_0^\infty = -\frac{1}{\lambda}\mathbf{P}(\zeta \in dz).$$

Hence by the equality $\psi(\lambda, \mu) = 1$ and from condition (5.2.5), we get $\mu \in (\mu_-^{(\zeta)},\ \mu_+^{(\zeta)})$, and

$$I_Y(\alpha, {}^>0, 0) = \frac{1 - \psi^{(\zeta)}(\mu)}{\lambda} < \infty$$

(if $\lambda = \lambda(\alpha) > 0$, then by inequality $I_Y(\alpha, 0, 0) > 0$ we have $\psi^{(\zeta)}(\mu) < 1$, and so condition (5.2.5) is satisfied automatically).

For $\lambda = 0$, letting $V(y) = y$, we get

$$M(0, dz) = \int_0^\infty \mathbf{P}(\zeta \in dz, \tau > y)\, dy = \int_0^\infty y\mathbf{P}(\zeta \in dz, \tau \in dy),$$

$$I_Y(\alpha, {}^{>}0, 0) = \int_{-\infty}^\infty \int_0^\infty e^{\mu z} y\mathbf{P}(\zeta \in dz, \tau \in dy) = \psi'_{(1)}(0, \mu).$$

This proves (5.2.14) and (5.2.15).

(ii) The proof of the second assertion is based on Theorem 5.2.1. For brevity, we denote by B_Y the event

$$B_Y = B_Y(x, \Delta, {}^{>}u, v) := \{Y(T) \in \Delta[x),\ B({}^{>}u, v)\}.$$

Note that

$$\mathbf{P}(B_Y) = \mathbf{P}(B_Y,\ \tau_1 > T) + \mathbf{P}(B_Y,\ \tau_1 \le T), \tag{5.5.1}$$

where

$$\mathbf{P}(B_Y,\ \tau_1 > T) = \mathbf{P}(\tau_1 > T + v,\ \zeta_1 \in \Delta[x)).$$

The second summand in (5.5.1) is

$$\mathbf{P}(B_Y,\ \tau_1 \le T) = \int_{-\infty}^\infty \mathbf{P}(Z(T) \in dz, B({}^{>}u, v, \Delta[x - z)),\ \tau_1 \le T). \tag{5.5.2}$$

We now approximate the integral on the right of (5.5.2) by the series

$$\sum := \sum_k \mathbf{P}(Z(T) \in \delta[z_k);\ B({}^{>}u, v, \Delta[x - z_k)),\ \tau_1 \le T), \tag{5.5.3}$$

where

$$z_k = x - s_k, \qquad s_k = k\delta, \qquad k = \dots, -1, 0, 1, \dots$$

Let Δ be fixed and let δ converges to zero as $T \to \infty$. Then

$$\begin{aligned}\mathbf{P}(B_Y,\ \tau_1 \le T) &= \sum_k \int_{\delta[z_k)} \mathbf{P}\big(Z(T) \in dz,\ B({}^{>}u, v, \Delta[x - z)),\ \tau_1 \le T\big)\\ &\le \sum_k \int_{\delta[z_k)} \mathbf{P}\big(Z(T) \in dz,\ B({}^{>}u, v, \Delta'[x' - z_k)),\ \tau_1 \le T\big),\end{aligned} \tag{5.5.4}$$

where

$$x' = x - \delta, \qquad \Delta' = \Delta + \delta, \tag{5.5.5}$$

so that

$$\mathbf{P}(B_Y,\ \tau_1 \le T) \le \sum_k \mathbf{P}\big(Z(T) \in \delta[z_k),\ B({}^{>}u, v, \Delta'[x' - z_k)),\ \tau_1 \le T\big).$$

A similar argument gives the lower estimate

$$\mathbf{P}(B_Y,\ \tau_1 \le T) \ge \sum_k \mathbf{P}\big(Z(T) \in \delta[z_k),\ B({}^{>}u, v, \Delta''[x - z_k)),\ \tau_1 \le T\big),$$

where

$$\Delta'' = \Delta - \delta. \tag{5.5.6}$$

Let us now find the asymptotics of the series $\sum$ in (5.5.3) and verify that it is insensitive to the substitutions (5.5.5) and (5.5.6) as $\delta \to 0$.

We first consider the subsum $\sum_1$ of the series (5.5.3) formed by the summands corresponding to s_k and such that $|s_k| < \varepsilon\sqrt{T}$, where $\varepsilon \to 0$ as $T \to \infty$, $\varepsilon\sqrt{T} \to \infty$. Setting

$$\alpha_k := \frac{z_k}{T} = \alpha - \frac{s_k}{T},$$

we have, as $T \to \infty$,

$$D(\alpha_k) = D(\alpha) - \frac{s_k}{T}\,\mu(\alpha) + O\Big(\Big(\frac{s_k}{T}\Big)^2\Big). \tag{5.5.7}$$

The functions $C(\alpha)$ and $\lambda(\alpha)$ from (5.2.4) are continuous with respect to α, so that

$$C(\alpha_k) \sim C(\alpha), \qquad \lambda(\alpha_k) \sim \lambda(\alpha) \quad \text{for} \quad |s_k| < \varepsilon\sqrt{T}, \quad T \to \infty.$$

Hence if δ converges to 0 sufficiently slowly as $T \to \infty$, then by Theorem 5.2.1 (see also (5.4.3)),

$$\sum\nolimits_1 \sim \frac{\delta}{\sqrt{T}}\,\psi_1 C(\alpha) e^{-TD(\alpha)} \sum_{k:\ |s_k| < \varepsilon\sqrt{T}} e^{\mu(\alpha)s_k} \times \int_u^\infty e^{\lambda(\alpha)y}\mathbf{P}(\tau > y + v,\ \zeta \in \Delta[s_k))\,dy, \tag{5.5.8}$$

where the sum on the right multiplied by δ is asymptotically equivalent to

$$\begin{aligned}
&\int_{-\infty}^\infty e^{\mu(\alpha)s} \int_u^\infty e^{\lambda(\alpha)y}\mathbf{P}(\tau > y + v,\ \zeta \in \Delta[s))\,dy\,ds \\
&\quad = \int_u^\infty e^{\lambda(\alpha)y} \int_{s=-\infty}^\infty e^{\mu(\alpha)s} \int_{z=s}^{s+\Delta} \mathbf{P}(\tau > y + v,\ \zeta \in dz)\,ds\,dy \\
&\quad = \int_u^\infty e^{\lambda(\alpha)y} \int_{z=-\infty}^\infty \mathbf{P}(\tau > y + v,\ \zeta \in dz) \int_{s=z}^{z+\Delta} e^{\mu(\alpha)s}\,ds\,dy \\
&\quad = \frac{e^{\mu(\alpha)\Delta} - 1}{\mu(\alpha)} \int_u^\infty e^{\lambda(\alpha)y}\mathbf{E}(e^{\zeta\mu(\alpha)};\ \tau > y + v)\,dy = \frac{e^{\mu(\alpha)\Delta} - 1}{\mu(\alpha)} I_Y(\alpha, u, v).
\end{aligned}$$

If we now replace x on the right by $x' = x - \delta$ (that is, we replace α by $\alpha' = \alpha - \frac{\delta}{T}$) and substitute $\Delta' = \Delta + \delta$ or $\Delta'' = \Delta - \delta$ for Δ, then, clearly, the new expression will be asymptotically equivalent to the original one. The same applies to the factor

$$\psi_1 C(\alpha) e^{-TD(\alpha)}$$

in (5.5.8). Hence the original expression for $\mathbf{P}(B_Y,\ \tau_1 \le T)$ in (5.5.4) is asymptotically equivalent to

$$\frac{e^{\mu(\alpha)\Delta}-1}{\mu(\alpha)\sqrt{T}}\,\psi_1 C(\alpha)e^{-TD(\alpha)}I_Y(\alpha,u,v).$$

It remains to estimate from above the subsums of the series (5.5.3) over the exterior of the interval $(-\varepsilon\sqrt{T},\varepsilon\sqrt{T})$, i.e., over $s_k \notin (-\varepsilon\sqrt{T},\varepsilon\sqrt{T})$. Let us briefly describe this procedure. First, consider the interval $\varepsilon(-T,-\sqrt{T})$, where ε is so small that $\frac{x}{T}-\varepsilon=\alpha-\varepsilon\in K$. Then the subsum $\sum_2$ of the series (5.5.3) over the interval $-\varepsilon T < s_k \le -\varepsilon\sqrt{T}$ is estimated from above as before, but instead of the asymptotic representation (5.5.7) one should use the inequality

$$D(\alpha_k) \ge D(\alpha) - \frac{s_k}{T}\,\mu(\alpha), \tag{5.5.9}$$

which follows from the convexity of the function $D(\alpha)$ and the equality $D'(\alpha)=\mu(\alpha)$. Hence, in place of (5.5.8), we obtain the inequality

$$\sum\nolimits_2 \le \frac{\delta}{\sqrt{T}}\,\psi_1 C(\alpha)e^{-TD(\alpha)}\sum_{s_k<-\varepsilon\sqrt{T}} e^{\mu(\alpha)s_k}\int_0^\infty e^{\lambda(\alpha)y}\mathbf{P}(\tau>y,\ \zeta\in\Delta[s_k))\,dy,$$

where the sum on the right multiplied by δ is asymptotically equivalent to

$$\int_{-\infty}^{-\varepsilon\sqrt{T}} e^{\mu(\alpha)s}\int_0^\infty e^{\lambda(\alpha)y}\mathbf{P}(\tau>y,\ \zeta\in\Delta[s))\,dy\,ds.$$

Integrating by parts, one can easily show that this integral has the same order of smallness as the integral

$$\int_{-\infty}^{-\varepsilon\sqrt{T}}\int_0^\infty e^{\lambda(\alpha)y+\mu(\alpha)s}\mathbf{P}(\tau\in dy,\ \ \zeta\in ds). \tag{5.5.10}$$

But this integral over the interval $s<-\varepsilon\sqrt{T}$ is negligible in comparison with the full integral, which is equal to $\mathbf{A}(\lambda(\alpha),\mu(\alpha))=1$. Thus, we have

$$\sum\nolimits_2 = o\left(\frac{\psi_1 C(\alpha)}{\sqrt{T}}\right)e^{-TD(\alpha)}. \tag{5.5.11}$$

The subsum $\sum_3$ of the series (5.5.3) over the interval $\varepsilon\sqrt{T}\le s_k<\varepsilon T$ is estimated similarly.

Next, consider the subsum $\sum_4$ of the series (5.5.3) over the interval $s_k>\varepsilon T$. Here the integro-local theorems for $\mathbf{P}(Z(T)\in\delta[z_k))$ for $z_k<x-\varepsilon T$ (see (5.5.3)), generally speaking, are not applicable. However, the probabilities of the events considered here are exponentially smaller than the principal part, and so it suffices to apply the local large deviation principle for $Z(T)$ (see Theorem 3.4.1), which implies, for $\alpha_k=\alpha-\frac{s_k}{T}<\alpha-\varepsilon$, $s_k>\varepsilon T$, that

$$\ln\mathbf{P}(Z(T)\in\delta[z_k)) = -T\widehat{D}(\alpha_k)+o(T)$$

as $T \to \infty$, where the function $\widehat{D}(\alpha)$ is convex and coincides with $D(\alpha)$ in K (see Theorem 3.5.9), so that, as in (5.5.9), we have

$$\widehat{D}(\alpha_k) \geq D(\alpha) - \frac{s_k}{T}\,\mu(\alpha).$$

Hence

$$\ln \mathbf{P}(Z(T) \in \delta[z_k)) \leq -TD(\alpha) + s_k\mu(\alpha) + o(T).$$

Since the analog of the integral (5.5.10) over the domain $s > \varepsilon T$ is exponentially small, it follows that $\sum_4$ can be estimated as in (5.5.11). The subsum $\sum_5$ over the domain $s_k < -\varepsilon T$ is treated similarly.

(iii) The claim that in the homogeneous case we can neglect the summand $P_1(v, \Delta[x))$ on the right of (5.2.16) is proved in Remark 5.2.7 (see (5.2.18)).

(iv) The admissible inhomogeneity condition $(\mathcal{A}_K) \subset (\mathcal{A}^{(st)})$ was verified in assertion (iii) of Theorem 5.2.1. The following assertion shows that the summand $P_1(v, \Delta[x))$ is small relative to the principal part of the right-hand side of (5.2.16).

Lemma 5.5.1 *Let $\alpha = \alpha_T \to \alpha^0 \in K$ as $T \to \infty$, and let*

$$(\lambda(\alpha^0), \mu(\alpha^0)) \in (\mathcal{A}_1), \qquad (0, \mu(\alpha^0)) \in (\mathcal{A}_1). \tag{5.5.12}$$

Then, for some $\varepsilon > 0$,

$$\mathbf{P}(\tau_1 > T,\ \zeta_1 \in \Delta[x)) = O(e^{-T(D(\alpha)+\varepsilon)}) \quad \textit{as} \quad T \to \infty. \tag{5.5.13}$$

The conditions (5.5.12) of Lemma 5.5.1 are satisfied, because

$$(0, \mu(\alpha)) \in (\mathcal{A}^{(st)}) \quad \text{for all} \quad \alpha \in K.$$

To verify this, we have to show that $\psi^{(st)}(0, \mu(\alpha))$ is finite. However, the function

$$\psi^{(st)}(\lambda, \mu(\alpha)) = \frac{\psi(\lambda, \mu(\alpha)) - \psi(0, \mu(\alpha))}{a_\tau \lambda}$$

has a removable singularity at the point $\lambda = 0$ and is equal to

$$\frac{1}{a_\tau}\frac{\partial \psi(\lambda, \mu(\alpha))}{\partial \lambda}\Big|_{\lambda=0}.$$

This value is finite because $(0, \mu(\alpha)) \in (\mathcal{A})$ by (5.2.5). Therefore, to verify Theorem 5.2.10 it remains to prove Lemma 5.5.1.

Proof of Lemma 5.5.1 We set

$$P_1 := \mathbf{P}(\tau_1 > T,\ \zeta_1 \in \Delta[x)).$$

To estimate the probability P_1 from above, we define in parallel with the vector (τ_1, ζ_1) the random vector $(\widehat{\tau}_1, \widehat{\zeta}_1)$ with the distribution

$$\mathbf{P}((\widehat{\tau}_1, \widehat{\zeta}_1) \in B) := \frac{1}{\psi_1(\lambda(\alpha), \mu(\alpha))}\mathbf{E}(e^{\lambda(\alpha)\tau_1 + \mu(\alpha)\zeta_1};\ (\tau_1, \zeta_1) \in B),$$

$B \subset \mathbb{R}^2$ (the Cramér transform of (τ_1, ζ_1)). We have

$$\begin{aligned}
P_1 &= e^{-T(\lambda(\alpha)+\mu(\alpha)\alpha)}\mathbf{E}(e^{\lambda(\alpha)\tau_1+\mu(\alpha)\zeta_1-\lambda(\alpha)(\tau_1-T)-\mu(\alpha)(\zeta_1-x)};\ \tau_1 \ge T,\ \zeta_1 \in \Delta[x)) \\
&= e^{-TD(\alpha)}\psi_1(\lambda(\alpha), \mu(\alpha))\mathbf{E}(e^{-\lambda(\alpha)(\widehat{\tau}_1-T)-\mu(\alpha)(\widehat{\zeta}_1-x)}; \widehat{\tau}_1 \ge T,\ \widehat{\zeta}_1 \in \Delta[x)) \\
&\le e^{-TD(\alpha)+|\mu(\alpha)|\Delta}\psi_1(\lambda(\alpha), \mu(\alpha))E(T),
\end{aligned}$$

where

$$E(T) := \mathbf{E}(e^{-\lambda(\alpha)(\widehat{\tau}_1-T)}; \widehat{\tau}_1 \ge T).$$

So, as $T \to \infty$, we have

$$P_1 = O(e^{-TD(\alpha)}E(T)). \tag{5.5.14}$$

Let $\lambda(\alpha^0) > 0$. Then $\lambda(\alpha) > 0$ for all sufficiently large T, and so,

$$E(T) \le \mathbf{P}(\widehat{\tau}_1 \ge T).$$

By the first condition in (5.5.12), the vector $\widehat{\tau}_1$ (whose distribution depends on $\alpha = \alpha_T$) satisfies condition [**C**] uniformly in T:

$$\mathbf{E}e^{h|\widehat{\tau}_1|} \le M \quad \text{for all} \quad T \in [T_0, \infty)$$

and some $h > 0$, $T_0 < \infty$, $M < \infty$. Therefore, for some $\varepsilon > 0$,

$$E(T) = O(e^{-\varepsilon T}) \quad \text{as} \quad T \to \infty.$$

Hence from (5.5.14) we arrive at (5.5.13) in the case $\lambda(\alpha^0) > 0$.

Now let $\lambda(\alpha^0) \le 0$. Then

$$\begin{aligned}
P_1 &= e^{-T(\lambda(\alpha)+\mu(\alpha)\alpha)+T\lambda(\alpha)}\mathbf{E}(e^{\mu(\alpha)\zeta_1-\mu(\alpha)(\zeta_1-x)};\ \tau_1 \ge T,\ \zeta_1 \in \Delta) \\
&\le e^{-TD(\alpha)+T\lambda(\alpha)+|\mu(\alpha)|\Delta}G(T), \qquad G(T) := \mathbf{E}(e^{\mu(\alpha)\zeta_1};\ \tau_1 \ge T).
\end{aligned}$$

Applying the Hölder inequality to estimate $G(T)$ from above (as in the proof of Lemma 5.3.5), we get, for any $p > 0$, $q > 0$, $p + q = 1$,

$$G(T) \le (\mathbf{E}(e^{\frac{1}{p}\mu(\alpha)\zeta_1})^p\mathbf{P}^q(\tau_1 \ge T).$$

By the second condition in (5.5.12), there exists $p_0 < 1$ such that

$$\overline{\lim_{T\to\infty}}\,(\mathbf{E}(e^{\frac{1}{p_0}\mu(\alpha)\zeta_1})^{p_0} < \infty.$$

Hence by condition [**C**] for τ_1, we have, for some $T_0 < \infty$, $c < \infty$, $\upsilon > 0$, and $T \ge T_0$,

$$G(T) \le ce^{-\upsilon T}, \qquad P_1 \le ce^{-TD(\alpha)+T\lambda(\alpha)+|\mu(\alpha)|\Delta}e^{-\upsilon T} \le ce^{-TD(\alpha)-\frac{\upsilon}{2}T}.$$

The last inequality is valid because in the case $\lambda(\alpha^0) \le 0$ we have

$$e^{T\lambda(\alpha)+|\mu(\alpha)|\Delta} = e^{o(T)} \quad \text{as} \quad T \to \infty.$$

Thus, (5.5.13) is also established in the case $\lambda(\alpha^0) \le 0$. Lemma 5.5.1, and therefore Theorem 5.2.10, are proved. □

5.5.2 Proof of Theorem 5.2.13 on Finite-Dimensional Distributions

The proof is based on Theorems 5.2.1 and 5.4.4. We split the proof into several steps.

1. We first refine the remainder term $O(e^{-T(D(\alpha)+\varepsilon)})$ in (5.2.4) using condition $[\widehat{\mathbf{P}}_K]$ (see (5.2.22)). To this end, we improve the estimate of the second part

$$J_2 := \int_{T(1-p)}^{T} \widetilde{H}((T-dy),\ \Delta[x])\mathbf{P}(\tau > y+v,\ \zeta \in \mathbf{w})$$

of the integral J in (5.4.3) (see (5.4.13)). By condition $[\widehat{\mathbf{P}}_K]$ (see (5.2.22), (5.2.23)), for all sufficiently large T we have

$$\begin{aligned} J_2 &\le \int_{T(1-p)}^{T} \widetilde{H}(T-dy,\ \Delta[x])\mathbf{P}(\tau > y)e^{\delta y}e^{-\delta y}\mathbf{P}(\tau > y+v,\ \zeta \in \mathbf{w} \mid \tau > y) \\ &\le c\widehat{P}((v,\infty),\ \mathbf{w}) \int_{T(1-p)}^{\infty} \widetilde{H}(T-dy,\ \Delta[x])e^{\delta y}\mathbf{P}(\tau > y). \end{aligned}$$

Here the integral can be estimated in the same way as the integral in (5.4.14), but now taking into account the inequality $\lambda(\alpha) < \lambda_+ - \delta$, $\delta > 0$. As a result, we get (see (5.4.18), (5.4.19))

$$J_2 \le c_1\widehat{P}((v,\infty),\ \mathbf{w})e^{-T(D(\alpha)+\varepsilon)}, \qquad c_1 = \text{const.} \tag{5.5.15}$$

2. By condition $[\widehat{\mathbf{P}}_K]$ (see (5.2.22) and (5.2.23)),

$$\begin{aligned} I_Z(\alpha, {}^{>}0, v, \mathbf{w}) &= \int_0^{\infty} e^{\lambda(\alpha)y}\mathbf{P}(\tau > y+v,\ \zeta \in \mathbf{w})\,dy \\ &\le c\int_0^{\infty} e^{(\lambda(\alpha)+\delta)y}\mathbf{P}(\tau > y)\widehat{P}((v,\infty),\ \mathbf{w}) \le c_2\widehat{P}((v,\infty),\ \mathbf{w}),\ \ c_2 = \text{const.} \end{aligned}$$

This together with (5.2.4) and (5.5.15) establishes

$$\mathbf{P}(Z(T) \in \Delta[x];\ B({}^{>}0, v, \mathbf{w})) \le \frac{c_3\Delta}{\sqrt{T}}\, e^{-TD(\alpha)}\widehat{P}((v,\infty),\ \mathbf{w}) \tag{5.5.16}$$

for all v, $\mathbf{w}$ and all sufficiently large T.

3. Consider the process

$$Z_T(t) := Z(T+t) - Z(T) \quad \text{for} \quad t \ge 0.$$

The initial vector $\boldsymbol{\xi}_{1,T}$ for the process $Z_T(t)$ is the vector

$$\boldsymbol{\xi}_{1,T} = (\chi(T), \zeta(T)). \tag{5.5.17}$$

Its distribution $P_{1,T,\alpha}$ under the condition $\{Z(T) \in \Delta[x]\}$ depends on T and $\alpha = \frac{x}{T}$ and by (5.2.4) and (5.5.16), it has the following property as $\alpha \to \alpha^0$

$$\begin{aligned} P_{1,T,\alpha}((v,\infty),\ \mathbf{w}) &= \frac{\mathbf{P}(Z(T) \in \Delta[x],\ B({}^{>}0, v, \mathbf{w}))}{\mathbf{P}(Z(T) \in \Delta[x])} \\ &\le c_4\widehat{P}((v,\infty),\ \mathbf{w}), \quad c_4 = \text{const} \end{aligned} \tag{5.5.18}$$

for all sufficiently large T. Besides, as $T \to \infty$,

$$P_{1,T,\alpha}\big((v,\infty),\ \mathbf{w}\big) \to \frac{I_Z(\alpha^0, {}^{>}0, v, \mathbf{w})}{I_Z(\alpha^0, {}^{>}0, 0, \mathbb{R})} =: P_{\alpha^0}\big((v,\infty),\ \mathbf{w}\big). \tag{5.5.19}$$

Hence we can apply Theorem 5.4.4 for the process $Z_T(t)$ on $[0, U]$, for $U \sim uT$, $u > 0$, $T \to \infty$ (condition $[\widehat{\mathbf{P}}_1]$ of this theorem (see (5.4.23)) is satisfied in view of (5.5.18) and (5.5.19)). From this theorem, for the conditional probability

$$\mathbf{P}(Z_T(U) \in \Delta[x')|\ Z(T) \in \Delta[x))$$

we have the integro-local theorem of the same form as Theorem 5.2.1, but with the right-hand side, in which T has to be replaced by U, $\alpha' := \frac{x'}{U}$ has to be substituted for α, and $\psi_{(\alpha^0)}(\lambda(\alpha'), \mu(\alpha'))$ has to be written instead of ψ_1, where $\psi_{(\alpha^0)}$ is the Laplace transform of the distribution P_{α^0}, which is defined in (5.5.19).

4. From the above we readily get the conclusion of Theorem 5.2.13 with $N = 2$ if T is replaced by u_1T, and $(1-u_1)T$ is substituted for U. For an arbitrary N, the result follows by induction as in Steps 1–3. Theorem 5.2.13 is proved. □

5.5.3 Proof of Theorem 5.2.14

The proof of Theorem 5.2.14 is similar to the proof above but simpler. As already pointed out, $\lambda(\alpha)$ is separated from λ_+ for all α, and hence, $\lambda(\alpha) < \lambda_+ - \varepsilon$ for some $\varepsilon > 0$. Therefore, by condition $[\mathbf{h}]$,

$$\begin{aligned} I_Z(\alpha, {}^{>}0, v, \mathbf{w}) &= \int_0^\infty e^{\lambda(\alpha)t}\mathbf{P}(\tau > t + v, \zeta \in \mathbf{w})\, dt \\ &= e^{-\lambda(\alpha)v} \int_v^\infty e^{\lambda(\alpha)t}\mathbf{P}(\tau > t, \zeta \in \mathbf{w})\, dt \\ &\le e^{-\lambda(\alpha)v} e^{\lambda(\alpha)v + h(v)}\mathbf{P}(\tau > v, \zeta \in \mathbf{w}) = e^{h(v)}\mathbf{P}(\tau > v, \zeta \in \mathbf{w}). \end{aligned}$$

A similar inequality holds for the integral J_2 (cf. (5.5.15)) with the correction factor $e^{-T(D(\alpha)+\varepsilon)}$ (cf. Step 1 of the proof of Theorem 5.2.13). Therefore, for the distribution $P_{1,T,\alpha}$ of the initial vector (5.5.17), in place of (5.5.18) we have

$$\begin{aligned} P_{1,T,\alpha}\big((v,\infty),\ \mathbf{w}\big) &= \frac{\mathbf{P}(Z(T) \in \Delta[x),\ B(0, v, \mathbf{w}))}{\mathbf{P}(Z(T) \in \Delta[x))} \\ &\le e^{h(v)}\mathbf{P}(\tau > v,\ \zeta \in \mathbf{w}) \end{aligned} \tag{5.5.20}$$

for all sufficiently large T. The compact set $\mathcal{A}_K$ lies in the domain of convergence of the integral

$$\int e^{\lambda v + \mu z + h(v)}\mathbf{P}(\tau > v, \zeta \in dz)\, dv,$$

and hence from (5.5.20), as before, the admissible inhomogeneity condition is met. The rest of the proof repeats in a simplified form the arguments in Steps 3 and 4 of the proof of Theorem 5.2.13. This completes the proof of Theorem 5.2.14. □

5.6 Exact Asymptotics of the Laplace Transform of the Distribution of a Compound Renewal Process and Related Problems

5.6.1 Main Result

In §3.7, based on LDPs for CRPs, we have found rough asymptotics of the Laplace transform, that is, the representation

$$\ln \mathbf{E}e^{\mu Z(T)} = T\widehat{A}(\mu)(1 + o(1)) \quad \text{as} \quad T \to \infty \tag{5.6.1}$$

in the entire domain $\mu \in (\mu^-, \mu^+)$ in which the fundamental function $\widehat{A}(\mu)$ is defined. Proceeding in a similar manner, we by using the integro-local Theorem 5.2.1 and arguing as in the proof of Theorem 3.7.1 find the *exact* asymptotics of $\mathbf{E}e^{\mu Z(T)}$ as $T \to \infty$, but now in a more narrow domain. Recall that in Remark 5.2.3 to the integro-local Theorem 5.2.1 it was noted that the compact set $K^>$ appearing in Theorem 5.2.1 can be defined using the fundamental function $\widehat{A}(\mu)$. We let $(\widehat{\mu}_-, \widehat{\mu}_+)$ denote the maximal domain of analyticity of the function $\widehat{A}(\mu)$, which contains the point $\mu = 0$, and denote

$$\widehat{\alpha}_- = A'(\widehat{\mu}_- + 0), \qquad \widehat{\alpha}_+ = A'(\widehat{\mu}_+ - 0).$$

Under this notation, the interval $(\widehat{\alpha}_-, \widehat{\alpha})$ is a domain of analyticity of the function $\widehat{D}(\alpha)$, which contains the point $\alpha = a$.

If $\lambda_+ > D(0)$, then $\widehat{\mu}_\pm = \mu_\pm$, $\widehat{\alpha}_\pm = \alpha_\pm$ and the compact sets $K^<$ and $K^>$ are indistinguishable by the definition (the forbidden set $[\beta_-, \beta_+]$ is empty).

It is clear that the condition $\lambda_+ > D(0)$ is met if $\lambda_+ = \infty$ or $a = 0$ (in this case, $D(0) = 0$). According to assertion (iv) of Theorem 3.3.2, this condition is also satisfied if the boundary $\partial\mathcal{A}^{\leq 0}$ in $\mathbb{R}^2$ is embedded to the open half-plane $\{\lambda < \lambda_+\}$ (this is always so if $\mathbf{A}(\lambda_+, \mu) > 0$ for all μ).

Below we will use the integro-local Theorem 5.2.1. This theorem involves the compact set $K^>$, which we now replace in an equivalent way by any closed interval $\widehat{K} \subset (\widehat{\alpha}_-, \widehat{\alpha}_+)$ containing the point $\alpha = a$. Let

$$\widehat{\mathcal{A}} = \left\{(\lambda(\alpha), \mu(\alpha)),\ \alpha \in \widehat{K}\right\}$$

be the analytic segment of the curve $\partial\mathcal{A}^{\leq}$ in $\mathbb{R}^2$. We let $Q(\alpha)$ denote the product

$$Q(\alpha) = \psi_1(\lambda(\alpha), \mu(\alpha))C(\alpha)I_Z(\alpha, {}^{>}0, 0, \mathbb{R})$$

on the right of (5.2.4) with $u = 0$, $v = 0$, $\mathbf{w} = \mathbb{R}$. For $\mu \in (\mu_-, \mu_+)$, consider the function $\alpha(\mu) := A'(\mu)$, which is the inverse function of $\mu(\alpha) = D'(\alpha)$, so that

$$A(\mu) = \sup_\alpha (\mu\alpha - D(\alpha)) = \mu\alpha(\mu) - D(\alpha(\mu)).$$

The following assertion holds.

Theorem 5.6.1 *Let (τ, ζ) be a nonlattice vector and let the "admissible inhomogeneity" conditions*

$$\mathcal{A}^{\leq 0} \subset [\mathcal{A}_1], \quad \widehat{\mathcal{A}} \subset (\mathcal{A}_1), \quad \mathbf{P}(\tau_1 > T) = o\Big(\frac{1}{\sqrt{T}} e^{-TD(0)}\Big) \quad as\ T \to \infty \tag{5.6.2}$$

be met. Then, for any $\mu \in (\widehat{\mu}_-, \widehat{\mu}_+)$,

$$\mathbf{E}e^{\mu Z(T)} = \frac{\sqrt{2\pi} Q(\alpha(\mu)) e^{-TA(\mu)}}{\sqrt{D''(\alpha(\mu))}} (1 + o(1)) \quad as \quad T \to \infty. \tag{5.6.3}$$

Remark 5.6.2 A similar assertion also holds for arithmetic ζ. If the component τ is arithmetic, then the integral $I_Z(\alpha, {}^{>}0, 0, \mathbb{R})$ should be replaced by the corresponding sum.

If the CRP $Z(t)$ is homogeneous, then conditions (5.6.2) are superfluous.

Theorem 5.6.1 shows that the approximation (5.6.1) in the domain $(\widehat{\mu}_-, \widehat{\mu}_+)$ is pretty accurate: the next term of the asymptotic expansion for $\ln \mathbf{E}e^{\mu Z(T)}$ turns out to be a "constant" (it does not depend on T).

Proof We set, as in §3.7,

$$E(B) = \mathbf{E}(e^{\mu Z(T)};\ Z(T)/T \in B),$$

and let, as before, $d_- < d_+$ be the boundaries of the set $\mathcal{D}_1^{<\infty} = \{\alpha : D(\alpha) < \infty\}$. Next, for large N and small $\varepsilon > 0$, we put

$$N_- = \max(-N, d_- + \varepsilon), \qquad N_+ = \min(N, d_+ + \varepsilon), \qquad B_N = [N_-, N_+).$$

From the proof of Theorem 3.7.1 it follows (see the proofs of Lemmas 3.7.2 and 3.7.3) that, for sufficiently large N and sufficiently small ε, there exists an $\varepsilon_1 > 0$ such that, for all sufficiently large T,

$$E(\overline{B}_N) \leq e^{-TA(\mu) - \varepsilon_1 T}. \tag{5.6.4}$$

We have

$$\mathbf{E}e^{\mu Z(T)} = E(\mathbb{R}) = E(B_N) + E(\overline{B}_N), \tag{5.6.5}$$

and hence by (5.6.4) it suffices to estimate $E(B_N)$. For some $h > 0$, we split the domain B_N into three parts:

$$B_1 = [N_-, \alpha(\mu) - h), \quad B_2 = [\alpha(\mu) - h, \alpha(\mu) + h), \quad B_3 = [\alpha(\mu) + h, N_+).$$

We first estimate $E(B_2)$, which gives the principal part of $E(B_N)$. It is clear that, for sufficiently small $h > 0$, one can choose a compact set $\widehat{K} \subset (\widehat{\alpha}_-, \widehat{\alpha}_+)$ such that the domain B_2 will wholly lie in $\widehat{K}$.

Given a small $\Delta > 0$, we set

$$x_k = T\alpha(\mu) + k\Delta, \qquad k = -M, -M+1, \ldots, M-1,$$

where for simplicity we assume that $M = \frac{hT}{\Delta}$ is integer. We represent the integral

$$E(B_2) = \int_{T(\alpha(\mu)-h)}^{T(\alpha(\mu)+h)} e^{\mu x} \mathbf{P}(Z(T) \in dx)$$

as the sums

$$E(B_2) = \sum_{k=-M}^{M-1} \int_{\Delta[x_k)} e^{\mu x} \mathbf{P}(Z(T) \in dx) = \sum_{k=-M}^{M-1} e^{\mu x_k} \mathbf{P}(Z(T) \in \Delta[x_k))(1 + o(1))$$

as $\Delta \to 0$.

To simplify the presentation, we first assume that $0 \notin B_2$. In this case, for

$$\alpha_k = \frac{x_k}{T}, \qquad k = -M, \ldots, M-1,$$

from Theorem 5.2.1 (its conditions are satisfied), we have, as $T \to \infty$ and $\Delta = \overline{o}(1)$,

$$E(B_2) = \sqrt{T} \sum_{k=-M}^{M-1} \frac{\Delta}{T} Q(\alpha_k) e^{T(\mu\alpha_k - D(\alpha_k))}(1 + o(1)). \tag{5.6.6}$$

Here the function $Q(\alpha)$ is continuous, and the analytic function $\mu\alpha - D(\alpha)$ attains its maximum $A(\mu)$ at the point $\alpha(\mu)$, and, for small h, $|\alpha - \alpha(\mu)| \le h$, admits the expansion

$$\mu\alpha - D(\alpha) = A(\mu) - \frac{(\alpha - \alpha(\mu))^2 D''}{2} + O(|\alpha - \alpha(\mu)|^3), \tag{5.6.7}$$

where, for brevity, we put $D'' = D''(\alpha(\mu))$. We have $\alpha_{k+1} - \alpha_k = \frac{\Delta}{T}$, and hence the sum in (5.6.6) is the integral sum for the integral

$$Q(\alpha(\mu)) \int_{\alpha(\mu)-h}^{\alpha(\mu)+h} e^{T(\mu\alpha - D(\alpha))} d\alpha (1 + o(1)).$$

Taking into account (5.6.7) and replacing $\sqrt{TD''}(\alpha - \alpha(\mu))$ by u, we find that this integral is equal to

$$\frac{Q(\alpha(\mu))e^{TA(\mu)}}{\sqrt{TD''}} \int_{-h\sqrt{TD''}}^{h\sqrt{TD''}} e^{-\frac{u^2}{2}} du (1 + o(1)) = \frac{\sqrt{2\pi}\, Q(\alpha(\mu))e^{TA(\mu)}}{\sqrt{TD''}} (1 + o(1)). \tag{5.6.8}$$

If $0 \in B_2$, then, in view of the embedding $\{\tau_1 > T\} \subset \{Z(T) \in \Delta[0)\}$, another summand $\mathbf{P}(\tau_1 > T)$ will appear on the right of (5.2.4) and in the sum (5.6.6). But by the third admissible inhomogeneity condition in (5.6.2) this summand has no effect on the final result in (5.6.8).

Thus, using (5.6.6) we have, by the above,

$$E(B_2) = \frac{\sqrt{2\pi}\, Q(\alpha(\mu))e^{TA(\mu)}}{\sqrt{D''}} (1 + o(1)) \quad \text{as} \quad T \to \infty. \tag{5.6.9}$$

Let us now estimate the term $E(B_3)$. We again take a small Δ and put

$$\alpha_l = \alpha(\mu) + h + l\Delta, \qquad l = 0, \ldots, L-1,$$

assuming, for simplicity, that $L = \frac{N_+ - \alpha(\mu) - h}{\Delta}$ is an integer number. Let for definiteness, $\mu \geq 0$. In this case, for any $l < L$ and $\Delta = \overline{o}(1)$ as $T \to \infty$, from the local LDP for the CRP $Z(T)$, we have

$$J_l := \int_{T\Delta[\alpha_l)} e^{\mu x}\mathbf{P}(Z(T) \in dx) \leq e^{T(\alpha_l+\Delta)\mu}\mathbf{P}(Z(T) \in T\Delta[\alpha_l)) = e^{T(\mu\alpha_l - D(\alpha_l))+o(T)}$$

(the conditions of Theorem 3.4.1 are satisfied in view of (5.6.2)). Since $D'' > 0$ for $\mu \in (\widehat{\mu}_-, \widehat{\mu}_+)$, the function $\mu\alpha - D(\alpha)$ is concave, its maximum $A(\mu)$ is attained at the point $\alpha = \alpha(\mu)$. Hence for $\alpha_l \geq \alpha(\mu) + h$, there exists an $h_1 > 0$ such that, for all $l < L$,

$$\mu\alpha_l - D(\alpha_l) \leq A(\mu) - h_1.$$

Therefore,

$$J_l \leq e^{T(A(\mu)-h_1)+o(T)},$$

$$E(B_3) = \sum_{l=0}^{L-1} J_l \leq L e^{T(A(\mu)-h_1)+o(T)}. \tag{5.6.10}$$

In a similar manner we estimate $E(B_1)$. Combining (5.6.4), (5.6.5), (5.6.9), and (5.6.10), we get (5.6.3). Theorem 5.6.1 is proved. □

5.6.2 Refinement of the Inequalities of Theorem 4.6.3 for the Distribution of $\overline{Z}(T)$

By an appeal to Theorem 5.6.1, one can refine, in a number of cases, Theorem 4.6.3 on inequalities for $\mathbf{P}(\overline{Z}(T) \geq x)$ (for example, by replacing $o(T)$ by $c(\alpha) + o(1)$ in inequalities (4.6.15) and (4.6.16)). If

$$A(\widehat{\mu}_+) > 0, \qquad \mu_0 = \sup\{\mu : A(\mu) \leq 0\},$$

then $\mu_0 = 0$ in the case $a \geq 0$, and $\mu_0 > 0$, $A(\mu_0) = 0$ in the case $a < 0$. We put $\alpha_0 = A'(\mu_0)$. The following assertion holds.

Theorem 5.6.3 *Let $Z(t)$ be a homogeneous CRP, let (τ, ζ) be a nonlattice vector, let $A(\widehat{\mu}_+) > 0$, and let $\alpha = x/T$. Then, as $T \to \infty$,*

$$\mathbf{P}(\overline{Z}(T) \geq x) \leq \begin{cases} e^{-\mu_0 x}(1 + o(1)) & \text{if } a < 0,\ \alpha \in [0, \alpha_0]; \\ e^{-TD(\alpha)}(1 + o(1)) & \text{if } \alpha \in [\alpha_0, \widehat{\alpha}_+). \end{cases} \tag{5.6.11}$$

Inequalities (5.6.11) are quite sharp. From the results of Chapter 6 it will follow that the first inequality in (5.6.11) (in the domain $\alpha \in [0, \alpha_0)$) is unimprovable up to a constant factor, and the second one (in the domain $\alpha \in [\alpha_0, \widehat{\alpha}_+)$) is sharp up to the factor $\frac{C}{\sqrt{T}}$.

If $Z(T)$ is a random walk ($\tau \equiv 1$), then $\widehat{\mu}_+ = \mu^+$, the inequalities in (5.6.11) become sharp (not the asymptotic one), that is, they hold for all T, and $o(1)$ can be replaced by 0.

Proof of Theorem 5.6.3 The proof is similar to that of Theorem 4.6.3, but now Theorem 5.6.1 is used instead of Theorem 3.7.1. By Theorem 5.6.1, for the homogeneous CRP $Z(t)$ and $\mu \in (\widehat{\mu}_-, \widehat{\mu}_+)$, we have

$$\mathbf{E}e^{\mu Z(T)} = ce^{TA(\mu)}(1 + \varepsilon_T), \tag{5.6.12}$$

where $\varepsilon_T \to 0$ as $T \to \infty$, and c depends only on μ. Since $\eta_Z(x) := \inf\{t : Z(t) \ge x\}$ is a Markov moment, from (5.6.12) for $\mu \in [0, \widehat{\mu}_+)$, $t < T$, it follows that

$$\mathbf{E}(e^{\mu Z(T)} \mid \eta_Z(x) = t) \ge \mathbf{E}e^{\mu x + \mu Z(T-t)} = ce^{\mu x + (T-t)A(\mu)}(1 + \varepsilon_{T-t}).$$

If we use inequality (4.6.19), then, in addition to (5.6.12), we get, for $\mu \in [0, \widehat{\mu}_+)$,

$$\mathbf{E}e^{\mu Z(T)} \ge c\int_0^T e^{\mu x + (T-t)A(\mu)}(1 + \varepsilon_{T-t})\mathbf{P}(\eta_Z(x) \in dt).$$

One can get from here that, as $T \to \infty$,

$$(1 + \varepsilon_T) \ge e^{\mu x}\int_0^T e^{-tA(\mu)}(1 + \varepsilon_t)\mathbf{P}(\eta_Z(x) \in dt),$$
$$1 \ge e^{\mu x}\int_0^T e^{-tA(\mu)}\mathbf{P}(\eta_Z(x) \in dt)(1 + o(1)).$$

For $A(\mu) \le 0$, this gives

$$\mathbf{P}(\overline{Z}(T) \ge x) = \mathbf{P}(\eta_Z(x) \le T) \le \int_0^T e^{-tA(\mu)}\mathbf{P}(\eta_Z(x) \in dt) \le e^{-\mu x}(1 + o(1)). \tag{5.6.13}$$

If $\alpha \in [0, \alpha_0]$, then $\mu(\alpha) \in [0, \mu_0]$. Putting $\mu = \mu_0$ in (5.6.13), we get the first inequality in (5.6.11).

For $A(\mu) \ge 0$, we have

$$\mathbf{P}(\overline{Z}(T) \ge x) \le \int_0^T e^{-tA(\mu)+TA(\mu)}\mathbf{P}(\eta_Z(x) \in dt) \le e^{-\mu x + TA(\mu)}(1 + o(1)). \tag{5.6.14}$$

If $\alpha = x/T \in [\alpha_0, \widehat{\alpha}_+)$, then $\mu(\alpha) \in [\mu_0, \widehat{\mu}_+)$, $A(\mu(\alpha)) \ge 0$, and so, putting $\mu = \mu(\alpha)$ in (5.6.14), we get

$$-\mu x + TA(\mu) = -T(\alpha\mu(\alpha) - A(\mu(\alpha))) = -TD(\alpha).$$

This proves the second inequality in (5.6.11). Theorem 5.6.3 is proved. □

It is clear that in the case $A(\widehat{\mu}_+) > 0$ for $\alpha = \alpha_0$ we have $\mu(\alpha) = \mu(\alpha_0) = \mu_0$, $A(\mu_0) = 0$, $D(\alpha_0) = \alpha_0\mu_0$, and the right-hand sides in (5.6.11) are equal. A more detailed analysis of the properties of the distribution of $\overline{Z}(T)$ will be carried over in Chapter 6.

5.6.3 Exact Asymptotics of the Moments of a CRP

As in the previous section, to simplify the statements and proofs, we will restrict ourselves to consideration of homogeneous CRPs only. In Theorem 1.2.1 it was shown that, for $\mathbf{E}\tau^2 < \infty$ and $\mathbf{E}|\tau\zeta| < \infty$,

$$\mathbf{E}Z(T) - aT = \frac{a_\zeta \mathbf{E}\tau^2}{2a_\tau^2} - \frac{\mathbf{E}\tau\zeta}{a_\tau} + o(1) = O(1) \quad \text{as} \quad T \to \infty.$$

If $\mathbf{E}\tau^2 < \infty$, $\xi = \zeta - a\tau$, and $\sigma_\xi = \mathbf{E}\xi^2 < \infty$, then

$$\mathbf{D}Z(T) = \frac{\sigma_\xi^2 T}{a_\tau} + o(T). \tag{5.6.15}$$

If, in addition, $\mathbf{E}\tau^3 < \infty$, then the remainder term $o(T)$ in (5.6.15) can be replaced by $O(\sqrt{T})$.

Estimation of the variable $Z(T) - aT$ of high order involves essential technical difficulties. However, under the Cramér condition for (τ, ζ), we can proceed as in §5.6.1 and employ the integro-local theorems to derive the exact asymptotics for the moments $\mathbf{E}|Z(T) - aT|^k$ for any $k > 0$ and to estimate the moments of $(Z(T) - aT)$ of odd integer orders $k \geq 3$.

Theorem 5.6.4 *Let $Z(t)$ be a homogeneous CRP, let a vector (τ, ζ) be nonlattice, and let w be a normally distributed random variable with parameters $(0, 1)$. Then, for any $k > 0$,*

$$\mathbf{E}|Z(T) - aT|^k = \sigma^k T^{k/2}\, \mathbf{E}|w|^k (1 + o(1)) \quad as \quad T \to \infty. \tag{5.6.16}$$

If $k \geq 3$ is an odd integer, then

$$\mathbf{E}[Z(T) - aT]^k = o(T^{k/2}) \quad as \quad T \to \infty.$$

Assertion (5.6.16) remains valid also in the case when at least one of the components τ or ζ is arithmetic. The conditions on the structure of the distribution of (τ, ζ) appeared in Theorem 5.6.4, because its proof uses the integro-local theorem for $Z(T)$. It is clear that these conditions are not connected to the essence of the matter in this section, but we have not succeeded to find an easy way to get rid of them.

If τ and ζ are independent, then (5.6.16) can be easily established via the inequality

$$\mathbf{P}\left(|z_T - w| > \frac{b \ln T}{\sqrt{T}} + \frac{y}{\sqrt{T}}\right) \leq ce^{-\lambda y}$$

for some $b > 0$, $c < \infty$, $\lambda > 0$ and all sufficiently large T, where $z_T = \frac{Z(T)-aT}{\sigma\sqrt{T}}$, and the processes $Z(tT)$ and $w(t)$ are constructed in an appropriate way on the same probability space. This inequality follows from the results of [56] (see also §7.1 below). If we set $h_T = \frac{b \ln T}{\sqrt{T}}$ and use the representation $z_T = w + r_T$, where $\mathbf{P}(|r_T| > h_T + \frac{y}{\sqrt{T}}) \leq ce^{-\lambda y}$, then we easily get

$$\mathbf{E}|z_T|^k = \mathbf{E}|w + r_T|^k = \mathbf{E}|w|^k + O(h_T).$$

Proof of Theorem 5.6.4 We represent $Z(T) - aT$ in the form

$$Z(T) - aT = Z_{\nu(T)} - aT_{\nu(T)} - a\gamma(T) = S_{\nu(T)} - a\gamma(T),$$

where, as before, $S_n = \sum_{k=1}^{n} \xi_k$, $\xi_k = \zeta_k - a\tau_k$, and $\mathbf{E}\xi_k = 0$. The process $S_{\nu(t)}$ is a CRP with zero mean. The variables $\gamma(T)$ for all T have uniformly bounded exponential moment (see §1.1), and by Theorem 1.2.1 we have $\mathbf{E}S^2_{\nu(T)} \to \infty$ as $T \to \infty$, so that $\mathbf{E}|S_{\nu(T)}|^k \to \infty$ for $k \geq 2$ and $T \to \infty$. One can get from here that

$$\mathbf{E}|Z(T) - aT|^k = \mathbf{E}|S_{\nu(T)}|^k(1 + o(1)) \quad \text{as} \quad T \to \infty.$$

Hence the problem of the asymptotic behavior of the moments $\mathbf{E}|Z(T) - aT|^k$ is reduced to the problem of the asymptotic behavior of the moments $\mathbf{E}|Z(T)|^k$ for the CRP $Z(t)$ with zero mean.

So, let $a = 0$. To prove (5.6.16) we will follow the scheme of the proof of Theorem 5.6.1. We set

$$E_k(\mu, B) = \mathbf{E}\left(|Z(T)|^k e^{\mu Z(T)};\ \frac{Z(T)}{T} \in B\right), \quad E_k(B) = E_k(0, B).$$

For some $h > 0$, we split the real line $\mathbb{R}$ into three parts as follows:

$$B_1 = (-\infty, -h), \qquad B_2 = [-h, h), \qquad B_3 = [h, \infty),$$

so that

$$E_k(\mathbb{R}) = \sum_{j=1}^{3} E_k(B_j). \tag{5.6.17}$$

Next, the power function x^k grows as $x \to \infty$ slower than any exponential function:

$$x^k \leq c_{k,\mu} e^{\mu x} \qquad \text{for} \quad x > 0, \ \ \mu > 0,$$

here $c_{k,\mu}$ depends only on k and μ. It follows that

$$E_k(B_3) \leq c_{k,\mu} E_0(\mu, B_3).$$

From the proof of Theorem 5.6.1 it follows that since the neighborhood of the point $a = 0$ lies in the compact set $\widehat{K}$, the neighborhood of the point $\mu = 0$ is contained in $(\widehat{\mu}_-, \widehat{\mu}_+)$, and hence by (5.6.4) and (5.6.10)

$$E_0(\mu, B_3) \leq e^{T(A(\mu)-\varepsilon)}$$

with some $\varepsilon > 0$ and all sufficiently large T. Since $A(\mu) \to 0$ as $\mu \to 0$, we can choose μ so as to have $A(\mu) = \varepsilon/2$. For such μ, we get

$$E_k(B_3) \leq c_{k,\mu} e^{-T\varepsilon/2} \tag{5.6.18}$$

for all sufficiently large T. The estimate for $E_k(B_1)$ is similar.

So, in order to estimate

$$\mathbf{E}|Z(T)|^k = E_k(\mathbb{R}),$$

it suffices to find the asymptotics of $E_k(B_2)$, which makes the main contribution to the asymptotics of $E_k(\mathbb{R})$. We choose $\Delta > 0$ and put

$$x_j = j\Delta, \qquad j = -M, \ldots, M-1,$$

where for simplicity we assume that $M = \frac{hT}{\Delta}$ is integer. We represent the integral

$$E_k(B_2) = \int_{-hT}^{hT} |x|^k \mathbf{P}(Z(T) \in dx)$$

as the sum

$$E_k(B_2) = \sum_{j=-M}^{M-1} \int_{\Delta[x_j)} |x|^k \mathbf{P}(Z(T) \in dx) = \Big[\sum_{j=-M}^{M-1} |x_j|^k \mathbf{P}(Z(T) \in \Delta[x_j))\Big](1+o(1))$$

as $\Delta \to 0$. By Theorem 5.2.1 ($\psi_1 = 1$ in the homogeneous case) and the remark at the end of the theorem to the effect that

$$\psi_1 C(\alpha) I_Z(\alpha, {}^>0, 0, \mathbb{R}) = \frac{1 + o(1)}{\sigma\sqrt{2\sigma}} \quad \text{as} \quad \alpha \to 0$$

for α near $a = 0$, we get

$$\begin{aligned} E_k(B_2) &= (1 + o(1)) \sum_{j=-M}^{M} \frac{\Delta}{\sqrt{T}} |\alpha_j T|^k \psi_1 C(\alpha_j) I_Z(\alpha_j, {}^>0, 0, \mathbb{R}) e^{-TD(\alpha_j)} \\ &= \frac{(1 + o(1))T^{k/2}}{\sqrt{2\pi}} \sum_{j=-M}^{M-1} \frac{\Delta}{\sigma\sqrt{T}} |\alpha_j \sqrt{T}|^k e^{-TD(\alpha_j)} \end{aligned} \tag{5.6.19}$$

as $T \to \infty$, $\Delta = \overline{o}(1)$, and for $\alpha_j = \frac{x_j}{T}$. We have

$$TD(\alpha_j) = \frac{T\alpha_j^2}{2\sigma^2} + O(|\alpha_j|^3 T)$$

as $\alpha_j \to 0$, and hence

$$TD(\alpha_j) = \frac{T\alpha_j^2}{2\sigma^2} + o(1) \quad \text{for} \quad \alpha_j = o(T^{-1/3})$$

and

$$TD(\alpha_j) = \frac{T\alpha_j^2}{2\sigma^2} (1 + o(1)) \quad \text{for} \quad \alpha_j = o(1), \quad T \to \infty.$$

Writing $y_j = \frac{\alpha_j \sqrt{T}}{\sigma}$ in (5.6.19), we get $y_{j+1} - y_j = \frac{\Delta}{\sigma\sqrt{T}}$, and

$$E_k(B_2) = \frac{(1 + o(1))\sigma^k T^{k/2}}{\sqrt{2\pi}} \sum_{j=-M}^{M-1} \Big(\frac{\Delta}{\sigma\sqrt{T}}\Big)|y_j|^k e^{-\frac{y_j^2}{2}}. \tag{5.6.20}$$

The sum in the right-hand side of (5.6.20) is the integral sum for the integral $\int_{-h\sqrt{T}}^{h\sqrt{T}} |y|^k e^{-y^2/2} dy$, and hence

$$E_k(B_2) = \sigma^k T^{k/2} \mathbf{E}|w|^k (1 + o(1)) \quad \text{as} \quad T \to \infty. \tag{5.6.21}$$

Comparing (5.6.21), (5.6.17), and (5.6.18), we get (5.6.16). The second assertion of the theorem can be derived from the above estimates and since $\mathbf{E}w^k = 0$ for odd integer k. Theorem 5.6.4 is proved. □

The conclusion of Theorem 5.6.4 can be extended to the convergence

$$\mathbf{E}f\left(\frac{Z(T) - aT}{\sigma\sqrt{T}}\right) \to \mathbf{E}f(w) \quad \text{as} \quad T \to \infty$$

for any continuous function $f(v) \geq 0$ that grows at infinity slower than any exponential function $((\ln f(v))^+ = o(|v|)$ as $|v| \to \infty)$. For alternating in sign functions f of the same form, one should use the relation

$$\mathbf{E}f\left(\frac{Z(T) - aT}{\sigma\sqrt{T}}\right) \to \mathbf{E}f^+(w) - \mathbf{E}f^-(w) \quad \text{as} \quad T \to \infty,$$

where f^+, f^- are, respectively, the positive and negative parts of f.

5.7 Integro-Local Theorems for Markov Additive Processes under the Cramér Conditions

The sums

$$\mathrm{X}_n = \sum_{k=1}^{n} \xi_k(\mathrm{x}_k)$$

of the random variables $\xi_k(\mathrm{x}_k)$ given on the states of the Markov chain $\{\mathrm{x}_k\}$, which were defined in §§1.8 and 2.5 (where the main notation was also introduced), have the form (see (1.8.9) in §1.8)

$$\mathrm{X}_n = Z(n) + \rho_n, \qquad \rho_n = \sum_{k=n-\gamma(n)+1} \xi_k(\mathrm{x}_k),$$

where $Z(n)$ is the "embedded" CRP constructed by cycles generated by returning of the chain in a positive atom e_0. As before, the increment of the sums $\mathrm{X}(n)$ on one cycle of length τ will be denoted by ζ, so that the vector (τ, ζ) is a "governing" vector for the CRP $Z(n)$. We assume that the vector (τ, ζ) satisfies the Cramér condition. As already pointed out, for each fixed $\gamma(n) = l$ the random variable ρ_n conditionally does not depend on $Z(n)$, and the distribution of $\rho_n = \rho(l)$ will not depend on n. We let $(\mu_-^{(\rho(l))}, \mu_+^{(\rho(l))})$ denote the domain of analyticity of the function $\mathbf{E}(e^{\mu\rho_n} \mid \gamma(n) = l)$.

Let $K^<$ be the compact set defined in (5.2.1).

Theorem 5.7.1 *Let the random variable ζ be nonlattice and let $\alpha = x/n \in K^< \cap (A'(\mu_-^{(\rho(l))}), A'(\mu_+^{(\rho(l))}))$. Then, for any fixed l and $\Delta = \overline{o}(1)$, as $n \to \infty$,*

$$\mathbf{P}(\mathrm{X}_n \in \Delta[x), \gamma(n) = l) = \frac{\Delta C(\alpha)}{\sqrt{n}} e^{-nD(\alpha)} \mathbf{P}(\tau > l) \mathbf{E}e^{\mu(\alpha)\rho(l)} (1 + o(1)),$$

where $C(\alpha)$ is defined in Theorem 5.2.1.

Proof We have

$$\mathbf{P}(X_n \in \Delta[x), \gamma(n) = l) = \mathbf{P}(Z(n) + \rho_n \in \Delta[x), \gamma(n) = l)$$
$$= \int \mathbf{P}(Z(n) \in \Delta[x - y), \gamma(n) = l)\mathbf{P}(\rho_n \in dy \mid \gamma(n) = l).$$

By Corollary 5.4.3,

$$\mathbf{P}(X_n \in \Delta[x), \gamma(n) = l)$$
$$= \frac{\Delta C(\alpha)}{\sqrt{n}} \mathbf{P}(\tau > l) \int e^{-nD(\alpha - y/n)} \mathbf{P}(\rho(l) \in dy)(1 + o(1)). \quad (5.7.1)$$

Here, for small y/n,

$$D(\alpha - y/n) = D(\alpha) - \frac{y}{n}\mu(\alpha) + o\Big(\frac{y}{n}\Big)$$

and for all y,

$$D(\alpha - y/n) \geq D(\alpha) - \frac{y}{n}\mu(\alpha).$$

Hence the integral in (5.7.1) is asymptotically equivalent to

$$e^{-nD(\alpha)}\mathbf{E}e^{\mu(\alpha)\rho(l)}, \quad (5.7.2)$$

where the Laplace transform on the right of (5.7.2) is finite for $\mu(\alpha) \in (\mu_-^{(\rho(l))}, \mu_+^{(\rho(l))})$ or, equivalently, for

$$\alpha \in (A'(\mu_-^{(\rho(l))}), A'(\mu_+^{(\rho(l))})).$$

Theorem 5.7.1 is proved. □

In view of Theorem 5.7.1 it is natural to expect that if α is such that

$$\ln \mathbf{E}e^{\mu(\alpha)\rho(l)} < \lambda_+ l \quad (5.7.3)$$

(recall that the left-hand side in (5.7.3), as well as $\mu(\alpha)$, is close to zero for small $|\alpha - a|$), then

$$\mathbf{P}(X_n \in \Delta[x)) = \frac{\Delta C(\alpha)}{\sqrt{n}} e^{-nD(\alpha)} \sum_{l=0}^{\infty} \mathbf{P}(\tau > l)\mathbf{E}e^{\mu(\alpha)\rho(l)}(1 + o(1)),$$

as $n \to \infty$, where the series on the right is finite.

Let us find conditions sufficient for (5.7.3) to hold for a finite Markov chain with states $e_0, \ldots, e_N$. Consider the random variable ξ^+ with the distribution $\mathbf{P}(\xi^+ > t) = \max_j \mathbf{P}(\xi(e_j) > t)$. It is clear that $\xi(e_j) \underset{p}{\leq} \xi^+$ and $\rho(l) \underset{p}{\leq} \sum_{k=1}^{l} \xi_k^+$, where ξ_k^+ are independent copies of ξ^+. Hence, for $\mu(\alpha) \geq 0$,

$$\mathbf{E}e^{\mu(\alpha)\rho(l)} \leq \big[\psi^{(\xi^+)}(\mu(\alpha))\big]^l, \qquad \psi^{(\xi^+)}(\mu) = \mathbf{E}e^{\mu\xi^+}, \quad (5.7.4)$$

and therefore, (5.7.3) is satisfied if

$$\ln \psi^{(\xi^+)}(\mu(\alpha)) < \lambda_+.$$

An analogous condition also holds for $\mu(\alpha) \le 0$ in terms of the random variable ξ^- with the distribution $\mathbf{P}(\xi^- > t) = \min_j \mathbf{P}(\xi(\mathrm{e}_j) > t)$.

If $\mu^{(+)}$ is the positive solution of the equation $\ln \psi^{(\xi^+)}(\mu) = \lambda_+$ (if such a solution exists), then (5.7.4) will be satisfied for $\mu(\alpha) \in [0, \mu^{(+)})$ or equivalently for $\alpha \in [a, A'(\mu^{(+)}))$. The range of $\alpha \le a$ for which (5.7.3) is satisfied is described similarly.

6

Exact Asymptotics in Boundary Crossing Problems for Compound Renewal Processes

6.1 Asymptotics of Distributions of the Maximal Value of a Compound Renewal Process with Linear Drift. First Passage Time of a High Level

Let, as before,

$$Z^{(q)}(t) = Z(t) + qt.$$

In this section, under rather broad conditions, we find the asymptotics of the distributions of the variable $\overline{Z}^{(q)} := \sup_{t\geq 0} Z^{(q)}(t)$ and of the time $\eta_{Z^{(q)}}(x) = \min\{t : Z^{(q)}(t) > x\}$ for *homogeneous* CRPs $Z(t)$. For an extension of some results given here to the inhomogeneous case, see Remark 6.1.6.

6.1.1 Preliminaries

The problem of the asymptotics of the distribution of $\overline{Z}^{(q)}$ can be reduced in many aspects to the similar problem for the maximum of sequential sums of random variables. Therefore, before proceeding to the formulations and proofs of the main results, it is worth recalling the basic facts concerning the distribution of the maximum value of a random walk. It will be convenient for us to present these results in terms of the random walks

$$Z_k^{(q)} := Z_k + qT_k = \sum_{j=1}^{k} \zeta_j^{(q)}, \qquad \zeta_j^{(q)} \underset{d}{=} \zeta^{(q)} = \zeta + q\tau,$$

under the assumption that

$$a_{\zeta^{(q)}} := \mathbf{E}\zeta^{(q)} = a_\zeta + qa_\tau < 0 \qquad (q < -a),$$

so that

$$\overline{Z}_\infty^{(q)} := \sup_{k\geq 0} Z_k^{(q)} < \infty$$

with probability 1.

Next, following [17], we give the definition of an upper power function.

Definition 6.1.1 A finite function C is called an *upper power function* if

$$\lim_{t\to\infty} \frac{G(t+b)}{G(t)} = 1 \tag{6.1.1}$$

for each b (that is, G is a locally constant function – such functions are also called long-tailed functions by many authors) and if, for any $0 < h \le 1$, it satisfies the inequality

$$0 < \frac{G(hx)}{G(x)} \le c(h) < \infty, \qquad x \ge 1,$$

where the function $c(h)$ is bounded on any closed interval $[h_1, 1]$ with some $h_1 > 0$.

It is easily seen, for example, that any function satisfying (6.1.1) and such that

$$c_1 x^m l(x) \le G(x) \le c_2 x^m l(x),$$

for some $0 < c_1 < c_2 < \infty$ and m, where $l(x)$ is a slowly varying at infinity function, is an upper power function.

It is also easy to see that an upper power function always majorizes some *power function*. Indeed, for $x = 2^N$, we have

$$G(x) \ge c^{-1}(1/2)G(x/2) \ge c^{-N}(1/2)G(x/2^N) = G(1)x^{-\gamma},$$
$$\gamma = \log_2 c(1/2).$$

This explains the term “upper power function.”

We need the following assertion.

Theorem 6.1.2 (i). *For all $x \ge 0$,*

$$\mathbf{P}(\overline{Z}_\infty^{(q)} > x) \le e^{-\mu_0 x}, \tag{6.1.2}$$

where

$$\mu_0 := \sup\{\mu : \psi_q(\mu) \le 1\}, \qquad \psi_q(\mu) := \mathbf{E}e^{\mu\zeta^{(q)}}$$

(*see Lemma* 2.1.1 *in* [38]).

(ii) *If $\mu_0 > 0$, $\psi_q(\mu_0) = 1$, $\psi_q'(\mu_0) < \infty$, then*

$$\mathbf{P}(\overline{Z}_\infty^{(q)} > x) = ce^{-\mu_0 x}(1 + o(1)) \quad \textit{as} \quad x \to \infty, \quad c = \text{const} \tag{6.1.3}$$

(*the classical Cramér–Lundberg theorem, see, e.g.,* [54], [5], [6, XIII, 5]).

Let $\widetilde{\chi}$ be the first overshoot over an infinite boundary by a random walk whose jumps have the distribution of the random variable $\widetilde{\zeta}^{(q)}$:

$$\mathbf{P}(\widetilde{\zeta}^{(q)} \in dt) = e^{\mu_0 t}\mathbf{P}(\zeta^{(q)} \in dt) \tag{6.1.4}$$

(*the Cramér transform of $\zeta^{(q)}$*), *and let $\mathbf{E}\widetilde{\zeta}^{(q)} = \psi_q'(\mu_0) < \infty$. Then $c = \mathbf{E}e^{-\mu_0\widetilde{\chi}} < 1$.* (*For other interpretations of the constant c, see, e.g.,* §21 in [17].)

If the distribution of $\zeta^{(q)}$ contains an absolutely continuous component or is arithmetic, then $o(1)$ in (6.1.3) *can be replaced by $O(e^{-hx})$ with some $h > 0$* (*Theorem* 21.11 *in* [17]).

(iii) *Let* $\psi_q(\mu_0) < 1$ *or* $\mu_0 = 0$, *and let the function*

$$F^I(x) := \int_x^\infty F(t)\,dt, \quad \text{where} \quad F(x) = \mathbf{P}(\zeta^{(q)} > x)$$

(*"the integrated tail" of the distribution of* $\zeta^{(q)}$) *be such that*

$$e^{\mu_0 x} F^I(x)$$

is an upper power function. Then

$$\mathbf{P}(\overline{Z}_\infty^{(q)} > x) \sim \frac{F^I(x)}{B(\mu_0)C(\mu_0)} \quad \text{as} \quad x \to \infty, \tag{6.1.5}$$

where

$$\begin{aligned} B(\mu) &= \frac{1 - \psi_q(\mu)}{\mu}, \\ C(\mu) &= \frac{1 - \mathbf{E}(e^{\mu\chi(+0)}; \eta(+0) < \infty)}{1 - \mathbf{P}(\eta(+0) < \infty)} = [\mathbf{E}e^{\mu\overline{Z}_\infty^{(q)}}]^{-1}, \end{aligned} \tag{6.1.6}$$

$\chi(t), \eta(t)$ *have the same meaning as in* §1.1, *but are related to the walk* $\{Z_k^{(q)}\}$, *and so, for* $\mu_0 = 0$,

$$B(0) = -a^{(q)} = -a_\zeta - q a_\tau > 0, \qquad C(0) = 1.$$

If the random variable $\zeta^{(q)}$ *is arithmetic, then x in* (6.1.3), (6.1.5) *should be chosen to be integer* (*Theorems* 21.11 *and* 21.12 *in* [17]).

(iv) *If* $\mu_0 = 0$, *the function* $F^I(x)$ *is subexponential, and if* $a_{\zeta^{(q)}} < 0$, *then*

$$\mathbf{P}(\overline{Z}_\infty^{(q)} > x) \sim \frac{1}{|a_{\zeta^{(q)}}|} F^I(x) \quad \text{as} \quad x \to \infty \tag{6.1.7}$$

(*see, for example, Theorem* 12.7.3 *in* [21]).

Note that in assertions (iii), (iv) the case $\mu_0 = 0$ is allowed (this means that the Cramér condition [**C**] is not met).

The relation

$$\mathbf{P}(\overline{Z}_\infty^{(q)} > x) \sim \frac{MF(x)}{1 - \psi_q(\mu_0)} \tag{6.1.8}$$

(which is close to (6.1.5)), where

$$M = \mathbf{E}e^{\mu_0 \overline{Z}_\infty^{(q)}},$$

was obtained in [121] in the case when

$$\mu_0 > 0 \qquad (\mathbf{E}\zeta^{(q)} < 0), \quad \psi_q := \psi_q(\mu_0) < 1, \tag{6.1.9}$$

and the convolution $F^{*2}(x) := F * F(x)$ of the distributions $F(x)$ has the property

$$F^{*2}(x) \sim 2\psi_q F(x) \quad \text{as} \quad x \to \infty. \tag{6.1.10}$$

The intermediate case $\mu_0 > 0, \psi_q(\mu_0) = 1, \psi_q'(\mu_0) = \infty$ (cf. assertion (ii) of Theorem 6.1.2) was studied in [80].

6.1.2 Distribution of the Maximal Value of a CRP with Drift

Let us now return to the CRPs. We set

$$\overline{Z}^{(q)} = \sup_{t\geq 0} Z^{(q)}(t).$$

Let, as before, $\zeta^{(q)} = \zeta + q\tau$, $\psi_q(\mu) = \mathbf{E}e^{\mu\zeta^{(q)}}$. If the equation

$$\psi_q(\mu) = 1 \tag{6.1.11}$$

has a positive solution μ_0 (this is possible only in the case $\mathbf{E}\zeta^{(q)} < 0$), then by $\widetilde{\zeta}^{(q)}$ we denote, as before, the Cramér transform at the point μ_0 of the random variable $\zeta^{(q)}$, that is, the random variable with distribution (6.1.4). By $\widetilde{\chi}$ we denote the overshoot over an infinite boundary by a random walk generated by sums of independent random variables with the same distribution as for $\widetilde{\zeta}^{(q)}$.

The following assertion holds.

Theorem 6.1.3 *Assume that there exists a solution $\mu_0 > 0$ of equation* (6.1.11) *such that $\psi_q'(\mu_0) < \infty$. Let $q \leq 0$. Then*

$$\mathbf{P}(\overline{Z}^{(q)} > x) = ce^{-\mu_0 x}(1 + o(1)) \quad as \quad x \to \infty, \tag{6.1.12}$$

where $c = \mathbf{E}e^{-\mu_0\widetilde{\chi}} < 1$.

If $q > 0$, τ and ζ are independent, then

$$\mathbf{P}(\overline{Z}^{(q)} > x) = c\psi^{(\tau)}(q\mu_0)e^{-\mu_0 x}(1 + o(1)) \quad as \quad x \to \infty. \tag{6.1.13}$$

If the distribution of $\zeta^{(q)}$ either contains an absolutely continuous component or is arithmetic, then $o(1)$ in (6.1.12), (6.1.13) *can be replaced by $O(e^{-hx})$ with some $h > 0$.*

If the random variable $\zeta^{(q)}$ is arithmetic, then x in (6.1.12), (6.1.13) *should be chosen to be integer.*

Proof Let, as before,

$$Z_k^{(q)} := \sum_{j=1}^{k} \zeta_j^{(q)},$$

where $\zeta_j^{(q)}$ are independent copies of $\zeta^{(q)}$, and $\overline{Z}_\infty^{(q)} := \sup_{k\geq 0} Z_k^{(q)}$. If $q \leq 0$, then $\sup_{t\geq 0} Z^{(q)}(t)$ can be attained only at the points T_k, $k = 0, 1, \ldots$. As a result,

$$\overline{Z}^{(q)} = \sup_{k\geq 0}(Z_k + qT_k) = \sup_{k\geq 0} Z_k^{(q)} = \overline{Z}_\infty^{(q)}.$$

It remains to apply assertion (ii) of Theorem 6.1.2, which implies (6.1.12).

If $q > 0$, then in the case of independent τ and ζ, we have

$$\overline{Z}^{(q)} = q\tau_1 + \sup_{k\geq 1}\{Z_k + q(T_{k+1} - \tau_1)\} \underset{d}{=} q\tau + \overline{Z}_\infty^{(q)},$$

where τ and $\overline{Z}_{\infty}^{(q)}$ are independent. Hence, making $x \to \infty$, we have

$$\mathbf{P}(\overline{Z}^{(q)} > x) = \int_0^{\infty} \mathbf{P}(\tau \in dt)\mathbf{P}(\overline{Z}_{\infty}^{(q)} > x - qt)$$
$$\sim ce^{-\mu_0 x} \int_0^{\infty} \mathbf{P}(\tau \in dt)e^{\mu_0 qt} = c\psi^{(\tau)}(q\mu_0)e^{-\mu_0 x}. \qquad (6.1.14)$$

The passage to the limit in (6.1.14) is justified by the dominated convergence theorem and the inequality $\mathbf{P}(\overline{Z}_{\infty}^{(q)} \geq x) \leq e^{-\mu_0 x}$ in (6.1.2). That $\psi^{(\tau)}(q\mu_0)$ is finite follows from the relation

$$1 = \psi_q(\mu_0) = \psi^{(\zeta)}(\mu_0)\psi^{(\tau)}(q\mu_0)$$

and since $\zeta = \zeta^{(q)} - q\tau < \zeta^{(q)}$. Theorem 6.1.3 is proved. □

Consider now the case when the solution μ_0 of equation (6.1.11) does not exist. By μ_0 we will now understand the more general characteristic:

$$\mu_0 = \sup\{\mu : \psi_q(\mu) \leq 1\}$$

(the case $\mu_0 = 0$, which means that the Cramér condition **[C]** is not met, is not excluded). We set, as before,

$$F(x) = \mathbf{P}(\zeta^{(q)} > x), \qquad F^I(x) = \int_x^{\infty} F(t)dt.$$

Theorem 6.1.4 *Let $\psi_q = \psi_q(\mu_0) < 1$ or $\mu_0 = 0$, and let $e^{\mu_0 x}F^I(x)$ be an upper power function. Then, for $q \leq 0$,*

$$\mathbf{P}(\overline{Z}^{(q)} > x) \sim \frac{F^I(x)}{B(\mu_0)C(\mu_0)} \quad as \quad x \to \infty, \qquad (6.1.15)$$

where $B(\mu)$, $C(\mu)$ are defined in (6.1.6), *so that, for $\mu_0 = 0$,*

$$B(0) = -a_{\zeta^{(q)}} = -a_{\zeta} - qa_{\tau}, \qquad C(0) = 1. \qquad (6.1.16)$$

If the distribution $F(x)$ satisfies the conditions $\mu_0 > 0$ and (6.1.10), *then* (6.1.15) *remains valid and can be written as*

$$\mathbf{P}(\overline{Z}^{(q)} > x) \sim \frac{\mu_0 M F^I(x)}{1 - \psi_q}. \qquad (6.1.17)$$

If the random variable $\zeta^{(q)}$ is arithmetic, then x in (6.1.15) *should be chosen to be integer.*

If $q \leq 0$, $\mu_0 = 0$, $a_{\zeta^{(q)}} < 0$, and if the function $F^I(x)$ is subexponential, then

$$\mathbf{P}(\overline{Z}^{(q)} > x) \sim \frac{F^I(x)}{|a_{\zeta^{(q)}}|} \quad as \quad x \to \infty. \qquad (6.1.18)$$

In the case $q > 0$, when τ and ζ are independent, and $\mu_0 > 0$, relation (6.1.15) *is preserved but with the additional factor $\psi^{(q)}(q\mu_0)$ in the right-hand side.*

If $q > 0$, $\mu_0 = 0$, *then we assume in addition that the function* F^I *is regularly varying and that, for any fixed* $h > 0$,

$$\mathbf{P}(\tau \geq hx) = o(F^I(x)) \quad as \quad x \to \infty. \tag{6.1.19}$$

Then (6.1.18) *holds.*

Remark 6.1.5 Condition (6.1.19) apparently is always met, but its proof is very cumbersome, because it requires consideration of various relations between the distributions of τ and ζ. This condition is satisfied, for example, if τ and ζ are independent and if, for any $h > 0$,

$$\mathbf{P}(\tau \geq hx) = o\Big(\int_x^\infty \mathbf{P}(\tau \geq t)\, dt\Big). \tag{6.1.20}$$

Indeed, assuming, for simplicity, that $p := \mathbf{P}(\zeta \geq 0) > 0$, and letting $q = 1$ without loss of generality, we get by (6.1.20),

$$F^I(x) = \int_x^\infty \mathbf{P}(\tau + \zeta \geq t)\, dt \geq p \int_x^\infty \mathbf{P}(\tau \geq t)\, dt \gg \mathbf{P}(\tau \geq hx)$$

as $x \to \infty$, which proves (6.1.19).

Violation of (6.1.20) means, roughly speaking, that the distribution of τ either has large lacunas or satisfies the Cramér condition. If $\mathbf{P}(\tau \geq x)$ decays at least exponentially, then $\mathbf{P}(\tau \geq hx) = o(\mathbf{P}(\zeta \geq x))$, because the function $F^I(x)$ is regularly varying. So (6.1.19) again holds.

Proof of Theorem 6.1.4 The argument is completely similar to that of Theorem 6.1.3 and is based on assertions (i), (iii) of Theorem 6.1.2 and on the results from [121] (see (6.1.8)). Explanations are needed only in the case $q > 0$, $\mu_0 = 0$. In this case, we estimate $\mathbf{P}(\overline{Z}^{(q)} \geq x)$ from below and above. Using (6.1.5), (6.1.16), we have

$$\mathbf{P}(\overline{Z}^{(q)} > x) = \mathbf{P}(q\tau + \overline{Z}_\infty^{(q)} > x) \geq \mathbf{P}(\overline{Z}_\infty^{(q)} > x) \sim \frac{F^I(x)}{|a_\zeta + qa_\tau|}.$$

On the other hand, by condition (6.1.19), for $h \in (0, 1)$, we find that

$$\begin{aligned}\mathbf{P}(\overline{Z}^{(q)} > x) &\leq \mathbf{P}(q\tau \geq hx) + \mathbf{P}(\overline{Z}_\infty^{(q)} \geq (1-h)x) \\ &= \mathbf{P}(q\tau \geq hx) + \frac{F^I((1-h)x)}{|a_\zeta + qa_\tau|}(1 + o(1)) = \frac{F^I((1-h)x)}{|a_\zeta + qa_\tau|}(1 + o(1))\end{aligned}$$

as $x \to \infty$. This inequality holds for any $h > 0$. The left-hand side of this inequality does not depend on h and since $F^I((1-h)x)/F^I(x) \to 1$ as $x \to \infty$, $h \to 0$, it follows that this inequality also holds for $h = 0$. This proves (6.1.18), and therefore, Theorem 6.1.4. □

Remark 6.1.6 Theorems 6.1.3 and 6.1.4 can be extended most easily to the inhomogeneous case for $q \leq 0$ by using the total probability formula with respect to the first (inhomogeneous) jump (τ_1, ζ_1). We set

$$\zeta_1^{(q)} = \zeta_1 + q\tau_1, \qquad \psi_{q,1}(\mu) = \mathbf{E}\exp\mu\zeta_1^{(q)}.$$

Then, clearly,

$$\mathbf{P}(\overline{Z}_1^{(q)} > x) = \mathbf{P}(\zeta_1^{(q)} > x) + \int_{-\infty}^{x} \mathbf{P}(\zeta_1^{(q)} \in dv)\mathbf{P}(\overline{Z}^{(q)} > x - v), \qquad (6.1.21)$$

where $\overline{Z}_1^{(q)}$ corresponds to an inhomogeneous CRP.

Let the conditions of Theorem 6.1.3 hold. Assume that $\psi_{q,1}(\mu_0) < \infty$. Then by (6.1.12),

$$\begin{aligned}\mathbf{P}(\overline{Z}_1^{(q)} > x) &= \mathbf{P}(\zeta_1^{(q)} > x) + c\psi_{q,1}(\mu_0)e^{-\mu_0 x}(1 + o(1))\\ &= c\psi_{q,1}(\mu_0)e^{-\mu_0 x}(1 + o(1)) \quad \text{as} \quad x \to \infty.\end{aligned}$$

If the conditions of Theorem 6.1.4 are satisfied, then from (6.1.21) and (6.1.17) we get

$$\mathbf{P}(\overline{Z}_1^{(q)} > x) = \mathbf{P}(\zeta_1^{(q)} > x) + \frac{\psi_{q,1}(\mu_0)MF^I(x)}{1 - \psi_q}(1 + o(1)) \quad \text{as} \quad x \to \infty.$$

Here for $\mu_0 > 0$ the finiteness of $\psi_{q,1}(\mu_0)$ does not imply that the summand $\mathbf{P}(\zeta_1^{(q)} > x)$ is negligible.

6.1.3 Distribution of the First Passage Time of a High Level

We denote by

$$Z^{[q]}(t) = Z_{\nu(t)}^{(q)} = \sum_{k=1}^{\nu(t)} \zeta_k^{(q)}$$

the "usual" CRP (without linear drift) such that

$$Z^{[q]}(T_k) = Z^{(q)}(T_k).$$

By using the CRP $Z^{[q]}(t)$ one can reduce, for $q \le 0$, the problems on the boundary functionals $\overline{Z}^{(q)}(T)$, $\eta_{Z^{(q)}}(x) = \inf\{t;\ Z^{(q)}(t) > x\}$ for processes with drift qt to analogous problems for "usual" CRPs, because

$$\begin{aligned}\mathbf{P}(\overline{Z}^{(q)}(t) > x) &= \mathbf{P}(\overline{Z}^{[q]}(t) > x),\\ \mathbf{P}(\eta_{Z^{(q)}}(x) < t) &= \mathbf{P}(\eta_{Z^{[q]}}(x) < t).\end{aligned} \qquad (6.1.22)$$

We set

$$\eta_{Z_\wedge^{(q)}}(x) = \min\{k : Z_k^{(q)} > x\}.$$

In the case $\mu_0 > 0$, $\psi_q(\mu_0) = 1$ from Theorem 3.4.3 in [23] we have the following result on the limit distribution of $\eta_{Z_\wedge^{(q)}}(x)$ under the condition $\overline{Z}_\infty^{(q)} > x$ and $x \to \infty$. We also set

$$\sigma_1^2 = (\ln \psi_q(\mu_0))''.$$

Theorem 6.1.7 *Let $\mu_0 > 0$, $\psi_q(\mu_0) = 1$, $\sigma_1 < \infty$, $q \le 0$. Then, as $x \to \infty$,*

$$\mathbf{P}\Big(\eta_{Z_\wedge^{(q)}}(x) < \frac{x}{\psi_q'(\mu_0)} + \frac{z\sigma_1\sqrt{x}}{(\psi_q'(\mu_0))^{3/2}} \,\Big|\, \overline{Z}_\infty^{(q)} > x\Big) = \Phi(z) + o(1),$$

where $\Phi(z)$ is the distribution function of the normal law.

Below in §6.4 we will apply (6.1.22) to prove an analog of Theorem 6.1.7 for the first passage time $\eta_{Z^{(q)}}(x)$ of a high level x by the process $Z^{(q)}(t)$ (see Corollary 6.4.3).

In significant contrast to this assertion is the behavior of the conditional distribution of $\eta_{Z_\wedge^{(q)}}(x)$ in the case when $\mu_0 > 0$, $\psi_q(\mu_0) < 1$ and the conditions of Theorem 6.1.4 are satisfied. We illustrate this by the following example, in which, for $t > 0$,

$$F(t) := \mathbf{P}(\zeta^{(q)} > t) = V_q(t)e^{-\mu_0 t}, \qquad \mu_0 > 0, \tag{6.1.23}$$

where $V_q(t)$ is a function regularly varying at infinity. It is easily seen that if

$$I_1 := \int_0^\infty V_q(t)\,dt < \infty,$$

then the distribution of $\zeta^{(q)}$ can be chosen so as to have $\psi_q < 1$, and hence, the conditions of Theorem 6.1.4, as well as conditions (6.1.9), (6.1.10) will be met, so that (6.1.17) holds. In this case, there exists the proper limit distribution of $\eta_{Z_\wedge^{(q)}}$ under the condition $\overline{Z}_\infty^{(q)} > x$.

Theorem 6.1.8 *Let (6.1.23) hold and let $I_1 < \infty$, $\psi_q < 1$. Then, as $x \to \infty$,*

$$\begin{aligned} &\mathbf{P}(\eta_{Z_\wedge^{(q)}}(x) = 1) = F(x), \\ &\mathbf{P}(\eta_{Z_\wedge^{(q)}}(x) = k) \sim c_k F(x), \qquad k \ge 2, \end{aligned} \tag{6.1.24}$$

where c_k for $k \ge 1$ satisfy the recurrence equations

$$\begin{aligned} c_1 = 1, \qquad c_{k+1} &= c_k\psi_q + \mu_0 I_k, \\ I_k := \int_0^\infty &\mathbf{P}(\eta_{Z_\wedge^{(q)}}(u) = k)e^{\mu_0 u}\,du < \infty. \end{aligned} \tag{6.1.25}$$

Besides, there exists the proper limit conditional distribution

$$\lim_{x\to\infty} \mathbf{P}(\eta_{Z_\wedge^{(q)}}(x) = k \,|\, \overline{Z}_\infty^{(q)} > x) = \frac{c_k(1-\psi_q)}{M}, \quad M = \mathbf{E}e^{\mu_0 \overline{Z}_\infty^{(q)}}. \tag{6.1.26}$$

Proof The first relation in (6.1.24) is clear.

Further, assume that, for $k \ge 1$ and $x \to \infty$,

$$\mathbf{P}(\eta_{Z_\wedge^{(q)}}(x) = k) \sim c_k F(x) \tag{6.1.27}$$

($c_1 = 1$). We will show by induction that (6.1.27) also holds for $k+1$. Indeed, we have

$$dF(t) = dV_q(t)e^{-\mu_0 t} - \mu_0 V_q(t)e^{-\mu_0 t}\,dt, \qquad (6.1.28)$$

and hence, for regularly varying functions $V_q(t)$, the integration of the second summand in (6.1.28) over $t \in (x/2, x)$, for which $V_q(t) \to 0$ as $x \to \infty$, gives the main contribution, and so by (6.1.27) we have

$$\mathbf{P}(\eta_{Z_\wedge^{(q)}}(x) = k+1) = -\int_{-\infty}^{x} dF(t)\mathbf{P}(\eta_{Z_\wedge^{(q)}}(x-t) = k)$$
$$\sim -c_k \int_{-\infty}^{x/2} dF(t)F(x-t) + \mu_0 \int_0^{x/2} \mathbf{P}(\eta_{Z_\wedge^{(q)}}(u) = k)F(x-u)\,du.$$

Here for $t \in (-\infty, x/2)$ we have $V_q(x-t) < cV_q(x)$ and $V_q(x-t) \sim V_q(x)$ for each fixed t and $x \to \infty$. Hence

$$\int_{-\infty}^{x/2} dF(t)F(x-t) \sim V_q(x)e^{-\mu_0 x}\int_{-\infty}^{x/2} dF(t)e^{-\mu_0 t} \sim F(x)\psi_q.$$

Next, for $u \in (0, x/2)$ we have $V_q(x-u) < cV_q(x)$ and $V_q(x-u) \sim V_q(x)$ for each fixed u. Therefore,

$$\int_0^{x/2} \mathbf{P}(\eta_{Z_\wedge^{(q)}}(u) = k)F(x-u)\,du \sim V_q(x)e^{-\mu_0 x}\int_0^{\infty} \mathbf{P}(\eta_{Z_\wedge^{(q)}}(u) = k)e^{\mu_0 u}\,du.$$

As a result, we get

$$\mathbf{P}(\eta_{Z_\wedge^{(q)}}(x) = k+1) = c_k\psi_q F(x) + \mu_0 F(x)\int_0^{\infty} \mathbf{P}(\eta_Z(u) = k)e^{\mu_0 u}\,du.$$

This proves (6.1.25).

Next,

$$\mathbf{P}(\eta_{Z_\wedge^{(q)}}(x) = k) \le F^{*k}(x) \sim k\psi_q^{k-1}F(x),$$

so that

$$\mathbf{P}(\eta_{Z_\wedge^{(q)}}(x) > k \,|\, \overline{Z}_\infty^{(q)} > x) \le ck\psi_q^k \quad \text{as} \quad x \to \infty, \quad c = \text{const},$$

and so the conditional distribution of $\eta_{Z_\wedge^{(q)}}(x)$ is majorized by an exponentially decaying function. This together with (6.1.27) proves (6.1.26), as well as the fact that the sum of the right-hand sides in (6.1.26) is 1. Theorem 6.1.8 is proved. □

For the distribution of $\eta_{Z_\wedge^{(q)}}(x)$, see also [65].

Remark 6.1.9 An assertion analogous to Theorem 6.1.8 can also be established in the case when $V_q(t)$ in (6.1.23) is an integrable semiexponential function. However, in this case the calculations in the proof are more cumbersome.

The nature of the conditional distribution of the time $\eta_{Z^{(q)}}(x)$ for the CRP $Z^{(q)}(t)$ under the condition $\overline{Z}^{(q)} > x$ is more complicated. Here, under rather broad conditions, we succeeded in proving only the uniform boundedness in probability of the random variable $\eta_{Z^{(q)}}(x)$ for all x on the set

$$\{\overline{Z}^{(q)} > x\} = \{\eta_{Z^{(q)}}(x) < \infty\}.$$

We will use the fact that

$$\eta_{Z^{(q)}}(x) = T_\eta, \quad \text{where} \quad \eta = \eta_{Z^{(q)}_{\wedge}}(x),$$

and consider only the case when τ and ζ are independent or linearly dependent.

Theorem 6.1.10 *Let $\zeta = h\tau + \omega$, where τ and ω are independent, let $q \le 0$, $h + q \le 0$, let $\psi_q < 1$, and let*

$$F^{(\omega)}(t) = \mathbf{P}(\omega > t) = V_0(t)e^{-\mu_0 t}, \tag{6.1.29}$$

where $V_0(t)$ is a function regularly varying at infinity such that $\int_0^\infty V_0(t)\,dt < \infty$.

Then condition (6.1.23) *holds,*

$$F(t) \sim c_q F^{(\omega)}(t) \quad as \quad t \to \infty, \tag{6.1.30}$$

where $c_q = \mathbf{E}e^{\mu_0(q+h)\tau} \le 1$, and

$$\limsup_{x\to\infty} \mathbf{P}(\eta_{Z^{(q)}}(x) > v \,|\, \overline{Z}^{(q)} > x) \to 0 \quad as \quad v \to \infty. \tag{6.1.31}$$

Apparently, the assertion of Theorem 6.1.10 remains valid also in the case when $V_q(t)$ in (6.1.23) is an integrable semiexponential function (cf. Remark 6.1.9).

Proof of Theorem 6.1.10 Since $\zeta^{(q)} = \zeta + q\tau = \omega + (q+h)\tau$, we may without loss of generality consider only the case of independent τ and ζ, in which we put $\zeta = \omega$ and replace the linear drift parameter by $q + h \le 0$.

We will first prove that (6.1.29) implies condition (6.1.23) and that (6.1.30) is met, that is, under the adopted agreement,

$$F(t) \sim c_q F^{(\zeta)}(t) \quad \text{as} \quad t \to \infty, \tag{6.1.32}$$

where $q \le 0$, $c_q = \mathbf{E}e^{\mu_0 q\tau} \le 1$ (cf. (6.1.30)).

If $q = 0$, then relation (6.1.32) becomes an identity. If $q < 0$, then we set, for simplicity, $q = -1$. Hence, for independent τ and ζ (that is, for $\zeta = \omega$), condition (6.1.29) becomes the condition $F^{(\zeta)}(t) = V_0(t)e^{-\mu_0 t}$, and so

$$\begin{aligned} F(t) = \mathbf{P}(\zeta^{(q)} > t) &= \int_0^\infty \mathbf{P}(\tau \in du)F^{(\zeta)}(t+u) \\ &= V_0(t)e^{-\mu_0 t}\int_0^\infty \mathbf{P}(\tau \in du)e^{-\mu_0 u}V_0(t+u)/V_0(t). \end{aligned} \tag{6.1.33}$$

Since $V_0(t)$ is a regularly varying integrable function, it follows that, for all sufficiently large t, there exists a constant $c_0 < \infty$ such that

$$\sup_{u\ge 0}\,[V_0(t+u)/V_0(t)] \le c_0.$$

Hence by the dominated convergence theorem, the integral in (6.1.33) converges as $t \to \infty$ to $c_q = \mathbf{E}e^{-\mu_0\tau}$, and besides, (6.1.32), (6.1.30), and (6.1.23) are satisfied.

The following lemma is needed for the proof of the main assertion (6.1.31).

Lemma 6.1.11 *Let the conditions of Theorem* 6.1.10 *hold. Then, for any fixed* $N > 0$,

$$\mathbf{P}(\eta_{Z_\wedge}(x) \le N) \sim b_N F^{(\zeta)}(x) \quad \textit{as} \quad x \to \infty, \quad b_N = \text{const}.$$

Proof Repeating the arguments used in the proof of (6.1.24), (6.1.25) in Theorem 6.1.8 in application to the sequence $Z_k^{(0)} = Z_k$, we get that

$$\mathbf{P}(\eta_{Z_\wedge}(x) = 1) = F^{(\zeta)}(x),$$
$$\mathbf{P}(\eta_{Z_\wedge}(x) = k) \sim C_k F^{(\zeta)}(x)$$

as $x \to \infty$, where C_k for $k \ge 1$ satisfies the recurrence relations $C_1 = 1$,

$$C_{k+1} = C_k\psi(\mu_0) + \mu_0 I_{k,0},$$
$$I_{k,0} = \int_0^\infty \mathbf{P}(\eta_{Z_\wedge}(u) = k)e^{\mu_0 u}\,du$$

(note that here $\psi(\mu_0) \ge 1$ for $\mathbf{E}\zeta \ge 0$). It follows that

$$\mathbf{P}(\eta_{Z_\wedge}(x) \le N) \sim b_N F^{(\zeta)}(x), \quad \text{where} \quad b_N = \sum_{j=1}^N C_j.$$

Lemma 6.1.11 is proved. □

Let us return to the proof of Theorem 6.1.10. For $\eta = \eta_{Z_\wedge^{(q)}}(x)$ we have

$$\begin{aligned}\mathbf{P}(\eta_{Z^{(q)}}(x) > v \,|\, \overline{Z}^{(q)} > x) &= \mathbf{P}(T_\eta > v \,|\, \overline{Z}^{(q)} > x) \\ &= \frac{\mathbf{P}(T_\eta > v;\ \{\eta \le N\} \cup \{N < \eta < \infty\})}{\mathbf{P}(\overline{Z}^{(q)} > x)} \le \frac{\mathbf{P}(T_N > v,\ \eta \le N)}{\mathbf{P}(\overline{Z}^{(q)} > x)} + \varepsilon_N(x),\end{aligned} \tag{6.1.34}$$

where

$$\varepsilon_N(x) \le \mathbf{P}(\eta > N \,|\, \eta < \infty),$$

and hence by Theorem 6.1.8,

$$\limsup_{x\to\infty} \varepsilon_N(x) \le \varepsilon_N \to 0 \quad \text{as} \quad N \to \infty.$$

Next, $Z_k^{(q)} \le Z_k^{(0)} = Z_k$ for $q \le 0$, and so $\eta_{Z_\wedge^{(q)}}(x) \ge \eta_{Z_\wedge}(x)$. Hence by Lemma 6.1.11,

$$\begin{aligned}\mathbf{P}(T_N > v,\ \eta \le N) &\le \mathbf{P}(T_N > v, \eta_{Z_\wedge}(x) \le N) \\ &= \mathbf{P}(T_N > v)\mathbf{P}(\eta_{Z_\wedge}(x) \le N) \sim \mathbf{P}(T_N > v)b_N F^{(\zeta)}(x) \quad \text{as} \quad x \to \infty.\end{aligned}$$

Now from (6.1.34) by Theorem 6.1.4 we obtain

$$\limsup_{x\to\infty} \mathbf{P}(\eta_{Z^{(q)}}(x) > v \,|\, \overline{Z}^{(q)} > x) \le c b_N \mathbf{P}(T_N > v) + \varepsilon_n, \qquad c = \text{const}.$$

For a given $\varepsilon > 0$, we can choose N so as to have $\varepsilon_N < \varepsilon/2$, and then choose v so that $cb_N\mathbf{P}(T_N > v) < \varepsilon/2$. For this v, the left-hand side in (6.1.31) is smaller than ε. This proves (6.1.31), and therefore, Theorem 6.1.10. □

Apparently, in addition to (6.1.31) the conditional distribution of $\eta_{Z^{(q)}}(x)$ also *converges*, under broad conditions, as $x \to \infty$ to a proper distribution.

Remark 6.1.12 If τ and ζ are linearly dependent ($\zeta = h\tau + \omega$ and $q + h > 0$), then the convergence (6.1.31) in Theorem 6.1.10 may fail to hold. The large jump $\zeta^{(q)} = \omega + (q+h)\tau$, which ensures the attainment of a level x, can be generated by the large jump τ, and the time $\eta_{Z^{(q)}}(x)$ will be comparable with x.

To complement Theorems 6.1.7–6.1.10, it remains to consider the case $\mu_0 = 0$, that is, the case of the so-called heavy tails of the distribution of $\zeta^{(q)}$ on the half-axis $t > 0$ or, what, for $q \le 0$, is the same, the case of heavy tails of the distribution of ζ.

Consider the regularly varying distributions of $\zeta^{(q)}$:

$$F(t) = \mathbf{P}(\zeta^{(q)} > t) = t^{-\alpha} l(t), \tag{6.1.35}$$

where $\alpha > 1$ and $l(t)$ is a function slowly varying at infinity.

Theorem 6.1.13 *Let* (6.1.35) *hold and let one of the following two conditions be met*:

either $\alpha \in (1,2)$, $\mathbf{P}(\zeta^{(q)} < -t) < cF(t)$, $TF(x) \to 0$ *as* $x \to \infty$;

or $\alpha > 2$, $d = \mathbf{D}\zeta^{(q)} < \infty$, $x \gg \sqrt{(\alpha-2)da_\tau^{-1}T\ln T}$.

Also let $a + q < 0$ *and* $q \le 0$. *Then*

$$\mathbf{P}(\eta_{Z^{(q)}}(x) < T) = (1 + o(1)) \int_0^T F(x - (a+q)t)\,dH(t) \tag{6.1.36}$$

as $T \to \infty$, *where* $H(t) = \mathbf{E}\nu(t)$;

$$\mathbf{P}(\eta_{Z^{(q)}}(x) < vx) = \frac{xF(x)[1 - (1 - (a+q)v)^{1-\alpha}]}{(\alpha-1)|a+q|a_\tau}(1 + o(1)), \tag{6.1.37}$$

$$\mathbf{P}(\eta_{Z^{(q)}}(x) > vx \,|\, \overline{Z}^{(q)} > x) \sim [1 + |a+q|v]^{1-\alpha} \quad \textit{as } x \to \infty. \tag{6.1.38}$$

Proof The main assertion (6.1.36) follows from Theorem 16.4.1 in [38] (see also Theorem 16.5.1 in [38]). Assertion (6.1.37) is a consequence of (6.1.36). Next, by Theorem 6.1.4 (see, for example, (6.1.18)) and since $|a+q|a_\tau = |a_\zeta + qa_\tau| = |a_{\zeta^{(q)}}|$ we conclude that

$$\mathbf{P}(\overline{Z}^{(q)} > x) = \frac{x^{1-\alpha}l(x)}{(\alpha-1)|a_{\zeta^{(q)}}|}(1 + o(1)) \quad \text{as} \quad x \to \infty.$$

Now from (6.1.3) we get (6.1.38). Theorem 6.1.13 is proved. □

With the help of (6.1.22) and the results of [79] assertion (6.1.36) can be extended to subexponential distributions F.

In the following sections we will as a rule assume that the Cramér condition is met.

6.2 Limit Theorems under the Cramér Condition for the Conditional Distribution of Jumps when the Trajectory Has a Fixed End

The content of this section is similar to that of §3.1 in [23].

6.2.1 Limit Conditional Distribution of Jumps

As in §5.2, we will assume that the normalized deviation $\alpha = x/T$ is contained in some compact set

$$K \subset (\alpha_-, \alpha_+) \setminus [\beta_-, \beta_+]$$

($\alpha_\pm$ are defined in Theorem 3.5.2, and $\beta_\pm$, in Lemma 3.5.8), by which we mean a closed interval containing the neighborhood of the point $\alpha = a$. To simplify the presentation, we shall consider only the *homogeneous case* $\boldsymbol{\xi}_1 \underset{d}{=} \boldsymbol{\xi} = (\tau, \zeta)$. By ${}^{\alpha}\boldsymbol{\xi}$ we denote the Cramér transform of $\boldsymbol{\xi}$ at the point $(\lambda(\alpha), \mu(\alpha))$, that is, the random vector with the distribution

$$\mathbf{P}({}^{\alpha}\boldsymbol{\xi} \in d\boldsymbol{v}) = e^{\lambda(\alpha)v_{(1)}+\mu(\alpha)v_{(2)}}\mathbf{P}(\boldsymbol{\xi} \in d\boldsymbol{v}), \quad \boldsymbol{v} = (v_{(1)}, v_{(2)}),$$

so that the density of the distribution of ${}^{\alpha}\boldsymbol{\xi}$ with respect to the distribution of $\boldsymbol{\xi}$ exists:

$${}^{\alpha}p(\boldsymbol{v}) := \frac{\mathbf{P}({}^{\alpha}\boldsymbol{\xi} \in d\boldsymbol{v})}{\mathbf{P}(\boldsymbol{\xi} \in d\boldsymbol{v})} = e^{\lambda(\alpha)v_{(1)}+\mu(\alpha)v_{(2)}} \tag{6.2.1}$$

(recall that $\psi(\lambda(\alpha), \mu(\alpha)) = 1$). Given an arbitrary fixed integer $m \geq 1$, we let K_{ξ} denote an arbitrary compact set in $\mathbb{R}^{2m}$. Further, let $\vec{\boldsymbol{v}} = (\boldsymbol{v}_1, \ldots, \boldsymbol{v}_m)$, $\boldsymbol{v}_i \in \mathbb{R}^2$ and let $1 \leq k_1 \leq \cdots < k_m$ be an arbitrary set of indexes.

Theorem 6.2.1 *Let ζ be a nonlattice random variable, let condition* [**C**] *be met, and let $\alpha = x/n$. Then the density*

$$p_{T,x}(\vec{\boldsymbol{v}}) = \frac{\mathbf{P}(\vec{\boldsymbol{\xi}} \in d\vec{\boldsymbol{v}} |\, Z(T) \in \Delta[x))}{\mathbf{P}({}^{\alpha}\vec{\boldsymbol{\xi}} \in d\vec{\boldsymbol{v}})} \tag{6.2.2}$$

of the conditional distribution of $\vec{\boldsymbol{\xi}} = (\boldsymbol{\xi}_{k_1}, \ldots, \boldsymbol{\xi}_{k_m})$ under the condition $\{Z(T) \in \Delta[x)\}$ with respect to the distribution of ${}^{\alpha}\vec{\boldsymbol{\xi}} = ({}^{\alpha}\boldsymbol{\xi}_1, \ldots, {}^{\alpha}\boldsymbol{\xi}_m)$, where ${}^{\alpha}\boldsymbol{\xi}_j$ are independent copies of ${}^{\alpha}\boldsymbol{\xi}$, tends to 1 as $T \to \infty$ uniformly in $\alpha \in K$, $\vec{v} \in K_{\xi} \subset \mathbb{R}^{2m}$, $\Delta \in [\Delta_T, \Delta_0]$. Here, $\Delta_T = \overline{o}(1)$ as $T \to \infty$ and $\Delta_0 > 0$ is any fixed number.

The same assertion also holds for the vector $\vec{\boldsymbol{\xi}}$ with random indexes $\nu(T) - k_1, \ldots, \nu(T) - k_m$ for $0 \leq k_1 < \cdots < k_m$.

It will be clear from the proof of the theorem that we can also take finite sets of other indexes between 1 and $\nu(T)$.

"Theorem 6.2.1 asserts that under condition $\{Z(T) \in \Delta[x)\}$ the conditional distribution $\vec{\xi}$ converges in variation to the distribution of ${}^{\alpha}\vec{\xi}$." In particular, from Theorem 6.2.1 we have

Corollary 6.2.2 *Let the conditions of Theorem* 6.2.1 *be met. Then, for any Borel sets* $B_1, \ldots, B_m$ *in* $\mathbb{R}^2$,

$$\mathbf{P}(\xi_1 \in B_1, \ldots, \xi_m \in B_m \mid Z(T) \in \Delta[x)) = \prod_{i=1}^{m} \mathbf{P}({}^{\alpha}\xi_i \in B_i) + \varepsilon_T, \tag{6.2.3}$$

where $\varepsilon_T \to 0$ *as* $T \to \infty$ *uniformly in* $\alpha \in K$, $\Delta \in [\Delta_T, \Delta_0]$.

In what follows, the principal part of the right-hand side of (6.2.3) will be called the "limit distribution" as $T \to \infty$, even though this name is not quite correct, because $\alpha \in K$ may depend on T. The same convention will be adhered to in some other analogous cases. Formally, the use of the term "limit distribution" in (6.2.3) is justified only in the cases when $\alpha \to \alpha_0 \in K$ as $T \to \infty$ and α on the right of (6.2.3) is replaced by α_0.

Instead, it would be more correct to speak about "close distributions" in (6.2.3). This more rigorous terminology was introduced in §3.3 of [23] and used to study similar problems for random walks. However, for the sake of brevity, below we will follow the above (not quite rigorous) convention about the meaning of the words "limit distribution".

Proof of Theorem 6.2.1 Setting

$$\sum_{i=1}^{m} \boldsymbol{v}_i = (t, z),$$

we have

$$\begin{aligned} p^0_{T,x}(\vec{v}) := \frac{\mathbf{P}(\vec{\xi} \in d\boldsymbol{v} \mid Z(T) \in \Delta[x))}{\mathbf{P}(\vec{\xi} \in d\boldsymbol{v})} &= \frac{\mathbf{P}(\vec{\xi} \in d\boldsymbol{v}, Z(T) \in \Delta[x))}{\mathbf{P}(\vec{\xi} \in d\boldsymbol{v})\mathbf{P}(Z(T) \in \Delta[x))} \\ &= \frac{\mathbf{P}(Z(T-t) \in \Delta[x-z))}{\mathbf{P}(Z(T) \in \Delta[x))}. \end{aligned} \tag{6.2.4}$$

By Theorem 5.2.1,

$$\mathbf{P}(Z(T) \in \Delta[x)) = \frac{\Delta}{\sqrt{T}} C(\alpha) e^{-TD(\alpha)} I_Z(\alpha, 0, 0, \mathbb{R})(1 + o(1)), \tag{6.2.5}$$

where $\alpha = x/T$, $C(\alpha)$ is a positive continuous function on K,

$$I_Z(\alpha, 0, 0, \mathbb{R}) = \int_0^{\infty} e^{\lambda(\alpha)y} \mathbf{P}(\tau > y)\, dy,$$

$\Delta = \Delta_T = \overline{o}(1)$ as $T \to \infty$, and the remainder term $o(1)$ is uniform in $\alpha \in K$.

Next, we set

$$\alpha_m := \frac{x - z}{T - t} = \alpha - \frac{z}{T} + \frac{\alpha t}{T} + O(T^{-2}).$$

Since $\alpha_m = \alpha + o(1)$ as $T \to \infty$ and since the function $C(\alpha)$ in Theorem 5.2.1 is a positive continuous function of α, an appeal of this theorem to the numerator and denominator in (6.2.4) shows that

$$p^0_{T,x}(\vec{v}) = \frac{I_Z(\alpha_m, 0, 0, \mathbb{R})}{I_Z(\alpha, 0, 0, \mathbb{R})} \exp\{-(T-t)D(\alpha_m) + TD(\alpha)\} \times (1 + o(1)). \quad (6.2.6)$$

Here

$$D(\alpha_m) = D\Big(\alpha - \frac{z}{T} + \frac{\alpha t}{T} + O(T^{-2})\Big) = D(\alpha) + \mu(\alpha)\Big(\frac{\alpha t}{T} - \frac{z}{T}\Big) + O(T^{-2});$$

and therefore,

$$p^0_{T,x} = \exp\{tD(\alpha) - \mu(\alpha)(\alpha t - z)\}(1 + o(1)). \quad (6.2.7)$$

We have $D(\alpha) = \alpha\mu(\alpha) - A(\mu(\alpha))$, and hence the expression in curly brackets in (6.2.7) is

$$z\mu(\alpha) + t(D(\alpha) - \alpha\mu(\alpha)) = z\mu(\alpha) - tA(\mu(\alpha)) = z\mu(\alpha) + t\lambda(\alpha).$$

In addition,

$$\frac{I_Z(\alpha_m, 0, 0, \mathbb{R})}{I_Z(\alpha, 0, 0, \mathbb{R})} \to 1 \quad \text{as} \quad T \to \infty. \quad (6.2.8)$$

Hence by (6.2.1), (6.2.6)

$$p^0_{T,x}(\vec{\boldsymbol{v}}) = \prod_{i=1}^{m} \frac{\mathbf{P}({}^{\alpha}\xi_i \in d\boldsymbol{v}_i)}{\mathbf{P}(\xi_i \in d\boldsymbol{v}_i)}(1 + o(1)) = \frac{\mathbf{P}({}^{\alpha}\vec{\xi} \in d\vec{\boldsymbol{v}})}{\mathbf{P}(\vec{\xi} \in d\vec{\boldsymbol{v}})}(1 + o(1)),$$

$$p_{T,x}(\vec{\boldsymbol{v}}) = p^0_{T,x}(\vec{\boldsymbol{v}}) \frac{\mathbf{P}(\vec{\xi} \in d\vec{\boldsymbol{v}})}{\mathbf{P}({}^{\alpha}\vec{\xi} \in d\vec{\boldsymbol{v}})} = 1 + o(1),$$

where the remainder term is uniform in $\alpha \in K$ and $\vec{\boldsymbol{v}} \in K_{\xi}$.

The proof of the second assertion of the theorem repeats the same arguments. Theorem 6.2.1 is proved. □

Remark 6.2.3 In view of Theorem 5.2.1, the limit conditional distribution of $\vec{\xi}$ under the condition $\{Z(T) \in \Delta[x), B(u, v, \mathbf{w})\}$ remains the same as in Theorem 6.2.1. Indeed, now $I_Z(\alpha, u, v, \mathbf{w})$ appears instead of the factor $I_Z(\alpha, 0, 0, \mathbb{R})$ in (6.2.5). If this factor is nonzero, then the ratio of these two factors

$$\frac{I_Z(\alpha_m, u, v, \mathbf{w})}{I_Z(\alpha, u, v, \mathbf{w})}$$

(this is an analog of (6.2.6)) tends, as in (6.2.8), to 1 as $T \to \infty$; all the arguments from the proof of Theorem 6.2.1 remain valid.

It is clear that the "limit" conditional distribution of $\vec{\xi}$ will be the same also in the case if we replace the event $B(u, v, \mathbf{w})$ (see Theorem 5.2.1) by the event that the vector

$(\gamma(T), \chi(T), \zeta(T))$ lies in an arbitrary parallelepiped of positive measure generated by the distribution

$$\frac{I_Z(\alpha, u, v, \mathbf{w})}{I_Z(\alpha, 0, 0, \mathbb{R})}.$$

Of course, the above also applies to the vectors $\vec{\xi}$ with random indexes specified in the second assertion of Theorem 6.2.1.

6.2.2 On the Distribution of the Vector ${}^{\alpha}\xi$

The Laplace transform of the distribution of ${}^{\alpha}\xi$ has the form

$${}^{\alpha}\mathbf{A}(\lambda, \mu) := \mathbf{E}e^{\langle {}^{\alpha}\xi, (\lambda,\mu)\rangle} = \mathbf{A}(\lambda + \lambda(\alpha), \mu + \mu(\alpha)).$$

The fundamental function ${}^{\alpha}A(\mu)$ corresponding to the jumps of ${}^{\alpha}\xi$ near the point $(0, 0)$ is the solution of the equation $\mathbf{A}(-\lambda + \lambda(\alpha), \mu + \mu(\alpha)) = 0$, that is,

$${}^{\alpha}A(\mu) = \lambda(\alpha) + A(\mu + \mu(\alpha)),$$

so that ${}^{\alpha}A(0) = 0$. The mean value of the process ${}^{\alpha}Z(t)$ with jumps ${}^{\alpha}\xi$ per unit time is asymptotically equivalent to

$${}^{\alpha}a := {}^{\alpha}A'(0) = A'(\mu(\alpha)) = \alpha \tag{6.2.9}$$

(see §1.1; recall that $\mu(\alpha)$ is the inverse function of $A'(\mu)$). The variance of ${}^{\alpha}Z(t)$ per unit time is asymptotically equivalent to

$${}^{\alpha}\sigma^2 = A''(\mu(\alpha)).$$

6.3 Integro-Local Theorems for the First Passage Time of a High Level by the Trajectory of a Compound Renewal Process

We set

$$\eta_Z(x) := \inf\{t : Z(t) > x\}, \qquad \chi_Z(x) = Z(\eta_Z(x)) - x.$$

In this section, we study the asymptotic behavior of the probability

$$\mathbf{P}(\eta_Z(x) \in \delta[T),\ \chi_Z(x) \in \Delta[v)) \quad \text{for} \quad \alpha = \frac{x}{T} > a + \varepsilon$$

and for some $\varepsilon > 0$.

To formulate the main assertion, we introduce the CRP ${}^{\alpha}Y(t)$ in the same way as the CRP $Y(t)$ (defined in §1.1) but with the jumps $({}^{\alpha}\tau, -{}^{\alpha}\zeta)$, where, as before, $({}^{\alpha}\tau, {}^{\alpha}\zeta) = {}^{\alpha}\xi$:

$${}^{\alpha}Y(t) = -{}^{\alpha}Z_{\eta(t)}, \qquad {}^{\alpha}Z_n = \sum_{j=1}^{n} {}^{\alpha}\zeta_j,$$

and ${}^{\alpha}\zeta_j$ are independent copies of ${}^{\alpha}\zeta$. By (6.2.9), we have

$$\mathbf{E}\,{}^{\alpha}Y(t) \sim -{}^{\alpha}at = -\alpha t \quad \text{as} \quad t \to \infty.$$

Hence for ${}^{\alpha}\overline{Y} = \sup_{t\geq 0} {}^{\alpha}Y(t)$ with $\alpha \geq \alpha_0 > 0$ we have

$$\mathbf{P}({}^{\alpha}\overline{Y} \leq -v) > 0$$

for $v > 0$ such that $\mathbf{P}(\zeta_1 < -v) > 0$ (recall that ${}^{\alpha}Y(0) = -{}^{\alpha}\zeta_1$).

Theorem 6.3.1 *Let $\boldsymbol{\xi} = (\tau, \zeta)$ be a nonlattice vector and let $\alpha = x/T > a + \varepsilon$ with some $\varepsilon > 0$. Then*

$$\begin{aligned}
&\mathbf{P}(\eta_Z(x) \in \delta[T),\ \chi_Z(x) \in \Delta[v)) \\
&\quad = \mathbf{P}(Z(T+\delta) \in \Delta[x+v),\ \gamma(T+\delta) < \delta)\mathbf{P}({}^{\alpha}\overline{Y} \leq -v)(1+o(1)) \\
&\quad = \frac{\Delta\delta C(\alpha)}{\sqrt{T}} e^{-TD(\alpha)-v\mu(\alpha)}\mathbf{P}({}^{\alpha}\overline{Y} \leq -v)(1+o(1)),
\end{aligned} \tag{6.3.1}$$

where $C(\alpha)$ is the positive continuous function defined in Theorem 5.2.1, *$\Delta = \Delta_T = \overline{o}(1)$, $\delta = \delta_T = \overline{o}(1)$ as $T \to \infty$, and the remainder term $o(1)$ is uniform in $\alpha \in K \cap \{\alpha \geq \alpha_0 > 0\}$.*

Proof We set

$$B_{Z,T} = \{\eta_Z(x) \in \delta[T),\ \chi_Z(x) \in \Delta[v)\}.$$

Lower estimate for $\mathbf{P}(B_{Z,T})$. We have

$$B_{Z,T,\gamma} := \Big\{\overline{Z}(T+\delta-\gamma(T+\delta)-0) \leq x,\ Z(T+\delta) \in x+\Delta[v),\ \gamma(T+\delta) < \delta\Big\} \subset B_{Z,T}. \tag{6.3.2}$$

Hence

$$\mathbf{P}(B_{Z,T}) \geq \mathbf{P}(B_{Z,T,\gamma}). \tag{6.3.3}$$

Upper estimate. The event

$$B_{Z,T} = \{Y(T) \in x + \Delta(v), \chi(T) < \delta,\ \overline{Z}(T + \chi(T) - 0) < x\} \tag{6.3.4}$$

lies in the σ-algebra generated by the process $Y(T)$ through the Markov moment $\eta(T)$ inclusively, so that the events $B_{Z,T}$ and the random variable $\tau_{\eta(T)+1}$ are independent. On the other hand, by (6.3.4)

$$\begin{aligned}
&B_{Z,T} \bigcap \{\tau_{\eta(T)+1} \geq \delta\} \\
&\subset \{Z(T+\delta) \in x + \Delta[v), \gamma(T+\delta) < \delta,\ \overline{Z}(T+\delta-\chi(T+\delta)-0) < x\} = B_{Z,T,\gamma};
\end{aligned}$$

$$\mathbf{P}(B_{Z,T})\mathbf{P}(\tau \geq \delta) \leq \mathbf{P}(B_{Z,T,\gamma}). \tag{6.3.5}$$

Since in what follows as δ we will take the sequence $\delta = \delta_T \to 0$ tending to 0 sufficiently slowly as $T \to \infty$, we have $\mathbf{P}(\tau \geq \delta) = 1 + o(1)$ as $T \to \infty$, and so by (6.3.3) and (6.3.5),

$$\mathbf{P}(B_{Z,T}) = \mathbf{P}(B_{Z,T,\gamma})(1+o(1)) \quad \text{as} \quad T \to \infty. \tag{6.3.6}$$

Next, we have

$$\mathbf{P}(B_{Z,T,\gamma}) = \mathbf{P}(Z(T+\delta) \in \Delta[x+v), \gamma(T+\delta) < \delta)$$
$$\mathbf{P}(\overline{Z}(T+\delta-\gamma(T+\delta)-0) \le x \mid Z(T+\delta) \in \Delta[x+v), \gamma(T+\delta) < \delta). \quad (6.3.7)$$

Further

$$I_Z(\alpha, 0, 0, \mathbb{R}) - I_Z(\alpha, \delta, 0, \mathbb{R}) = \int_0^{\delta} e^{\lambda(\alpha)y}\mathbf{P}(\tau > y)\, dy \sim \delta$$

as $\delta \to 0$ and

$$TD\left(\frac{x+v}{T}\right) = TD(\alpha) + \mu(\alpha)v + o(1)$$

as $T \to \infty$, and so by Theorem 5.2.1 the first factor on the right of (6.3.7) has the same asymptotic behavior as the product

$$\frac{\Delta\delta C(\alpha)}{\sqrt{T}} e^{-TD(\alpha)-v\mu(\alpha)}$$

(cf. (6.3.1)).

Let us now show that the second factor on the right of (6.3.7) is asymptotically equivalent to $\mathbf{P}({}^{\alpha}\overline{Y} \le -v)$. Our presentation of the proof will sometimes be partly schematic, because (a) it is very transparent and follows the same line of arguments as in §3.3 [23] in the proof of the analogous result for random walks and (b) the transition to a detailed exposition presents no significant difficulties.

By Theorem 6.2.1 and Remark 6.2.3, the conditional distribution of the jumps $\xi_{\nu(T+\delta)}, \xi_{\nu(T+\delta)-1}, \ldots, \xi_{\nu(T+\delta)-m-1}$ under the condition

$$\{Z(T+\delta) \in \Delta[x+v), \gamma(T+\delta) < \delta\}$$

has the "limit" conditional distribution, which coincides with the distribution of $({}^{\alpha}\xi_1, \ldots, {}^{\alpha}\xi_m)$, where ${}^{\alpha}\xi_j$ are independent. If we consider the process $Z(t)$ in the opposite direction from the point $T+\delta-\gamma(T+\delta)$ (this is the process $Z(T+\delta-\gamma(T+\delta)-u)$, $u>0$), then on the event $\{{}^{\alpha}\overline{Y} \le -v\}$ it will begin with the negative jump $-{}^{\alpha}\zeta_1 \le -v$. On the first m "steps," the oppositely directed limit process ${}^{\alpha}Y(u)$ has the form of the process $Y(u)$, but with the jumps $({}^{\alpha}\tau_1, -{}^{\alpha}\zeta_1), \ldots, ({}^{\alpha}\tau_m, -{}^{\alpha}\zeta_m)$. By choosing a large m, the probability $\mathbf{P}({}^{\alpha}\overline{Z}_m \le -v)$, where

$${}^{\alpha}\overline{Z}_m = \max_{k \le m} \sum_{j=1}^{k} (-{}^{\alpha}\zeta_j),$$

can be made to be arbitrarily close to $\mathbf{P}({}^{\alpha}\overline{Y} \le -v)$, where ${}^{\alpha}\overline{Y} = \sup_{u \ge 0} {}^{\alpha}Y(u)$. In other words, by the choice of N the probability $\mathbf{P}({}^{\alpha}\overline{Y}(N) \le -v)$, where ${}^{\alpha}\overline{Y}(N) = \max_{u \le N} {}^{\alpha}Y(u)$, can be made to be arbitrarily close to $\mathbf{P}({}^{\alpha}\overline{Y} \le -v)$.

It remains to verify that the conditional probability of the process $Z(t)$ to cross the boundary x on the time interval $(0, T+\delta-M)$ under the condition $\{Z(T+\delta) \in$

$\Delta[x+v), \gamma(T+\delta) < \delta\}$ is negligibly small for large M. More precisely, it suffices to show that the conditional probability

$$P^{\rm cond} := \mathbf{P}(\eta_Z(x) < T - M \mid Z(T) \in \Delta[x+v), \gamma(T) < \delta) \\ = \frac{\mathbf{P}(\eta_Z(x) < T - M, Z(T) \in \Delta[x+v), \gamma(T) < \delta)}{\mathbf{P}(Z(T) \in \Delta[x+v), \gamma(T) < \delta)} \tag{6.3.8}$$

is negligibly small for large T, M and $T - M$. We set $t_k = kh$, $k = 0, 1, \ldots, N$; $h[t) := [t, t+h)$, and assume that $N = \frac{T-M}{h}$ is an integer number. Since $\eta_Z(x)$ is a Markov moment, the numerator on the right of (6.3.8) is

$$\Sigma := \sum_{k=0}^{N} \int_{t_k}^{t_{k+1}} \mathbf{P}(\eta_Z(x) \in dt)\mathbf{P}(Z^*(T-t) \in \Delta[x+v), \gamma(T-t) < \delta), \tag{6.3.9}$$

where the process Z^* has the same structure as Z and does not depend on Z. By Theorem 5.2.1, the second factor in (6.3.9) can be replaced by

$$\mathbf{P}(Z(T-t_k) \in \Delta[x+v), \gamma(T-t_k) < \delta)(1+\varepsilon_h),$$

where $\varepsilon_h \to 0$ as $h \to 0$, so that the resulting expression differs from $\mathbf{P}(Z(T-t_k) \in \Delta[x+v))$ by the factor $\delta(1+o(1))$ for small δ. This remark also applies to the denominator in (6.3.8). As a result, we see that, for small h and δ,

$$P^{\rm cond} = \sum_{k=0}^{N} \frac{\mathbf{P}(\eta_Z(x) \in h[t_k))\mathbf{P}(Z(T-t_k) \in \Delta[x+v))(1+o(1))}{\mathbf{P}(Z(T) \in \Delta[x+v))}. \tag{6.3.10}$$

The numerators in (6.3.8) and (6.3.10) mean the probability of the bundle of "curved" trajectories (passing above the point $(T-M, x)$) connecting the point $(0,0)$ with the neighborhood of the point $(T, x+v)$. The denominator in these ratios is equal to the probability of all trajectories connecting the above domains, including the "straight" (the most probable) trajectories (see §4.7.1). Since the function D is strictly convex, this means that the numerator in (6.3.8) and (6.3.10) is exponentially smaller) with respect to M) than the denominator, so that the probability $P^{\rm cond}$ is small for large M. To obtain an explicit estimate for $P^{\rm cond}$, we note that since $\eta_Z(x)$ is a Markov moment, the next horizontal jump τ^* does not depend on the prehistory and is distributed as τ. Therefore,

$$\mathbf{P}(\eta_Z(x) \in h[t_k), \tau^* > h) = \mathbf{P}(\eta_Z(x) \in h[t_k))\mathbf{P}(\tau > h).$$

On the other hand,

$$\{\eta_Z(x) \in h[t_k), \tau^* > h\} \subset \{Z(t_{k+1}) > x\}$$

and hence, for h such that $\mathbf{P}(\tau > h) > 1/2$, we get

$$\mathbf{P}(\eta_Z(x) \in h[t_k)) \le \frac{\mathbf{P}(Z(t_{k+1}) > x)}{\mathbf{P}(\tau > h)} \le 2\mathbf{P}(Z(t_{k+1}) > x).$$

Hence, using (6.3.10), this gives

$$P^{\mathrm{cond}} \le \sum_{k=0}^{N} \frac{2\mathbf{P}(Z(t_{k+1}) > x)\mathbf{P}(Z(T-t_k) \in \Delta[x+v))}{\mathbf{P}(Z(T) \in \Delta[x+v))}(1+o(1)). \qquad (6.3.11)$$

It remains to apply the asymptotic representations for all terms of the right of (6.3.11), which are given by Theorem 5.2.1. We leave this to the reader.

The arguments showing that P^{cond} is small may also be found in the proof of Theorem 6.4.1 below (see the estimates of the sums $\sum_{T,x}$ and $R_{T,x}$ in (6.4.10) and (6.4.11)). Theorem 6.3.1 is proved. □

We set

$$P_\alpha(v) = \mathbf{P}({}^{\alpha}\overline{Y} \le v), \qquad J_\alpha(v) = \int_0^v e^{-u\mu(\alpha)} P_\alpha(-u)\,du. \qquad (6.3.12)$$

In Theorem 6.3.1, we put $v = v_k = k\Delta$, $k = 0, 1, \ldots$, and sum the resulting relations with respect to k from 0 to $N = v/\Delta$ (assuming that N is integer). In the left-hand side, we get the probability $\mathbf{P}(\eta_Z(x) \in \delta[T), \chi_Z(x) < v)$, and on the right we have

$$\frac{\delta C(\alpha)}{\sqrt{T}} e^{-TD(\alpha)} \sum_{k=0}^{N} \Delta e^{-v_k\mu(\alpha)} P_\alpha(-v_k)(1+o(1)),$$

where the integral sum corresponds to the interval $J_\alpha(v)$. Hence we reach the following.

Corollary 6.3.2 *Let the conditions of Theorem* 6.3.1 *be met. Then*

$$\mathbf{P}(\eta_Z(x) \in \delta[T),\ \chi_Z(x) < v) = \frac{\delta C(\alpha)}{\sqrt{T}} e^{-TD(\alpha)} J_\alpha(v)(1+o(1)),$$

where $\delta = \overline{o}(1)$ *as* $T \to \infty$ *and the remainder term is uniform in* α *from* $K \cap \{\alpha \ge \alpha_0 > 0\}$.

The following result is of special interest in view of Corollary 6.3.2.

Lemma 6.3.3 *The integral* $J_\alpha(\infty)$ *converges for all* $\alpha \in K$.

Proof Note that

$$P_\alpha(u) < \mathbf{P}({}^{\alpha}\zeta \ge u).$$

Hence

$$\begin{aligned} J_\alpha(\infty) &< \int_0^\infty e^{-u\mu(\alpha)} \int_u^\infty \mathbf{P}({}^{\alpha}\zeta \in dt)\,du \\ &= \int_{t=0}^\infty \int_{u=0}^t e^{-u\mu(\alpha)} du \mathbf{P}({}^{\alpha}\zeta \in dt) = \int_0^\infty \frac{(1-e^{-t\mu(\alpha)})}{\mu(\alpha)} \mathbf{P}({}^{\alpha}\zeta \in dt) \\ &= \frac{1}{\mu(\alpha)}(\mathbf{P}({}^{\alpha}\zeta > 0) - \mathbf{E}(e^{-\mu(\alpha){}^{\alpha}\zeta};\ {}^{\alpha}\zeta \ge 0), \end{aligned} \qquad (6.3.13)$$

where

$$\mathbf{E}(e^{-\mu(\alpha)^\alpha\zeta}; \zeta \ge 0) \le \mathbf{E}e^{-\mu(\alpha)^\alpha\zeta} = \mathbf{A}(\lambda(\alpha), 0) < \infty$$

since $\alpha \in K$ (that is, $\alpha \notin [\beta_-, \beta_+]$), and therefore, $\lambda(\alpha) < \lambda_+$.

If α is separated from a (and so, $|\mu(\alpha)| > \varepsilon$ for some $\varepsilon > 0$), then, clearly, $J_\alpha(\infty) < \infty$. If $\alpha \to a$ as $T \to \infty$, then $\mu(\alpha) \to 0$ and

$$J_\alpha(\infty) \to \mathbf{E}({}^\alpha\zeta; {}^\alpha\zeta \ge 0) < \infty,$$

inasmuch as ${}^\alpha\zeta$ satisfies the Cramér condition. Lemma 6.3.3 is proved. □

6.4 Integral Theorems for the Distribution of $\overline{Z}(T) = \max_{t\le T} Z(t)$

To begin with we note that in the case of independent τ and ζ the asymptotics of the distributions of

$$\mathbf{P}(\overline{Z}(T) > x,\ \chi_Z(x) > u), \quad \mathbf{P}(\overline{Z}(T) < x, Z(T) < x - y)$$

and of several other distributions was treated in detail in [49] by *analytic methods* and the factorization technique. The constant factors in the asymptotics of the probabilities were given in [49] in terms of the components of the factorization of the function $1 - z\psi^{(\zeta)}(\mu)$. This requires an explanation, which, however, would lead us too far afield from the main course of our presentation. So, we will not dwell on the results of [49] and pay most attention to the study of the general case with the help of *probabilistic* approaches.

6.4.1 The Case $a < 0$, $\alpha > 0$

We set

$$K_\varepsilon := \{\beta_+^* + \varepsilon, \alpha_+ - \varepsilon\} =: [\alpha_{(\varepsilon)}, \alpha^{(\varepsilon)}],$$

where

$$\beta_+^* = \begin{cases} \beta_+ & \text{if } \lambda_+ < D(0), \\ 0 & \text{if } \lambda_+ \ge D(0) \end{cases}$$

and, as before,

$$\alpha_+ = A'(\mu_+ - 0), \qquad \beta_+ = A'(m_+),$$

μ_+ is defined in (3.5.8) and m_+ is the greatest solution of the equation $A(\mu) = -\lambda_+$ (it exists if $\lambda_+ < D(0)$). In the case $a < 0$, $A(\mu_+) > 0$, the function $A(\mu)$ has the greatest zero $\mu_0 > 0$: $A(\mu_0) = 0$, and the value

$$\alpha_0 = A'(\mu_0)$$

is defined. We let $A_{T,x,v}$ denote the event

$$A_{T,x,v} := \{\overline{Z}(T) > x,\ \chi_Z(x) \in \Delta[v]\}.$$

Theorem 6.4.1 *Let ξ be a nonlattice vector, let $a < 0$, $A(\mu_+) > 0$, $\alpha = \frac{x}{T}$, and let $c(\alpha, v) = C(\alpha)P_\alpha(-v)$, where $C(\alpha)$ and $P_\alpha(v)$ are defined in Theorem 5.2.1 and in (6.3.12), respectively. Assume that $\Delta = \overline{o}(1)$ as $T \to \infty$. Then:*

(i) *if $\alpha > \alpha_0$ and $\alpha - \alpha_0$ is separated from 0, then*

$$\mathbf{P}(A_{T,x,v}) = \frac{\Delta}{\sqrt{T}} \frac{c(\alpha, v)}{\alpha L'(\alpha)} e^{-TD(\alpha)-v\mu(\alpha)}(1 + o(1)), \tag{6.4.1}$$

where

$$L(\alpha) = \frac{D(\alpha)}{\alpha};$$

(ii) *if $\alpha = \alpha_0 + \rho$, where $\rho = o(1)$ as $T \to \infty$, then, as $T \to \infty$,*

$$\mathbf{P}(A_{T,x,v}) = \frac{\Delta\sqrt{2\pi}\, c(\alpha_0, v)}{\alpha_0\sqrt{D''}} e^{-x\mu_0 - v\mu(\alpha_0)}\Phi(\rho\sqrt{xL''})(1 + o(1)), \tag{6.4.2}$$

where $\Phi(z)$ is the distribution function of the standard normal law, $L'' = L''(\alpha_0)$ and $D'' = D''(\alpha_0) = \alpha_0 L''(\alpha_0)$;

(iii) *if $\alpha < \alpha_0$ and $\alpha_0 - \alpha$ is separated from 0, then*

$$\mathbf{P}(A_{T,x,v}) = \frac{\Delta\sqrt{2\pi} c(\alpha_0, v)}{\alpha_0\sqrt{D''}} e^{-x\mu_0 - v\mu(\alpha_0)}(1 + o(1)) \tag{6.4.3}$$

as $T \to \infty$.

In assertions (i), (iii), the remainder term $o(1)$ is uniform in α, where α is taken from the intersections of the domains for α given in these sections with the compact set $K_{2\varepsilon}$.

Summing (6.4.2) and (6.4.3) with respect to $v = v_k = k\Delta$, $k = 0, 1, \ldots$, and approximating the result by an integral, we arrive at the following assertion.

Corollary 6.4.2 *If $\alpha = \alpha_0 - \rho$, $\rho\sqrt{x} \to \infty$ as $x \to \infty$, then*

$$\mathbf{P}(\overline{Z}(T) > x) = \mathbf{P}(\overline{Z} > x)(1 + o(1)) = \frac{\sqrt{2\pi}C(\alpha_0)P_{\alpha_0}}{\alpha_0\sqrt{D''}} e^{-x\mu_0}(1 + o(1)), \tag{6.4.4}$$

where

$$P_{\alpha_0} = \int_0^\infty P_{\alpha_0}(-v)e^{-v\mu(\alpha_0)}\, dv.$$

The conditional limit distribution of $\chi_Z(x)$ under the condition $\overline{Z} > x$ has the density

$$\frac{P_{\alpha_0}(-v)e^{-v\mu(\alpha_0)}}{P_{\alpha_0}}.$$

We have $\overline{Z} = \sup_{k\geq o} Z_k$, and hence the second equality in (6.4.4) is one of the forms of the classical Cramér–Lundberg theorem on the distribution of $\overline{Z}$. Since $A^{(\zeta)}(\mu_0) = \mathbf{A}(0, \mu_0) = \mathbf{A}(-A(\mu_0), \mu_0) = 0$, it follows that $\mu_0 = \mu_0^{(\zeta)} = \max\{\mu : A^{(\zeta)}(\mu) = 0\}$. Hence there is no need to verify assertion (iii) of the theorem, because the asymptotics of the right-hand side of (6.4.2) as $\rho \to \infty$ is the same, up to a (very complicated) constant factor, as the asymptotics of $\mathbf{P}(\overline{Z} > x)$ in Theorem 6.1.3. This implies that the constant factors in (6.1.12) and (6.4.4) are equal.

Corollary 6.4.3 *The conditional distribution of the variable $\eta_Z(x) = \inf\{t : Z(t) > x\}$ under the condition $\{\overline{Z} > x\}$ has the property*

$$\mathbf{P}\Big(\frac{\eta_Z(x) - x/\alpha_0}{\sigma_{\eta Z}\sqrt{x}} < z \,|\, \overline{Z} > x\Big) \Rightarrow \Phi(z) \quad as \quad x \to \infty, \tag{6.4.5}$$

where

$$\sigma_{\eta Z}^{-1} = \alpha_0^2 \sqrt{L''} = \alpha_0^{3/2}\sqrt{D''}.$$

The proof of Corollary 6.4.3 will be given below.

As already pointed out, a rather complete study of the distribution of the maximum $\overline{Z}_n$ in the large deviation zone for random walks was carried out in [13] (see also [23, §3.9]) by analytic methods with the use of factorization. A considerable part of the results from [13] was carried over in [49], for independent τ and ζ, to CRPs and the distribution of $\overline{Z}(T)$. The machinery used in these studies is very subtle and goes beyond the scope of the present monograph. A more transparent probabilistic approach to the study of the distribution of $\overline{Z}_n$ for random walks was proposed in [41]. Theorem 6.3.1 uses the same approach. For CRPs, this approach is based on the existence of limit conditional distributions (for a fixed end of the trajectory) for the jumps τ and ζ, which was established in Theorem 6.2.1. The essential deficiency of this approach is that it assumes that the normalized deviation of $\alpha = \frac{x}{T}$ is separated from zero. In this case, there exist the conditional limit distributions of τ and ζ described in Theorem 6.2.1. In the case $\alpha \to 0$ as $T \to \infty$, we have recourse to the results of [49].

There is a different approach, which can be applied in the case $\alpha = o(T^{-1/3})$, based on the extension of the invariance principle to the large deviation zone. This approach will be presented in Chapter 7.

The convergence of the conditional distribution of the first high-level passage time to the normal law for random walks with negative drift (cf. Theorem 6.4.1 and Corollary 6.4.3) was established in [13]. The same convergence for CRPs in the case of independent τ and ζ was proved in [49]. As already pointed out, both these works used the analytic technique based on factorization.

Proof of Corollary 6.4.3 We have

$$\mathbf{P}(\eta_Z(x) < x/\alpha_0 + y\sqrt{x} \,|\, \overline{Z} > x) = \mathbf{P}(\overline{Z}(x/\alpha_0 + y\sqrt{x}) > x \,|\, \overline{Z} > x)$$
$$= \frac{\mathbf{P}(\overline{Z}(x/\alpha_0 + y\sqrt{x}) > x)}{\mathbf{P}(\overline{Z} > x)}. \tag{6.4.6}$$

Here, for $T = x/\alpha_0 + y\sqrt{x}$,

$$\alpha = \frac{x}{x/\alpha_0 + y\sqrt{x}} = \alpha_0\Big(1 - \frac{y\alpha_0}{\sqrt{x}} + O\Big(\frac{y^2}{x}\Big)\Big) = \alpha_0 - \rho,$$

where $\rho = \frac{y\alpha_0^2}{\sqrt{x}} + O(\frac{y^2}{x})$. Hence by Theorem 6.4.1 we can complete equality (6.4.6) with the value

$$\Phi\Big(\frac{y\alpha_0^2\sqrt{xL''}}{\sqrt{x}}\Big)(1 + o(1)).$$

Putting $z = y\alpha_0^2\sqrt{L''}$, we arrive at (6.4.5). Corollary 6.4.3 is proved. □

The following assertion will be needed in the proof of Theorem 6.4.1.

Lemma 6.4.4 *Let* $A(\mu^+) > 0$ $(\mu^+ = \sup\{\mu : A(\mu) < \infty\})$. *Then:*

(i) *The maximal solution of the equation* $A(\mu) = 0$ *is given by*

$$\mu_0 = \max\{\mu : A(\mu) = 0\} \begin{cases} > 0 & \textit{for} \quad a < 0, \\ = 0 & \textit{for} \quad a \geq 0. \end{cases}$$

If $A(\mu_+) > 0$, *then the function* $A(\mu)$ *is analytic at the point* μ_0;

(ii) *The function*

$$L(\alpha) = \frac{1}{\alpha} D(\alpha)$$

is convex on the half-axis $(0, \infty)$; *in the case* $A(\mu_+) > 0$ *attains its minimum at the point*

$$\alpha_0 := A'(\mu_0), \qquad L(\alpha_0) = \mu_0; \quad \textit{here} \;\; \alpha_0 \in K_\varepsilon$$

for a sufficiently small $\varepsilon > 0$,

$$L''(\alpha_0) = \frac{\mu'(\alpha_0)}{\alpha_0} = \frac{D''(\alpha_0)}{\alpha_0} > 0$$

(in the case $a \geq 0$, *we have* $\alpha_0 = a$ *and* $L(\alpha)$ *attains its minimum on* $[a, \infty)$ *at the point* $\alpha_0 = a$, $L(\alpha_0) = \mu_0 = 0$*).*

Proof (i) The first assertion of the lemma is clear, since the function $A(\mu)$ is convex and since $A'(0) < 0$. If $A(\mu_+) > 0$, then $[0, \mu_0] \subset [0, \mu_+)$ lies in the domain of analyticity of the function $A(\mu)$.

(ii) Like $D(\alpha)$, the function $L(\alpha)$ is differentiable a.e., and

$$\begin{aligned} L'(\alpha) &= -\alpha^{-2} D(\alpha) + \alpha^{-1}\mu(\alpha) = -\alpha^{-2}(D(\alpha) - \alpha\mu(\alpha)) \\ &= \alpha^{-2} A(\mu(\alpha)) = -\alpha^{-2}\lambda(\alpha). \end{aligned} \tag{6.4.7}$$

In Lemma 3.5.8 it was shown that $\lambda(\alpha)$ decreases on $(0, \infty)$. Hence the function $\alpha^{-2}\lambda(\alpha)$ is also decreasing and $L'(\alpha)$ is increasing. This proves the convexity of $L(\alpha)$.

In view of (6.4.7), the equation $L'(\alpha) = 0$ for the minimum point of $L(\alpha)$ is equivalent to the equation

$$A(\mu(\alpha)) = 0,$$

which in turn is equivalent to the equation $\mu(\alpha) = \mu_0$, or, equivalently,

$$\alpha = \alpha_0 = A'(\mu_0)$$

($\mu(\cdot)$ is the inverse function of $A'(\cdot)$).

Next,

$$L(\alpha_0) = \frac{1}{\alpha_0} D(\alpha_0) = \frac{1}{\alpha_0}(\alpha_0 \mu(\alpha_0) - A(\mu(\alpha_0))) = \mu(\alpha_0) = \mu_0.$$

In the case $\lambda_+ < D(0)$ we have $\beta_+ < \mu_0$, and hence $\mu_0 > \beta_+^*$, $\mu_0 \in (\beta_+^*, \alpha_+)$, $\mu_0 \in K_\varepsilon$ for sufficiently small $\varepsilon > 0$.

Differentiating equality (6.4.7), we get

$$L''(\alpha) = -2\alpha^{-3} A(\mu(\alpha)) + \alpha^{-2} A'(\mu(\alpha))\mu'(\alpha),$$
$$L''(\alpha_0) = \alpha_0^{-1} \mu'(\alpha_0) = \alpha_0^{-1} D''(\alpha_0) > 0,$$

proving Lemma 6.4.4. □

Proof of Theorem 6.4.1 Let, as before,

$$B_{Z,T} = \{\eta_Z(x) \in \delta[T), \chi_Z(x) \in \Delta[v)\}.$$

According to Theorem 6.3.1,

$$\mathbf{P}(B_{Z,T}) = \frac{\Delta \delta C(\alpha) P_\alpha(-v)}{\sqrt{T}} e^{-TD(\alpha) - v\mu(\alpha)}(1 + o(1)), \tag{6.4.8}$$

where

$$P_\alpha(-v) := \mathbf{P}({}^\alpha\overline{Y} \le -v),$$

$C(\alpha)$ is defined in Theorems 5.2.1 and 6.3.1, $\Delta = \Delta_T = \overline{o}(1)$, $\delta = \delta_T = \overline{o}(1)$ as $T \to \infty$, and the remainder term is uniform in $\alpha \in K_\varepsilon \subset K \bigcap \{\alpha \ge \alpha_0 > 0\}$.

Suppose for definiteness that $N = \frac{T}{\delta}$ is an integer number. Then, for $\alpha \in K_{2\varepsilon}$, $t_k = k\delta$, $\theta_k = \frac{t_k}{T}$, $k = 0, 1, \ldots, N-1$, we have

$$\mathbf{P}(A_{T,x,v}) = \mathbf{P}(\overline{Z}(T) > x, \chi_Z(x) \in \Delta[v)) = \sum_{k=0}^{N-1} \mathbf{P}(B_{Z,t_k})$$
$$= \sum_{k=0}^{N-1} \mathbf{P}(\eta_Z(x) = \delta[t_k), \chi_Z(x) \in \Delta[v)) = \Sigma_{T,x} + R_{T,x}, \tag{6.4.9}$$

where

$$\Sigma_{T,x} := \sum_{\theta_k \in \left[\frac{\alpha}{\alpha(\varepsilon)}, 1\right)} \frac{\Delta \delta c(\alpha/\theta_k, v)}{\sqrt{T\theta_k}} e^{-T\theta_k D(\alpha/\theta_k) - v\mu(\alpha/\theta_k)}(1 + o(1)), \tag{6.4.10}$$

$c(\alpha, v) = C(\alpha)P_\alpha(-v)$, and hence by (6.3.6),

$$R_{T,x} := \sum_{\theta_k < \frac{\alpha}{\alpha^{(\varepsilon)}}} \mathbf{P}(B_{Z,t_k}) = \sum_{\theta_k < \frac{\alpha}{\alpha^{(\varepsilon)}}} \mathbf{P}(B_{Z,t_k,\gamma})(1 + o(1))$$
$$\leq \sum_{\theta_k < \frac{\alpha}{\alpha^{(\varepsilon)}}} \mathbf{P}(Z(t_k + \delta) \in \Delta[x + v))(1 + o(1)). \tag{6.4.11}$$

Here $\frac{\alpha}{\alpha^{(\varepsilon)}} \leq \frac{\alpha^{(\varepsilon)} - \varepsilon}{\alpha^{(\varepsilon)}} < 1$ for $\alpha \in K_{2\varepsilon}$.

Let us return to the sum (6.4.10). Since $\theta_{k+1} - \theta_k = \frac{\delta}{T}$, the sum $\Sigma_{T,x}$ can be approximated by the integral:

$$\Sigma_{T,x} = J_{T,x}(1 + o(1)),$$

where

$$J_{T,x} := \Delta\sqrt{T} \int_{\alpha/\alpha^{(\varepsilon)}}^{1} c\Big(\frac{\alpha}{\theta}, v\Big)\theta^{-1/2} e^{-T\theta D(\alpha/\theta) - v\mu(\alpha/\theta)} d\theta, \tag{6.4.12}$$

and the remainder term is still uniform, but now with respect to $\frac{\alpha}{\theta} \in K_\varepsilon$.

The logarithm of the principal exponential part of the integrand in (6.4.12) is

$$-T\theta D\Big(\frac{\alpha}{\theta}\Big) = -xL\Big(\frac{\alpha}{\theta}\Big), \tag{6.4.13}$$

where $L(\alpha) = \frac{1}{\alpha}D(\alpha)$, $\frac{\alpha}{\theta} \in [\alpha, \alpha^{(\varepsilon)})$. By Lemma 6.4.4, $\alpha_0 = A'(\mu_0) \in K_\varepsilon$. In addition, $L(\alpha_0) = \mu_0$ and $L''(\alpha_0) = \mu'(\alpha_0)\alpha_0^{-1} > 0$.

Consider the following main cases.

(i) $\alpha > \alpha_0$ and $\alpha - \alpha_0$ is separated from 0. In (6.4.12), we change the variables as follows: $\theta = \frac{1}{1+t}$, $t = \frac{1}{\theta} - 1$. Hence $\frac{\alpha}{\theta} = \alpha + \alpha t$, $d\theta = -\frac{dt}{(1+t)^2}$, and

$$L\Big(\frac{\alpha}{\theta}\Big) = L(\alpha) + \alpha t L'(\alpha) + O(t^2) \quad \text{as} \quad t \to 0, \quad xL(\alpha) = TD(\alpha),$$
$$\int_{\alpha/\alpha^{(\varepsilon)}}^{1} e^{-T\theta D(\alpha/\theta)} d\theta = e^{-TD(\alpha)} \int_0^{\alpha^{(\varepsilon)}/\alpha - 1} \frac{1}{(1+t)^2} e^{-\alpha t T L'(\alpha) + O(t^2 T)} dt$$
$$= e^{-TD(\alpha)} \frac{1}{T\alpha L'(\alpha)} (1 + o(1)).$$

The above shows that the principal part of the integral in (6.4.12) is governed by the integral over a neighborhood of the point $t = 0$ ($\theta = 1$), and hence (see (6.4.12))

$$J_{T,x} = \frac{\Delta c(\alpha, v)}{\sqrt{T}\alpha L'(\alpha)} e^{-TD(\alpha) - v\mu(\alpha)}(1 + o(1)).$$

Consider now the remainder term $R_{T,x}$ in (6.4.9) and in (6.4.11).

The function $L(s)$ is convex, and hence $L(s) > L(\alpha)$ for $s > \alpha > \alpha_0$. By the local large deviation principle, for $\theta_k < \frac{\alpha}{\alpha^{(\varepsilon)}}$ and $T \to \infty$,

$$\begin{aligned} \ln \mathbf{P}(Z(t_k+\delta) \in \Delta[x+v)) &< -TD\Big(\frac{x}{t_k}\Big) = -xL\Big(\frac{\alpha}{\theta_k}\Big) \\ &< -xL(\alpha^{(\varepsilon)}) < -x(L(\alpha)+\varepsilon_1) = -TD(\alpha) - x\varepsilon_1, \qquad (6.4.14) \\ R_{T,x} &= O(e^{-TD(\alpha)-T\varepsilon_2}) \end{aligned}$$

for some $\varepsilon_1 > 0$, $\varepsilon_2 > 0$. Therefore, returning to (6.4.9), we find that

$$\mathbf{P}(A_{T,x,v}) = -\frac{\Delta}{\sqrt{T}} \frac{c(\alpha, v)}{\alpha L'(\alpha)} e^{-TD(\alpha)-v\mu(\alpha)}(1+o(1)),$$

where the remainder term is uniform in $\alpha \in K_{2\varepsilon}$. This proves (6.4.1).

(ii) $\alpha = \alpha_0 - \rho$, where $\rho = o(1)$ as $T \to \infty$. In this case,

$$\begin{aligned} \frac{\alpha}{\theta} &= (1+t)(\alpha_0 - \rho) = \alpha_0 - \rho + t(\alpha_0-\rho), \\ L\Big(\frac{\alpha}{\theta}\Big) &= L(\alpha_0) + \frac{1}{2}(\rho - t(\alpha_0-\rho))^2 L''(\alpha_0) + O(|\rho|+t)^3. \end{aligned}$$

With the change of variables

$$u = \big[t(\alpha_0 - \rho) - \rho\big]\sqrt{xL''},$$

where for brevity we put $L'' = L''(\alpha_0)$, we get

$$xL\Big(\frac{\alpha}{\theta}\Big) = xL(\alpha_0) + \frac{xu^2}{2} L'' + O\bigg(|\rho| + \frac{|u|}{\sqrt{x}}\bigg)^3.$$

Returning to (6.4.12), (6.4.13) and taking it into account that, as before, the principal part of the integral $J_{T,x}$ is governed by the integral over the neighborhood of the point $t = 0$ (with the interval $\theta = 1$), and since $d\theta = -\frac{du}{\alpha_0\sqrt{xL''}}(1+o(1))$ as $T \to \infty$, $\rho \to 0$, we have

$$\begin{aligned} J_{T,x} &= \frac{\Delta\sqrt{T}\, c(\alpha, v)}{\alpha_0\sqrt{xL''}} e^{-x\mu_0 - v\mu(\alpha)} \int_{-\rho\sqrt{xL''}}^{\infty} e^{-\frac{u^2}{2}} du(1+o(1)) \\ &= \frac{\Delta\sqrt{2\pi}\, c(\alpha_0, v)}{\alpha_0^{3/2}\sqrt{L''}} e^{-x\mu_0 - v\mu(\alpha_0)} \Phi(\rho\sqrt{xL''})(1+o(1)), \qquad (6.4.15) \end{aligned}$$

since $\alpha \to \alpha_0$ as $T \to \infty$.

The remainder term $R_{T,x}$ is estimated similarly. This proves (6.4.2).

(iii) $\alpha < \alpha_0$ and $\alpha_0 - \alpha$ is separated from 0. Under these conditions, the maximum of the integrand in (6.4.12) is attained inside the integration interval at the point $\theta = \frac{\alpha}{\alpha_0} < 1$ (so that $\frac{\alpha}{\theta} = \alpha_0$), where

$$\frac{\alpha}{\alpha_0} > \frac{\alpha}{\alpha^{(\varepsilon)}}, \quad \text{because} \quad \alpha_0 < \alpha^{(\varepsilon)},$$

so that $\theta \in \big(\frac{\alpha}{\alpha^{(\varepsilon)}}, 1\big)$. In this case, the principal contribution into the asymptotics of $J_{T,x}$ comes from the integral over the neighborhood of the point $\theta = \alpha/\alpha_0$, and so use should be made of the standard Laplace method for integrals of the form (6.4.12). Let us expand $L(\frac{\alpha}{\theta})$ near the point $\frac{\alpha}{\theta} = \alpha_0$:

$$L\left(\frac{\alpha}{\theta}\right) = L(\alpha_0) + \frac{1}{2}\left(\frac{\alpha}{\theta} - \alpha_0\right)^2 L'' + O\left(\left|\frac{\alpha}{\theta} - \alpha_0\right|^3\right)$$

and change the variables

$$u = \sqrt{xL''}\left(\frac{\alpha}{\theta} - \alpha_0\right), \quad \theta = \alpha\left(\frac{u}{\sqrt{xL''}} + \alpha_0\right)^{-1}, \quad d\theta \sim -\frac{\alpha\, du}{\alpha_0^2\sqrt{xL''}}$$

as $u \to 0$. This gives us

$$\begin{aligned} J_{T,x} &= \Delta\sqrt{T}c(\alpha_0, v)\sqrt{\frac{\alpha_0}{\alpha}}\, e^{-x\mu_0 - v\mu(\alpha_0)} \frac{\alpha\sqrt{2\pi}}{\alpha_0^2\sqrt{xL''}}(1 + o(1)) \\ &= \frac{\Delta c(\alpha_0, v)\sqrt{2\pi}}{\alpha_1^{3/2}\sqrt{L''}}\, e^{-x\mu_0 - v\mu(\alpha_0)}(1 + o(1)). \end{aligned} \tag{6.4.16}$$

The remainder term $R_{T,x}$ is estimated as in (6.4.14). This proves (6.4.3). It remains to note that in (6.4.15), (6.4.16)

$$L'' = L''(\alpha_0) = \alpha_0^{-1}D''(\alpha_0),$$

so that

$$\alpha_0^{3/2}\sqrt{L''} = \alpha_0 D''(\alpha_0) = \alpha_0 D''.$$

Theorem 6.4.1 is proved. □

6.4.2 The Case $\alpha > a \geq 0$

By Lemma 6.4.4, $L'(\alpha) = -\alpha^{-2}\lambda(\alpha) > 0$ for $\alpha > a$, and therefore, the function $L(\alpha) = \frac{1}{\alpha} D(\alpha)$ is increasing for $\alpha > a$ and the function $L(\frac{\alpha}{\theta})$ attains its maximum with respect to $\theta \in [0, 1]$ at the point $\theta = 1$. Hence here the three situations considered in Theorem 6.4.1 do not happen and we have to consider only the analog of case (i). We will use the novation of Theorem 6.4.1, but now as the compact set K_ε we take the closed interval $K_\varepsilon = [a + \varepsilon, \alpha_+ - \varepsilon]$.

Theorem 6.4.5 *Let ξ be a nonlattice vector and let $0 \leq a < \alpha = \frac{x}{T} \in K_\varepsilon$. Then the asymptotic behavior of the probability $\mathbf{P}(A_{T,x,v})$ is given by (6.4.1),*

$$\mathbf{P}(\overline{Z}(T) > x) = \frac{C(\alpha)P_\alpha}{\sqrt{T}\alpha L'(\alpha)}\, e^{-TD(\alpha)}(1 + o(1)), \tag{6.4.17}$$

where

$$P_\alpha = \int_0^\infty P_\alpha(-v)e^{-v\mu(\alpha)}\, dv,$$

and the remainder term $o(1)$ is uniform in $\alpha \in K_\varepsilon$.

The proof of Theorem 6.4.5 repeats that of assertion (i) in Theorem 6.4.1. Here, as in assertion (i) of Theorem 6.4.1, the conditional distribution of $\eta_Z(x)$ under the condition $\overline{Z}(T) > x$ is concentrated in a neighborhood of the point T and is exponential.

Corollary 6.4.6 *Suppose that either $a < 0$, $\alpha > \alpha_0$, and $\alpha - \alpha_0$ is separated from zero or $\alpha > a \ge 0$ and $\alpha - a$ is separated from zero. Then*

$$\mathbf{P}(T - \eta_Z(x) > t \mid \overline{Z}(T) > x) = e^{-tb}(1 + o(1)), \tag{6.4.18}$$

where $b = \alpha^2 L'(\alpha) > 0$, and the remainder term is uniform in $\alpha \in K_\varepsilon$.

Proof We have

$$TD\Big(\frac{x}{T}\Big) - (T-t)D\Big(\frac{x}{T-t}\Big) = x\Big[L(\alpha) - L\Big(\frac{\alpha}{\theta}\Big)\Big] \tag{6.4.19}$$

for

$$\theta = \frac{T-t}{T} = 1 - \frac{t}{T}, \qquad \frac{1}{\theta} = 1 + \frac{t}{T} + O(T^{-2}), \qquad T \to \infty.$$

Hence by (6.4.1), (6.4.17), and (6.4.19),

$$\mathbf{P}(T - \eta_Z(x) > t \mid \overline{Z}(T) > x) = \frac{\mathbf{P}(\overline{Z}(T-t) > x)}{\mathbf{P}(\overline{Z}(T) > x)} = \exp\Big\{x\Big[L(\alpha) - L\Big(\frac{\alpha}{\theta}\Big)\Big]\Big\}(1 + o(1)). \tag{6.4.20}$$

Next,

$$L\Big(\frac{\alpha}{\theta}\Big) = L(\alpha) + \alpha\Big(\frac{1}{\theta} - 1\Big)L'(\alpha) + O(T^{-2}),$$

$$x\Big[L(\alpha) - L\Big(\frac{\alpha}{\theta}\Big)\Big] = -x\Big[\frac{\alpha t}{T}L'(\alpha) + O(T^{-2})\Big] = -\alpha^2 L'(\alpha)t + O(T^{-1}).$$

So, the right-hand side in (6.4.20) is $\exp\{-\alpha^2 L'(\alpha)t\}(1 + o(1))$. This proves (6.4.18), and therefore, Corollary 6.4.6. □

Relation (6.4.17) means that, under the conditions of Theorem 6.4.5, the asymptotics of the probabilities $\mathbf{P}(Z(T) > x)$ and $\mathbf{P}(\overline{Z}(T) > x)$ coincide up to a constant factor (with a fixed α). Indeed, from (5.2.13) for large Δ it easily follows that

$$\mathbf{P}(Z(T) > x) = \frac{C(\alpha)}{\sqrt{T}\mu(\alpha)} I_Z(\alpha, 0, 0, \mathbb{R}) e^{-TD(\alpha)}(1 + o(1)).$$

Hence by (6.4.17),

$$\frac{\mathbf{P}(\overline{Z}(T) > x)}{\mathbf{P}(Z(T) > x)} = \frac{P_\alpha \mu(\alpha) I_Z(\alpha, 0, 0, \mathbb{R})}{\alpha L'(\alpha)}(1 + o(1)).$$

6.4.3 The Case $a > 0$, $\alpha \sim a$ as $T \to \infty$

In this case, the limit theorems for $\overline{Z}(T)$ and $Z(T)$ in a broad domain of deviations of x will be the same.

Let the process $Y^*(t)$ be defined in the same way as the process $Y(t)$ but has jumps $(\tau_j, -\zeta_j)$, $j = 1, 2, \ldots$ Since $\mathbf{E}Y^*(t) \sim -at$ as $t \to \infty$, the proper random variable $\overline{Y}^* = \sup_{t>0} Y^*(t)$ is defined. From the results of §6.1, we get

$$\mathbf{P}(\overline{Y}^* > y) < e^{-cy}$$

for some $c > 0$ and $y \to \infty$.

Theorem 6.4.7 *Let ξ be a nonlattice vector and let $a > 0$, $\alpha - a = o(1)$ as $T \to \infty$. Then:*

(i)
$$\overline{Z}(T) = Z(T) + Y_T^*, \tag{6.4.21}$$
where the distribution of Y_T^* converges as $T \to \infty$ to the distribution of $\overline{Y}^*$ and $\mathbf{P}(Y_T^* > y) < \mathbf{P}(\overline{Y}^* > y)$ for all y.

(ii) *If $x - aT = o(T^{2/3})$, then*

$$\mathbf{P}(\overline{Z}(T) > x) = \mathbf{P}(Z(T) > x)(1 + o(1)) = \left[1 - \Phi\left(\frac{x - aT}{\sigma\sqrt{T}}\right)\right](1 + o(1)) \tag{6.4.22}$$

as $T \to \infty$, where Φ is the distribution function of the normal law.

Proof (i) It is clear that (6.4.21) holds with

$$Y_T^* = \sup_{u>0} (Z(T - u) - Z(T))$$

and that the distribution of $Y_T^* \underset{p}{\leq} \overline{Y}^*$ converges as $T \to \infty$ to the distribution of $\overline{Y}^*$.

(ii) Setting

$$A_y = \{\overline{Y}^* < y\},$$

we have

$$\mathbf{P}(\overline{Z}(T) > x) \geq \mathbf{P}(Z(T) > x),$$
$$\mathbf{P}(\overline{Z}(T) > x) \leq \mathbf{P}(\overline{Z}(T) > x, A_y) + O(e^{-cy}) \leq \mathbf{P}(Z(T) > x - y) + O(e^{-cy}), \tag{6.4.23}$$

as $y \to \infty$. Let us estimate the probabilities $\mathbf{P}(Z(T) > x)$ and $\mathbf{P}(Z(T) > x - y)$ using sums of integro-local probabilities. We set

$$v_k = k\Delta, \qquad k \leq N, \qquad N\Delta = o(T^{2/3}).$$

Hence, for $x - aT = o(T^{2/3})$ and $x_k = x + v_k$,

$$x_k - aT = x - aT + v_k = o(T^{2/3}), \tag{6.4.24}$$

and besides, for $\alpha_k = x_k/T$ from Theorem 5.2.1 we have

$$\mathbf{P}(Z(T) \in \Delta[x_k)) = T^{-1/2}\Delta C(\alpha_k) I_Z(\alpha_k, 0, 0, \mathbb{R}) e^{-TD(\alpha_k)}, \tag{6.4.25}$$

where the functions $C(\alpha)$ and $I_Z(\alpha, 0, 0, \mathbb{R})$ are continuous. So, $C(\alpha_k) \sim C(a)$ and $I_Z(\alpha_k, 0, 0, \mathbb{R}) \sim I_Z(a, 0, 0, \mathbb{R})$ for all $k \le N$. By the integro-local theorems, for $\Delta_T = \frac{\Delta}{\sigma\sqrt{T}}$, $x = aT$ we have

$$\mathbf{P}(Z(T) - aT \in \Delta[0)) = \mathbf{P}\Big(\frac{Z(T) - aT}{\sigma\sqrt{T}} \in \Delta_T[0)\Big) \sim \frac{\Delta}{\sigma\sqrt{2\pi T}},$$

hence

$$C(a) I_Z(a, 0, 0, \mathbb{R}) = \frac{1}{\sigma\sqrt{2\pi}}.$$

Besides, for $x_k - aT = o(T^{2/3})$,

$$TD(\alpha_k) = TD''(a)\frac{(\alpha_k - a)^2}{2} + o(1) = \frac{1}{2\sigma^2}\frac{(x_k - aT)^2}{T} + o(1). \tag{6.4.26}$$

Similarly, for $y = T^{1/3}$

$$\begin{aligned} \mathbf{P}(Z(T) \in \Delta[x_k - y)) &\sim \frac{T^{-1/2}\Delta}{\sigma\sqrt{2\pi}} e^{-TD(\alpha_k - y/T)}, \\ TD\Big(\alpha_k - \frac{y}{T}\Big) &= \frac{(x_k - aT - y)^2}{2\sigma^2 T} + o(1) \\ &= \frac{(x_k - aT)^2}{2\sigma^2 T} - \frac{(x_k - aT)y}{\sigma^2 T} + o(1) \\ &= \frac{(x_k - aT)^2}{2\sigma^2 T} + o(1) = TD(\alpha_k) + o(1). \end{aligned} \tag{6.4.27}$$

Here the last relation follows from the fact that, as $x_k - aT = o(T^{2/3})$ and for $y = T^{1/3}$,

$$\frac{(x_k - aT)y}{T} = o(1) \quad \text{as} \quad T \to \infty.$$

Next, in view of (6.4.25) and (6.4.26) the principal part of the asymptotics of $\mathbf{P}(Z(T) > x)$ is given by

$$\mathbf{P}(Z(T) > x) \sim \sum_{k=0}^{N} \mathbf{P}(Z(T) \in \Delta[x_k)) \sim \frac{C(\alpha) I_Z(\alpha, 0, 0, \mathbb{R})}{\sqrt{T}} \sum_{k=0}^{N} \Delta e^{-TD(\alpha_k)}. \tag{6.4.28}$$

Here the sum can be naturally approximated by an integral. We leave this to the reader

Similarly, in view of (6.4.27) the probability $\mathbf{P}(Z(T) > x - y)$ is also asymptotically equivalent to the right-hand side of (6.4.28). Therefore, returning to (6.4.23), we find that

$$\mathbf{P}(\overline{Z}(T) > x) \le \mathbf{P}(Z(T) > x)(1 + o(1)) + O(e^{-cy}), \tag{6.4.29}$$

where $cy = cT^{1/3}$, and now an appeal to (6.4.24) shows that

$$TD(\alpha_k) \sim \frac{1}{2\sigma^2}\frac{(x_k - aT)^2}{T} = o(T^{1/3}).$$

This means that the remainder term on the right of (6.4.29) is negligible with respect to $\mathbf{P}(Z(T) > x)$. This proves (6.4.22).

Now the asymptotics of $\mathbf{P}(Z(T) > x)$ can be found by approximating the sum (6.4.28) by the integral. As a result, we get

$$\mathbf{P}(Z(T) > x) = \mathbf{P}\Big(\frac{Z(T) - aT}{\sigma\sqrt{T}} > \frac{x - aT}{\sigma\sqrt{T}}\Big) \sim 1 - \Phi\Big(\frac{x - aT}{\sigma\sqrt{T}}\Big)$$

as $T \to \infty$. Theorem 6.4.7 is proved. □

6.4.4 Asymptotics of the Probability that the Trajectory of a CRP Does Not Cross a High Level x for $\alpha = \frac{x}{T} < a$

In this section, we will consider the "boundary crossing problem of the second kind" with respect to the trajectories of $Z(t)$, $t \le T$, with horizontal boundary, that is, we will study the asymptotics of the probability $\mathbf{P}(\overline{Z}(T) \le x)$ (complementary to the probabilities from §6.4.2) in the case $\alpha < a$, when this probability tends to 0. Our analysis will be based on the same approaches as in §§6.4.1 and 6.4.2.

Let

$$P_\alpha(v) = \mathbf{P}({}^\alpha\overline{Y} \le v), \qquad Q_\alpha = \int_0^\infty e^{\mu(\alpha)u} P_\alpha(u)\, du.$$

Theorem 6.4.8 *Let $\boldsymbol{\xi}$ be a nonlattice vector and let $a > 0$, $\alpha < a$, $\alpha = \frac{x}{T} \in K_\varepsilon$. Then*

$$\mathbf{P}(\overline{Z}(T) \le x, Z(T) \in x - \Delta[v)) = \frac{\Delta C(\alpha) P_\alpha(v) I_Z(\alpha, 0, 0, \mathbb{R})}{\sqrt{T}} e^{-TD(\alpha) + v\mu(\alpha)}(1 + o(1)), \tag{6.4.30}$$

where $\Delta = \overline{o}(1)$ as $T \to \infty$, the remainder term $o(1)$ is uniform in $\alpha \in K_\varepsilon$, $\lambda(\alpha) < 0$, the function $C(\alpha)$ is defined in Theorem 5.2.1, and

$$I_Z(\alpha, 0, 0, \mathbb{R}) = \int_0^\infty e^{\lambda(\alpha)y} \mathbf{P}(\tau > y)\, dy.$$

Setting $v = v_k = k\Delta$, $k = 0, 1, \ldots$ and summing the probabilities on the left of (6.4.30) for $v = v_k$ with respect to k, we arrive at the following.

Corollary 6.4.9 *If the conditions of Theorem 6.4.8 are satisfied, then $\mu(\alpha) < 0$ and*

$$\mathbf{P}(\overline{Z}(T) \le x) = \frac{C(\alpha) Q_\alpha I_Z(\alpha, 0, 0, \mathbb{R})}{\sqrt{T}} e^{-TD(\alpha)}(1 + o(1)), \tag{6.4.31}$$

where the remainder term $o(1)$ is uniform in $\alpha \in K_\varepsilon$.

As before, here the asymptotics of $\mathbf{P}(\overline{Z}(T) \le x)$ is the same, up to a constant factor (with a fixed α), as that of $\mathbf{P}(Z(T) \le x)$.

Proof of Theorem 6.4.8 Consider the probabilities

$$p(u) := \mathbf{P}\big(Z(T) \in x - \Delta[v), \gamma(T) \in \delta[u), \overline{Z}(T - \gamma(T) - 0 \le x)\big).$$

As in the proof of Theorem 6.3.1, we get

$$p(u) = \mathbf{P}(Z(T) \in x - \Delta[v), \gamma(T) \in \delta[u))\mathbf{P}({}^{\alpha}\overline{Y} \le v)(1 + o(1)). \tag{6.4.32}$$

Next,

$$I_Z(\alpha, u, 0, \mathbb{R}) - I_Z(\alpha, u + \delta, 0, \mathbb{R}) = \int_u^{u+\delta} e^{\lambda(\alpha)y}\mathbf{P}(\tau > y)\, dy \sim \delta e^{\lambda(\alpha)u}\mathbf{P}(\tau > y)$$

for $\delta = \bar{o}(1)$, $T \to \infty$ (assuming for simplicity that the function $\mathbf{P}(\tau > y)$ is continuous), and hence by Theorem 5.2.1 the asymptotics of the first factor on the right of (6.4.32) coincides with the asymptotics of the product

$$\frac{\Delta\delta C(\alpha)}{\sqrt{T}} e^{-TD(\alpha)+v\mu(\alpha)+u\lambda(\alpha)}\mathbf{P}(\tau > u)$$

(see (6.3.1)). The second factor on the right of (6.4.32) does not depend on u, and hence, letting $u = u_k\delta$, $k = 0, 1, \ldots$, and summing $p(u_k)$ with respect to k, we get

$$\sum_{k=0}^{\infty} p(u_k) = \frac{\Delta C(\alpha)P_\alpha(v)}{\sqrt{T}} e^{-TD(\alpha)+v\mu(\alpha)} I_Z(\alpha, 0, 0, \mathbb{R})(1 + o(1)).$$

This proves (6.4.30). That $o(1)$ is uniform is verified as before. Theorem 6.4.8 is proved. □

Remark 6.4.10 In the study of boundary crossing problems in §§6.3–6.4, the asymptotics of the distributions under consideration remains incomplete in the case $\alpha = o(1)$ as $T \to \infty$, because in this setting, the probabilistic approaches involving the Cramér transform of jumps and the consideration of the "inverse trajectories" do not work. In this case, the values of $P_\alpha(-v)$ in the asymptotics in question tends to zero as $\alpha \to 0$. This apparently means a change in the nature of these asymptotics as $\alpha \to 0$. As already pointed out, for independent τ and ζ, a rather complete asymptotic analysis of boundary crossing problems with rectilinear boundaries in the domains of large and normal deviations was given in [49], where the factorization method was applied for finding double integral transforms of the distributions in question with subsequent inversion of these transforms. The exposition in [49] is quite technical and is beyond the scope of the present monograph – however, it "covers" the case $\alpha = o(1)$, $x \to \infty$ as $T \to \infty$. For example, from the results of [49] we get:

Theorem 6.4.11 *Let τ and ζ be independent, let τ be nonlattice, let the distribution of ζ have an absolutely continuous component, and let $x \to \infty$, $x = o(T)$ as $T \to \infty$. Then*

$$\mathbf{P}(\overline{Z}(T) \le x, Z(T) > x - v) = \frac{\alpha c(v)e^{-TD(\alpha)}}{\sqrt{T}}(1 + o(1)) \tag{6.4.33}$$

as $T \to \infty$, where the constant factor $c(v)$ depends on v and on the distribution of (τ, ζ).

Comparing this theorem with Corollary 6.4.9, we see that, as $\alpha \to 0$, the factor α, which tends 0, appears in the asymptotic formulas for the left-hand sides of (6.4.33), (6.4.32).

6.5 Integro-Local Theorems in Boundary Crossing Problems for Compound Renewal Processes

6.5.1 Integro-Local Theorems for the First Boundary Crossing Problem

It is clear from the proof of Theorem 6.3.1 that it can be extended to the time

$$\eta_{Z,g} = \min\{t : Z(t) > g_T(t)\} \tag{6.5.1}$$

of the first crossing by the CRP trajectory of an arbitrary remote boundary, where $g_T(t)$ depends on an increasing parameter T so that $x := g_T(T) \to \infty$ as $T \to \infty$. For example, the function

$$g_T(t) = Tg(t/T) \tag{6.5.2}$$

has this property, where $g(u)$ is a fixed continuous function such that $g(1) > 0$.

Without loss of generality we shall study the asymptotics of the event that the boundary g_T is crossed in a neighborhood of the point $t = T$. We shall assume that $g_T(t)$ satisfies the following conditions.

(a) The function $g_T(t)$ is asymptotically linear in a neighborhood of $t = T$, that is,

$$g_T(t) = x + (t - T)g' + o(1) \quad \text{for } |t - T| \le N, \quad \alpha = \frac{x}{T} \in K, \tag{6.5.3}$$

as $T \to \infty$, for any fixed $N > 0$ and some $g' < \alpha$.

If, for example, the function $g_T(t)$ is of the form (6.5.2) and $g(t)$ is differentiable at the point $t = 1$, then $g' = g'(1)$.

(b) In the domain $t \le T - N$, for some $r < g'$,

$$g_T(t) \ge x + r(t - T) > 0. \tag{6.5.4}$$

Note that if these conditions $g' < 0$, then the qualitatively new possibility arises: the process $Z(t)$ can cross the boundary $g_T(t)$ not only with a vertical jump, but also with a horizontal jump.

We let $A^{\uparrow}$ ($A^{\rightarrow}$, respectively) denote the event that the process $Z(T)$ crosses the boundary $g_T(t)$ with a vertical (horizontal) jump. We shall find the asymptotics of the probabilities

$$\mathbf{P}(\eta_{Z,g} \in \Delta[T), A^{\uparrow}) \quad \text{and} \quad \mathbf{P}(\eta_{Z,g} \in \Delta[T), A^{\rightarrow}) \tag{6.5.5}$$

and therefore, the asymptotics of the probabilities

$$\mathbf{P}(\eta_{Z,g} \in \Delta[T)) = \mathbf{P}(\eta_{Z,g} \in \Delta[T), A^{\uparrow}) + \mathbf{P}(\eta_{Z,g} \in \Delta[T), A^{\rightarrow})$$

as $T \to \infty$. In fact, we will study the probabilities of more informative events than in (6.5.5).

(1) *Let us start with the probabilities of crossing the boundary with a vertical jump.* We set

$$\chi_{Z,g} := Z(\eta_{Z,g}) - g_T(\eta_{Z,g}).$$

Theorem 6.5.1 *Let ξ be a nonlattice vector, let $\alpha = \frac{x}{T} > a + \varepsilon$, and let conditions* (a), (b) *be met for any $g' < \alpha - \varepsilon$ and some $\varepsilon > 0$. Then*

$$\begin{aligned}\mathbf{P}(\eta_{Z,g} \in \delta[T), A^{\uparrow}, \chi_{Z,g} \in \Delta[v)) \\ &= \mathbf{P}(Z(T+\delta) \in \Delta[x+v),\ \gamma(T+\delta) < \delta) P_{\alpha,g'}(-v)(1+o(1)) \\ &= \frac{\Delta\delta C(\alpha)}{\sqrt{T}} e^{-TD(\alpha)-v\mu(\alpha)} P_{\alpha,g'}(-v)(1+o(1)),\end{aligned} \tag{6.5.6}$$

where

$$P_{\alpha,g'}(-v) := \mathbf{P}\Big(\sup_{t\geq 0}\,({}^{\alpha}Y(t) + g't) \leq -v\Big), \tag{6.5.7}$$

$\Delta = \Delta_T = \overline{o}(1)$, $\delta = \delta_T = \overline{o}(1)$ *as $T \to \infty$, and the remainder term $o(1)$ is uniform in $\alpha \in K \bigcap \{\alpha \geq \varepsilon\}$, $\varepsilon > 0$.*

If $g' \geq 0$, then $\mathbf{P}(A^{\uparrow} \mid \eta_{Z,g} \in \Delta[T)) = 1$ and the event $A^{\uparrow}$ under the probability sign in (6.5.6) can be omitted. Putting $v = v_k = k\Delta$ and summing the probabilities (6.5.6) over the half-open intervals $\Delta[v_k)$, for $k = 0, 1, \ldots$, we establish:

Corollary 6.5.2 *Under the conditions of Theorem* 6.5.1,

$$\mathbf{P}(\eta_{Z,g} \in \delta[T)) = \frac{\delta C(\alpha) P_{\alpha,g'}}{\sqrt{T}} e^{-TD(\alpha)}(1+o(1)),$$

where

$$P_{\alpha,g'} = \int_0^{\infty} e^{-v\mu(\alpha)} P_{\alpha,g'}(-v)dv. \tag{6.5.8}$$

Proof of Theorem 6.5.1 The conditions of Theorem 6.5.1 are such that they enable us to get the required assertion by a slight modification of the arguments in the proof of Theorem 6.3.1. We leave this to the reader. In some relations in the proof of Theorem 6.3.1 we need to replace the process $Z(t)$ by the process $Z(t) + g't$ with drift $g't$. For example, in the definition of the event $B_{Z,T,\gamma}$ in (6.3.2), the event

$$\{\overline{Z}(T+\delta-\gamma(T+\delta)-0) \leq x\}$$

should be replaced by the event

$$\{\sup_{t<T+\delta-\gamma(T+\delta)} (Z(t) + g't) \leq x\}.$$

The fact that the conditional probability

$$\mathbf{P}(\eta_{Z,g} < T - M \mid Z(T) \in \Delta[x+v))$$

is small for large M (cf. (6.3.8) and (6.3.10)) follows from condition (b) and the inclusion

$$\{\eta_{Z,g} < T - M\} \in \{\eta_Z^{(r)}(x - rT) < T - M\},$$

where $\eta_Z^{(r)}(x - rT)$ is the first passage time of the boundary $x - rT$ by the process $Z(t) - rt$. The remaining estimates are obtained similarly to (6.3.8)–(6.3.10), but now for the process $Z(t) - rt$ and the boundary $x - rT$. □

(2) *Crossing the boundary with a horizontal jump.* Let us now consider the case $g' < 0$, when the boundary $g_x(t)$ can be crossed with a horizontal jump.

Theorem 6.5.3 *Let the conditions of Theorem* 6.5.1 *hold and let* $g' < 0$, $\lambda_+ > D(0)$. *Then*

$$\mathbf{P}(\eta_{Z,g} \in \delta[T), A^{\rightarrow},\ \chi(T) > v)$$
$$= \frac{\delta|g'|}{\sqrt{T}} C(\alpha) e^{-TD(\alpha)} \int_0^\infty e^{\lambda(\alpha)u} \mathbf{P}(\tau > u + v) P_{\alpha,g'}(-g'u)\, du(1 + o(1)), \quad (6.5.9)$$

where $\delta = \delta_T = \overline{o}(1)$ *as* $T \to \infty$, *the remainder term is uniform in* $\alpha \in K \cap \{\alpha \geq \varepsilon\}$, $\varepsilon > 0$, *and the probability* $P_{\alpha,g'}(-v)$ *is defined in* (6.5.7).

The (unrestrictive) condition $\lambda_+ > D(0)$ was introduced to simplify the proof.

Proof of Theorem 6.5.3 For $g' < 0$, we have

$$g_T(T) - g_T(T + \delta) \sim -\delta g' > 0 \quad \text{as} \quad \delta \to 0. \quad (6.5.10)$$

Setting

$$\Delta_g(x] = (g_T(T + \delta), g_T(T) = x],$$

we have

$$\mathbf{P}(\eta_{Z,g} \in \delta[T), A^{\rightarrow},\ \chi(T) > v)$$
$$= \int_0^T \mathbf{P}(Z(T) \in \Delta_g(x],\ \gamma(T) \in du,\ \chi(T) > v,\ B_T(u)) + P_{\delta[T)}, \quad (6.5.11)$$

where

$$B_T(u) = \Big\{ \sup_{t \leq T-u} (Z(t) - g_T(t)) \leq 0 \Big\},$$
$$P_{\delta[T)} \leq \mathbf{P}(Z(T + \delta) \in \Delta_g[x),\ \gamma(T + \delta) < \delta).$$

For $v > 0$ and $\delta < v$, the summand $P_{\delta[T)}$ on the right of (6.5.11) disappears, because the event $\{\chi(T) > \delta, \gamma(T + \delta) < \delta\}$ is empty for $\delta < v$.

We estimate the first summand on the right of (6.5.11) in several steps.

I. As before, we find that the conditional probability of the event $B_T(u)$ under the condition $\{Z(T) \in \Delta_g(x],\ \gamma(T) \in du,\ \chi(T) > v\}$ for each fixed u and v is asymptotically equivalent as $T \to \infty$ to the probability $P_{\alpha,g'}(-g'u)$ (see (6.5.7); $g_T(T-u) - x - g'u \sim -g'u$ for small u). Hence, for a fixed M,

$$J_0^M := \int_0^M \mathbf{P}(Z(T) \in \Delta_g(x],\ \gamma(T) \in du,\ \chi(T) > v, B_T(u))$$
$$\sim \int_0^M \mathbf{P}(Z(T) \in \Delta_g(x],\ \gamma(T) \in du,\ \chi(T) > v) P_{\alpha,g'}(-g'u). \qquad (6.5.12)$$

Next, the function $P_{\alpha,g'}(-g'u)$ is increasing in u, and hence we can estimate the integral from below by the sums

$$\sum_{u_k \le M} \mathbf{P}(Z(T) \in \Delta_g(x],\ \gamma(T) \in \delta[u_k),\ \chi(T) > v) P_{\alpha,g'}(-g'u_k), \qquad (6.5.13)$$

where $u_k = k\delta$, $k = 0, \ldots, N = T/\delta$ (we assume that N is an integer number). By Theorem 5.2.1 and using (6.5.10), for $u_k < M$ and large fixed M, we have

$$\mathbf{P}(Z(T) \in \Delta_g(x], \gamma(T) \in \delta[u_k),\ \chi(T) > v)$$
$$= \frac{\delta|g'|C(\alpha)}{\sqrt{T}} e^{-TD(\alpha)} \big[I_Z(\alpha, u, v, \mathbb{R}) - I_Z(\alpha, u+\delta, v, \mathbb{R})\big](1 + o(1)),$$

where

$$I_Z(\alpha, u, v, \mathbb{R}) - I_Z(\alpha, u+\delta, v, \mathbb{R}) \sim \delta e^{\lambda(\alpha)u} \mathbf{P}(\tau > u + v)$$

as $\delta \to 0$ (we assume for simplicity that the function $\mathbf{P}(\tau > u)$ is continuous).

Returning to (6.5.12) and (6.5.13), we find that

$$J_0^M \ge \frac{\delta|g'|C(\alpha)}{\sqrt{T}} e^{-TD(\alpha)} \int_0^M e^{\lambda(\alpha)u} \mathbf{P}(\tau > u+v) \times P_{\alpha,g'}(-g'u)\, du (1 + o(1)). \qquad (6.5.14)$$

If in (6.5.13) we replace $P_{\alpha,g'}(-g'u_k)$ by $P_{\alpha,g'}(-g'u_{k+1})$, then we get the upper estimate for J_0^M, which gives the estimate with the same right-hand side as in (6.5.14). It is also clear that by choosing M we can make the integral on the right of (6.5.14) to be arbitrarily close to the integral from 0 to ∞. So for large M the right-hand side in (6.5.14) is asymptotically equivalent to that of (6.5.9).

II. Let us now estimate from above the integral

$$\int_M^{T^*} \mathbf{P}(Z(T) \in \Delta_g(x], \gamma(T) \in du, \chi(T) > v, B_T(u))$$

(see (6.5.11)), where T^* is such that the value $\alpha^* := \frac{x}{T^*} \sim c^*\alpha$, $c^* > 1$, $\alpha^* \in K$, still satisfies all the conditions of Theorem 5.2.1 for α. This integral is majorized by the integral

$$J_M^{T^*} := \int_M^{T^*} \mathbf{P}(Z(T) \in \Delta_g(x],\ \gamma(T) \in du, \chi(T) > v), \qquad (6.5.15)$$

which, as before, we approximate by the sums

$$\sum_{u_k \in [M,T^*]} \mathbf{P}(Z(T) \in \Delta_g(x],\ \gamma(T) \in \delta[u_k),\ \chi(T) > v), \tag{6.5.16}$$

where for the kth summand P_k we have

$$\begin{aligned} P_k &:= \mathbf{P}(Z(T) \in \Delta_g(x],\ \gamma(T) \in \delta[u_k),\ \chi(T) > v) \\ &= \mathbf{P}(Z(T-u_k) \in \Delta_g(x],\ \gamma(T-u_k) < \delta,\ \chi(T-u_k) > u_k + v). \end{aligned}$$

Setting

$$A_{T,k} = \{Z(T-u_k) \in \Delta_g(x], \gamma(T-u_k) < \delta\}, \tag{6.5.17}$$

we have

$$P_k = \mathbf{P}(A_{T,k})\mathbf{P}(\chi(T-u_k) > u_k + v \mid A_{T,k}),$$

where the second factor on the right is

$$\mathbf{P}(\chi(T-u_k) > u_k + v \mid \gamma(T-u_k) < \delta) \le \mathbf{P}(\tau > u_k + v). \tag{6.5.18}$$

Hence, by Theorem 5.2.1 for $\alpha_k = \frac{x}{T-u_k}$ we have

$$P_k \le \frac{|g'|\delta^2 C(\alpha_k)}{\sqrt{T-u_k}}\, e^{-(T-u_k)D(\alpha_k)}\mathbf{P}(\tau > u_k + v)(1+o(1)).$$

Here

$$(T-u_k)D(\alpha_k) = T\mathbf{D}(1-\rho_k, \alpha) \quad \text{as} \quad \rho_k = \frac{u_k}{T}.$$

Since the function $\mathbf{D}(t,\alpha)$ is convex (see Lemma 3.5.7), we have

$$\mathbf{D}(1-\rho_k, \alpha) \ge D(\alpha) - \rho_k \lambda(\alpha) \tag{6.5.19}$$

and, therefore, for all k,

$$\begin{aligned} &(T-u_k)D(\alpha_k) \ge TD(\alpha) - u_k\lambda(\alpha), \\ &\sum_{u_k \in [M,T^*]} P_k < \frac{|g'|\delta e^{-TD(\alpha)}}{\sqrt{T-T^*}} \sum_{u_k \in [M,T^*]} \delta C(\alpha_k) e^{u_k\lambda(\alpha)}\mathbf{P}(\tau > u_k + v)(1+o(1)) \\ &\qquad = \frac{c|g'|\delta e^{-TD(\alpha)}}{\sqrt{T-T^*}} \int_M^\infty e^{u\lambda(\alpha)}\mathbf{P}(\tau > u + v)\, du(1+o(1)) \end{aligned} \tag{6.5.20}$$

with some c. Since $\lambda_+ > D(0)$, we have $\lambda(\alpha) < \lambda_+$, and the integral in (6.5.20) converges to 0 as $M \to \infty$. It follows that

$$J_M^{T^*} = o(J_0^M) \quad \text{as} \quad M \to \infty,\ T \to \infty.$$

III. Let us now estimate

$$J_{T^*}^{(1-\varepsilon)T} = \int_{T^*}^{(1-\varepsilon)T} \mathbf{P}(Z(T) \in \Delta_g(x],\ \gamma(T) \in du, \chi(T) > v)$$

for small fixed $\varepsilon > 0$. As in Step II of the proof, we approximate the integral by sums and use inequality (6.5.18), from which we have (see (6.5.17) and (6.5.18))

$$P_k \le \mathbf{P}(A_{T,k})\mathbf{P}(\tau > u_k + v).$$

To obtain the required upper estimate for

$$\sum_{u_k \in [(1-\varepsilon)T, T^*]} P_k,$$

we use rougher approaches than in Step II. By the LDP for CRPs, we get, as $T \to \infty$,

$$\ln \mathbf{P}(A_{T,k}) \le \ln \mathbf{P}(Z(T-u_k) \in \Delta_g(x]) \le -(T-u_k)\widehat{D}\Big(\frac{x}{T-u_k}\Big)(1+o(1)).$$

We have $\lambda_+ > D(0)$, and hence $\widehat{D}(\cdot) = D(\cdot)$. So by (6.5.19), we get, as before,

$$\ln \mathbf{P}(A_{T.k}) \le -T(1-\rho_k)D\Big(\frac{\alpha}{1-\rho_k}\Big) = T\mathbf{D}(1-\rho_k, \alpha) \le -TD(\alpha) + u_k\lambda(\alpha),$$

$$\sum P_k \le \frac{c}{\delta}\, e^{-TD(\alpha)} \int_{T^*}^{(1-\varepsilon)T} e^{u\lambda(\alpha)}\mathbf{P}(\tau > u + v)\, du$$

with some $c < \infty$. Next, $\lambda(\alpha) < \lambda_+ - \varepsilon$ for some $\varepsilon > 0$, and hence

$$\sum P_k \le \frac{ce^{-TD(\alpha)}}{\delta} e^{-\varepsilon_1 T}$$

with some $\varepsilon_1 > 0$. Therefore, the value of the integral $J_{T^*}^{(1-\varepsilon)T}$ is negligibly small compared to J_0^M.

IV. Let us now estimate

$$J_{(1-\varepsilon)T}^T = \int_{(1-\varepsilon)T}^T \mathbf{P}(Z(T) \in \Delta_g(x], \gamma(T) \in du) \le \mathbf{P}(\overline{Z}(\varepsilon T) > x)\mathbf{P}(\tau > (1-\varepsilon)T). \tag{6.5.21}$$

By Lemma 4.6.2 (see (4.6.8)), we have $\frac{\mathfrak{l}_\alpha(t)}{t} \to \infty$ as $t \to 0$. Hence $\mathfrak{l}'_\alpha(\varepsilon) > 0$ for all $\alpha > 0$ and a sufficiently small ε. Hence by Theorem 4.6.4,

$$\mathbf{P}(\overline{Z}(\varepsilon T) > x) \le \exp\Big\{-\varepsilon TD\Big(\frac{x}{\varepsilon T}\Big) + o(1)\Big\}$$

as $T \to \infty$, where by (6.5.19),

$$\varepsilon TD\Big(\frac{\alpha}{\varepsilon}\Big) = T\mathbf{D}(\varepsilon, \alpha) \ge TD(\alpha) - T\lambda(\alpha)(1-\varepsilon).$$

Together with (6.5.21) this gives

$$J_{(1-\varepsilon)T}^T \le e^{-TD(\alpha)+\lambda(\alpha)(1-\varepsilon)T+o(T)}\mathbf{P}(\tau > (1-\varepsilon)T) \le e^{-TD(\alpha)-\varepsilon_1 T}$$

with some $\varepsilon_1 > 0$. This again proves that the integral is negligibly small in comparison with J_0^M.

V. It remains to estimate

$$P_{\delta[T)} \le \mathbf{P}(Z(T+\delta) \in \Delta_g(x],\ \gamma(T+\delta) < \delta)$$

(see (6.5.11). By Theorem 5.2.1,

$$P_{\delta[T)} \leq \frac{c\delta^2}{\sqrt{T}}\, e^{-TD(\alpha)} = o(J_0^M)$$

as $\delta \to 0$ and some $c < \infty$.

Now (6.5.9) follows from the above estimates. Theorem 6.5.3 is proved. □

6.5.2 Integro-Local Theorems for the Ruin Probability Problem

In the example given in the introduction it was mentioned that, for an insurance company with initial surplus z, claim payouts $\zeta_1, \zeta_2, \ldots$ and the "average rate" q of incoming insurance premiums, the probability of ruin prior to time T is

$$\mathbf{P}\Big(\sup_{u\leq T}(Z(t) - qt) > z\Big).$$

This corresponds to the probability of crossing the boundary

$$g_T(t) = z + qt = x + q(t - T) \quad \text{for} \quad x = z + qT \tag{6.5.22}$$

by the CRP $Z(t)$. It is clear that such boundary satisfies conditions (a), (b) with

$$g' = q > 0, \qquad \alpha = \frac{x}{T} = \alpha_z + q > q, \qquad \alpha_z = \frac{z}{T},$$

which excludes the possibility of crossing the boundary by a horizontal jump. This is the classical run problem, which has been extensively studied (see, for example, the books [4, 6, 9, 63]).

In this section, we will obtain an integro-local result for the time $\eta_{Z,g}$ of first crossing of the boundary (6.5.22). To get a meaningful result, we will assume that $\alpha_z = z/T$ and $\alpha - a$ are separated from 0. (In what follows, an appropriate estimate for the probability $\mathbf{P}\eta_{Z,g} > T)$ will be obtained when deriving integral theorems in the case $T \gg z$.)

The next assertion follows from Theorems 6.3.1 and 6.5.1.

Corollary 6.5.4 *Let ξ be a nonlattice vector, let $q > a$, and let $\alpha_z = z/T > \varepsilon$ with some $\varepsilon > 0$. Then*

$$\mathbf{P}(\eta_{Z,g} \in \delta[T),\ \chi_{Z,g} \in \Delta[v)) = \frac{\Delta\delta C(\alpha)}{\sqrt{T}}\, e^{-TD(\alpha)-v\mu(\alpha)} P_{\alpha,q}(-v)(1 + o(1)),$$

where $\Delta = \Delta_T = \overline{o}(1)$, $\delta = \delta_T = \overline{o}(1)$ as $T \to \infty$. Besides, the remainder term $o(1)$ is uniform in $\alpha \in K \bigcap \{\alpha \geq \varepsilon\}$.

Corollary 6.5.5 *Under the conditions of Corollary* 6.5.4,

$$\mathbf{P}(\eta_{Z,g} \in \delta[T)) = \frac{\delta C(\alpha) P_{\alpha,q}}{\sqrt{T}}\, e^{-TD(\alpha)}(1 + o(1)),$$

where the factor $P_{\alpha,q}$ is defined in (6.5.8).

6.6 Integral Theorems in Boundary Crossing Problems

6.6.1 Integral Theorems in the First Boundary Crossing Problem

From Theorems 6.5.1, 6.5.3 one can find the asymptotics of the "integral" probability $\mathbf{P}(\eta_{Z,g} < T)$ in the first general boundary crossing problem for a boundary of the form

$$g_T(t) = Tg(t/T),$$

where $g(u)$, $u \in [0, 1]$, is a function not depending on T (see §4.6). For solving this problem, we should, as was done in [23, §3.7] for random walks, consider the family of level lines $\mathfrak{l}_\alpha(t)$ (see §4.6) and fix the value $\alpha = \alpha^g$ at which the level line $\mathfrak{l}_\alpha(t)$ touches for the first time the curve $g(t)$ and also fix the point t^g at which the level line touches the curve (we assume that it is a unique tangent point). Here we should assume that at the point t^g the curve and the level line touch smoothly (the second derivatives of the functions $\mathfrak{l}_\alpha(t)$ and $g(t)$ at the point t^g exist and are different; and the first derivatives are equal).

To simplify the presentation, we consider in detail only the case when $t^g \in (0, 1)$ and the boundary $g_T(t)$ can be crossed only with a vertical jump ($\mathfrak{l}'_{\alpha^g}(t^g) > 0$; this is always satisfied if $a \geq 0$). Other "regular" kinds of tangency of $\mathfrak{l}_{\alpha^g}(t)$ and $g_T(t)$ will be considered briefly.

We set $\rho_g = g(t^g)/t^g$.

Theorem 6.6.1 *Let $\xi = (\tau, \zeta)$ be a nonlattice vector, let $\alpha^g \in K \cap \{\alpha \geq \varepsilon\}$, $\varepsilon > 0$, $t^g \in (0, 1)$, let the function $g(t)$ be twice continuously differentiable in the neighborhood of the point t^g and be separated from the function $\mathfrak{l}_{\alpha^g}(t)$ outside this neighborhood, and let $q := g'(t^g) = \mathfrak{l}'_{\alpha^g}(t^g) > 0$, $b := g''(t^g) - \mathfrak{l}''_{\alpha^g}(t^g) > 0$. Then*

$$\mathbf{P}(\eta_{Z,g} < T) = \frac{C(\rho_g) P_{\rho_g, q}\sqrt{2\pi}}{\sqrt{b\mu(\rho_g)t^g}}\, e^{-TD(\alpha^g)}(1 + o(1)) \tag{6.6.1}$$

as $T \to \infty$, where the function $P_{\alpha,q}$ is defined in (6.5.7), (6.5.8) (*with g' replaced by q*).

Proof Let us first find the asymptotics of the probability

$$\mathbf{P}(\eta_{Z,g} \in (Tt^g)_{\varepsilon T})$$

with some $\varepsilon > 0$. To this end, we shall use the integro-local Theorem 6.5.1 and Corollary 6.5.2 and consider the partition of the neighborhood of the point t^g into small half-open intervals $\delta[t_k)$, $t_k = t^g \pm \frac{k\delta}{T}$, $k = 0, 1, \ldots$ The sought-for probability $\mathbf{P}(\eta_{Z,g} \in (Tt^g)_{\varepsilon T})$ will be the sum over $t_k \in (t^g)_\varepsilon$ of the integro-local probabilities $\mathbf{P}(\eta_{Z,g} \in \delta[Tt_k))$ found with the use of Theorem 6.5.1, whose conditions are satisfied (here small additional stipulations are needed related to the choice of a small $\varepsilon > 0$ and the passage from the time T (in Theorem 6.5.1) to the time Tt_k).

So, consider the sum

$$\Sigma_{Z,g} = \sum_{k=-N}^{N} \mathbf{P}(\eta_{Z,g} \in \delta[Tt_k]), \tag{6.6.2}$$

where $N = \frac{T\varepsilon}{\delta}$ is assumed to be integer. We set

$$T_{(k)} = Tt_k, \quad \alpha_k = \frac{Tg(t_k)}{T_{(k)}} \to \frac{g(t^g)}{t^g} = \rho_g \tag{6.6.3}$$

for each k and $T \to \infty$, so that $g'(t_k) \to q$. By Theorem 6.5.1 and Corollary 6.5.2 we have

$$\mathbf{P}(\eta_{Z,g} \in \delta[T_{(k)}]) = \frac{\delta C(\alpha_k) P_{\alpha_k,q}}{\sqrt{T}} e^{-T_{(k)} D(\alpha_k)} (1 + o(1)) \tag{6.6.4}$$

as $T \to \infty$, where the function $P_{\alpha,q}$ is defined in (6.5.8), $P_{\alpha_k,q} \to P_{\rho_g,q}$, $C(\alpha_k) \to C(\rho_g)$, and $T_{(k)} \sim t^g T$ as $T \to \infty$.

In (6.6.4),

$$T_{(k)} D(\alpha_k) = T\Big(t^g + \frac{k\delta}{T}\Big) D\Big(\frac{g(t^g + \frac{k\delta}{T})}{t^g + \frac{k\delta}{T}}\Big), \tag{6.6.5}$$

$$g\Big(t^g + \frac{k\delta}{T}\Big) = g(t^g) + g'(t^g)\frac{k\delta}{T} + \frac{g''(t^g)}{2}\Big(\frac{k\delta}{T}\Big)^2 (1 + o(1)), \tag{6.6.6}$$

$$\Big(t^g + \frac{k\delta}{T}\Big)^{-1} = (t^g)^{-1} - \frac{k\delta}{Tt^g} + \Big(\frac{k\delta}{Tt^g}\Big)^2 (1 + o(1)) \tag{6.6.7}$$

as $T \to \infty$. Now one can easily derive the asymptotic expansion for $t_k D(\alpha_k)$ with the principal term $t^g D(\frac{g(t^g)}{t^g}) = t^g D(\frac{l_{\alpha^g}(t^g)}{t^g}) = D(\alpha^g)$. We next note that if in (6.6.5) and in what follows we substitute $l_{\alpha^g}(t)$ for $g(t)$, then we get analogous asymptotic expansion. Since $l_{\alpha^g}(t^g) = g(t^g)$, $l'_{\alpha^g}(t^g) = g'(t^g)$, the first three terms of the expansion differ only in that they contain $\mu(\rho_g) l''_{\alpha^g}(t^g)$ in place of $D'(\rho_g) g''(t^g) = \mu(\rho_g) g''(t^g)$. Besides, by the definition of a level line, for $\beta_k = \frac{l_{\alpha^g}(t_k)}{t_k}$ we have

$$t_k D(\beta_k) = D(\alpha^g).$$

Hence

$$t_k D(\alpha_k) = t_k D(\alpha_k) \pm t_k D(\beta_k) = D(\alpha^g) + \frac{\mu(\rho_g)}{2} (g''(t^g) - l''_{\alpha^g}(t^g)) \Big(\frac{k\delta}{T}\Big)^2 (1 + o(1))$$

and so using (6.6.4),

$$\mathbf{P}(\eta_{Z,g} \in \delta[T_k]) = \frac{\delta C(\rho_g) P_{\rho_g,q}}{\sqrt{Tt^g}} e^{-TD(\alpha^g) - \frac{b\mu(\rho_g)}{2} \frac{(k\delta)^2}{T}} (1 + o(1)).$$

Now by applying (6.6.4)) and (6.6.2) we obtain for the sum $\Sigma_{Z,g}$ the integral sum with the partition step of length $\frac{\delta}{\sqrt{T}}$ corresponding to the integral

$$\frac{C(\alpha^{(g)}) P_{\rho_g,q}}{\sqrt{t^g}} e^{-TD(\alpha^g)} \int_{-\sqrt{T}\varepsilon}^{\sqrt{T}\varepsilon} e^{-\frac{b\mu(\rho_g) z^2}{2}} dz (1 + o(1)) \tag{6.6.8}$$

as $T \to \infty$. This proves (6.6.1).

Next, for any $\varepsilon > 0$,

$$\mathbf{P}(\eta_{Z,g} < T) = \mathbf{P}(\eta_{Z,g} \in (Tt^g)_{\varepsilon T}) + \mathbf{P}(\eta_{Z,g} \in [0,T] \setminus (Tt^g)_{\varepsilon T}), \tag{6.6.9}$$

where by Theorem 4.6.3 the second summand in the right-hand side of (6.6.9) is $O(e^{-T(D(\alpha)+\varepsilon_1)})$ with some $\varepsilon_1 > 0$. Hence representation (6.6.8) remains valid also for the probability $\mathbf{P}(\eta_{Z,g} < T)$. Theorem 6.6.1 is proved. □

In the case

$$t_g = 1, \qquad g'(1) = \mathrm{l}'_{\alpha^g}(1), \qquad b = g''(1) - \mathrm{l}''_{\alpha^g}(1) > 0 \tag{6.6.10}$$

the argument is the same as in the proof of Theorem 6.6.1, but in (6.6.2) the sum $\Sigma_{Z,g}$ should be taken only over nonnegative values of k. As a result, we find that $\mathbf{P}(\eta_{Z,g} < T)$ is equal to the halved right-hand side in (6.6.1) with $t^g = 1$, $\alpha^g = g(1)$.

Let us now briefly discuss the case

$$t_g = 1 \quad (\alpha^g = g(1)), \qquad g'(1) < l'_{\alpha^g}(1) = -\frac{\lambda(\alpha^g)}{\mu(\alpha^g)}$$

(see Lemma 4.5.4). In this case, the asymptotics of the probability $\mathbf{P}(\eta_{Z,g} < T)$ has the form

$$\mathbf{P}(\eta_{Z,g} < T) = \frac{c}{\sqrt{T}} e^{-TD(\alpha)}(1 + o(1)),$$

where the constant c is obtained by summing the constants in the right-hand sides of (6.5.6) with respect to $v = v_k = k\Delta$ and $T = T_{(k)} = T - k\delta$, $k = 0, 1, \ldots$ This will not change the principal part of the asymptotics in (6.5.6).

Using Theorem 6.5.1 one can also easily find the asymptotics of the probability $\mathbf{P}(\eta_{Z,g} < T, \chi_{Z,g} < v)$.

The "transition domain" involving the normal distribution (similar to the domain appearing for the distribution of $Z(T)$ in assertion (ii) in Theorem 6.4.1) will also appear in this problem if in (6.6.10) we consider $\mathbf{P}(\eta_{Z,g} < T - v\sqrt{T})$, rather than $\mathbf{P}(\eta_{Z,g} < T)$. The normal distribution will appear also in this case when we consider $\mathbf{P}(\eta_{Z,g} < T - v\sqrt{T})$, for example, if α^g is not a fixed value satisfying (6.6.10), but if it only converges to it with the rate $c/\sqrt{T}$.

6.6.2 On the Second Boundary Crossing Problem

Recall that the second boundary crossing problem for CRPs is concerned with the probability that the trajectory $Z(t)$ on the interval $[0,T]$ will not cross a boundary $g_T(t)$, whereas the straight line $z = at$ will cross this boundary on $[0,T]$ (we consider the problem with one boundary). In this section, we will assume that the boundary $g_T(t)$ satisfies conditions (a), (b) of §6.5.1 (see (6.5.3), (6.5.4)) and the condition

(c) $a > \alpha + \varepsilon$ with some $\varepsilon > 0$.

The more general case where condition (b) is not satisfied is very difficult and its asymptotic analysis is hardly possible within the framework of the present monograph (cf. the study of the second boundary crossing problem for random walks in [23]).

Under the above conditions, we have the following assertion. We set, as before,

$$P_{\alpha,g'}(v) = \mathbf{P}\big(\sup_{t\ge 0}({}^{\alpha}Y(t) + g't) \le v\big).$$

Theorem 6.6.2 *Let conditions* (a), (b) *of* §6.5.1 *be met and let* $g' \le 0$. *Then*

$$\mathbf{P}(\eta_{Z,g} > T, Z(T) \in \Delta[x - v))$$
$$= \frac{\Delta C(\alpha)}{\sqrt{T}} e^{-TD(\alpha)+\mu(\alpha)v} P_{\alpha,g'}(v) \int_0^{\infty} e^{\lambda(\alpha)u} \mathbf{P}(\tau > u)\, du \qquad (6.6.11)$$

uniformly in $\alpha \in K \cap (\varepsilon, a - \varepsilon)$ *with some* $\varepsilon > 0$.

If $g' > 0$, *then*

$$\mathbf{P}(\eta_{Z,g} > T, Z(T) \in \Delta[x - v))$$
$$= \frac{\Delta C(\alpha)}{\sqrt{T}} e^{-TD(\alpha)+\mu(\alpha)v} \int_0^{v/g'} e^{\lambda(\alpha)u} \mathbf{P}(\tau > u) P_{\alpha,g'}(v - ug')\, du (1 + o(1)) \qquad (6.6.12)$$

uniformly in α *from the same domain.*

The integral in (6.6.11) is convergent, because $\lambda(\alpha) < \lambda_+$ for $\alpha \in K$.

Proof If $g' \le 0$, then the asymptotic behavior of the principal parts of the probabilities in (6.6.11) can be determined in the same way (with minor modifications) as in the proof of Theorem 6.5.3, in which (following the proof of this theorem) the probability $\mathbf{P}({}^{\alpha}\overline{Y} \le v)$ should be replaced by $\mathbf{P}(\sup_{t\ge 0}({}^{\alpha}Y(t) + g't) \le v)$.

The estimate for the probabilities to cross the boundary $g_T(t)$ on the time interval $(0, T - M)$ for large M is obtained as in Theorems 6.3.1 and 6.5.1.

If $g' > 0$, then the principal part of the asymptotics is obtained in a slightly different way. We have

$$\mathbf{P}(\eta_{Z,g} > T, Z(T) \in \Delta[x - v))$$
$$= \int_0^{v/g'} \mathbf{P}\Big(Z(T) \in \Delta[x - v),\ \gamma(T) \in du,\ \max_{t\le T-u}(Z(t) - g_T(t)) \le 0\Big)$$
$$= \int_0^{v/g'} \mathbf{P}\Big(Z(T) \in \Delta[x - v), \gamma(T) \in du\Big)$$
$$\times \mathbf{P}\big(\max_{t\le T-u}(Z(t) - g_T(t)) \le 0 \,|\, Z(T) \in \Delta[x - v),\ \gamma(T) \in du\big). \qquad (6.6.13)$$

As in the proofs of Theorems 6.3.1 and 6.5.1, we get

$$\mathbf{P}\Big(\max_{t\le T-u}(Z(t) - g_T(t)) \le 0 \,|\, Z(T) \in \Delta[x - v), \gamma(T) \in du\Big)$$
$$\sim \mathbf{P}\Big(\sup_{t\ge 0}({}^{\alpha}Y(t) - g't) \le v - ug'\Big) = P_{\alpha,g'}(v - ug').$$

Approximating the integral in (6.6.13) by sums of the probabilities over the events $\gamma(T) \in \delta[u_k)$, $u_k = k\delta$, for k from 0 to $N = \frac{v}{g'\delta}$ (N is assumed to be integer), we get

$$\sum_{k=0}^{N} \mathbf{P}(Z(T) \in \Delta[x-v), \gamma \in \delta[u_k)) P_{\alpha,g'}(v - u_k g')(1+o(1)).$$

Now by using Theorem 5.2.1, one can easily arrive at (6.6.12). Theorem 6.6.2 is proved. □

As in the above, from Theorem 6.6.2 we get

Corollary 6.6.3 *If $g' \le 0$, then*

$$\mathbf{P}(\eta_{Z,g} > T, Z(T) \ge x - v) = \frac{C(\alpha)}{\sqrt{T}} e^{-TD(\alpha)} \int_0^v e^{\mu(\alpha)t} P_{\alpha,g'}(t)\, dt \int_o^\infty e^{\lambda(\alpha)u} \mathbf{P}(\tau > u)\, du (1+o(1)). \tag{6.6.14}$$

If $g' > 0$, then

$$\mathbf{P}(\eta_{Z,g} > T, Z(T) \ge x - v) = \frac{C(\alpha)}{\sqrt{T}} e^{-TD(\alpha)} \int_0^v \int_0^{t/g'} e^{\lambda(\alpha)u + \mu(\alpha)t} \mathbf{P}(\tau > u) P_{\alpha,g'}(t - ug')\, du\, dt (1+o(1)) \tag{6.6.15}$$

uniformly in α in the same domains as in Theorem 6.6.2.

The integral in (6.6.14) for $v = \infty$ converges, because $\mu(\alpha) < 0$ for $\alpha < a$ and $\lambda(\alpha) < \lambda_+$ for $\alpha \in K$.

6.7 Applications to the Ruin Probability Problem for Insurance Companies

In the introduction, we described the classical ruin probability problem for an insurance company. This problem is posed as follows. Let $T_1 = \tau_1$, $T_2 = \tau_1 + \tau_2, \ldots$ be the times of significant claim payouts for insured cases and let $\zeta_1, \zeta_2, \ldots$ be the claim payments. Next, let r be the premium rate (the amount received by the company from insured customers per time unit). If x is the initial surplus of the company, then its surplus at time t is equal to $x + rt - Z(t)$. This means that if $\inf_{t \le T} (x + rt - Z(t)) < 0$, then the company will go bankrupt by time T. In other words, the probability of ruin prior to time T is equal to

$$\mathbf{P}\Big(\sup_{t \le T} (Z(t) - rt) > x\Big) = \mathbf{P}(\overline{Z}^{(q)}(T) > x) \text{ for } q = -r < 0, \tag{6.7.1}$$

where, as before,

$$Z^{(q)}(t) = Z(t) + qt, \qquad \overline{Z}^{(q)}(T) = \max_{t \le T} Z^{(q)}(t).$$

The problem of approximation for probabilities (6.7.1) has been extensively studied (see, e.g., the books [9], [4], [63], and [6]).

The ruin probability must be small, and hence the initial surplus x must be large. The probability can be approximately evaluated if one knows the asymptotics of probability (6.7.1) as $x \to \infty$. It is also interesting to know the distribution of the time $\eta_{Z^{(q)}}(x)$ until the moment of ruin.

The process $Z^{(q)}(t)$ in (6.7.1) is a CRP with *linear drift*. There exist two approaches capable of reducing the above problems of the distribution of $Z^{(q)}(T)$ and $\eta_{Z^{(q)}}(x)$ to the corresponding problems for "usual" CRPs (without drift).

The first approach We consider the CRP $Z^{[q]}(t)$ as a "usual" CRP governed by the vector $(\tau, \zeta^{(q)})$, $\zeta^{(q)} = \zeta + q\tau$, so that the values of the processes $Z^{(q)}$ and $Z^{[q]}$ are equal at times T_k: $Z^{(q)}(T_k) = Z^{[q]}(T_k)$, $k = 0, 1, \ldots$ It is easily seen that, for $q \le 0$,

$$\overline{Z}^{(q)}(T) = \overline{Z}^{[q]}(T) = \max_{t \le T} Z^{[q]}(t), \quad \eta_{Z^{(q)}}(x) = \eta_{Z^{[q]}}(x),$$

and besides, $\overline{Z}^{(q)} = \overline{Z}^{[q]} = \sup_{k \ge 0} Z_k^{(q)}$, $Z_k^{(q)} = Z_k + qT_k$. It now remains to apply the results of §§6.1, 6.4 to the processes $Z^{[q]}$. But for this purpose one should know the characteristics of the process $Z^{[q]}$; we give them the superscript (q). The most important characteristic here is the fundamental function $A^{(q)}(\mu)$, which in the analyticity domain is the solution of the equation

$$\mathbf{A}^{(q)}(-\lambda, \mu) = \ln \mathbf{E} e^{\lambda\tau + \mu\zeta^{(q)}} = 0$$

with respect to λ. We have $\mathbf{A}^{(q)}(-\lambda, \mu) = \mathbf{A}(-\lambda + q\mu, \mu)$, and hence one can easily show that

$$A^{(q)}(\mu) = q\mu + A(\mu),$$

because $\mathbf{A}^{(q)}(-A^{(q)}(\mu), \mu) = \mathbf{A}(-q\mu - A(\mu) + q\mu, \mu) = 0$.

For the deviation function $D^{(q)}(\alpha)$ of the processes $Z^{(q)}$ and $Z^{[q]}$, we similarly find that $D^{(q)}(\alpha) = D(\alpha - q)$ (this follows from the results of Chapter 3 on LDPs for CRPs). The remaining characteristics of the CRP $Z^{[q]}$ are determined similarly.

The second approach It assumes the problem of evaluation of probabilities (6.7.1) as the first boundary crossing problem for the original process $Z(t)$ with $q = -r$ and the boundary $g_T(t) = x + rt = Tg(t/T)$, where $g(u) = \alpha + ru$, $u \in [0, 1]$, $\alpha = x/T$. This problem was studied in §§6.5 and 6.6. Here the characteristics of the CRP remain the same, but now the defining role is played by the parameter $\alpha = \alpha^g$ for which the level line $\mathfrak{l}_\alpha(t)$ touches for the first time the curve $g(t) = \alpha + rt$ and by the point t^g at which they are tangent.

We will mainly stick to the first approach. Let, as before, $\psi_q(\mu) = \mathbf{E}e^{\mu\zeta^{(q)}}$. We set

$$\mu_0^{(q)} = \sup\{\mu : \psi_q(\mu) \le 1\}. \tag{6.7.2}$$

We put $\mu_0^{(q)} = \mu_0$ in this section, to simplify the notation. From (6.7.2) it follows that, for $\mathbf{E}\zeta^{(q)} < 0$,

$$\mu_0 = \sup\{\mu : A^{(q)}(\mu) \le 0\}. \tag{6.7.3}$$

Indeed, $A^{(q)}(\mu)$ is a convex function, $(A^{(q)})'(0) = \frac{\mathbf{E}\zeta^{(q)}}{a_\tau} < 0$, $\mathbf{A}^{(q)}(0,\mu_0) = \ln\psi_q(\mu_0) = 0$. The last relation here implies that $A^{(q)}(\mu_0) = 0$ (recall that $\mathbf{A}^{(q)}(-A^{(q)}(\mu_0),\mu) = 0$). This expression coincides with the right-hand side of (6.7.3).

By considering the results of §§6.1, 6.4, 6.6 one can single out three classes of problems which are of principal interest and are defined in terms of the governing vector (τ,ζ) and the value $q = -r$. To these problems there correspond three different types of limit laws for the boundary functionals $\overline{Z}^{(q)}(T)$, $\overline{Z}^{(q)} = \overline{Z}^{(q)}(\infty)$ and $\eta_{Z^{(q)}}(x)$.

Class (A) $\mu_0 > 0, \psi_q(\mu_0) = 1, \psi'_q(\mu_0) < \infty$ (the last condition is always fulfilled if $\psi_q(\mu_0 + 0) < \infty$).

Class (B) $\mu_0 > 0, \psi_q(\mu_0) < 1$.

Class (C) $\mu_0 = 0$ (the Cramér condition [**C**] is not met).

In cases (B), (C) we should impose in addition some conditions of regular behavior at infinity of the function $F(x) = \mathbf{P}(\zeta^{(q)} > x)$. For example, in case (C) the function $F(x)$ should vary regularly, or more generally, to be a subexponential function.

Let us briefly recall the results of §§6.1, 6.4, 6.6 on the asymptotic analysis in classes (A)–(C), which are concerned with the boundary functionals $\overline{Z}^{(q)}(T)$, $\overline{Z}^{(q)}$ and $\eta_{Z^{(q)}}(x)$. As already pointed out, to each of these classes there corresponds its own type of limit laws.

Class (A) *We set* $\alpha_0 = (A^{(q)})'(\mu_0) = q + A'(\mu_0)$. *Then, for* $q \le 0$, $T > \frac{x}{\alpha_0} + \rho$, *where* $\rho \gg \sqrt{x}$,

$$\mathbf{P}(\overline{Z}^{(q)}(T) > x) = \mathbf{P}(\overline{Z}^{(q)} > x)(1 + o(1)) = ce^{-\mu_0 x}(1 + o(1))$$

as $x \to \infty$, $c < 1$ (see Corollary 6.4.2). *If* $T < \frac{x}{\alpha_0} - \rho$, $\rho \gg \sqrt{x}$, $x \to \infty$, *then*

$$\mathbf{P}(\overline{Z}^{(q)}(T) > x) = \mathbf{P}(\eta_{Z^{(q)}}(x) \le T) \sim \frac{c}{\sqrt{T}} e^{-TD(\alpha - q)}, \quad c = \text{const.}$$

The conditional distribution of the ruin time $\eta_{Z^{(q)}}(x)$ *under the condition* $\overline{Z}^{(q)} > x$ *is asymptotically normal with parameters* $(\frac{x}{\alpha_0}, \sigma_{\eta_{Z[q]}}\sqrt{x})$, *where* $\sigma_{\eta_{Z[q]}}$ *is defined in* §6.4.1 (see Corollary 6.4.3).

Thus, if a ruin occurs, it happens at a time close to x/α_0 (up to a quantity of order $\sqrt{x}$). In §6.4 and in subsection 6.6.2 we have also obtained quite complete results about the distributions of $\eta_{Z[q]}(x)$ in the entire Cramér spectrum of deviations and also about the debt formed after the ruin (the variable $\chi_{Z[q]}(x)$).

Class (B) Here we have the following results.

(a) *We set* $F(x) = \mathbf{P}(\zeta^{(q)} > x)$, $F^I(x) = \int_x^\infty F(t)\,dt$. *Assume that* $q \le 0$, $\psi_q := \psi_q(\mu_0) < 1$, *and that* $e^{\mu_0 x} F^I(x)$ *is an upper power function. Then*

$$\mathbf{P}(\overline{Z}^{(q)} > x) = \frac{\mu_0 M F^I(x)}{1-\psi_q}(1+o(1)), \quad \text{as} \quad x \to \infty, \tag{6.7.4}$$
$$M = \mathbf{E}^{\mu_0 \overline{Z}^{(q)}}, \quad q = -r < 0.$$

(b) *If* $F(x)e^{\mu_0 x}$ *is subexponential or, equivalently, the function* $F(x)$ *is such that* $F{*}F(x) \sim 2\psi_q F(x)$ *as* $x \to \infty$, *then*

$$\mathbf{P}(\overline{Z}^{(q)} > x) = \frac{MF(x)}{1-\psi_q}(1+o(1)), \quad \text{as} \quad x \to \infty \tag{6.7.5}$$

(see Theorem 6.1.2 and [121]).

If in assertion (a) we assume in addition that

$$\frac{F(x+v)}{F(x)} \to e^{-\mu_0 v} \quad \text{as} \quad x \to \infty \quad \text{for any fixed } v,$$

then

$$\mu_0 F^I(x) = \mu_0 \int_0^\infty F(x+v)\,dv \sim \mu_0 F(x) \int_0^\infty e^{-\mu_0 v} dv = F(x)$$

and the right-hand sides in (6.7.4) and (6.7.5) are equal.

Under additional rather broad conditions, the ruin times $\eta_{Z^{(q)}}(x)$ *on the set* $Z^{(q)} > x$ *are bounded in probability,*

$$\limsup_{x\to\infty} \mathbf{P}(\eta_{Z^{(q)}}(x) > v \mid Z^{(q)} > x) \to 0 \quad \textit{as} \quad v \to \infty$$

(this is true, for example, for independent or linearly dependent τ and ζ for $h+q \le 0$; see Theorem 6.1.10).

This means that if the ruin happens, then it happens "almost immediately" after the start of work under the condition $\overline{Z}^{(q)} > x$ (we have $\eta_{Z^{(q)}}(x)/x \xrightarrow[p]{} 0$ as $x \to \infty$).

The "intermediate" case between (A) and (B), when $\mu_0 > 0$, $\psi_q(\mu_0) = 1$, and $\psi_q'(\mu_0) = \infty$, was studied in [80]; see also [17, §21.5].

Class (C) *Suppose that* $F^I(x) = \int_x^\infty F(t)\,dt$ *is an upper power function. Let* $a + q = a - r < 0$. *Then*

$$\mathbf{P}(\overline{Z}^{(q)} > x) = \frac{F^I(x)}{|a+q|a_\tau}(1+o(1)) \quad \textit{as} \ x \to \infty, \tag{6.7.6}$$
$$|a+q|\,a_\tau = |a_\zeta - r a_\tau|.$$

If $F(x) = \mathbf{P}(\zeta^{(q)} > x) = t^{-\alpha} l(x)$, *where* $\alpha > 1$, $l(x)$ *is a slowly varying function, then*

$$\mathbf{P}(\eta_{Z^{(q)}}(x) < T) = (1 + o(1))a_\tau^{-1} \int_0^T F(x - (a+q)t)\,dt, \tag{6.7.7}$$

$$\lim_{x\to\infty} \mathbf{P}(\eta_{Z^{(q)}}(x) > vx \mid \overline{Z}^{(q)} > x) = [1 + |a+q|v]^{1-\alpha}$$

(see Theorem 6.1.13). *The conditional ruin time $\eta_{Z^{(q)}}(x)$ divided by x converges not to a constant value (as in cases* (A), (B)), *but it converges in distribution to a proper random variable θ with power distribution.*

In [79], assertion (6.7.7) was extended to the case when $F(x)$ and $F^I(x)$ are subexponential functions.

The above means that in cases (A)–(C) the trajectories of the processes $z(v), v > 0$, which are limit processes as $x \to \infty$ for the processes

$$z_x(v) = \frac{Z^{(q)}(vx)}{x}, \qquad v > 0,$$

have under the conditions $\overline{Z}^{(q)} > x$, $a + q < 0$, the forms shown in Figure 6.1.

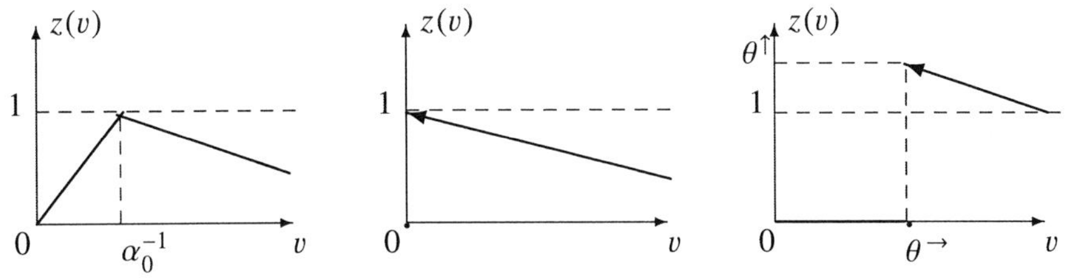

Figure 6.1. *The limit trajectories of the processes $z(v)$ under the condition $\overline{Z}^{(q)} > x$ and some additional conditions in cases* (B), (C).

Here $\theta^{\rightarrow}, \theta^{\uparrow}$ are random variables with the distributions

$$\mathbf{P}(\theta^{\rightarrow} > v) = (1 + |a+q|v)^{1-\alpha}, \qquad \mathbf{P}(\theta^{\uparrow} > 1 + v) = (1+v)^{-\alpha}, \quad v > 0.$$

The limit behavior of the sequence $\{Z_k\}$ under the conditions $\overline{Z}_\infty > x$ and $\mathbf{E}\zeta < 0$ was studied in [65].

Note that, according to the opinion of experts, in case (C) the accuracy of approximation (6.7.6) is not always satisfactory in real-world applications. In this regard, in Theorem 7.5.3 in [38] and in [20], a much sharper second-order asymptotics for the distribution of $\overline{Z}^{(q)}$ was found: *if $F(x)$ is regularly varying or a subexponential function on* $(0, \infty)$, *if the random variable $\zeta^{(q)}$ is nonlattice, and if $a_{\zeta^{(q)}} = \mathbf{E}\zeta^{(q)} < 0$ and $b := \mathbf{E}(\zeta^{(q)})^2 < \infty$, then*

$$\mathbf{P}(Z^{(q)} > x) = \frac{F^I(x)}{|a_{\zeta^{(q)}}|} + cF(x)(1 + o(1)) \quad \text{as} \quad x \to \infty,$$

where

$$c = \frac{b}{2(a_{\zeta^{(q)}})^2} + \frac{\mathbf{E}\overline{Z}^{(q)}}{a_{\zeta^{(q)}}}.$$

This assertion is also true in the arithmetic case for a somewhat different value of c.

In [24] and [29, §2], a second-order approximation for $\mathbf{P}(Z^{(q)} > x)$ was derived in the case $b = \infty$.

If $a_{\zeta^{(q)}} < 0$ is small and if $b < \infty$, then, for x comparable with $|a_{\zeta^{(q)}}|^{-1}$, in all three classes (A)–(C) the probability $\mathbf{P}(\overline{Z}^{(q)} > x)$ will be close to some proper value (the so-called transition phenomena):

$$\lim_{a_{\zeta^{(q)}} \to 0} \mathbf{P}\Big(\overline{Z}^{(q)} > \frac{v}{|a_{\zeta^{(q)}}|}\Big) = e^{-\frac{2v}{b}} \tag{6.7.8}$$

(see [77], [100], [17, §24]; see also the precise statement of the problem there). If, in addition, $\mathbf{E}|\zeta^{(q)}|^3 < \infty$, then a second-order approximation for (6.7.8) was found in [28] and [29, §1].

Note also that we can consider the more general statement of the problem when the surplus $x = x(t)$ and the drift $q = q(t)$ are not constant but depend on time t (the insurance company can have other sources of income, and the insurance intensity $r = r(t)$ may change with time). In this case, we obtain the first boundary crossing problem for the trajectory $Z(t)$ and the boundary $x(t)+R(t)$, $R(t) = \int_0^t r(u)\,du$. If we assume that this boundary has the form $Tg(t/T)$ on the interval $[0, T]$, where $g(u) > 0$ does not depend on x and T, and $g(0) = \alpha = x(0)/T$, then in case (A) we get the boundary crossing problem considered in §6.6 (see Theorem 6.6.1). In case (C), we get the boundary crossing problem, whose solution is described by Theorem 16.4.1 in [38].

7

Extension of the Invariance Principle to the Zones of Moderately Large and Small Deviations

The results of this chapter (see §§7.2–7.5) are new also for random walks, which are special cases of CRPs.

7.1 Strong Approximation of a CRP by a Wiener Process

By a strong approximation of a sequence of processes we mean a definition of the processes under study and of an approximating ("limit") process on the common probability space such that the distance between the trajectories of these processes can be estimated in some metric. For some definitions, the resulting approximation turns out to be quite sharp.

The well-known Komlós–Major–Tusnády inequality (see [78]) establishes a strong approximation of the trajectories of random walks by the trajectories of the Wiener process. Using this approximation and the results of [55], Csörgö, Deheuvels, and Horváth [56] constructed a similar approximation of the trajectories of a CRP with independent τ and ζ.

The main assertion on strong approximation of normalized renewal processes under the Cramér condition is as follows:

$$y_T(t) = \frac{Y(tT) - atT}{\sigma\sqrt{T}}, \qquad 0 \le t \le 1.$$

As in §2.1, we say that ζ and τ are linearly dependent if

$$\zeta = h\tau + \omega, \qquad h = \text{const} \neq 0,$$

where τ and ω are independent.

Theorem 7.1.1 *Let τ and ζ be either independent or linearly dependent and satisfy the Cramér condition* [**C**]. *Then the processes $y_T(t)$ and the standard Wiener process $w(t)$, $0 \le t \le 1$, can be defined on a common probability space in such a way that,* *for some positive constants b, c, λ, λ_1, and all $y > 0$, for all sufficiently large T, the following inequality holds:*

$$\mathbf{P}\Big(\sup_{t\in[0,1]} |y_T(t) - w(t)| > \frac{b\ln T}{\sqrt{T}} + \frac{y}{\sqrt{T}}\Big) \le ce^{-\lambda y} + e^{-\lambda_1 T} \tag{7.1.1}$$

where $\lambda_1 = \infty$ if τ and ζ are independent. Besides,

$$\rho(y_T, w) = O\Big(\frac{\ln T}{\sqrt{T}}\Big) \quad a.s. \quad as \quad T \to \infty, \tag{7.1.2}$$

where ρ is the uniform metric.

It is an open question whether this result remains valid for an arbitrary nondegenerate distribution of a vector (τ, ζ) under the Cramér condition [**C**].

Proof For independent ζ and τ, the conclusion of Theorem 7.1.1 was established in [56]. Let us prove (7.1.1) for linearly dependent ζ and τ. We set

$$\Omega_n = \sum_{k=1}^{n} \omega_k, \qquad \omega_T(t) := \frac{\Omega_{\eta(tT)} - (a_\omega tT)/a_\tau}{\sigma\sqrt{T}}, \qquad t \in [0, 1],$$

where ω_k are independent copies of ω. Since

$$Z_{\eta(t)} = hT_{\eta(t)} + \Omega_{\eta(t)},$$

we have

$$Z_{\eta(t)} - at = hT_{\eta(t)} + \Omega_{\eta(t)} - at = \Omega_{\eta(t)} - (a-h)t + h\chi(t), \tag{7.1.3}$$

where $\chi(t) = T_{\eta(t)} - t$,

$$a_\zeta = ha_\tau + a_\omega, \qquad h = \frac{a_\zeta - a_\omega}{a_\tau}, \qquad a - h = \frac{a_\zeta}{a_\tau} - \frac{a_\zeta - a_\omega}{a_\tau} = \frac{a_\omega}{a_\tau}.$$

If, without loss of generality, we put $a_\tau = 1$, then by (7.1.3) we get

$$y_T(t) = \frac{Z_{\eta(tT)} - atT}{\sigma\sqrt{T}} = \omega_T(t) + \frac{h\chi(tT)}{\sigma\sqrt{T}}. \tag{7.1.4}$$

By the assertion of Theorem 7.1.1 for independent τ and ω (see [56]), there exists a Wiener process $w(t)$ such that

$$\mathbf{P}\Big(\rho(\omega_T, w) \ge \frac{b\ln T + y}{\sqrt{T}}\Big) \le ce^{-\lambda y}, \tag{7.1.5}$$

where ρ is the uniform metric. Next,

$$\rho(y_T, w) \le \rho(y_T, \omega_T) + \rho(\omega_T, w), \qquad \rho(y_T, \omega_T) = h\max_{t\le T} \frac{\chi(t)}{\sigma\sqrt{T}}. \tag{7.1.6}$$

Hence in order to prove (7.1.1), it suffices in view of (7.1.5) to check that, for some c_1, λ_1, λ^*,

$$\mathbf{P}\Big(\rho(y_T, \omega_T) \ge \frac{b\ln T + y}{\sqrt{T}}\Big) \le c_1 e^{-\lambda^* y} + e^{-\lambda_1 T}. \tag{7.1.7}$$

In view of (7.1.6), the left-hand side of this inequality is majorized by

$$\mathbf{P}(\eta(T) > 2T) + \mathbf{P}(\max_{k\le 2T} \tau_k > \sigma(b\ln T + y)/h)$$
$$\le \mathbf{P}(T_{[2T]} < T) + 2T\mathbf{P}(\tau > \sigma(b\ln T + y)/h), \tag{7.1.8}$$

where, for some $\lambda_1 > 0$, $\mu > 0$,

$$\mathbf{P}(T_{[2T]} < T) \le e^{-\lambda_1 T}, \quad \mathbf{P}\Big(\tau > \frac{\sigma(b\ln T + y)}{h}\Big) \le me^{-\mu\sigma(b\ln T + y)/h},$$
$$m = \mathbf{E}e^{\mu\tau} < \infty.$$

If b is chosen so as to have $\mu\sigma b/h > 1$, then the second summand on the right of (7.1.8) does not exceed $e^{-y\mu\sigma/h}$ for sufficiently large T. This proves (7.1.7), and therefore, (7.1.1).

Assertion (7.1.2) now follows from (7.1.4), assertion (7.1.2) for independent τ and ζ, and from the fact that

$$\sup_{t\le T} \frac{\chi(t)}{\sqrt{t}} \underset{\text{a.s.}}{\to} 0 \quad \text{as} \quad T\to\infty.$$

Theorem 7.1.1 is proved. □

Note that assertion (7.1.2) is contained in [113].

A similar theorem holds for the normalized processes

$$z_T(t) = \frac{Z(tT) - atT}{\sigma\sqrt{T}}.$$

Theorem 7.1.2 *Under the conditions of Theorem 7.1.1, there exist positive constants b, c, λ, λ_1 such that, for all y and all sufficiently large T,*

$$\mathbf{P}\Big(\sup_{t\in[0,1]} |z_T(t) - w(t)| > \frac{b\ln T}{\sqrt{T}} + \frac{y}{\sqrt{T}}\Big) \le ce^{-\lambda y} + e^{-\lambda_1 T} \tag{7.1.9}$$

(the constants b, c, λ may differ from those in Theorem 7.1.1).

Besides,

$$\rho(z_T, w) = O\Big(\frac{\ln T}{\sqrt{T}}\Big) \quad \text{a.s. as } T\to\infty.$$

Proof We have

$$\rho(z_T, w) \le \rho(z_T, y_T) + \rho(y_T, w), \tag{7.1.10}$$

and so in view of Theorem 7.1.1 it suffices to check that, for some b, c, λ, λ_1, and all sufficiently large T,

$$\mathbf{P}\Big(\rho(z_T, y_T) > \frac{b\ln T + y}{\sqrt{T}}\Big) \le ce^{-y} + e^{-\lambda_1 T}. \tag{7.1.11}$$

By the definition of the processes Y and Z we have

$$\rho(z_T, y_T) \le \frac{\max_{t\le T} |\zeta_{\eta(t)}|}{\sigma\sqrt{T}}. \tag{7.1.12}$$

Now to prove (7.1.9) we follow the same lines as in (7.1.8), where τ_k under the maximum sign should be replaced by $|\zeta_k|$. Theorem 7.1.2 is proved. □

If the Cramér condition is not met, then the strong approximation for CRPs has the following form.

Theorem 7.1.3 *Let τ and ζ be either independent or linearly dependent and let $\mathbf{E}\tau^r < \infty$ and $\mathbf{E}|\zeta|^r < \infty$ for some $r > 2$. Then the processes y_T and w can be constructed on a common probability space to satisfy the inequality*

$$\mathbf{P}\Big(\sup_{t\in[0,1]} |y_T(t) - w(t)| > \frac{y}{\sqrt{T}}\Big) \le \delta_T T y^{-r} \tag{7.1.13}$$

for all y from the interval $(c_1 T^{1/r}, c_2\sqrt{T\ln T})$, $\delta_T = o(1)$ as $T \to \infty$, and some positive constants c_1, c_2.

Besides,

$$\rho(y_T, w) = o(T^{1/r}) \quad \text{a.s. as } T \to \infty.$$

Proof of Theorem 7.1.3 For independent ζ and τ, Theorem 7.1.3 was proved in [56]. Its extension to the case of linearly dependent ζ and τ proceeds as above (see the proof of Theorem 7.1.1) with the use of the representation

$$Y(t) - at = \Omega_{\eta(t)} - \frac{a_\omega t}{a_\tau} + h\chi(t)$$

and estimates of $\chi(t)$. Since τ and ω are independent, it suffices to check that

$$\mathbf{P}\Big(\max_{t\le\eta(T)} \chi(t) > y\Big) = o(Ty^{-r}). \tag{7.1.14}$$

Indeed, by (7.1.8) for $a_\tau = 1$ we have

$$\mathbf{P}\Big(\max_{t\le\eta(T)} \chi(t) > y\Big) \le \mathbf{P}(T_{[2T]} < T) + \mathbf{P}\Big(\max_{j\le 2T} \tau_j > y\Big) \le e^{-\lambda_1 T} + 2To(y^{-r}).$$

This proves (7.1.14). Since (7.1.6) holds, it also proves Theorem 7.1.3. □

For the processes z_T from Theorem 7.1.3 we have the following assertion.

Theorem 7.1.4 *There exist a constant $\lambda_1 > 0$ and $\delta_T = o(1)$ as $T \to \infty$, such that, for the same y as in Theorem 7.1.3,*

$$\mathbf{P}\Big(\rho(z_T, w) > \frac{y}{\sqrt{T}}\Big) \le e^{-\lambda_1 T} + \delta_T T y^{-r}.$$

Proof The argument is the same as in Theorem 7.1.2. Without loss of generality we put $a_\tau = 1$ and then use inequalities (7.1.10), (7.1.12) and Theorem 7.1.3. As a result, we have

$$\mathbf{P}\Big(\rho(z_T, w) > \frac{y}{\sqrt{T}}\Big) \le e^{-\lambda_1 T} + \mathbf{P}\Big(\max_{k \le 2T} |\zeta_k| > \sigma y\Big),$$

where the second summand on the right in view of Chebyshev's inequality is majorized by $2To(y^{-r})$. Theorem 7.1.4 is proved. □

7.2 Extension of the Invariance Principle to the Zone of Moderately Large Deviations

The invariance principle for CRPs in the normal deviation zone was established in [55, 112]; see also §1.5.

Assume that condition [**C**] is met. In §4.8.3 it was shown that if a set $B \subset \mathfrak{B}(\mathbb{D}, \rho)$ (ρ is the uniform metric) satisfies the conditions

$$\begin{aligned} \mathrm{I}_0((B)) &= \mathrm{I}_0([B]) > 0, \\ \mathfrak{W}((vB)) &= \mathfrak{W}([vB]) > 0 \quad \text{for any fixed } v > 0, \end{aligned}$$

where $\mathfrak{W}(\cdot)$ is the measure corresponding to the Wiener process $w(t)$, $t \in [0, 1]$, then the asymptotic equivalence

$$\ln \mathbf{P}(y_T \in xB) \sim \ln \mathfrak{W}(xB) \quad \text{as} \quad T \to \infty$$

holds as $T \to \infty$ for all

$$x \ge x_0 > 0, \qquad x = o(\sqrt{T}). \tag{7.2.1}$$

This is an extension of the "rough" (logarithmic) invariance principle to the zone of deviations (7.2.1), which includes the normal and moderately large deviations $x \to \infty$, $x = o(\sqrt{T})$ as $T \to \infty$.

Now let us find under what conditions on a set B and the deviations x the asymptotic equivalence

$$\mathbf{P}(y_T \in xB) \sim \mathbf{P}(w \in xB) = \mathfrak{W}(xB) \quad \text{as} \quad T \to \infty \tag{7.2.2}$$

holds for the *sharp* (rather than "rough") asymptotics and, in addition, the same equivalence holds for the process $z_T(t)$. The following assertion holds.

Theorem 7.2.1 *Let ζ and τ be either independent or linearly dependent, let*

$$0 < x_0 \le x = o(T^{1/6}) \quad \textit{as} \quad T \to \infty$$

and let a set $B \in \mathfrak{B}(\mathbb{D}, \rho)$ satisfy the conditions

$$\mathrm{I}_0(B) := \inf_{g \in B} \int_0^1 (g'(t))^2 dt < \infty,$$

and

$$[\mathbf{B}_\infty] \qquad \mathfrak{W}((x\partial B)_\varepsilon) = o(\mathfrak{W}(xB)) \text{ as } \varepsilon = o(1/x) \text{ and } T \to \infty. \tag{7.2.3}$$

Then (7.2.2) *holds.*

The same assertion holds for the processes z_T.

Proof If $x > 0$ is fixed, then condition $[\mathbf{B}_\infty]$ means that as $\varepsilon \to 0$

$$\mathfrak{W}((x\partial B)_\varepsilon) = o(1).$$

This is equivalent to saying that $\mathfrak{W}(x\partial B) = 0$ and now (7.2.2) follows from the invariance principle for CRPs.

Assume now that $x \to \infty$ as $T \to \infty$. By Theorem 7.1.1, for any sequence $\varepsilon = \varepsilon_T \gg \frac{\ln T}{\sqrt{T}}$ as $T \to \infty$ for some $\lambda > 0$, one has

$$\mathbf{P}(\rho(y_T, w) \geq \varepsilon) \leq e^{-\lambda\varepsilon\sqrt{T}} \tag{7.2.4}$$

(here one should put $\frac{y}{\sqrt{T}} = \varepsilon$ in (7.1.1) for $y \gg \ln T$ and slightly reduce λ).

Using this, we can obtain an *upper estimate* for $\mathbf{P}(y_T \in xB)$. Indeed, by (7.2.4) and in view of condition $[\mathbf{B}_\infty]$, we have

$$\begin{aligned}\mathbf{P}(y_T \in xB) &\leq \mathbf{P}(y_T \in xB,\ \rho(y_T, w) < \varepsilon) + \mathbf{P}(\rho(y_T, w) \geq \varepsilon)\\ &\leq \mathbf{P}(w \in (xB)_\varepsilon) + e^{-\lambda\varepsilon\sqrt{T}} \leq \mathbf{P}(w \in xB)(1 + o(1)) + e^{-\lambda\varepsilon\sqrt{T}} \quad \text{as} \quad T \to \infty.\end{aligned} \tag{7.2.5}$$

We set

$$x = \delta T^{1/6}, \quad \varepsilon = \delta/x \quad \text{as} \quad \delta = \delta_T \to 0, \quad T \to \infty$$

(so that $\varepsilon \gg \frac{\ln T}{\sqrt{T}}$). In this case, in view of (7.2.3) and using the moderately large deviation principle for the process w with $c = \frac{I_0(B)}{2} < \infty$, we get

$$\ln \mathbf{P}(w \in xB) \sim -\frac{x^2}{2} I_0(B) = -c\delta^2 T^{1/3}. \tag{7.2.6}$$

On the other hand, in (7.2.5)

$$\lambda\varepsilon\sqrt{T} = \frac{\lambda\delta\sqrt{T}}{x} = \lambda T^{1/3} \gg \delta^2 T^{1/3} \quad \text{as} \quad T \to \infty. \tag{7.2.7}$$

Hence in view of (7.2.5) and (7.2.6),

$$\mathbf{P}(y_T \in xB) \leq \mathbf{P}(w \in xB)(1 + o(1)). \tag{7.2.8}$$

Let us now derive a *lower estimate*. Setting $(B)_{-\varepsilon} = B \setminus (\partial B)_\varepsilon$, we have, under condition $[\mathbf{B}_\infty]$,

$$\begin{aligned}\mathbf{P}(y_T \in xB) &\geq \mathbf{P}(y_T \in xB,\ \rho(y_T, w) < \varepsilon)\\ &\geq \mathbf{P}(w \in (xB)_{-\varepsilon}) - \mathbf{P}(\rho(y_T, w) \geq \varepsilon) \geq \mathbf{P}(w \in xB)(1 + o(1)) - e^{-\lambda\varepsilon\sqrt{T}}.\end{aligned} \tag{7.2.9}$$

Therefore, similarly to the above (see (7.2.6) and (7.2.7)), we obtain

$$\mathbf{P}(y_T \in xB) \geq \mathbf{P}(w \in xB)(1 + o(1)).$$

This together with (7.2.8) gives (7.2.2).

For the processes z_T one should apply the above arguments with the use of Theorem 7.1.2. Theorem 7.2.1 is proved. □

It is unknown at present whether in Theorem 7.1.3 one can relax condition [**C**] to condition [$\mathbf{C}_V$] (to answer this question one needs to find an analog of Theorem 7.1.1 under condition [$\mathbf{C}_V$]).

If $\tau \equiv 1$, then relation (7.2.2) extends the invariance principle for the random walks $\{Z_k\}_{k=1}^n$ to the zone of moderately large deviations $x \to \infty$, $x = o(n^{1/6})$ as $n \to \infty$. In general, the boundary $o(n^{1/6})$ is the best possible, as follows from the consideration of the distribution $z_n(1) = \frac{Z_n - a_\zeta n}{\sigma_\zeta \sqrt{n}}$ in the case when $m_3 := \mathbf{E}(\zeta - a_\zeta)^3 \neq 0$. In this case,

$$\mathbf{P}(z_n(1) > n^{1/6}) \sim [1 - \Phi(n^{1/6})] \exp\left\{-\frac{m_3}{6\sigma_\zeta^4}\right\} \quad \text{as} \quad n \to \infty$$

(see (9.1.3), (9.3.7), (9.3.12), and (9.3.13) in [21]).

7.3 First Boundary Crossing Problem in the Zone of Moderately Large Deviations

Recall that the first boundary crossing problem with a function $g(t) \in \mathbb{D}(0,1)$, $g(t) > 0$, is concerned with the asymptotic behavior of the probabilities $\mathbf{P}(y_T \in xB_g)$, $\mathbf{P}(z_T \in xB_g)$ as $x \to \infty$, where

$$xB_g = \left\{f \in \mathbb{D}(0,1) : \sup_{t \leq 1} (f(t) - xg(t)) > 0\right\}.$$

In Chapter 6, the asymptotics of probabilities of this type were studied in the zone of "properly large" deviations when $x \sim c\sqrt{T}$ as $T \to \infty$. We now consider the case of moderately large deviations $x = o(T^{1/6})$.

To be able to use Theorem 7.2.1 in these circumstances, we need to check whether the sets xB_g satisfy condition [$\mathbf{B}_\infty$]. As in §4.6, we consider the level lines $\mathbf{l}_x(t)$, but now only for the Wiener process

$$\mathbf{l}_x(t) = x\sqrt{t}, \qquad t > 0$$

($\ln \mathbf{P}(w(t) > \mathbf{l}_x(t)) \sim -\frac{(\mathbf{l}_x(t))^2}{2t} = -\frac{x^2}{2}$ does not depend on t). Let v_g be the value of v at which the function $v\sqrt{t}$ first touches the curve $g(t)$ as v increases:

$$v_g := \sup\left\{v : \sup_{t \leq 1} (g(t) - v\sqrt{t}) > 0\right\}.$$

Theorem 7.3.1 *Let τ and ζ be either independent or linearly dependent, let $0 < x_0 < x = o(T^{1/6})$ as $T \to \infty$, and let at least one of the following (interrelated) conditions be satisfied:*

(a) *The function $v_g\sqrt{t}$ touches the curve $g(t)$ at a unique point t_g lying inside the interval $(0, 1)$. For sufficiently small Δ, the function $g(t)$ satisfies, in the Δ-neighborhood of the tangent point t_g of these curves, the Lipschitz condition. Besides, $g(t) > v_g\sqrt{t} + \delta$, $\delta = \delta(\Delta) > 0$, for $t \notin (t_g)_\Delta$, for sufficiently small Δ.*

(b) *The asymptotic formula*

$$\mathbf{P}(w \in xB_g) \sim V(x)\,\mathrm{e}^{-\frac{x_g^2}{2}} \quad \text{for} \quad x_g = v_g x, \quad x \to \infty \tag{7.3.1}$$

holds, where $V(x)$ is a regularly varying function (for example, a power-like function).

Then the set xB_g satisfies condition $[\mathbf{B}_\infty]$ of Theorem 7.2.1, and

$$\mathbf{P}(y_T \in xB_g) \sim \mathbf{P}(w \in xB_g) \sim \mathbf{P}(z_T \in xB_g) \quad as \quad x \to \infty.$$

Remark 7.3.2 From the proof of Theorem 7.3.1 we shall see that condition (a) can be extended to the case of several tangent points when the interval $(0, 1)$ can be split into a finite number of intervals or half-open intervals on each of which there is one tangency point and condition (a) holds

A sufficient condition for (b) to hold is the existence of the second derivative $g''(t)$ near the tangency point. Indeed, let us use the fact that the problem on the asymptotic behavior of $\mathbf{P}(w \in xB_g)$ can be "reduced" to the same problem for the random walk $w_n(t) = w\left[\frac{[nt]}{n}\right]$ with normally distributed jumps. Namely,

$$\mathbf{P}(\rho(w, w_n) > \varepsilon) \le 2n\,\mathbf{P}(w(1) > \varepsilon\sqrt{n}) \le 2ne^{-\frac{\varepsilon^2 n}{2}}. \tag{7.3.2}$$

Next, according to [14] for $x_g = v_g x$, $x = o(n)$ we have

$$\mathbf{P}(w_n \in xB_g) \sim V(x)e^{\frac{x_g^2}{2}} \quad \text{as} \quad x \to \infty, \tag{7.3.3}$$

where the function $V(x)$ has a power-law behavior, and the power exponent depends on the number of equal derivatives of the functions $g(t)$ and $v_g\sqrt{t}$ at the point t_g. Hence if ε and n are chosen so that

$$\varepsilon = o\left(\frac{1}{x}\right), \qquad \varepsilon^2 n \gg x^2 \quad \text{as} \quad x \to \infty, \tag{7.3.4}$$

then from (7.3.3) and (7.3.2) we get

$$\mathbf{P}(w \in xB_g) \le \mathbf{P}(w \in xB_g,\ \rho(w, w_n) < \varepsilon) + r_{x,n} \le \mathbf{P}(w_n \in (xB_g)_\varepsilon) + r_{x,n},$$

where

$$r_{x,n} \ll V(x)e^{-\frac{x_g^2}{2}}.$$

Next, we have $xg(t) - \varepsilon \geq (x - \varepsilon/\underline{g})g(t)$, where $\underline{g} = \min_{t\in[0,1]} g(t)$, and so for $\varepsilon = o(1/x)$ from (7.3.3) we obtain

$$\mathbf{P}(w_n \in (xB_g)_\varepsilon) \leq V\Big(x - \frac{\varepsilon}{\underline{g}}\Big) \exp\Big\{-\frac{(x_g - \varepsilon/\underline{g})^2}{2}\Big\} \sim V(x)e^{-\frac{x_g^2}{2}}$$

as $x \to \infty$. The proof of the lower estimate is similar. This proves (7.3.1).

Proof of Theorem 7.3.1 Let condition (a) be met. Setting

$$\eta_{xg} = \inf\{t : w(t) > xg(t)\},$$

we show that

$$\mathbf{P}(\eta_{xg} \notin (t_g)_\Delta, \eta_{xg} \leq 1) = o(\mathbf{P}(\eta_{xg} \in (t_g)_\Delta) \quad \text{as} \quad x \to \infty. \tag{7.3.5}$$

Let

$$l(t) = v_g\sqrt{t} + \delta, \qquad A = [0, 1] \setminus (t_g)_\Delta.$$

It is clear that

$$\mathbf{P}(\eta_{xg} \notin (t_g)_\Delta, \eta_{xg} \leq 1) = \mathbf{P}(\eta_{xg} \in A) \leq \mathbf{P}(\eta_{xl} \in A) \leq \mathbf{P}(\eta_{xl} \leq 1).$$

By the LDP for processes with independent increments (see [23, §5.3]),

$$\begin{aligned} \ln \mathbf{P}(\eta_{xl} \leq 1) &\sim -\frac{x^2(v_g + \delta)^2}{2}, \\ \ln \mathbf{P}(\eta_{xg} \leq 1) &\sim -\frac{x^2 v_g^2}{2} \end{aligned} \qquad \text{as} \quad x \to \infty. \tag{7.3.6}$$

This proves (7.3.5). Let us now show that condition $[\mathbf{B}_\infty]$ is satisfied. We have

$$\begin{gathered} \mathbf{P}(w \in (xB_g)_{-\varepsilon}) = \int_{t_g-\Delta}^{t_g+\Delta} \mathbf{P}(\eta_{xg} \in dt)\mathbf{P}\Big(\sup_{u\leq 1-t} (w(t+u) - xg(t+u)) > \varepsilon\Big) + r_x, \\ r_x \leq \mathbf{P}(\eta_{xg} \in A). \end{gathered} \tag{7.3.7}$$

To verify that the second factor in the integrand in (7.3.7) converges to 1 as $x \to \infty$, we need the following assertion, which will be proved a little later.

Lemma 7.3.3 *If $b > 0$, then*

$$\mathbf{P}\Big(\sup_{u\geq 0} (w(u) - bu) > \varepsilon\Big) = e^{-2b\varepsilon}.$$

For $t \in (t_g)_\Delta$, $|u| < \Delta$, and a sufficiently small Δ, we have by the Lipschitz condition

$$|g(t + u) - g(t)| \leq qu, \qquad q < \infty.$$

Besides, for $\eta_{xg} \in dt$ we have $w(t+u) \underset{p}{=} xg(t) + w(u)$. Hence by Lemma 7.3.3 for the chosen t and Δ,

$$\begin{aligned}
&\mathbf{P}\Big(\sup_{u\le 1-t}\,(w(t+u) - xg(t+u)) > \varepsilon\Big) \\
&\quad\ge \mathbf{P}\Big(\sup_{u\le\Delta}\,\{w(u) - x[g(t+u) - g(t)]\} > \varepsilon\Big) \ge \mathbf{P}\Big(\sup_{u\le\Delta}\,(w(u) - xqu) > \varepsilon\Big) \\
&\qquad = e^{-2\varepsilon xq} - \mathbf{P}\Big(\sup_{u\ge\Delta}\,(w(u) - xqu) > \varepsilon\Big), \qquad (7.3.8)
\end{aligned}$$

where q is the coefficient from the Lipschitz condition. Since $\varepsilon = o(1/x)$, the first summand on the right of (7.3.8) converges to 1 as $x \to \infty$. It is clear that the second summand converges to 0 for any fixed $\Delta > 0$ and $x \to \infty$ (for example, by the LDP).

So, in view of (7.3.5), (7.3.7) we have

$$\mathbf{P}(w \in (xB_g)_{-\varepsilon}) = \mathbf{P}(\eta_{xg} \in (t_g)_\Delta)(1 + o(1)) = \mathbf{P}(w \in xB_g)(1 + o(1)),$$
$$\mathbf{P}(w \in xB_g) = \mathbf{P}(w \in (xB_g)_\varepsilon)(1 + o(1)).$$

as $x \to \infty$. This proves that condition $[\mathbf{B}_\infty]$ is satisfied and therefore establishes Theorem 7.3.1 under condition (a).

The assertion of the theorem under condition (b) clearly holds, because

$$\mathbf{P}(w \in (xB_g)_\varepsilon) \le V(x - \varepsilon\overline{g})e^{-\frac{(x-\varepsilon\overline{g})^2 v_g^2}{2}}, \quad \text{where} \quad \overline{g} = \max_{t\in[0,1]} g(t),$$

$$\mathbf{P}(w \in xB_g) \sim V(x)e^{-\frac{x_g^2}{2}} \quad \text{as} \quad x \to \infty.$$

Hence condition $[\mathbf{B}_\infty]$ is satisfied, since

$$\mathbf{P}(w \in (\partial xB_g)_\varepsilon) = o(\mathbf{P}(w \in xB_g)) \quad \text{for} \quad \varepsilon = o(1/x), \quad x \to \infty.$$

Theorem 7.3.1 is proved. □

It now remains to prove Lemma 7.3.3.

Proof of Lemma 7.3.3 The process $w(t) - bu$ is upper continuous, and hence for $b > 0$,

$$\mathbf{P}\Big(\sup_{t\ge 0}\,(w(t) - bu) > \varepsilon\Big) = e^{-\lambda(b)\varepsilon}$$

for some $\lambda(b) > 0$. The Laplace transform of this distribution is

$$\lambda(b)\int_0^\infty e^{\lambda v - \lambda(b)v}\,dv = \frac{\lambda(b)}{\lambda(b) - \lambda}.$$

Using the factorization identities in [103], we conclude that the pole $\lambda = \lambda(b)$ of this transform coincides with the rightmost pole $\lambda = 2b$ of the function $\frac{1}{\psi(\lambda)-1}$, where

$$\psi(\lambda) = \mathbf{E}e^{\lambda(w(1)-b)} = \exp\Big\{\frac{\lambda^2}{2} - b\lambda\Big\}.$$

Hence $\lambda(b) = 2b$, proving Lemma 7.3.3. □

7.4 Extension of the Invariance Principle for Lipschitz Functionals to the Zone of Moderately Large Deviations

Consider the class $\mathcal{F}$ of functionals $F(f)$, $f \in \mathbb{D}(0, 1)$, measurable with respect to $\mathfrak{B}(\mathbb{D}, \rho)$ and satisfying the following conditions:

(a) $|F(f) - F(g)| < c\rho(f, g)$ (the Lipschitz condition);

(b) $P(x) := \mathbf{P}(F(w) > x) \geq e^{-cx^2}$ for some $c > 0$, $x \to \infty$;

(c) $P(x + o(1/x)) \sim P(x)$ as $x \to \infty$.

For example, the class $\mathcal{F}$ contains the functionals

$$\max_{t\in[0,1]} L(f(t)), \quad \max_{t\in[0,1]} \frac{f(t)}{\max(1, L(f(t)))}, \quad \int_0^1 L(f(t))dG(t) \quad \text{etc.,} \tag{7.4.1}$$

where G is a function of bounded variation, a function L satisfies the Lipschitz condition, and $L(v) \sim bv$, $b > 0$ as $v \to \infty$.

Theorem 7.4.1 *Let τ and ζ be either independent or linearly dependent, let condition* [**C**] *be met, and let $F \in \mathcal{F}$. Then, for $x \geq x_0 > 0$, $x = o(T^{1/6})$ as $T \to \infty$,*

$$\mathbf{P}(F(y_T) > x) \sim \mathbf{P}(F(w) > x). \tag{7.4.2}$$

Proof The *upper estimate*. By Theorem 7.1.1,

$$\begin{aligned}\mathbf{P}(F(y_T) > x) &\leq \mathbf{P}(F(y_T) > x,\ \rho(y_T, w) < \varepsilon) + \mathbf{P}(\rho(y_T, w) \geq \varepsilon)\\ &\leq \mathbf{P}(F(w) > x - c\varepsilon) + O(e^{-\lambda\varepsilon\sqrt{T}}).\end{aligned} \tag{7.4.3}$$

Here by conditions (b), (c) the first summand on the right of (7.4.3) is asymptotically equivalent to

$$P(x) \geq e^{-cx^2}.$$

On the other hand, for an appropriately chosen (see the proof of Theorem 7.2.1) sequence $\varepsilon = o(T^{1/6})$, $x \gg \frac{\ln T}{T}$, we have

$$e^{-\lambda\varepsilon\sqrt{T}} \ll e^{-cx^2}. \tag{7.4.4}$$

It follows from (7.4.4) that

$$\mathbf{P}(F(y_T) > x) \leq P(x)(1 + o(1)) \quad \text{as} \quad T \to \infty.$$

The *lower estimate*. A similar analysis shows that

$$\mathbf{P}(F(y_T) > x) \geq \mathbf{P}(F(y_T) > x,\ \rho(y_T, w) < \varepsilon) \geq \mathbf{P}(F(y_T) > x + c\varepsilon) + O(e^{-\lambda\varepsilon\sqrt{T}}).$$

Repeating the above arguments and using (7.4.4), we get (7.4.2). Theorem 7.4.1 is proved. □

If the function $L(v)$ in (7.4.1) grows slower than bv as $v \to \infty$, then the order of smallness of ε, and therefore the zone of admissible deviations of x, should be corrected.

7.5 Extension of the Invariance Principle to the Zone of Moderately Small Deviations

Consider now the problem on the asymptotic behavior of *small* deviations, that is, the probabilities

$$\mathbf{P}(y_T \in xB) \to 0 \quad \text{as} \quad x \to 0.$$

For this setting, at present there are no general results of LDP type. Only some partial results are known, which show that the asymptotics of these probabilities can differ depending on the nature of the sets B. Keeping in mind, first of all, the boundary crossing problems of the second type, we single out two types of sets.

I. The first-type sets are such that

$$\ln \mathbf{P}(w \in xB) > -\frac{c}{x^2} \quad \text{with some } c > 0 \quad \text{and } x \to 0. \tag{7.5.1}$$

According to [86], [87], this type of sets B includes the sets in boundary crossing problems of the form

$$B = \{f \in \mathbb{D}(0,1) : g_1(t) < f(t) < g_1(t)\}, \qquad t \in [0,1],$$

where $g_i \in \mathbb{C}(0,1)$, $i = 1, 2$; $g_2(0) < 0 < g_1(0)$. In this case, in (7.5.1) we have

$$c = \frac{\pi^2}{2} \int_0^1 (g_1(t) - g_2(t))^{-2}.$$

II. The second-type sets, which describe "one-sided small deviations," include "one-sided" sets in boundary crossing problems, for example, the sets

$$B = \{f \in \mathbb{D}[0,1) : \sup_{t\in[0,1]} (f(t) - g(t)) < 0\}, \quad g \in \mathbb{C}(0,1), \quad g(t) > 0 \text{ for } t \le t_0 > 0$$

(here we have constraints only on the positive values of $f(t)$). For such sets, for $x < x_0 = \text{const}$,

$$\mathbf{P}(w \in xB) > cx, \qquad c = \text{const}. \tag{7.5.2}$$

For example, for $B = \{f : \sup_{t\in[0,1]} f(t) < 1\}$ we have

$$\mathbf{P}(w \in xB) \sim \sqrt{\frac{2}{\pi}}\, x \quad \text{as} \quad x \to 0. \tag{7.5.3}$$

7.5.1 Extension of the Invariance Principle for the First-Type Sets

Theorem 7.5.1 *Let τ and ζ be either independent or linearly dependent, let condition* [**C**] *and relation* (7.5.1) *hold, and let, for some $x_0 > 0$,*

$$x_0 > x \gg T^{-1/6} \quad as \quad T \to \infty.$$

Next, assume that a set B satisfies the condition

$$[\mathbf{B}_0] \qquad \mathfrak{W}((x\partial B)_\varepsilon) = o(\mathfrak{W}(xB)) \quad as \quad \frac{\varepsilon}{x} \to 0. \tag{7.5.4}$$

Then

$$\mathbf{P}(y_T \in xB) \sim \mathfrak{W}(xB) \quad as \quad T \to \infty. \tag{7.5.5}$$

The same assertion holds for the processes z_T as well.

One can note that the conclusion of Theorem 7.5.1 features a certain "inverse symmetry" with respect to Theorem 7.2.1 ($T^{1/6}$ is replaced by $T^{-1/6}$, and $\varepsilon = o(1/x)$ is replaced by $\varepsilon = o(x)$). Besides, the asymptotics e^{-cx^2} in (7.3.1) is replaced by the asymptotics $e^{-cx^{-2}}$ in (7.5.1).

Proof of Theorem 7.5.1 If x is fixed, then the required result is given by Theorem 7.2.1. Assume that $x \to 0$.

I. The *upper estimate*. If $x \to 0$, $\varepsilon \gg \frac{\ln T}{\sqrt{T}}$, then relation (7.2.5) holds, where it is now assumed that $\varepsilon = o(x)$ (to be able to use condition $[\mathbf{B}_0]$). We set

$$x = NT^{-1/6}, \qquad N \to \infty.$$

Then in relation (7.2.5) we have, in view of (7.5.1),

$$\mathbf{P}(w \in xB) > \exp\{-cx^{-2}\} = \exp\left\{-\frac{cT^{1/3}}{N^2}\right\}. \tag{7.5.6}$$

On the other hand, if $\varepsilon = T^{-1/6} = o(x)$, then by (7.5.6) we have

$$e^{-\lambda\varepsilon\sqrt{T}} = e^{-\lambda T^{1/3}} \ll \mathbf{P}(w \in xB) \quad \text{as} \quad T \to \infty. \tag{7.5.7}$$

In view of (7.2.5), this means that

$$\mathbf{P}(y_T \in xB) \le \mathbf{P}(w \in xB)(1 + o(1)) \quad \text{as} \quad T \to \infty. \tag{7.5.8}$$

The *lower estimate*, like the upper one, is similar to the estimate in Theorem 7.2.1. By condition $[\mathbf{B}_0]$, (7.2.9) remains valid with $\varepsilon = o(x)$. From (7.2.9) in view of (7.5.6) and (7.5.7) we have

$$\mathbf{P}(y_T \in xB) \ge \mathbf{P}(w \in xB)(1 + o(1)).$$

To carry over the theorem to the processes z_T one should apply Theorem 7.1.2. Theorem 7.5.1 is proved. □

7.5.2 Extension of the Invariance Principle to the Zone of Small Deviations for Second-Type Sets

Theorem 7.5.2 *Let τ and ζ be either independent or linearly dependent, and let a second-type set B (see (7.5.2)) satisfy condition $[\mathbf{B}_0]$. Then*

I. *If the Cramér condition* $[\mathbf{C}]$ *is met, then, for $x \le x_0 > 0$, $x \gg \frac{\ln T}{\sqrt{T}}$ as $T \to \infty$,*

$$\mathbf{P}(y_T \in xB) \sim \mathfrak{W}(xB). \tag{7.5.9}$$

II. *If, for some* $r > 2$,

$$\mathbf{E}\tau^r < \infty, \qquad \mathbf{E}|\zeta|^r < \infty, \tag{7.5.10}$$

then there exists a sequence $\gamma = \gamma_T \to 0$ *as* $T \to \infty$ *such that* (7.5.9) *holds for*

$$x_0 \geq x > \gamma T^{\frac{2-r}{2(r+1)}}.$$

Assertions I, II *hold for the processes* z_T *as well.*

For example, (7.5.9) always holds for $r = 3$ and $x_0 \geq x \geq T^{-1/8}$. For $r = 4$, this relation holds for $x \geq T^{-1/5}$, and so on.

Proof I. The *upper estimate* under condition [**C**]. We set

$$x \leq x_0, \qquad x = \frac{N \ln T}{\sqrt{T}}, \quad N \to \infty \quad \text{as} \quad T \to \infty.$$

Then for

$$\varepsilon = \sqrt{N} \frac{\ln T}{\sqrt{T}} = o(x)$$

by (7.2.5) we have

$$\mathbf{P}(w \in xB) > cx, \tag{7.5.11}$$

$$e^{-\lambda\varepsilon\sqrt{T}} = e^{-\lambda\sqrt{N}\ln T} = T^{-\lambda\sqrt{N}} \ll \frac{N \ln T}{\sqrt{T}} = x. \tag{7.5.12}$$

Hence (7.5.8) holds.

The *lower estimate* follows from (7.2.9), (7.5.11), and (7.5.12). This proves (7.5.9).

II. The *upper estimate* under (7.5.10). Using Theorem 7.1.3 with $\varepsilon = \frac{y}{\sqrt{T}}$ in (7.1.13), we get

$$\mathbf{P}(\rho(y_T, w) > \varepsilon) \leq \delta_T T^{1-r/2} \varepsilon^{-r} \tag{7.5.13}$$

with

$$\varepsilon \in (c_1 T^{\frac{2-r}{2r}}, c_2\sqrt{\ln T}). \tag{7.5.14}$$

Further, similarly to (7.2.9) by condition [$\mathbf{B}_0$] we have

$$\begin{aligned} \mathbf{P}(y_T \in xB) &\leq \mathbf{P}(y_T \in xB,\ \rho(y_T, w) < \varepsilon) + \mathbf{P}(\rho(y_T, w) \geq \varepsilon) \\ &\leq \mathbf{P}(w \in (xB)_\varepsilon) + \delta_T T^{1-r/2} \varepsilon^{-r} \\ &\leq \mathbf{P}(w \in xB)(1 + o(1)) + \delta_T T^{1-r/2} \varepsilon^{-r}. \end{aligned} \tag{7.5.15}$$

We set

$$\varepsilon = \delta_T^{\frac{1}{2r}} x = o(1) \quad \text{as} \quad T \to \infty. \tag{7.5.16}$$

Then for

$$x_0 \geq x \geq \gamma T^{\frac{2-r}{2(r+1)}}, \qquad \gamma = \delta_T^{\frac{1}{2(r+2)}}, \tag{7.5.17}$$

the right-hand side in (7.5.13) equals

$$\sqrt{\delta_T}T^{1-r/2}x^{-r} \le \sqrt{\delta_T}\gamma^{-r}T^{\frac{2-r}{2(r+1)}} \ll x \quad \text{as} \quad T \to \infty, \tag{7.5.18}$$

and so, in view of (7.5.15),

$$\mathbf{P}(y_T \in xB) \le \mathbf{P}(w \in xB)(1 + o(1)).$$

It is clear that the so-chosen value of x is consistent with (7.5.16) and (7.5.14).

The *lower estimate* under (7.5.10). Similarly to (7.2.9), we have

$$\mathbf{P}(y_T \in xB) \ge \mathbf{P}(w \in xB)(1 + o(1)) - \delta_T T^{1-r/2}\varepsilon^{-r}. \tag{7.5.19}$$

The required lower bound with the above chosen values of ε and x (see (7.5.16), (7.5.17)) follows from (7.5.19) and (7.5.18).

To carry over these results to the process z_T one should employ Theorems 7.1.2 and 7.1.4. Theorem 7.5.2 is proved. □

7.5.3 Second Boundary Crossing Problem in the Small Deviation Zone

In the second boundary crossing problem *with two boundaries*, the asymptotics of $\mathbf{P}(y_T \in xB_{g_1g_2})$ for the process y_T in the zone of small deviations is studied, where

$$xB_{g_1g_2} = \{f \in \mathbb{D}(0,1) : xg_1(t) < f(t) < xg_2(t) \ \text{ for all } t \in [0,1]\}$$

as $x \to 0$. We set

$$g(t) = g_2(t) - g_1(t).$$

Theorem 7.5.3 *Let τ and ζ be either independent or linearly dependent, and let functions of bounded variation $g_2(t) > g_1(t)$ on $[0,1]$ satisfy the following conditions:*

(1) $g_2(0) > 0 > g_1(0)$, $\inf_{t\in[0,1]} g(t) > 0$;

(2) *for some function $\delta(\Delta) \downarrow 0$ as $\Delta \downarrow 0$ the moduli of continuity $\omega_1(\Delta)$, $\omega_2(\Delta)$ of the functions g_1, g_2 satisfy the inequality*

$$\omega_k(\Delta) \le \delta(\Delta)\sqrt{\Delta}, \qquad k = 1, 2. \tag{7.5.20}$$

Then, for $x \le x_0$ with some $x_0 > 0$, $x \gg T^{-1/6}$ as $T \to \infty$,

$$\mathbf{P}(y_T \in xB_{g_1g_2}) \sim \mathbf{P}(w \in xB_{g_1g_2}), \tag{7.5.21}$$

where, as $x \to 0$,

$$\begin{aligned} \mathbf{P}(w \in xB_{g_1g_2}) &\sim c\exp\Big\{-\frac{\pi^2}{2x^2}\int_0^1 g^{-2}(u)\,du\Big\}, \\ c &= \frac{4}{\pi}\sqrt{\frac{g(1)}{g(0)}}\cos\Big(\frac{\pi}{2}\,\frac{(g_1(0)+g_2(0))}{g(0)}\Big). \end{aligned} \tag{7.5.22}$$

The same assertion holds for the processes z_T as well.

Proof Formula (7.5.22) under conditions (1), (2) was established in [86]. The asymptotic equivalence (7.5.21) follows from the fact that

$$P(x) := \mathbf{P}(w \in xB_{g_1 g_2}) \sim ce^{-bx^{-2}}, \qquad b = \text{const},$$

as $x \to 0$, $\varepsilon = o(x)$ (the set $B_{g_1 g_2}$ is a first-type set; see (7.5.1)) and since condition $[\mathbf{B}_0]$ is met. Indeed,

$$(x+\varepsilon)^{-2} = x^{-2}\left(1+\frac{\varepsilon}{x}\right)^{-2} = x^{-2} - \frac{2\varepsilon}{x} + o\left(\frac{\varepsilon}{x}\right) = x^{-2} + o(1)$$

as $x \to 0$, $\varepsilon = o(x)$. This means that $P(x+\varepsilon) \sim P(x)$. Therefore, condition $[\mathbf{B}_0]$ of Theorem 7.5.1 is satisfied. Hence (7.5.21) also holds. Theorem 7.5.3 is proved. □

Consider now the second boundary crossing problem with *one* boundary, namely, let us study the asymptotic behavior of $\mathbf{P}(y_T \in xB_g)$, where

$$xB_g = \left\{ f \in \mathbb{D}(0,1) : \sup_{t\in[0,1]} (f(t) - xg(t)) < 0 \right\}, \qquad g(0) > 0.$$

These sets are second-type sets: for such sets the inequality $\mathbf{P}(w \in xB_g) > cx$ holds as $x \to 0$ (see (7.5.2)).

Theorem 7.5.4 *Let τ and ζ be either independent or linearly dependent, let a function $g > 0$ satisfy the Lipschitz condition, $g(0) > 0$, and let*

$$P(x) := \mathbf{P}(w \in xB_g) \sim cx \quad \textit{as} \quad x \to 0. \tag{7.5.23}$$

Then the following assertions hold.

I. *If the Cramér condition* $[\mathbf{C}]$ *is met, then*

$$\mathbf{P}(y_T \in xB_g) \sim \mathbf{P}(w \in xB_g) \quad \textit{for} \quad x \gg \frac{\ln T}{\sqrt{T}}, \qquad T \to \infty. \tag{7.5.24}$$

II. *If $\mathbf{E}\tau^r < \infty$ and $\mathbf{E}|\zeta|^r < \infty$, $r > 2$, then there exists a sequence $\gamma = o(1)$ as $x \to 0$ for which (7.5.24) is true with $x \gg \gamma T^{\frac{2-r}{2(r+1)}}$ as $T \to \infty$.*

Proof I. The first assertion of the theorem follows from assertion I of Theorem 7.5.2. Condition $[\mathbf{B}_0]$ (see (7.5.4)) is satisfied because $xg(t) + \varepsilon \le (x + \varepsilon/\underline{g})g(t)$, where $\underline{g} = \min_{t\in[0,1]} g(t)$, and hence, with $\varepsilon = o(x)$ in view of (7.5.23) we have

$$\left.\begin{aligned} \mathbf{P}(w \in (xB_g)_\varepsilon) &\le P(x + \varepsilon/\underline{g}) \sim P(x), \\ \mathbf{P}(w \in (\partial xB_g)_\varepsilon) &\le P(x + \varepsilon/\underline{g}) - P(x) = o(P(x)) \end{aligned}\right\} \quad \text{as } x \to \infty.$$

II. The second assertion follows similarly from assertion II of Theorem 7.5.2. Theorem 7.5.4 is proved. □

As was already pointed out, condition (7.5.23) is satisfied for the function $g(t) = g = \text{const}$ (see (7.5.3)). It is quite certain that condition (7.5.23) is also satisfied for smooth boundaries $g(t)$ (for $c = \sqrt{\frac{2}{\pi}} \int_0^1 g(t)\,dt$); however, we are not aware of any suitable results.

Appendix A

On Boundary Crossing Problems for Compound Renewal Processes when the Cramér Condition Is Not Fulfilled

In Chapters 3–6, we constructed the theory of large deviations of CRPs for which the distribution of jumps satisfies Cramér's moment condition. If the Cramér condition is not satisfied, then the nature of probabilities of large deviations of CRPs becomes quite different – this is well known from the example of random walks. In this case, we need to impose some conditions on the regular behavior of distributions of jumps at infinity. Under these conditions, the probabilities of large deviations for CRPs were studied in Chapter 16 of the book [38], where a rather detailed bibliography can also be found.

In this chapter, we will always assume that the CRPs under consideration are homogeneous. Finding conditions of admissible inhomogeneity does not present any substantial difficulty.

In §A.1, we present results on the asymptotics of the probabilities of large deviations of $\overline{Z}^{(q)} = \sup_{t\geq 0} Z^{(q)}(t)$, $Z^{(q)}(t) = Z(t) + qt$. These results can be easily obtained from the results of [38] on the asymptotic of the distribution of the maximum value of a *random walk* under analogous conditions. In the case when the distributions of jumps vary regularly at infinity, the asymptotics of the distributions of $Z(T)$ and $\overline{Z}^0(T) = \max_{t\leq T}(Z(t) - at)$ will be studied in §A.2. The asymptotics of probabilities of large deviations in the first boundary crossing problem will be considered in §A.3.

Many results of §§A.1–A.3 are given without proofs, since they are either contained in the monograph [38] or published in papers, but their proofs are beyond the scope of the main course of this book.

A.1 Distribution of the Maximal Value on the Whole Half-Axis of a Compound Renewal Process with Drift

A.1.1 Distribution of the Maximal Value of a Compound Renewal Process when the Cramér Condition Is Not Met

In this section, as in §6.1, we consider the asymptotics of the distribution of the variable $\overline{Z}^{(q)} = \sup_{t\geq 0} Z^{(q)}(t)$. The next assertion follows from Theorem 6.1.4.

Corollary A.1.1 *Assume that*

$$Q^{(q)}(x) = \int_x^\infty \mathbf{P}(\zeta + q\tau \geq t)\,dt \tag{A.1.1}$$

is an upper power function (see Definition 6.1.1) *and let* $a_\zeta + qa_\tau < 0$. *In the case* $q > 0$, *we assume in addition that*

$$\mathbf{P}(\tau > x) = o(Q^{(q)}(x)) \quad as \quad x \to \infty. \tag{A.1.2}$$

Then

$$\mathbf{P}(\overline{Z}^{(q)} > x) \sim \frac{Q^{(q)}(x)}{|a_\zeta + qa_\tau|} \quad as \quad x \to \infty. \tag{A.1.3}$$

For condition (A.1.2), see Remark 6.1.5.

Let us indicate a broader class of distributions of $\zeta + q\tau$ for which (A.1.3) holds.

Definition A.1.2 The distribution $F^{(\gamma)}(t) = \mathbf{P}(\gamma \geq t)$ of a random variable $\gamma \geq 0$ is called *subexponential* if the convolution

$$F^{(\gamma)*}(x) = -\int_0^x dF^{(\gamma)}(t) F^{(\gamma)}(x-t)$$

has the property

$$F^{(\gamma)*}(x) \sim 2F^{(\gamma)}(x) \quad \text{as} \quad x \to \infty.$$

The distribution of a random variable γ on $(-\infty, \infty)$ is called subexponential if the distribution of the variable $\max(0, \gamma)$ is also subexponential. A nonincreasing function $G(t)$ on $[0, \infty]$ is called subexponential if $G(t) \sim cF(t)$ as $t \to \infty$ for some $c < \infty$, where $F(t)$ is a subexponential distribution.

It can be shown that semiexponential and upper power distributions (as well as regularly varying ones) are subexponential. For more details on subexponential distributions see, for example, [11] and [38].

Theorem A.1.3 *Let*

$$Q^{(q)}(x) = \int_x^\infty \mathbf{P}(\zeta + q\tau \geq t)dt, \qquad x > 0$$

be a subexponential function and let $a_\zeta + qa_\tau < 0$. *In the case* $q > 0$, *we additionally assume that condition* (A.1.2) *is met. Then* (A.1.3) *holds.*

Proof As in Theorems 6.1.3 and 6.1.4, the proof is based of the results of the distribution of the maximal value of a random walk. Consider, for example, the random walk $\{Z_k\}$ and define

$$\overline{Z}_\infty = \sup_{k \geq 1} Z_k, \qquad Q^{(0)}(x) = \int_x^\infty \mathbf{P}(\zeta \geq t)dt.$$

In this case, the following result holds (see, for example, Theorem 17.7.3 in [21]).

Theorem A.1.4 *Let a function $Q^{(0)}(x)$ be subexponential and let $a_\zeta < 0$. Then*

$$\mathbf{P}(\overline{Z}_\infty \geq x) \sim \frac{Q^{(0)}(x)}{|a_\zeta|} \quad \text{as} \quad x \to \infty.$$

Let us return to the proof of Theorem A.1.3. We first consider the case $q \leq 0$. As already pointed out in §6.1, in this case

$$\overline{Z}^{(q)} \underset{d}{=} \overline{Z}_\infty^{(q)} = \sup_{k \geq 0}(Z_k + qT_k)$$

and so to derive (A.1.3) it suffices to employ Theorem A.1.4 with ζ replaced by $\zeta + q\tau$.

If $q > 0$, then

$$\overline{Z}^{(q)} \underset{d}{=} \tau + \overline{Z}_\infty^{(q)},$$

and (A.1.3) is also preserved if (A.1.2) holds (see the proof of Theorem 6.1.4). □

A.1.2 Second-Order Approximation for the Distribution of the Maximal Value of the CRP $Z^{(q)}(t)$

In insurance theory and risk theory, the ruin probability of an insurance company (see §6.7) can be described in terms of the probability $\mathbf{P}(\overline{Z}^{(q)} \geq x)$ (with corresponding definition of a governing sequence $\{\tau_j, \zeta_j\}$ and the values q, x). The same can be said about the probability of memory block overflow in satellite communication systems, and so on. Experts working in this field have repeatedly noted that the approximation of $\mathbf{P}(\overline{Z}^{(q)} > x)$ via (A.1.3) is not always satisfactory and is usually lower than the real value. This suggests the problem of finding more exact approximations for the probabilities under consideration, or, equivalently, of the second-order asymptotics of the probability $\mathbf{P}(\overline{Z}^{(q)} > x)$ with estimation of the remainder term.

If the Cramér condition is met, then, as pointed out in Theorem 6.1.3, under some additional conditions the correction factor $(1+o(1))$ as $x \to \infty$ on the right of (6.1.5) and (6.1.6) can be replaced by $(1 + O(e^{-hx}))$. This means that the approximation of $\mathbf{P}(\overline{Z}^{(q)} \geq x)$ is rather precise.

Assume now that the Cramér condition is not met. We set

$$\zeta^{(q)} := \zeta + q\tau.$$

From Theorem 7.5.3 in [38] (see also [55]) we have the following assertion.

Theorem A.1.5 *Let the distribution of $\zeta^{(q)}$ be nonlattice and be either regularly varying or semiexponential. Assume also that $a_{\zeta^{(q)}} := \mathbf{E}\zeta^{(q)} < 0$, $q \leq 0$, and $\mathbf{E}(\zeta^{(q)})^2 < \infty$. Then as $x \to \infty$*

$$\mathbf{P}(\overline{Z}^{(q)} > x) = \frac{Q^{(q)}(x)}{|a_{\zeta^{(q)}}|} + c\mathbf{P}(\zeta^{(q)} > x)(1 + o(1)),$$

where

$$c = \frac{\mathbf{E}(\zeta + q\tau)^2}{2(a_{\zeta^{(q)}})^2} - \frac{\mathbf{E}\overline{Z}^{(q)}}{a_{\zeta^{(q)}}},$$

and the function $Q^{(q)}(x)$ *is defined in* (A.1.1).

Since $c > 0$, this confirms the assumption that the first-order approximation (A.1.3) gives the value lower that the real one.

In the case $\mathbf{E}(\zeta^{(q)})^2 = \infty$, the correction term has a much more complex nature. In this way, rather complete results were obtained in [24]. We consider here only the most simple case in which $\mathbf{E}(\zeta_-^{(q)})^2 < \infty$, where $\zeta_-^{(q)} = \min(0, \zeta^{(q)})$. We will write $\zeta^{(q)} \in [\mathbf{R}_\alpha]$ if $\mathbf{P}(\zeta^{(q)} > x)$ is a regularly varying function at infinity with parameter $-\alpha$.

Theorem A.1.6 *Let* $\mathbf{E}\zeta^{(q)} < 0$, $q \le 0$, $\zeta^{(q)} \in \mathbf{R}_\alpha$, $\alpha \in (1,2)$, *and let* $\mathbf{E}(\zeta_-^{(q)})^2 < \infty$. *Then as* $x \to \infty$

$$\mathbf{P}(\overline{Z}^{(q)} > x) = \frac{Q^{(q)}(x)}{|a_{\zeta^{(q)}}|} + \frac{2(\alpha - 1)I(\alpha)(Q^{(q)}(x))^2}{(a_{\zeta^{(q)}})^2}(1 + o(1)),$$

where $I(\alpha) = \int_0^{1/2} s^{1-\alpha}(1-s)^{-\alpha}\,ds < \infty$ *for* $\alpha \in (1,2)$.

The second-order asymptotics for $\mathbf{P}(\overline{Z}^{(q)} \ge x)$ in the case $\mathbf{E}(\zeta_-^{(q)})^2 = \infty$ can be obtained via Theorem 1.2 in [24].

A.1.3 Transient Phenomena for CRPs. First- and Second-Order Asymptotics

In queueing theory, when studying the work of a queuing system in heavy traffic, the problem arises of approximating the distribution of the queue length or the waiting time when the loading is close to critical. This corresponds to the problem on the distribution of $\overline{Z}^{(q)}$ in the case when $a_\zeta + qa_\tau$ is negative, but small. The value $\overline{Z}^{(q)}$ increases unboundedly in probability with decreasing $|a_\zeta + qa_\tau|$. This leads us to the problem on the limit distribution of $\overline{Z}^{(q)}$ arising under these conditions.

Let us give more precise statement of this problem.

Assume that the random variable $\zeta^{(q)} = \zeta + q\tau$ depends on the parameter ε (in this case, we write $\zeta^{(q)} = {}^{(\varepsilon)}\zeta$). We also assume that

$$^{(\varepsilon)}a := \mathbf{E}^{(\varepsilon)}\zeta \to 0, \qquad ^{(\varepsilon)}a < 0 \quad \text{as} \quad \varepsilon \to 0. \tag{A.1.4}$$

In this case, we denote by ${}^{(\varepsilon)}Z(t)$ the processes $Z(t)$ corresponding to the vector (τ, ζ, q). We also set

$$^{(\varepsilon)}\overline{Z} = \sup_{t\ge0} {}^{(\varepsilon)}Z(t), \qquad ^{(\varepsilon)}Z_k = \sum_{j=1}^{k} {}^{(\varepsilon)}\zeta_j, \qquad ^{(\varepsilon)}\overline{Z}_\infty = \sup_{k\ge0} {}^{(\varepsilon)}Z_k,$$

where ${}^{(\varepsilon)}\zeta_j$ are independent copies of ${}^{(\varepsilon)}\zeta$. In the case of transient phenomena, the problem here is to study the limit distribution of ${}^{(\varepsilon)}\overline{Z}$ as $\varepsilon \to 0$. As we have already seen,

$$ {}^{(\varepsilon)}\overline{Z} \stackrel{d}{=} \begin{cases} {}^{(\varepsilon)}\overline{Z}_\infty & \text{for} \quad q \le 0, \\ q\tau + {}^{(\varepsilon)}\overline{Z}_\infty & \text{for} \quad q > 0, \end{cases} $$

where the second equality holds if τ and ζ are independent.

The following assertion was obtained in [77], [100], [17, Theorem 4.18].

Theorem A.1.7 *Let* ${}^{(\varepsilon)}a \to 0$ *and* $\mathbf{D}^{(\varepsilon)}\zeta \to \sigma^2 > 0$ *as* $\varepsilon \to 0$ *and let the following uniform convergence condition hold*:

$$ \int_{|t|>N} t^2 \mathbf{P}({}^{(\varepsilon)}\zeta \in dt) \to 0 \quad as \quad N \to \infty \ \textit{uniformly in } \varepsilon \tag{A.1.5} $$

(*the convergence* ${}^{(\varepsilon)}\zeta \Rightarrow \zeta^{(q)}$ *is not assumed*). *Then, for each fixed* $v \ge 0$,

$$ \lim_{\varepsilon\to 0} \mathbf{P}\Big({}^{(\varepsilon)}\overline{Z}_\infty > \frac{v}{|{}^{(\varepsilon)}a|}\Big) = e^{-2v/\sigma^2}. $$

Condition (A.1.5) is satisfied if $\mathbf{E}|{}^{(\varepsilon)}\zeta|^{2+\delta} < c$ for some $c < \infty$ and $\delta > 0$.

A *complete asymptotic expansion* of $\mathbf{P}({}^{(\varepsilon)}\overline{Z}_\infty \ge v/\varepsilon)$ in powers of $\varepsilon = -{}^{(\varepsilon)}a$ was obtained in [17] in the special case when

$$ {}^{(\varepsilon)}\zeta = \zeta^{(q)} - \varepsilon, \qquad \mathbf{E}\zeta^{(q)} = 0, $$

the distribution of ζ has an absolutely continuous component, and

$$ \mathbf{E}e^{\mu\zeta} < \infty \tag{A.1.6} $$

for sufficiently small $|\mu|$ (see also Theorem 4.19 in [17]).

From Theorem A.1.7 we have the following analog of the Kingman–Prokhorov theorem. For simplicity, we identify the parameter ε with $-{}^{(\varepsilon)}a$.

Theorem A.1.8 *Assume that conditions* (A.1.4), (A.1.5) *are met and*

$$ \mathbf{D}^{(\varepsilon)}\zeta \to \sigma^2 > 0 \quad as \quad \varepsilon \to 0. $$

If $q \le 0$ *or*

$$ q > 0, \quad \tau \textit{ and } \zeta \textit{ are independent}, \quad \varepsilon q\tau \xrightarrow[p]{} 0 \quad as \quad \varepsilon \to 0 $$

(*q and τ may depend on ε*), *then*

$$ \lim_{\varepsilon\to 0} \mathbf{P}\Big({}^{(\varepsilon)}\overline{Z} > \frac{v}{\varepsilon}\Big) = e^{-2v/\sigma^2}. \tag{A.1.7} $$

In applications, the most interesting case is $q \le 0$. For this reason and also for simplicity of presentation, we will consider only the case $q \le 0$ when dealing with the second-order approximation for the distribution of ${}^{(\varepsilon)}\overline{Z}$. In this case, from Theorem 3.2 in [28] we have:

Theorem A.1.9 *Assume that* ${}^{(\varepsilon)}\zeta \Rightarrow {}^{(0)}\zeta$, *conditions* (A.1.4) *are met, and*

$$\mathbf{E}|{}^{(\varepsilon)}\zeta|^3 \to \mathbf{E}|{}^{(0)}\zeta|^3, \qquad \mathbf{E}({}^{(\varepsilon)}\zeta)^2 = \mathbf{E}^{(0)}\zeta^2 + o(\varepsilon) \quad as \quad \varepsilon \to 0.$$

Assume in addition that the distributions of ${}^{(\varepsilon)}\zeta$ *and* ${}^{(0)}\zeta$ *have density with uniformly in* ε *bounded variation. Then*

$$\mathbf{P}(\varepsilon^{(\varepsilon)}\overline{Z} > v) = \exp\left\{-\frac{2v}{\sigma^2} + \frac{4v\varepsilon m_3}{3\sigma^6}\right\}\left[1 + \frac{2\varepsilon m_-}{\sigma^2} + \frac{2\varepsilon m_3}{3\sigma^4} + o(\varepsilon)\right],$$

where

$$m_3 = \mathbf{E}^{(0)}\zeta^3, \qquad m_- = \mathbf{E}\chi_-,$$

and χ_- *is the overshoot over an infinite negative barrier by the random walk* $\{{}^{(0)}Z_k\}$ *with jumps distributed as the variable* ${}^{(0)}\zeta$.

A.2 Asymptotics of the Distributions of $Z^0(T) = (Z(T) - aT)$ and $\overline{Z}^0(T) = \max_{t \le T} Z^0(t)$ when the Distributions of Jumps Are Regularly Varying

For $t > 0$, we put

$$F^{(\zeta)}(t) = \mathbf{P}(\zeta \ge t), \qquad F_-^{(\zeta)}(t) = \mathbf{P}(\zeta < -t), \qquad F^{(\tau)}(t) = \mathbf{P}(\tau \ge t).$$

We will need the following conditions.

$[\mathbf{R}_\alpha]$ We write $F \in [\mathbf{R}_\alpha]$ if $F = F(t)$ is a regularly varying function on $[0, \infty)$ with exponent $-\alpha$, that is, $F(t) = t^{-\alpha} l(t)$, where $\alpha > 1$, $l(t)$ is a slowly varying function (SVF).

Further, let us introduce two more conditions, which we shall use in what follows.

$[\mathbf{Q}_T]$ $F^{(\zeta)} \in [\mathbf{R}_\alpha]$ *for* $\alpha \in (1, 2)$ *and at least one of the following two conditions holds:*

(i) *either* $F_-^{(\zeta)}(t) \le cF^{(\zeta)}(t)$, $t > 0$ *and* $TF^{(\zeta)}(x) \to 0$,
(ii) *or* $F_-^{(\zeta)}(t) \le W(t) \in [\mathbf{R}_\gamma]$ *with some* $\gamma > 1$ *and*

$$T[F^{(\zeta)}(x/\ln x) + W(x/\ln x)] < c \quad \text{as } x \to \infty.$$

$[\mathbf{\Phi}_T]$ $F^{(\zeta)} \in [\mathbf{R}_\alpha]$ *for* $\alpha > 2$, $\sigma_\zeta^2 := \mathbf{D}\xi < \infty$ *and* $x \to \infty$ *in such a way that, for some* $c > 1$,

$$x > c\sigma_\zeta\sqrt{(\alpha - 2)a_\tau^{-1} T \ln T}, \quad \text{as} T \to \infty.$$

We set $H(t) = \mathbf{E}\nu(t)$.

Theorem A.2.1 (see Theorem 16.2.1 in [38]) *Let* τ *and* ζ *be independent, let either condition* $[\mathbf{Q}_T]$ *or condition* $[\mathbf{\Phi}_T]$ *be met, and let* $\delta > 0$ *be an arbitrary fixed number.*

I. *The case* $a \ge 0$.

(i) *If* $x \to \infty$, *then*

$$\mathbf{P}(Z^0(T) \geq x) \sim H(T)F^{(\zeta)}(x) \tag{A.2.1}$$

uniformly in T *such that* $x \geq \delta T$. *If* $T \to \infty$, *then* $H(T)$ *on the right of* (A.2.1) *can be replaced by* T/a_τ.

(ii) *If* $F^{(\tau)} \leq W_\tau \in [\mathbf{R}_\gamma]$ *for* $\gamma \in (1,2)$, *then* (A.2.1) *holds as* $x \to \infty$ *uniformly in the range of* T *satisfying, in addition to* $[\mathbf{Q}_T]$ *or* $[\mathbf{\Phi}_T]$, *the condition* $x \geq T^{1/\gamma}l(T)$ *for an appropriate SVF* l.

(iii) *If* $\alpha \in (1,2)$ *and* $F^{(\tau)}(t) = o(F^{(\zeta)}(t))$ *as* $t \to \infty$, *or if* $\alpha > 2$ *and* $\mathbf{E}\tau^2 < \infty$, *then* (A.2.1) *holds without any additional conditions on* x *and* T *(excluding the conditions mentioned in* $[\mathbf{Q}_T]$ *and* $[\mathbf{\Phi}_T]$).

II. *The case* $a < 0$.

(i) *Relation* (A.2.1) *holds as* $x \to \infty$ *uniformly in* T *from the range given by the inequality*

$$x_* := x + aT \geq \delta T.$$

(ii) *Let condition* $F^{(\tau)} \in [\mathbf{R}_\gamma]$ *hold. If*

$$T \to \infty, \quad x_* \to \infty \quad \text{and} \quad x_* = o(T),$$

then, for $\alpha \in (1,2)$,

$$\mathbf{P}(Z^0(T) \geq x) \sim \frac{T}{a_\tau}F^{(\zeta)}(x),$$

and for $\alpha > 2$,

$$\mathbf{P}(Z^0(T) \geq x) \sim \frac{T}{a_\tau}F^{(\zeta)}(x) + \frac{x_*^2 F^{(\zeta)}(x_*)F^{(\tau)}(T)}{a^2 a_\tau^2(\alpha-1)(\alpha-2)}.$$

(iii) *If* $F^{(\tau)}(t) = o(F^{(\zeta)}(t))$ *as* $t \to \infty$, *then assertions* I(i), *and* I(ii) *hold without any additional conditions on* x *and* T *(excluding the conditions mentioned in* $[\mathbf{Q}_T]$ *and* $[\mathbf{\Phi}_T]$ *and the assertions* I(i), I(ii)).

(iv) *Let the condition* $F_\tau \in [\mathbf{R}_\gamma]$ *be met with* $\gamma \in (1,2) \cup (2,\infty)$ *and let*

$$x_* = x + aT \to -\infty, \qquad x \geq T^{1/\gamma}\, l(T)$$

for an appropriate SVF l *(for its choice, see Lemma* 16.2.3 *in* [38]; *for* $\gamma > 2$, *the last inequality always holds by conditions* $[\mathbf{Q}_T]$ *or* $[\mathbf{\Phi}_T]$). *Then*

$$\mathbf{P}(Z^0(T) \geq x) \sim \frac{1}{a_\tau}\left[TF^{(\zeta)}(x) + (T + x/a)F^{(\tau)}(x/q)\right] \quad \text{as } T \to \infty.$$

Some remarks about the second part of Theorem A.2.1 are worth making. As already pointed out, case II (the presence of the positive linear drift $-at$) is more difficult than case I. In assertion II(i), the level x is *much higher* than the point $-aT$, near which the value $Z^0(T)$ may appear to be at time T due to the linear drift and in view of very long renewal time interval. In this case, the asymptotics of the probability

$\mathbf{P}(Z^0(T) \geq x)$ remains the same as in Part I of the theorem: roughly speaking, the crossing will occur if, as before, there is a large jump ζ_j on the time interval $[0, T]$.

In case II(ii), the level of x is still above $-aT$, but the difference between these values is relatively small: $x_* = x + aT = o(T)$. In this transient case, the form of the asymptotics of the probability $\mathbf{P}(Z^0(T) \geq x)$ is determined by the "thickness" of the right-hand tail of the distribution of the jumps ζ_j. In the case of a "thick" tail (when $\alpha \in (1, 2)$), the asymptotics remains the same as before and does not depend on the distribution of the lengths of the renewal intervals (we require that $F^{(\tau)} \in [\mathbf{R}_\gamma]$ for this distribution). If the tail of the distribution of the random variables ζ_j is "thinner" (that is $\alpha > 2$), then an additional term (which might either be dominating or negligible) will appear in the asymptotics, but now this term depends on the distribution of the random variable τ. The appearance of this term stems from the following possibility: one of the renewal intervals at the very beginning of the time interval $[0, T]$ turns out to be very long (so, it covers the point $t = T$), and besides, the sum of jumps ζ_j "accumulated" on the initial time interval (before this renewal interval) appears to be large enough for the process to overshot the "gap" between $-aT$ and x. The last event in turn is also a "large deviation" and takes place as a result of a single "moderately large" jump ζ_j, which exceeds x_* (for this reason, the factor $F^{(\zeta)}(x_*)$ appears). The appearance of the factor x_*^2 can be explained, roughly speaking, by fact that the number of such jumps corresponding to these largest values of τ_k and ζ_j at the beginning of the process trajectory has the order x_*, and besides, the sequences of random variables $\{\tau_k\}$ and $\{\zeta_j\}$ are independent. Hence the probability of the required combination of the events has the order of the variable $x_* F^{(\zeta)}(x_*) \times x_* F^{(\tau)}(T)$.

In case II(iv), the level x is now markedly lower than $-aT$ (the condition $x + aT \to -\infty$ is satisfied), and hence the probability of its exceedance is increased due to the contribution corresponding to the presence of one very large τ_k in the initial part of the interval $[0, T]$. The form of this additional term can be explained as follows. Roughly speaking, in the absence of large deviations in the walk $\{Z_n\}$, in order for the trajectory to be above the level x at time T it suffices that one of the first (approximately $(T + x/a)/a_\tau$) renewal intervals should be large ($> -x/a$). In this case, the trajectory $Z(t)$ will oscillate near the origin until the beginning of this long renewal interval, and then it will move along the line of slope $-a > 0$ within this interval (and will exceed the level x on this interval), after which (if $t < T$ still holds) it will again oscillate at approximately the same level (and, therefore, it will be still above the level x by the time T).

The very narrow transient case $x_* = O(1)$ ($-aT$ and x are nearly equal) is very difficult to analyse. This case is not considered here and is not covered by Theorem A.2.1.

Let us now consider the distribution of $\overline{Z}^0(t)$.

As for the usual random walks, the first-order asymptotics for the probabilities $\mathbf{P}(\overline{Z}^0(T) \geq x)$ of large deviations of the maximum of the process is the same as for

$\mathbf{P}(Z^0(T) \geq x)$. The reason here is the same: if the process $\{Z^0(t)\}$ crosses a high level x somewhere inside the interval $[0, T]$ (say, as a result of the jump ζ_j), then it will remain, with high probability, "near" the point $Z^0(T_j)$ until the end of the interval $[0, T]$ (recall that this process has zero average drift). Hence the events $\{Z^0(T) \geq x\}$ and $\{\overline{Z}^0(T) \geq x\}$ are "nearly equivalent." The level x may also be crossed in the interval of linear growth, but the above also remains valid in this case.

Theorem A.2.2 (see Theorem 16.2.2 in [38]) *The conclusions of Theorem* A.2.1 *remain valid under the corresponding conditions for the probability* $\mathbf{P}(\overline{Z}^0(T) \geq x)$ *as well.*

A.3 First Boundary Crossing Problem in the Case of Regular Variation of Jump Distributions

In this section, we consider the problem of crossing of an arbitrary boundary $\{g_T(t);\, t \in [0, T]\}$ by the trajectory of the process $\{Z^0(t)\}$ in the case when $\inf_{t \leq T} g(t)$ goes to infinity sufficiently rapidly, so that the event

$$B_{T,g} := \Big\{ \sup_{t \leq T} (Z^0(t) - g_T(t)) \geq 0 \Big\}$$

corresponds to the large deviation zone. Note that the assumption that the process $Z^{(q)}$ under consideration has zero drift ($q = -a$) does not impair the generality.

One of the important factors in the following considerations is the possibility of implementing the event $B_{T,g}$ of interest due to a very large renewal interval. To avoid cumbersome calculations, this possibility can be excluded. For $q \leq 0$, there is no such possibility. For $q > 0$, we can assume that the condition

$$\inf_{0 \leq t \leq T} (g_T(t) - qt) > \delta T \tag{A.3.1}$$

is satisfied with some fixed $\delta > 0$. In this case, it is impossible to cross the boundary $g_T(t)$ by one large jump τ (with some margin), while the presence of two large jumps τ_j and ζ_j is very unlikely.

Hence, a high boundary can actually be crossed, like for random walks, only with one large jump ζ_j, and besides, at the moment of jump it suffices to cross, roughly speaking, the level

$$f_T(t) := \inf_{t \leq s \leq T} g_T(s),$$

because after crossing the level $f_T(t)$ the trajectory of the process $\{Z^{(q)}(t)\}$ for $q = -a$ will again "follow" the (horizontal) line of mean drift and at some time will be higher than the boundary, which by this time will "decrease" down to the level $f_T(t)$.

Let us now formulate the main assertion of this section. We denote by $\mathcal{G}_{(x,N)}$ the class of all measurable boundaries $g_T(t)$ such that

$$x \le \inf_{0 \le t \le T} g_T(t) \le Nx, \quad x > 0, \quad N \in (1, \infty).$$

Theorem A.3.1 (see Theorem 16.4.1 in [38]) *Let the distribution of the jumps ζ_j satisfy either condition $[\mathbf{Q}_T]$ or condition $[\mathbf{\Phi}_T]$, and let $\delta > 0$ and $K > 1$ be arbitrary fixed numbers.*

I. *The case $a \ge 0$. The relation*

$$\mathbf{P}(B_{T,g}) = (1 + o(1)) \int_0^T F^{(\zeta)}(f_T(t))\, dH(t), \quad x \to \infty \tag{A.3.2}$$

holds uniformly in $g_T \in \mathcal{G}_{(x,N)}$ for $x \ge \delta T$.

If, in addition, the distribution of the random variable τ satisfies the condition $F^{(\tau)}(t) \le W_\tau(t) \Subset [\mathbf{R}_\gamma]$ for $\gamma \in (1, 2)$, then (A.3.2) *holds as $x \to \infty$ uniformly in the range of T satisfying (in addition to $[\mathbf{Q}_T]$ or $\mathbf{\Phi}_T]$) the condition $x \ge T^{1/\gamma} l(T)$ with an appropriate SVF l (the choice of $l(t)$ is indicated in Lemma* 16.2.3 *of* [38]).

II. *The case $a < 0$.*

(i) *If* (A.3.1) *is met, then* (A.3.2) *holds uniformly in $g_T \in \mathcal{G}_{(x,N)}$.*

(ii) *If $F^{(\tau)}(t) = o(F^{(\zeta)}(t))$ as $t \to \infty$, then all the assertions from Part* I *of the theorem remain valid.*

It is easily seen that, as $T \to \infty$,

$$\int_0^T F^{(\zeta)}(f_T(t))\, dH(t) \sim \frac{1}{a_\tau} \int_0^T F^{(\zeta)}(f_T(t))\, dt \tag{A.3.3}$$

uniformly in $g \in \mathcal{G}_{(x,N)}$, so that in this case the integral on the right of (A.3.2) can be replaced by the expression in the right-hand side of (A.3.3).

The following assertion is immediate from Theorem A.3.1 and the definition of a regularly varying function.

Corollary A.3.2 *Let $g(s)$, $s \in [0, 1]$, be a given measurable function with values in the interval $[c_1, c_2] \subset (0, \infty)$. Then, for the boundary*

$$g_T(t) := xg(t/T), \quad 0 \le t \le T,$$

the relation

$$\mathbf{P}(B_{T,g}) \sim \frac{TF^{(\zeta)}(x)}{a_\tau} \int_0^1 f^{-\alpha}(s)\, ds \quad \text{as } T \to \infty,$$

where $f(s) := \inf_{s \le u \le 1} g(u)$, $0 \le s \le 1$, holds under the corresponding conditions on x from Theorem A.3.1 *in the case $a \ge 0$ or under the condition*

$$\inf_{0 \le s \le 1} (xf(s) + asT) \ge \delta T$$

in the case $a < 0$.

Indeed, it is necessary only to note that

$$F^{(\zeta)}(f_T(t)) = \frac{F^{(\zeta)}(xf(t/T))}{F^{(\zeta)}(x)} F^{(\zeta)}(x) = f^{-\alpha}(t/T)\frac{l(xf(t/T))}{l(x)} F^{(\zeta)}(x)$$
$$= (1 + o(1)) f^{-\alpha}(t/T) F^{(\zeta)}(x)$$

uniformly on $[0, T]$ by Theorem 1.1.1 in [38] on uniform convergence of SVFs to 1 in the last formula.

For a more detailed description of the case of linear boundaries of the form $g_T(t) = x + ct$, $c =$ const, see Chapter 16 in [38].

Basic Notation

The order of the symbols in the list does not always correspond to the order in which they appear in the book.

Random variables and processes

$\boldsymbol{\xi} = (\tau, \zeta)$, $\boldsymbol{\xi}_j = (\tau_j, \zeta_j)$, $j = 1, 2, \ldots$ – the governing vector of a CRP and its independent copies

$\boldsymbol{\xi}^*$ – a column vector

$T_n = \sum_{j=0}^{n} \tau_j$, $T_0 = 0$

$Z_n = \sum_{j=0}^{n} \zeta_j$, $Z_0 = 0$

$\mathbf{S}_n = \sum_{j=0}^{n} \boldsymbol{\xi}_j$, $\mathbf{S}_0 = (0, 0)$

$\xi = \zeta - a\tau$, ξ_j – independent copies of ξ

$S_n = \sum_{j=0}^{n} \xi_j$, $S_0 = 0$

$\eta(t) = \min\{k : T_k > t\}$ – a simple renewal process

$\nu(t) = \eta(t) - 1 = \max\{k : T_k \leq t\}$ – a simple renewal process

$Z(t) = Z_{\nu(t)}$ – a compound renewal process (CRP)

$Y(t) = Y_{\eta(t)}$ – a compound renewal process (CRP)

$\chi(t) = T_{\eta(t)} - t$ – the overshoot

$\overline{\chi}(t) = \max_{k \leq t} \chi(k)$

$\gamma(t) = t - T_{\nu(t)}$ – the undershoot

$\zeta(t) = \zeta_{\eta(t)}$

$\overline{\zeta}(t) = \max_{u \leq t} \zeta(u)$

$\tau(t) = \tau_{\eta(t)}$

$\overline{\tau}(t) = \max_{u \leq t} \tau(u)$

$Z^{(q)}(t) = Z(t) + qt$ – a CRP with linear drift q

$Z^{[q]}(t)$ – a CRP with jumps $(\tau, \zeta + q\tau)$

$Z^{(\mathrm{st})}(t)$ – a CRP with stationary increments

$(\tau^{(\mathrm{st})}, \zeta^{(\mathrm{st})})$ – the initial jump of $Z^{(\mathrm{st})}(t)$

$\overline{Z}(t) = \max_{u \leq t} Z(u)$

$\overline{Z}^{(q)}(t) = \max_{u \leq t} Z^{(q)}(u)$

$\overline{Z}^{[q]}(t) = \max_{u \leq t} Z^{[q]}(u)$

$\overline{Z} = \overline{Z}(\infty)$

$\overline{Z}^{(q)} = \overline{Z}^{(q)}(\infty)$

$\overline{Z}^{[q]} = \overline{Z}^{[q]}(\infty)$

$\overline{Z}_n = \max_{k \le n} Z_k$

$\zeta^{(q)} = \zeta + q\tau$

$Z_n^{(q)} = \sum_{j=1}^{n} \zeta_j^{(q)} = Z_n + qT_n$

$$y(t) = \begin{cases} \frac{Y(t)-at}{\sigma\sqrt{t}} & \text{for} \quad \sigma < \infty, \\ \frac{Y(t)-at}{\sigma(t)} & \text{under condition} \quad [\mathbf{R}_{\alpha,\beta}] \end{cases}$$

$$z(t) = \begin{cases} \frac{Z(t)-at}{\sigma\sqrt{t}} & \text{for} \quad \sigma < \infty, \\ \frac{Z(t)-at}{\sigma(t)} & \text{under condition} \quad [\mathbf{R}_{\alpha,\beta}] \end{cases}$$

ω – a random variable independent of τ (in the representation $\zeta = h\tau + \omega$ for linearly dependent τ and ζ)

$\Omega_n = \sum_{j=1}^{n} \omega_j, \quad \Omega_0 = 0$

$$y_T(t) = \begin{cases} \frac{Y(tT)}{x}, & t \in [0,1]\ x \sim T \text{ as } T \to \infty \text{ (in Chapter 3)} \\ \frac{Y(tT)-atT}{\sigma\sqrt{T}}, & t \in [0,1] \quad \text{(in Chapter 7)} \end{cases}$$

$$z_T(t) = \begin{cases} \frac{Z(tT)}{x}, & t \in [0,1]\ x \sim T \text{ as } T \to \infty \text{ (in Chapters 3, 4)} \\ \frac{Z(tT)-atT}{\sigma\sqrt{T}}, & t \in [0,1] \quad \text{(in Chapter 7)} \end{cases}$$

$\eta_g = \inf\{t : z_T(t) \ge g(t)\}$ (in Chapter 4)

$w(t), t \in [0,1]$ – a standard Wiener process

$w_{\alpha,\beta}(t), t \in [0,1]$ – a stable process with parameters (α, β)

$\eta_Z(x) = \inf\{t : Z(t) > x\}$

$\chi_Z(x) = Z(\eta_Z(x)) - x$

$\eta_{Z,g} = \inf\{t : Z(t) > g_T(t)\}$ (in Chapter 6)

${}^{\alpha}\boldsymbol{\xi} = ({}^{\alpha}\tau, {}^{\alpha}\zeta)$ – the Cramér transform (in distribution) of a random vector $\boldsymbol{\xi}$ at a point $(\lambda(\alpha), \mu(\alpha))$ (§6.2)

${}^{\alpha}Y(t)$ – the CRP constructed from the governing vector $({}^{\alpha}\tau, -{}^{\alpha}\zeta)$

${}^{\alpha}\overline{Y} = \sup_{t>0} {}^{\alpha}Y(t)$

Parameters

$a_\tau = \mathbf{E}\tau$

$a_\zeta = \mathbf{E}\zeta$

$a = a_\zeta / a_\tau$

$\sigma_\xi^2 = \mathbf{E}\xi^2$

$\sigma^2 = \frac{\mathbf{E}(\zeta - a\tau)^2}{a_\tau} = \frac{\mathbf{E}\xi^2}{a_\tau}$

$\boldsymbol{\sigma}^2 = \mathbf{E}\boldsymbol{\xi}^*\boldsymbol{\xi}$

$\lambda_+ = \sup\{\lambda : \mathbf{E}e^{\lambda\tau} < \infty\}$

Functions

$F_-(t) = \mathbf{P}(\xi \le -t)$

$F_+(t) = \mathbf{P}(\xi \geq t)$
$F(t) = F_-(t) + F_+(t) = \mathbf{P}(|\xi| \geq t)$
$F^{(-1)}(u) = \inf\{t : F(u) < u\}$ – the inverse function of F
$\left.\begin{array}{l}\sigma_\xi(n) = F^{(-1)}(1/n) \\ \sigma(t) = \sigma_\xi(t)a_\tau^{-1/\alpha}\end{array}\right\}$ if $F(t) = t^{-\alpha}l(t)$, $l(t)$ is a slowly varying function
$\Phi(t)$ – the distribution function of the normal law
$\phi(t)$ – density of the normal distribution
$\Phi_{\alpha,\beta}(t)$ – the distribution function of a stable law with parameters (α, β)
$\psi(\lambda, \mu) = \mathbf{E}e^{\lambda\tau+\mu\zeta}$
$\psi^{(\tau)}(\lambda) = \psi(\lambda, 0) = \mathbf{E}e^{\lambda\tau}$
$\psi^{(\zeta)}(\mu) = \psi(0, \mu) = \mathbf{E}e^{\mu\zeta}$
$\mathbf{A}(\lambda, \mu) = \ln\psi(\lambda, \mu)$
$A^{(\tau)}(\lambda) = \ln\psi^{(\tau)}(\lambda)$
$A^{(\zeta)}(\mu) = \ln\psi^{(\zeta)}(\mu)$
$A(\mu) = -\sup\{\lambda : \mathbf{A}(\lambda, \mu) \leq 0\}$ – the fundamental function under the broad condition $[\boldsymbol{\lambda}_+]$
$\mathbf{A}_1(\lambda, \mu) = \ln\mathbf{E}e^{\lambda\tau_1+\mu\zeta_1}$
$\mathbf{A}^{(\mathrm{st})}(\lambda, \mu) = \ln\mathbf{E}e^{\lambda\tau^{(\mathrm{st})}+\mu\zeta^{(\mathrm{st})}}$
$\boldsymbol{\Lambda}(t, \alpha) = \sup_{(\lambda,\mu)}(\lambda t + \mu\alpha - \mathbf{A}(\lambda, \mu))$ – the deviation function of $\{T_n, Z_n\}$
$H(t) = \sum_{n=0}^{\infty}\mathbf{P}(T_n \leq t)$ (in Chapter 1)
$H_1(t) = \sum_{n=1}^{\infty}\mathbf{P}(T_n \leq t)$ (in Chapter 1)
$H(B) = \sum_{n=0}^{\infty}\mathbf{P}(\mathbf{S}_n \in B)$ (in Chapter 3)
$\mathbf{D}_\Lambda(t, \alpha) = \inf_{v>0} v\boldsymbol{\Lambda}(\frac{t}{v}, \frac{\alpha}{v})$
$\mathbf{D}_\Lambda(B) = \inf_{(t,\alpha)\in B}\mathbf{D}_\Lambda(t, \alpha)$
$\mathbf{D}(t, \alpha) = \lim_{\varepsilon\to 0}\mathbf{D}_\Lambda((t, \alpha)_\varepsilon)$
$D(\alpha) = \mathbf{D}(1, \alpha) = \sup_\mu(\mu\alpha - A(\mu))$ – the deviation function of $Z(t)$ under the broad condition $[\boldsymbol{\lambda}_+]$
$\widehat{D}(\alpha) = \begin{cases}\inf[\mathbf{D}(t, \alpha) - (1-t)\lambda_+] & \text{for } \lambda_+ < \infty, \\ D(\alpha) & \text{for } \lambda_+ = \infty,\end{cases}$ – the deviation function of $Z(t)$ in the general case
$\widehat{A}(\mu) = \max(A(\mu), -\lambda_+) = \sup_\alpha(\alpha\mu - \widehat{D}(\alpha))$
$\mu(\alpha) = D'(\alpha)$ – the point at which $\sup_\mu(\mu\alpha - A(\mu))$ is attained
$\lambda(\alpha) = -A(\mu(\alpha))$
$\boldsymbol{\mu}(\theta, \alpha), \boldsymbol{\lambda}(\theta, \alpha)$ are defined in (5.3.2)
$L(t) = \sigma\sqrt{2t\ln\ln t}$ (in Chapter 1)
$L(r) = r\boldsymbol{\Lambda}(\frac{\theta}{r}, \frac{\alpha}{r})$ (in Chapter 3)
$L(\alpha) = D(\alpha)/\alpha$ (in Chapter 6)
$I_Z(\alpha, {}^{<}u, v, \mathbf{w}) = \int_0^u e^{\lambda(\alpha)y}\mathbf{P}(\tau > y + v, \zeta \in \mathbf{w})dy$
$I_Z(\alpha, {}^{>}u, v, \mathbf{w}) = \int_u^\infty e^{\lambda(\alpha)y}\mathbf{P}(\tau > y + v, \zeta \in \mathbf{w})dy$
$P_{T,\alpha}^{\lessgtr} = \mathbf{P}(Z(T) \in \Delta[x); B({}^{\lessgtr}u, v, \mathbf{w}))$, where $\alpha = x/T$
$P_\alpha(v) = \mathbf{P}({}^\alpha\overline{Y} \leq v)$
$J_\alpha(v) = \int_0^v e^{-u\mu(\alpha)}P_\alpha(-u)du$

$$I(f) = \begin{cases} \int_0^1 D(f'(t))dt & \text{if} \quad f \in \mathbb{C}_a, \quad f(0) = 0 \\ \infty & \text{otherwise} \end{cases}$$

$$I_0(f) = \begin{cases} \frac{1}{2}\int_0^1 (f'(t))^2 dt & \text{if} \quad f \in \mathbb{C}_a, \quad f(0) = 0 \\ \infty & \text{otherwise} \end{cases}$$

Domains, events

$\mathbb{C} = \mathbb{C}(0,1)$ – the space of continuous functions on $[0,1]$
$\mathbb{C}_a = \mathbb{C}_a(0,1)$ – the space of absolutely continuous functions on $[0,1]$
$\mathbb{D} = \mathbb{D}(0,1)$ – the space of functions on $[0,1]$ without discontinuities of the second kind
$\Delta[x) = [x, x+\Delta)$
$B_t(u,v) = \{\gamma(t) \ge u,\ \chi(t) \ge v\}$ (in Chapter 2)
$B(^{\lessgtr}u, v, \mathbf{w}) = \{\gamma(T) \lessgtr u,\ \chi(T) \ge v,\ \zeta(T) \in \mathbf{w}\}$ (in Chapter 5)
$B_g = \left\{f \in \mathbb{D} : \sup_{t\in[0,1]}(f(t) - g(t)) > 0\right\}$
$B_{g_1 g_2} = \{f \in \mathbb{D} : g_1(t) \le f(t) \le g_2(t) \quad \text{for all} \quad t \in [0,1]\}$
$\mathcal{A} = \{(\lambda,\mu) : \mathbf{A}(\lambda,\mu) < \infty\}$
$\mathcal{A}_1 = \{(\lambda,\mu) : \mathbf{A}_1(\lambda,\mu) < \infty\}$
$\mathcal{A}^{\le 0} = \{(\lambda,\mu) : \mathbf{A}(\lambda,\mu) \le 0\}$
$\mathcal{A}^{(\mathrm{st})} = \{(\lambda,\mu) : \mathbf{A}^{(\mathrm{st})}(\lambda,\mu) < \infty\}$
(μ^-, μ^+) – the effective domain of a fundamental function $A(\mu)$
(μ_-, μ_+) – the analyticity domain of a fundamental function $A(\mu)$ containing the point $\mu = 0$
$(\widehat{\mu}_-, \widehat{\mu}_+)$ – the analyticity domain of a function $\widehat{A}(\mu)$ containing the point $\mu = 0$
$(\alpha_-, \alpha_+) = (A'(\mu_- + 0), A'(\mu_+ - 0))$ – the analyticity domain of the deviation function $D(\alpha)$ containing the point $\alpha = a$
$(\widehat{\alpha}_-, \widehat{\alpha}_+) = (\widehat{A}'(\widehat{\mu}_- + 0), \widehat{A}'(\widehat{\mu}_+ - 0))$ – the analyticity domain of the deviation function $\widehat{D}(\alpha)$ containing the point $\alpha = a$
(β_-, β_+) – the interval in which $\lambda(\alpha) > \lambda_+$
$\mathcal{R} = \mathbb{R} \setminus (\beta_-, \beta_+)$
$K^<$ – the compact set in (α_-, α_+) containing the point $\alpha = a$
$K^>$ – the compact set in $(\alpha_-, \alpha_+) \setminus [\beta_-, \beta_+]$ containing the point $\alpha = a$
$\mathcal{A}_K^{\lessgtr} = \{(\lambda(\alpha), \mu(\alpha)) : \alpha \in K^{\lessgtr}\}$
$\mathcal{L}$ – the analyticity domain of $\mathbf{\Lambda}(\theta,\alpha)$ (in Chapter 5)
$\mathcal{D}$ – the open cone of analyticity of the function $\mathbf{D}(\theta,\alpha)$
$\mathbf{K}_{\mathbf{\Lambda}}$ – a compact set embedded in $\mathcal{L}$
$\mathbf{K}_{\mathcal{D}} \in \mathbb{R}^2$ – a compact set embedded in $\mathcal{D}$ and separated from the point $(0,0)$
$\mathcal{A}_{\mathbf{K}} = \{(\lambda(\frac{\alpha}{\theta}), \mu(\frac{\alpha}{\theta})) : (\theta,\alpha) \in \mathbf{K}_{\mathcal{D}}\}$

Conditions

$[\mathbf{R}_{\alpha,\beta}]$ – the attraction condition to a stable law with parameters (α,β) (see §1.6)
$[\mathbf{C}]$ – the Cramér moment condition: $\zeta \in [\mathbf{C}]$ if $\mathbf{E}e^{\lambda\zeta} < \infty$ in the neighborhood of the point $\lambda = 0$

$[\mathbf{C}_\infty]$ – the strong Cramér moment condition: $\zeta \in [\mathbf{C}_\infty]$ if $\mathbf{E}e^{\lambda\zeta} < \infty$ for all λ
$[\mathbf{C}_\varphi]$ – the Cramér condition on the characteristic function
$[\mathbf{C}^V]$ – the condition from §4.5.1
$[\lambda_+]$ – the condition $\lambda_+ \geq D(0)$
$[\overline{\lambda}_+]$ – the condition $\lambda_+ < D(0)$, $\ln \mathbf{P}(\tau > y) \sim -\lambda_+ t$ as $t \to \infty$
$[\widehat{\mathbf{P}}_{\mathbf{K}}]$ – the condition from §5.2.3
$[\mathbf{h}]$ – the condition from §5.2.3

Symbols

$\overline{o}(1)$; the equality $\Delta = \overline{o}(1)$ as $T \to \infty$ means that $\Delta = \Delta_T \to 0$ sufficiently slowly as $T \to \infty$
$o_p(1)$; the equality $\gamma = \gamma_T = o_p(1)$ as $T \to \infty$ means that $\gamma_T \underset{p}{\to} 0$ as $T \to \infty$
$\rho_{\mathbb{C}}$ – the uniform metric
ρ_S – the Skorokhod metric
$\rho_{\mathbb{D}}$ – the metric defined in §1.6
$\Rightarrow$ – the weak convergence in distribution
$\underset{\mathbb{C}}{\Rightarrow}$ – the $\mathbb{C}$-convergence (defined in §1.5.2)
$\underset{\mathbb{D}}{\Rightarrow}$ – the $\mathbb{D}$-convergence (defined in §1.6.3)
$\underset{S}{\Rightarrow}$ – the convergence in the Skorokhod metric (see §1.6.1)
$\Leftrightarrow$; the relation $\xi_n \Leftrightarrow \mathbf{F}$ means the weak convergence in distribution ξ_n to the distribution $\mathbf{F}$ as $n \to \infty$
$\in$; the relation $\xi \in [\mathbf{C}_\infty]$ means that the distribution of ξ satisfies condition $[\mathbf{C}_\infty]$
$(v)_\varepsilon$ – the ε-neighborhood of a point v
$(f)_\varepsilon$ – the ε-neighborhood of a function f (as a rule, in the uniform metric)
$\langle \boldsymbol{\lambda}, \boldsymbol{\xi} \rangle$ – the inner product of vectors $\boldsymbol{\lambda}$ and $\boldsymbol{\xi}$

References

[1] Aleskjavicene, A.: Large deviations for homogeneous Markov chains (1965), *Litovsk. Math. Sb.*, 5, 199–209.

[2] Anscombe, C.J.: Large sample theory of sequentional estimation (1952), *Proc. Cambridge Philos. Soc.*, 48, 600–607.

[3] Araujo, A., Giné, E.: The central limit theorem for real and Banach valued random variables (1980), Wiley.

[4] Asmussen, S.: Approximations for the probability of ruin within finite time (1984), *Scand. Actuarial J.*, 31–57.

[5] Asmussen, S.: Applied Probability and Queues Stochastic Modelling and Applied Probability (2003), 2nd ed., in: Applications of Mathematics (New York), vol. 51, Springer.

[6] Asmussen, S., Albrecher, H.: Ruin Probabilities. (2010), 2nd ed. Advanced Series on Statistical Science & Applied Probability 14. World Scientific.

[7] Athreya, K.B., Ney, P.E.: A new approach to the limit theory of recurrent Markov chains (1978), *Trans. Amer. Math. Soc.*, 245, 493–501.

[8] Azencott, R.: Grandes deviations et applications (1980), École d'Été de Probabilités de Saint-Flour VIII, 1978. Lecture Notes in Math. vol. 774, pp. 1–176. Springer.

[9] von Bahr, B.: Ruin probabilities expressed in terms of ladder height distributions (1974), *Scand. Actuarial J.*, 190–204.

[10] Billingsley, P.: Convergence of Probability Measures (1999), 2nd ed., John Wiley & Sons.

[11] Bingham, N.H., Goldie, C.H., and Teugels, J.L.: Regular Variations (1987), Cambridge University Press.

[12] Borovkov A.A.: Asymptotic Analysis of Random Walks. Light Tailed Jump Distributions (2020), Cambridge University Press.

[13] Borovkov A.A.: New limit theorems in boundary-value problems for sums of independent terms (1962), *Sibirsk. Mat. Zh.*, 3(5), 645–694.

[14] Borovkov A.A.: Analysis of large deviations in boundary-value problems with arbitrary boundaries I, II (1964), *Sibirsk. Mat. Zh.*, 5(2); *Sibirsk. Mat. Zh.*, 5(4), 750–767.

[15] Borovkov A.A.: Boundary problems for random walks and large deviations in functional spaces (1967), *Theory Probab. Appl.,* 12(4), 575–595.

[16] Borovkov A.A.: The convergence of tributions of functionals on stochastic processes (1972), *Russian Math. Surveys,* 27(1), 1–42.

[17] Borovkov A.A.: Stochastic Processes in Queueing Theory (1976), Springer.

[18] Borovkov A.A.: Asymptotic Methods in Queueing Theory (1984), Wiley.

[19] Borovkov A.A.: Ergodicity and Stability of Stochastic Processes (1998), Wiley.

[20] Borovkov A.A.: On subexponential distributions and asymptotics of the distribution of the maximum of sequential sums (2002), *Siberian Math. J.,* 43(6), 995–1022.

[21] Borovkov A.A.: Probability Theory (2013), Springer.

[22] Borovkov A.A.: Integro-local and local theorems for normal and large deviations of sums of nonidentically distributed random variables in the scheme of series (2009), *Theory Probab. Appl.,* 54(4), 571–587.

[23] Borovkov A.A.: Asymptotic Analysis of Random Walks: Rapidly Decreasing Distributions of Jumps (2013), Fizmatlit [in Russian].

[24] Borovkov A.A.: Second order approximation for distribution of maximum of random walk with negative drift and infinite variance (2014), *Theory Probab. Appl.,* 59(1), 3–22.

[25] Borovkov A.A.: Integral theorems for the first passage time of an arbitrary boundary by a compound renewal process (2015), *Siberian Math. J.*, 56(5), 765–782.

[26] Borovkov A.A.: Large deviation principles in boundary problems for compound renewal processes (2016), *Siberian Math. J.*, 57(3), 442–469.

[27] Borovkov A.A.: On the distribution of the first passage time of an arbitrary remote boundary by random walk (2016), *Theory Probab. Appl.,* 61(2), 235–254.

[28] Borovkov A.A.: Stability theorems and the second-order asymptotics in threshold phenomena for boundary functionals of random walks (2016), *Siberian Adv. Math.*, 26(4), 231–246.

[29] Borovkov A.A.: Some Boundary Problems of Probability Theory (2016), Saarbrucken: Palmarium Academic Publishing.

[30] Borovkov A.A.: Integro-local limit theorems for compound renewal processes (2017), *Theory Probab. Appl.*, 62(2), 175–195.

[31] Borovkov A.A.: Functional limit theorems for compound renewal processes (2019), *Siberian Math. J.*, 60(1), 27–40.

[32] Borovkov A.A.: Moderately large deviation principles for trajectories of compound renewal processes (2019), *Theory Probab. Appl.,* 64(2), 324–333.

[33] Borovkov A.A.: Extension of the invariance principle for compound renewal processes to the zones of moderately large and small deviations (2021), *Theory Probab. Appl.,* 65(4), 651–670.

[34] Borovkov A.A.: On large deviation principles for compound renewal processes (2019), *Math. Notes,* 106(6), 864–871.

[35] Borovkov, A.A.: Integro-local theorems in boundary problems for compound renewal processes (2019), *Sibirsk. Mat. Zh.*, 60(6), 1229–1246.

[36] Borovkov, A.A.: Boundary problems for compound renewal processes (2020), *Sibirsk. Mat. Zh.,* 61(1), 29–59.

[37] Borovkov, K.A.: Rate of convergence in the invariance principle for generalized renewal processes (1983), *Theory Probab. Appl.,* 27(3), 461–471.

[38] Borovkov, A.A., Borovkov, K.A.: Asymptotic Analysis of Random Walks. Vol. I: Slowly Decaying Jump Distributions (2008), Moscow: Fizmatlit (in Russian).

[39] Borovkov, A.A., Mogul'skii, A.A.F.: The second rate function and the asymptotic problems of renewal and hitting the boundary for multidimensional random walks (1996), *Siberian Math. J.*, 37(4), 647–682.

[40] Borovkov, A.A., Mogul'skii, A.A.F.: Integro-local limit theorems including large deviations for sums of random vectors (1999, 2001), *Theory Probab. Appl.*, 43(1), 1–12; *Theory Probab. Appl.*, 45(1), 3–22.

[41] Borovkov, A.A., Mogul'skii, A.A.: Limit theorems in the boundary hitting problem for a multidimensional random walk (2001), *Siberian Math. J.*, 42(2), 245–270.

[42] Borovkov, A.A., Mogul'skii, A.A. Integro-local and integral theorems for sums of random variables with semiexponential distributions (2006), *Siberian Math. J.*, 47(6), 990–1026.

[43] Borovkov, A.A., Mogul'skii, A.A.F.: On large deviation principles for random walk trajectories. I, II, III (2011, 2012, 2013), *Theory Probab. Appl.*, 56(4), 538–561; *Theory Probab. Appl.*, 58(1), 1–27; *Theory Probab. Appl.*, 58(1), 25–37.

[44] Borovkov, A.A., Mogul'skii, A.A.: Large deviation principles for sums of random vectors and the corresponding renewal functions in the inhomogeneous case (2015), *Siberian Adv. Math.*, 25(4), 255–267.

[45] Borovkov, A.A., Mogul'skii, A.A.: Large deviation principles for the finite-dimensional distributions of compound renewal processes (2015), *Siberian Math. J.*, 56(1), 28–53.

[46] Borovkov, A.A., Mogul'skii, A.A.: Large deviation principles for trajectories of compound renewal processes, I, II (2016), *Theory Probab. Appl.,* 60(2), 207–221; *Theory Probab. Appl.*, 60(3), 349–366.

[47] Borovkov, A.A., Mogul'skii, A.A.F.: Integro-local limit theorems for compound renewal processes under Cramér's condition. I, II (2018), *Siberian Math. J.*, 59(3), 383–402, *Siberian Math. J.*, 59(4), 578–597.

[48] Borovkov, A.A., Mogul'skii, A.A.F., Prokopenko, E.I.: Properties of the deviation rate function and the asymptotics for the Laplace thansform of the distribution of a compound renewal process (2019), *Theory Probab. Appl.*, 64(4), 499–512

[49] Borovkov, A.A., Rogozin, B.A.: Boundary problems in some two-dimensional random walks (1964), *Theory Probab. Appl.,* 9(3), 361–388.

[50] Chung, K.L.: A Course in Probability Theory (1974), 2nd ed., Academic Press.

[51] Çinlar, E.: Markov additive processes. I (1972), *Z. Warsch. verw. Gebiete*, 24, 84–94.

[52] Çinlar, E.: Markov additive processes. II (1972), *Z. Warsch. verw. Gebiete*, 24, 94–121.

[53] Cox D.R., Smith, W.L. Renewal Theory (1962), Wiley.

[54] Cramér, H.: Collective Risk Theory (1955), Stockholm: Erselte.

[55] Csörgö, M., Horváth, L., and Steinebach, J.: Invariance principles for renewal processes (1987), *Ann. Probability*, 15(4), 1441–1460.

[56] Csörgö, M., Deheuvels, P., and Hervatt, L.: An approximation of stopped sums with applications in queueing theory (1987), *Adv. Appl. Probability*, 19, 674–690.

[57] Dembo, A., Zeitouni, O.: Large Deviations Techniques and Applications (1998), 2nd ed., Springer.

[58] Doney, R.A.: A large deviation local limit theorem (1989), *Math. Proc. Cambridge Phil. Soc.*, 105(3), 575–577.

[59] Donsker, M.D. and Varadhan, S.R.S.: Asymptotic evaluation of certain Markov process expectations for large time. I, (1975) *Comm. Pure Math.*, 28, 1–47; (1975) 28(1), 279–301; (1976) 29, 389–461; (1983) 36, 183–212.

[60] Donsker, M.D., and Varadhan, S.R.S.: Asymptotic evaluation of certain Markov process expectations for large time. II // *Comm. Pure Math.*, 1975. V. 28(2). P. 279–301.

[61] Donsker, M.D., and Varadhan, S.R.S.: Asymptotic evaluation of certain Markov process expectations for large time. III // *Comm. Pure Math.*, 1976. V. 29(4). P. 389–461.

[62] Donsker, M.D., and Varadhan, S.R.S.: Asymptotic evaluation of certain Markov process expectations for large time. IV // *Comm. Pure Math.*, 1983. V. 36(2). P. 183–212.

[63] Drekic, S., and Willmot, G.E.: On the density and moments of the time of ruin with exponential claims (2003), *ASTIN Bull.*, 33(1), 11–21.

[64] Feng, J., Kurtz, T.G.: Large Deviations for Stochastic Processes (2006), Mathematical Surveys and Monographs, v. 131, AMS.

[65] Foss S. G., Puhalskii A. A.: On the limit law of a random walk conditioned to reach a high level (2011), *Stoch. Proc. and Appl.*, 121(2), 288–313.

[66] Frolov, A.N.: Limit theorems for increments of compound renewal processes (2007), *J. Math. Sci. (N. Y.)*, 152(6), 944–957.

[67] Gamkrelidze, N.G.: On local limit theorem for integer random vectors (2015), *Theory Probab. Appl.*, 59(3), 494–499.

[68] Gikhman, I.I., Skorokhod, A.V.: Theory of Random Processes. I (1974), Springer.

[69] Gnedenko, B.V., Kolmogorov, A.N.: Limit Distributions for Sums of Independent random Variables (1954), Addison-Wesley.

[70] Gnedenko, B.V., Sherif, A.: Limit theorems for extreme terms of a variational series (1983), *Dokl. Akad. Nauk SSSR*, 270(3), 523–525.

[71] Gudinas, P.P.: Large deviations for sums of random variables connected in a Markov chain (1986), + *Litovsk. Math. Sb.,* 26(2), P.246–258.

[72] Gut, Allan: Stopped Random Walks. Limit Theorem and Applications (2009), 2nd ed., Springer.

[73] Hervé, M.: Several Complex Variables (1963), Oxford University Press.

[74] Keilson J., Wishart, D.: Addenda to processes defined on a finite Markov chain (1964), *Proc. Cambridge Philos. Soc.*, 63 (1), 187–195.

[75] Keilson, J., Wishart, D.: A central limit theorem for processes defined on a Markov chain (1964), *Proc. Cambridge Philos. Soc.*, 60(3), 547–567.

[76] Khintchine, A.: Zwei Sätze über stochastische Prozesse mit stabilen Verteilungen (1938), *Rec. Math.* [*Mat. Sbornik*] *N.S.*, 3(45), 577–583.

[77] Kingman, F. G.: On queues in heavy traffic (1962), *J.R. Statist. Soc., Ser. B.*, 24(2), 383–392.

[78] Komlós, J., Major, P., and Tusnády: An approximation of partial sums of independent R.V.'s and the sample D.F. I (1976), *Z. Warsch. Verw. Gebiete*, 32(1), 111–131; (1976) 34. 33–58.

[79] Korshunov, D.: On subexponential tails for the maxima of negatively driven compound renewal and Lévy processes (2018), *Stochastic Processes and their Applications*, 128(4), 1316–1332.

[80] Korshunov, D.A.: The critical case of the Cramer–Lundberg theorem on the asymptotic tail behavior of the maximum of a negative drift random walk (2005), *Siberian Math. J.*, 46(6), 1077–1081.

[81] Loéve, M.: Probability Theory (1960), Van Nostrand Co.

[82] Macci, C.: Large deviations for compound Markov renewal processes with dependent jumps sizes and jumps waiting times (2007), *Bull. Belg. Math. Soc. Simon Stevin*, 14(2), 213–228.

[83] Meyn, S.P., Tweedie, R.L.: Markov Chains and Stochastic Stability (1993), Springer.

[84] Miller, H.D.: A convexity property in the theory of random variables defined on a finite Markov chain (1961), *Annals Math. Statist.*, (32), 1260–1270.

[85] Mitalauskas, A.: On a multidimensional limit theorem for lattice distributions (1960), *Trudy Lit. SSR Ser.* B, 2 3–14.

[86] Mogul'skii, A.A.: Small deviations in the sample function space (1975), *Theory Probab. Appl.,* 19(1), 726–736.

[87] Mogul'skii, A.A.: The Fourier method for finding the asymptotic behavior of small deviations of a Wiener process (1982), *Siberian Math. J.*, 23(3), 420–431.

[88] Mogul'skii, A.A.: The large deviation principle for a compound Poisson process (2017), *Siberian Adv. Math.*, 2(3), 160–186.

[89] Mogul'skii, A.A.: On a property of the Legendre transform (2018), *Siberian Adv. Math.*, 28(1), 65–73.

[90] Mogul'skii, A.A.: The extended large deviation principle for a process with independent increments (2017), *Siberian Math. J.*, 58(3), 515–524.

[91] Mogul'skii, A.A.: Local theorems for arithmetic compound renewal processes when Cramer's condition holds (2019), *Sib. Èlektron. Mat. Izv.*, 16, 21–41.

[92] Mogul'skii, A.A., Prokopenko, E. I.: Integro-local theorems for multidimensional compound renewal processes, when Cramer's condition holds I, II, III (2018), *Sib. Èlektron. Mat. Izv.*, 15, 475–502; *Sib. Èlektron. Mat. Izv.*, 15, 503–527; *Sib. Èlektron. Mat. Izv.*, 15, 528–553.

[93] Mogul'skii, A.A., Prokopenko, E.I.: Local theorems for multivariatearithmetic compound renewal processes under the Cramér condition (2019), *Mat. Tr.*, 22(2), 106–133.

[94] Nagaev A.V., Cramer large deviations when the extreme conjugate distribution is heavy-tailed (1999), *Theory Probab. Appl.*, 43(3), 405–421.

[95] Ney, P., Nummelin, T.: Markov additive processes. I. Eigenvalue properties and limit theorems (1987), *Annals of Probability*, 15(2), 561–592.

[96] Ney, P., Nummelin, T.: Markov additive processes. II. Large deviations (1987), *Annals of Probability*, 15(2), 593–609.

[97] Nummelin, E.: General Irreducible Markov Chains and Nonnegative Operators (1984), Cambridge University Press.

[98] Presman, E.L.: Factorization methods and boundary problems for sums of random variables given on Markov chains (1969), *Math. USSR-Izv.*, 3(4), 815–852.

[99] Prokhorov, Yu. V.: On a local limit theorem for lattice distributions (1954), *Dokl. Akad. Nauk SSSR, n. Ser.,* 98, 535–538.

[100] Prokhorov, Yu. V.: Transient phenomena in processes of mass service, *Lit. Matem. Sb.*, 3, 199–206.

[101] Raudelyunas, A.K.: On a multidimensional local limit theorem (1964), *Lit. Matem. Sb.*, 4, 141–145.

[102] Renyi, A.: On the asymptotic distribution of the sum of a random number of independent random variables (1957), *Acta Math. Acad. Sci. Hungar.*, 8, 193–199.

[103] Rogozin, B.A.: The distribution of the maximum of a process with independent increments (1969), *Siberian Math. J.*, 10(6), 989–1010.

[104] Rozanov, Yu. A.: On a local limit theorem for lattice distributions (1957), *Theory Probab. Appl.*, 2(2), 260–265.

[105] Rozovskii, L.V.: A lower bound for large deviation probabilities of sum of independent random variables with finite variances (2002), *J. Math. Sci. (New York)*, 109(6), 2192–2209.

[106] Rockafellar R.: Convex Analysis (1970), Princeton University Press.

[107] Rvacheva,E. L.: On domain of attraction of multidimensional distributions (1960), *Select Transl. Math. Statist. and Probability.* vol. 2, pp. 183–205.

[108] Saulis L., Statulevicius, V.: Limit Theorems for Large Deviations (1991), Kluwer.

[109] Shepp, L.A.: A local limit theorem (1964), *Ann. Math. Statist.,* 35, 419–423.

[110] Skorokhod, A.V.: Limit theorems for stochastic processes (1956), *Theory Probab. Appl.*, 1(3), 261–290.

[111] Smirnov N.V., Limit distributions for the terms of a variational series (1949), *Trudy Mat. Inst. Steklov., 25, Acad. Sci. USSR, Moscow–Leningrad*, 25, 3–60.

[112] Steinebach, J.: Invariance principles for renewal processes when only moments of low order exist (1988), *J. Multivariate Analysis*, 26(2), 166–183.

[113] Steinebach, J.: On the optimality of strong approximation rates for compound renewal processes (1988), *Statistics and Probability Letters*, 6(4), 263–267.

[114] Stone, C.: A local limit theorems for nonlattice multidimensional distribution functions (1965), *Ann. Math. Statist.,* 36(2), 546–551.

[115] Stone, C.: On local and ratio limit theorems (1966), Proc. Fifth Berkeley Symposium of Math Statistics and Probability. Berkeley and Los Angeles, University of California. Vol. II, Part II, pp. 217–224.

[116] Tkachuk, S.G.: Local limit theorems with admissible large deviations for stable limit laws (1973), *Izv. Akad. Nauk UzSSR, Ser. fiz.-Mat. Nauk*, 17 (21), 30–33.

[117] Torrang, I.: The law of the iterated logarithm–cluster points of deterministic and random subsequences (1987), *Probab. Math. Statist.*, VIII, 133–141.

[118] Varadhan, S.R.S.: Asymptotic probabilities and differential equations (1966), *Comm. Pure Appl. Math.*, 19(3), 261–286.

[119] Volkov, I.S.: On the distribution of sums of random variables defined on a homogeneous Markov chain with a finite number of states (1958), *Theory Probab. Appl.*, 3(4), 384–399.

[120] Volkov, I.S.: Analysis of some limit theorems for large deviations (1961), *Theory Probab. Appl.*, 6(3), 302–304.

[121] Zachary, S., Foss, S.G.: On the exact distributional asymptotics for the supremum of a random walk with increments in a class of light-tailed distributions (2006), *Siberian Math. J.*, 47(6), 1034–1041.

Index